Energy Transition, Climate Action and Sustainable Agriculture

Robert Ugochukwu Onyeneke ·
Chukwuemeka Chinonso Emenekwe ·
Chinedum Uzoma Nwajiuba
Editors

Energy Transition, Climate Action and Sustainable Agriculture

Perspectives and Strategies for Africa

Springer

Editors
Robert Ugochukwu Onyeneke
Department of Agricultural Economics
Alex Ekwueme Federal University
Ndufu-Alike
Ikwo, Nigeria

Chukwuemeka Chinonso Emenekwe
Department of Economics
and Development Studies
Alex Ekwueme Federal University
Ndufu-Alike
Ikwo, Nigeria

Chinedum Uzoma Nwajiuba
Department of Agribusiness
and Management
Michael Okpara University of Agriculture
Umudike
Umuahia, Nigeria

ISBN 978-3-031-83167-6 ISBN 978-3-031-83165-2 (eBook)
https://doi.org/10.1007/978-3-031-83165-2

This Springer imprint is published by the registered company Springer Nature Switzerland AG
The registered company address is: Gewerbestrasse 11, 6330 Cham, Switzerland

Preface

This book explores current thinking and presents the main opportunities and challenges associated with energy transition, climate action and sustainable agriculture especially in the context of developing nations. The book introduces evidence from empirical studies, projects and systematic literature review which show the drivers of energy transition, climate action and sustainable agriculture. The book further provides other issues associated with climate change, which is important to the achievement of energy transition and sustainable agriculture in developing countries. The ambition is that this book will become a leading publication on the topic of energy transition, climate action and sustainable agriculture.

The urgency of tackling climate change and the imperative for sustainable energy practices are an important issue in global sustainable development. The topic is aligned with the objectives of the Paris Agreement, focusing on equitable approaches to climate action and the challenges and opportunities associated with sustainable energy and agriculture reforms.

In regions where the impacts of climate change are most acutely felt, the challenges are immense. Developing countries, often the least responsible for global emissions, face the greatest burdens—economic, social, and environmental. In Nigeria for instance, a significant percentage of the population lacks access to electricity, and erratic power supply severely hampers economic growth. Meanwhile, the reliance on fossil fuels continues to pose serious environmental and health risks, complicating our path toward sustainable development. Yet, within these challenges lie vast opportunities for transformative growth and innovation. By embracing renewable energies and sustainable practices, Nigeria and other developing nations can leapfrog traditional polluting industrial processes and create new pathways for economic development that also address urgent environmental concerns.

It is against this background that the *Energy Transition, Climate Action and Sustainable Agriculture: Perspectives and Strategies for Africa* is being published. The book delved into challenges while also spotlighting the substantial opportunities that transitioning to renewable energy sources presents to developing nations' economies and society at large. The book featured papers that are peer-reviewed and selected based on their relevance and impact. It offers valuable insights and forward-thinking perspectives that are crucial for policy formulation and implementation. The papers addressed innovative strategies, practical solutions, and rigorous analyses that highlight the challenges and successes in climate action, sustainable agriculture and energy transition initiatives.

Ikwo, Nigeria

Robert Ugochukwu Onyeneke
Chukwuemeka Chinonso Emenekwe
Chinedum Uzoma Nwajiuba

Acknowledgments

The editors are grateful to the reviewers for their comments, which significantly improved the contributions. Also, we wish to express our sincere appreciation to everyone who contributed to the success of the maiden conference on "Climate Justice and Just Energy Transition in Developing Countries" in commemoration of the 60th birthday of Prof. Chinedum Uzoma Nwajiuba. Most of the contributions in this book were presented at this conference. The production editor and his team are also appreciated for their work in enhancing the quality of the contributions.

Contents

Part I
Energy

Chapter 1
Introduction

Robert Ugochukwu Onyeneke, Chukwuemeka Chinonso Emenekwe, and Chinedum Uzoma Nwajiuba

Africa faces a convergence of pressing challenges and transformative opportunities at the nexus of energy, climate, and agriculture. As the continent grapples with the multifaceted impacts of climate change, energy transitions, and food security, there is an urgent need for innovative, evidence-based solutions to guide policymakers, practitioners, and researchers in fostering sustainable development.

This book, Energy Transition, Climate Action, and Sustainable Agriculture: Perspectives and Strategies for Africa, offers a comprehensive exploration of these critical issues. By synthesizing contributions from leading scholars, the book provides actionable insights into the interdependencies among energy systems, climate resilience, and agricultural sustainability. Organized into three thematic sections—Energy, Climate, and Agriculture—the book draws on empirical studies, theoretical analyses, and policy frameworks to address Africa's unique challenges while identifying scalable opportunities for development.

R. U. Onyeneke (✉) · C. U. Nwajiuba
Department of Agricultural Economics, Alex Ekwueme Federal University Ndufu-Alike, Ikwo, Nigeria
e-mail: robert.onyeneke@funai.edu.ng

C. U. Nwajiuba
e-mail: chnwajiuba@yahoo.de

C. C. Emenekwe
Department of Economics and Development Studies, Alex Ekwueme Federal University Ndufu-Alike, Ikwo, Nigeria
e-mail: chukwuemeka.emenekwe@funai.edu.ng

C. U. Nwajiuba
Department of Agribusiness and Management, Michael Okpara University of Agriculture Umudike, Umuahia, Nigeria

R. U. Onyeneke et al. (eds.), *Energy Transition, Climate Action and Sustainable Agriculture*, https://doi.org/10.1007/978-3-031-83165-2_1

1.1 Energy: Transitioning to Sustainable Systems (Chapters 1–7)

The first section examines the transformative potential of energy transitions in Africa. It highlights how clean energy systems, financial mechanisms, and policy frameworks can catalyse economic growth, mitigate climate change, and improve public health.

- This chapter investigates the employment effects of renewable energy adoption in Nigeria, emphasizing the critical role of education and economic growth in achieving just energy transitions.
- Chapter 2 explores how foreign direct investment, development assistance, and institutional quality foster renewable energy consumption across Africa, while financial development presents challenges.
- Chapter 3 compares energy flexibility measures in Nigeria and the United Kingdom, identifying opportunities to enhance Nigeria's energy infrastructure for a just transition.
- Chapter 4 assesses the socio-economic implications of fuel subsidy removal in Nigeria, offering policy recommendations to mitigate adverse effects.
- Chapter 5 examines the relationship between clean energy transitions and public health outcomes, demonstrating the dual benefits of reducing carbon emissions and improving health indicators.
- Chapter 6 investigates blockchain technology as a tool for tracking carbon footprints in the energy sector, emphasizing its potential for enhancing transparency.
- Chapter 7 explores determinants of energy choices among rural households in Nigeria, shedding light on socio-economic and cultural factors influencing energy access.

These chapters collectively illustrate the opportunities and complexities of transitioning to sustainable energy systems in Africa, highlighting the need for inclusive and adaptive policy frameworks.

1.2 Climate: Building Resilience Through Policy and Practice (Chapters 8–20)

The second section focuses on Africa's response to climate change, addressing adaptation strategies, policy frameworks, and the socio-economic drivers of resilience. These chapters delve into the complex interplay between climate action and sustainable development.

- Chapter 8 examines the adoption of climate-smart agricultural techniques among smallholder farmers in Nigeria, emphasizing gender-sensitive solutions.
- Chapter 9 reviews climate-smart water conservation strategies, showcasing innovative approaches to managing water resources.

- Chapter 10 investigates the feasibility of carbon farming as a tool for carbon trading, presenting its potential to reverse climate change.
- Chapter 11 evaluates inclusive climate communication strategies, highlighting the role of public engagement in climate resilience.
- Chapter 12 analyses the risks and adaptation strategies of cassava value chain actors in Southeast Nigeria, emphasizing cooperative networks.
- Chapter 13 explores socio-economic drivers of climate resilience, linking education and infrastructure to adaptive capacity.
- Chapter 14 assesses flood insurance adoption among cassava farmers, presenting financial mechanisms to reduce vulnerability.
- Chapter 15 investigates land use dynamics and food security challenges in Nigeria, offering policy recommendations for sustainable agriculture.
- Chapter 16 examines perceptions of climate change impacts on agriculture and health, drawing lessons for community-based adaptation.
- Chapter 17 analyses gender and media coverage of climate change in Nigeria, advocating for equitable and inclusive narratives.
- Chapter 18 evaluates climate change risks in the oil and gas sector, offering insights into industry-specific adaptation strategies.
- Chapter 19 explores artificial intelligence in media coverage of climate change, emphasizing its role in shaping public awareness.
- Chapter 20 examines the green energy transition and its implications for public health in Nigeria, showcasing the benefits of sustainable development.

These chapters underscore the importance of integrating policy innovation, stakeholder engagement, and technological advancements to address climate change and enhance resilience in Africa.

1.3 Agriculture: Ensuring Food Security and Livelihoods (Chapters 21–26)

The third section highlights the pivotal role of agriculture in Africa's economic and social fabric. The chapters explore innovative farming practices, financial mechanisms, and adaptive strategies to secure food systems and strengthen livelihoods.

- Chapter 21 examines the financial risks faced by cassava farmers in Nigeria, offering strategies for managing climate-induced vulnerabilities.
- Chapter 22 evaluates anaerobic digestion technologies for renewable energy and waste management, presenting solutions for reducing greenhouse gas emissions.
- Chapter 23 investigates the relationship between energy transition and food security, highlighting the interdependencies of these sectors.
- Chapter 24 explores agroecological practices among smallholder farmers in Kenya, emphasizing their role in promoting sustainable agriculture.

- Chapter 25 examines the drivers of adoption for renewable energy technologies in farming communities, offering insights into policy interventions.
- Chapter 26 investigates the role of digital tools in enhancing agricultural resilience, highlighting the potential of technology in transforming food systems.

These chapters demonstrate how sustainable agricultural practices can address food insecurity, enhance climate resilience, and support Africa's socio-economic development.

This book contributes to the global discourse on sustainability by emphasizing Africa's role in addressing interconnected challenges across energy, climate, and agriculture. Key messages include:

- Energy Transitions: The transformative potential of renewable energy systems in fostering economic growth, reducing emissions, and improving public health.
- Climate Resilience: The importance of inclusive policies, community engagement, and adaptive technologies in enhancing resilience to climate change.
- Agricultural Sustainability: The centrality of agriculture in ensuring food security, supporting livelihoods, and mitigating climate risks.

Readers of this book will gain:

- A nuanced understanding of Africa's energy-climate-agriculture nexus.
- Evidence-based recommendations for policy and practice.
- Insights into emerging technologies and sustainable development strategies.
- Knowledge of how cross-sectoral approaches can address Africa's developmental challenges.

Chapter 2
Towards Just Energy Transition: An Assessment of the Employment Effects of Renewable Energy Consumption and Human Capital Development in Nigeria

Chukwuemeka Chinonso Emenekwe, Obioma Emenekwe, Chibuzo Emenekwe, and Chukwuma Emenekwe

Abstract This study investigates the impact of renewable energy consumption on employment in Nigeria, employing the Kernel-Based Regularized Least Squares (KRLS) method to capture the complex, non-linear relationships between key variables. Using annual data from authoritative sources, the analysis focuses on renewable energy consumption, fossil energy consumption, educational attainment levels, GDP per capita, and CO_2 emissions. The findings reveal a significant positive relationship between renewable energy consumption and employment, with a 1% increase in renewable energy consumption leading to a 0.098–0.191% increase in employment. Fossil energy consumption also positively impacts employment, though to a lesser extent. The study highlights the critical role of tertiary education in enhancing employment outcomes and underscores the importance of economic growth as a determinant of job creation. The analysis suggests potential adverse impacts of CO_2 emissions on employment, reinforcing the need for environmentally conscious policies. Based on these findings, the study recommends policies to enhance renewable energy adoption, invest in education, promote economic growth,

C. C. Emenekwe (✉)
Department of Economics and Development Studies, Alex Ekwueme Federal University Ndufu-Alike, Ikwo, Nigeria
e-mail: chukwuemeka.emenekwe@funai.edu.ng

O. Emenekwe
Department of Clinical and Health Services, Alex Ekwueme Federal University Ndufu-Alike, Ikwo, Nigeria
e-mail: emenekwe.obioma@funai.edu.ng

C. Emenekwe · C. Emenekwe
Department of Civil Engineering, Alex Ekwueme Federal University Ndufu-Alike, Ikwo, Nigeria
e-mail: emenekwec@yahoo.com

C. Emenekwe
e-mail: chuemenekwe@gmail.com

R. U. Onyeneke et al. (eds.), *Energy Transition, Climate Action and Sustainable Agriculture*, https://doi.org/10.1007/978-3-031-83165-2_2

and support industries transitioning from fossil fuels. Despite some limitations, such as data constraints and the generalization of employment impacts, the study provides valuable insights for policymakers aiming to foster sustainable development and climate justice in Nigeria.

Keywords Renewable energy consumption · Just energy transition · Employment · Climate justice · Machine learning · Kernel-based regularized least squares (KRLS) · Nigeria

2.1 Introduction

Energy accounts for two-thirds of the global greenhouse (GHG) emissions (IEA, 2021). The Paris Agreement, signed in December 2015, sets the goal of limiting "the increase in the global average temperature to well below 2 °C above pre-industrial levels and to pursue efforts to limit the temperature increase to 1.5 °C" (Article 2.1) (Emenekwe et al., 2022a). Net carbon emissions must fall to zero to meet this target, which will require drastic changes in the energy system.

Renewable energy sources (RES) have been pointed out as one of the most cost-effective mitigation strategies to decarbonize the energy system (Emenekwe et al., 2022a; IEA, 2021; Wang et al., 2021). This transition to renewable energy systems has the potential to affect the entire economy—from food and agriculture, ecosystems and climate, poverty, and health, to employment and economic growth, etc. A growing body of research literature is examining the interlinkages of the sustainable and clean energy goal (Sustainable Development Goal (SDG 7)) that aims to increase the share of renewables to ensure affordable, reliable, and sustainable sources of energy, with other SDG indicators. In particular, a large body of literature points to the positive impact of renewable energy on economic activities and emission reduction (Emenekwe et al., 2022a, 2022b, 2023; Le Treut et al., 2021; Su et al., 2021; Wang et al., 2021). Wang et al. (2021) show that renewable energy consumption reduce the coupling of carbon dioxide (CO_2) emissions and economic growth. Given the evidence of the positive impacts of renewable energy economy and environment, the energy transition process can be considered crucial for all countries.

Countries have always been advocating policies that promote an energy transition to renewable—beyond its role in reducing greenhouse gas (GHG) emissions—as a means of rationing energy supplies, encouraging technological advancement, and facilitating new jobs, as well as creating regional development opportunities (Proença & Fortes, 2020). In particular, creating jobs through renewable energy transition is a priority for many governments (Arvanitopoulos & Agnolucci, 2020; Cameron & van der Zwaan, 2015; Emenekwe et al., 2022c; Garrett-Peltier, 2017; Ram et al., 2022; UNDP & ILO, 2021). However, there appears to be no consensus among scholars and policymakers regarding the employment benefits of renewable energy. In a meta-analysis of the impacts of renewable energy development, Stavropoulos and Burger (2020) find that there is disagreement among empirical studies (academic research and policy reports) regarding the job-creating effects of renewable energy.

Further, the optimistic vision of renewable energy consumption has not been unanimous in the literature, with some authors arguing that its contribution to employment is not as large as envisioned and may even be negative (Almutairi et al., 2018; Böhringer et al., 2013; Cai et al., 2017; Mu et al., 2018). They claim that the effects of renewable energy consumption on employment are highly dependent on the financing mechanisms, the type of technologies and the scope of employment impacts (direct, indirect and induced). On the other hand, other studies find positive, beneficial employment effects of renewable energy (Proença & Fortes, 2020; Ram et al., 2022; Saboori et al., 2022; UNDP & ILO, 2021).

The existing ambiguity in the empirical literature centres on the types of renewable energies, the sample of countries, the employed methodologies, and the time intervals. Several methods have been applied to assess the employment effects associated with the penetration of RES in the energy system and particularly. These can be distinguished between the ones that analyse the linkage between renewable technologies and employment considering future energy scenarios and explicit methodologies, such as: input–output (I-O) matrixes or computable general equilibrium (CGE) models, which include the interrelationships between all sectors of the economy, or simpler analytical methods sustained by employment factors, i.e. job intensities. The other ones study a relationship between renewable technologies and employment using historical data and econometric methods.

Nigeria has made significant strides in promoting a transition from fossil fuels to renewable energy, aligning with its commitments under the Paris Agreement (Department of Climate Change, 2021; National Climate Change Council, 2023; Okereke et al., 2024). The country's updated Nationally Determined Contribution (NDC) aims for an ambitious 45% reduction in greenhouse gas emissions by 2030 with international support, highlighting key measures such as achieving 30% of grid-connected electricity from renewable sources (Department of Climate Change, 2021). This policy landscape is complemented by the Energy Transition Plan (ETP), which envisions a net-zero emissions target by 2060, necessitating substantial investments in solar energy and other renewables (Energy Transition Office, 2022).

By fostering renewable energy development, Nigeria is poised to generate substantial employment opportunities across various sectors, from construction and installation to operations and maintenance and innovation in green technologies, thereby contributing to Nigeria's economic diversification efforts (Emenekwe et al., 2024). The integration of renewables is expected to enhance energy security, reduce reliance on fossil fuels, and drive technological advancements, positioning Nigeria as a leader in Africa's green transition.

Although there is a vast literature focusing in the ex-ante linkages between renewable technologies and employment opportunities (Emenekwe et al., 2022c; Energy Transition Office, 2022; National Climate Change Council, 2023; Okereke et al., 2024; UNDP & ILO, 2021), fewer have been analysing the confirmed outcomes of this relation. Ex-post analysis is mainly focused on providing empirical evidence of the relationship between renewable energy consumption and economic development nexus. They tend to study its overall effect on GDP, neglecting the social face of renewables associated with employment creation. Moreover, there is a dearth of ex-post analysis investigating this relationship for Nigeria. Using Toda-Yamamoto causal analysis, Musa and Maijama'a (2020) find a significant bi-directional causality between renewable energy consumption and unemployment in Nigeria during the 1991–2015, while Lawan and Gidigbi (2019) find no causal relation from renewable consumption to employment in Nigeria during the 1991–2014 period.

Moreover, traditional econometric methods such as Ordinary Least Squares (OLS) and Generalized Linear Models (GLMs) remain prevalent in ex-post analyses of the relationship. These conventional techniques impose rigid functional form assumptions, potentially missing the complex, non-linear, and interactive effects within the data (Ferwerda et al., 2015; Hainmueller & Hazlett, 2014). As a result, these models often produce biased estimates and incorrect inferences about the elasticity between socio-economic and related phenomena (Ferwerda et al., 2015; Hainmueller & Hazlett, 2014). This methodological limitation highlights a significant gap in the literature, emphasizing the need for more sophisticated approaches.

Accurately estimating the elasticity between renewable energy consumption and employment is crucial for calibrating Computable General Equilibrium (CGE) models and other ex-ante modelling tools (Wang et al., 2021). Kernel-Based Regularized Least Squares (KRLS), proposed by Hainmueller and Hazlett (2014), offers a robust solution by accommodating complex, non-linear relationships without prespecifying the functional form. KRLS reduces the risk of model misspecification and enhances the accuracy of estimated marginal effects through its flexible, machine learning-based approach (Ferwerda et al., 2015; Hainmueller & Hazlett, 2014). Furthermore, KRLS mitigates overfitting and improves model stability, making it well-suited for generating reliable inputs for CGE models and other predictive tools. Adopting KRLS can bridge the current methodological gap, providing a more accurate and comprehensive understanding of the renewable energy-employment nexus and informing effective energy policies.

Consequently, to address the ambiguity in the literature on the link between renewable energy and employment, due to either because of short time period, or methodological limitation, the present study uses a long-term data for Nigeria and applies a more efficient machine learning estimator. Specifically, this paper aims to contribute to the ex-post literature by examining the empirical relationship between renewable energy consumption and employment in Nigeria, over the period 1980–2019, and using the KRLS estimator. This research paper answers the question: how does renewable energy consumption contribute to overall employment in Nigeria?

The remainder of this paper is organized as follows: Sect. 2.2 describes the employed KRLS machine learning method, the empirical model and the dataset. Section 2.3 presents and discusses the empirical results of the estimated model. Section 2.5 provides the concluding remarks and the possible implications of the outcomes.

2.2 Data and Methodology

This section is dedicated to the description of the empirical methods and dataset. Specifically, detail information and statistical properties of the dataset are provided alongside relevant guideline for the deployment of the empirical methods.

2.2.1 Model

This study extend the modelling analysis of Swain et al. (2022) by considering effects of human capital. Specifically, this study focuses on education component of human capital attainment by investigating school enrolment across primary, secondary, and tertiary levels. The functional form of the base model is given as follows:

$$Employ = f\,(Renewable, Fossil, HC, RGDPpc, MtCO_2pc) \tag{2.1}$$

where *Employ* represents the number of people employed in Nigeria each year in the sample. *Renewable* represents annual renewable energy consumption, *Fossil* represent fossil fuel consumption in Nigeria, *HC* represents human capital attainment in Nigeria, *RGDPpc* represents annual per capita GDP/income, and $MtCO_2pc$ represents per capita carbon emissions in million tonnes.

We further state the empirical model and decompose the human capital variable into the distinct levels of interest. Additionally, we convert the levels of the levels to natural logarithms to enhance linearity and address issues of heteroscedasticity (see Eqs. 2.2–2.4).

$$\begin{aligned}\ln Employ_t = {} & \beta_0 + \beta_1 \ln Renewable_t + \beta_2 \ln Fossil_t + \beta_3 \ln Primary_t \\ & + \beta_4 \ln RGDPpc_t + \beta_4 \ln MtCO_2pc_t + \varepsilon_t\end{aligned} \tag{2.2}$$

$$\begin{aligned}\ln Employ_t = {} & \beta_0 + \beta_1 \ln Renewable_t + \beta_2 \ln Fossil_t + \beta_3 \ln Secondary_t \\ & + \beta_4 \ln RGDPpc_t + \beta_4 \ln MtCO_2pc_t + \varepsilon_t\end{aligned} \tag{2.3}$$

$$\begin{aligned}\ln Employ_t = {} & \beta_0 + \beta_1 \ln Renewable_t + \beta_2 \ln Fossil_t + \beta_3 \ln Tertiary_t \\ & + \beta_4 \ln RGDPpc_t + \beta_4 \ln MtCO_2pc_t + \varepsilon_t\end{aligned} \tag{2.4}$$

where t is the time period, ln is indicates that variables have been converted to natural logarithmic forms, and ε_t is the idiosyncratic error term (iid). Further, Eqs. (2.2)–(2.4) are henceforth regarded as models I, II, and III, respectively.

2.2.2 Methodology

This study aims to examine the impact of renewable and non-renewable energy consumption on employment, considering the intricate relationships between energy, the economy, and the environment. Additionally, we assess the potential contributions of human capital, represented by school enrollment rates, per capita income, and CO_2 emissions, to employment variations. To achieve this, we utilize the Kernel-Based Regularized Least Squares (KRLS) method, as developed by Hainmueller and Hazlett (2014). The KRLS approach integrates machine learning algorithms with econometric principles, offering several advantages over traditional econometric techniques.

KRLS is particularly effective in deriving pointwise derivatives and mean marginal effects, conducting hypothesis testing, and producing reliable estimates. This method excels in addressing misspecification biases that often challenge other machine learning techniques. KRLS also provides interpretable and variable parameters for both classification and regression problems, even when the functional forms are uncertain. Essentially, KRLS identifies the underlying structure of the data series, protecting analysts from specification biases.

This method is highly beneficial for a range of analytical purposes, including understanding data generation processes, conducting causal inquiries driven by models, forecasting, and imputing missing data (Ferwerda et al., 2015; Hainmueller & Hazlett, 2014; Sarkodie & Owusu, 2020). Detailed procedural guidance on the KRLS method is provided by Hainmueller and Hazlett (2014), which supports its application in the four models examined in this study.

Prior to applying the KRLS method, essential preliminary tests, including stationarity and normality tests, were conducted. These tests ensure the appropriateness of the KRLS method for handling non-normal, non-stationary, and non-linear data series. This rigorous methodological approach ensures the robustness and validity of our findings.

2.2.3 Data Description

Table 2.1 shows the definition of variables employed and their sources. The data for this study were obtained from several authoritative sources to ensure accuracy and reliability. Employment data, measured as the number of persons engaged in millions, and the Human Capital Index (HCI) were sourced from the Penn World Table version 10 (2021). Data on territorial fossil CO_2 emissions per capita (in million tonnes of CO_2) were obtained from Friedlingstein et al. (2022). Information on GDP per capita (constant 2015 US \$), school enrollment rates at the primary, secondary, and tertiary levels (percent gross), and total population were sourced from the World Bank's World Development Indicators (WDI) database. Additionally, data on total energy consumption from renewable sources and other sources, as well as from fossil

Table 2.1 Definition of variables and data sources

Variable	Description	Source
Employ	Number of persons engaged (in millions)	Penn World Table v10 (2021)
$MtCO_2pc$	Territorial fossil CO_2 emissions (in million tonnes of CO_2) per capita	Friedlingstein et al. (2022)
RGDPpc	GDP per capita (constant 2015 US $)	WB WDI Database
Renew	Total energy consumption from renewables and other (MMBtu)	Energy Information Agency
Fossil	Total energy consumption from fossil sources (MMBtu)	Energy Information Agency
Primary	School enrolment, primary (% gross)	WB WDI Database
Secondary	School enrolment, secondary (% gross)	WB WDI Database
Tertiary	School enrolment, tertiary (% gross)	WB WDI Database
HCI	Human capital index	Penn World Table v10 (2021)
Pop	Population, total	WB WDI Database

Note MMBtu stands for million British thermal units. WB WDI stands for World Bank's World Development Indicators database

sources (measured in million British thermal units, MMBtu), were collected from the Energy Information Agency.

2.3 Empirical Results

2.3.1 Preliminary Results

Table 2.2 shows that summary statistics of the variables used in our analysis, elucidating the general characteristics of our dataset. The average number of persons engaged in employment across the study period is approximately 41.53 million, with a notable standard deviation of 13.69 million, indicating significant variability in employment levels over time. The range of employment spans from a minimum of 26.44 million to a maximum of 73.02 million, reflecting the dynamic nature of the Nigerian labour market.

Renewable energy consumption averages 18.6 million British thermal units (MMBtu), with a substantial standard deviation of 6.18 million MMBtu. This wide dispersion suggests considerable fluctuations in the adoption and utilization of renewable energy sources. The minimum and maximum values of 6.01 million MMBtu and 29.1 million MMBtu, respectively, underscore this variability. In contrast, fossil energy consumption is markedly higher, averaging 895 million MMBtu with a standard deviation of 376 million MMBtu. The consumption ranges from 390 million

Table 2.2 Summary statistics

Variable	Mean	Std. dev.	Min.	Max.	Obs.
Employ	41.528	13.692	26.435	73.021	40
Renewable	18,600,000	6,181,540	6,008,458	29,100,000	40
Fossil	895,000,000	376,000,000	390,000,000	1,710,000,000	40
Primary	91.865	9.166	76.463	113.146	40
Secondary	31.340	9.283	13.937	54.883	40
Tertiary	7.148	3.267	1.873	12.043	40
HCI	1.498	0.266	1.193	1.974	40
RGDPpc	1877.400	454.319	1408.209	2679.554	40
$MtCO_2pc$	0.685	0.140	0.408	0.933	40
Population	128,000,000	38,900,000	73,000,000	203,000,000	40

MMBtu to an upper limit of 1710 million MMBtu, highlighting the predominant reliance on fossil fuels within the energy mix of Nigeria.

Education indicators reveal primary school enrolment at a high average of 91.87%, indicating strong participation at the primary education level, albeit with some variation. Secondary and tertiary enrolment rates average 31.34% and 7.15%, respectively, reflecting lower participation rates in higher education stages, which could have implications for human capital development. The Human Capital Index (HCI) has a mean value of 1.50, with values ranging from 1.19 to 1.97, suggesting moderate human capital accumulation with room for improvement.

Economic indicators show that GDP per capita, adjusted to constant 2015 US dollars, averages $1877.40. The per capita CO_2 emissions, measured in million tonnes, average 0.685 with a standard deviation of 0.140, indicating relatively low but variable emissions per capita. The total population exhibits a mean of 128 million, ranging widely from 73 to 203 million, showcasing significant population growth over the study period.

Table 2.3 presents the Variance Inflation Factor (VIF) test results, which provide crucial insights into the potential multicollinearity among the independent variables in our models. Multicollinearity is a condition where independent variables are highly correlated, which can inflate the standard errors of the coefficients and make the results of the regression analysis less reliable. The general rule of thumb is that VIF values less than 10 indicate no severe multicollinearity issues. The mean VIF for all three model are well below the threshold of 10, indicating that multicollinearity is not a significant problem in these models.

Table 2.4 reports the Shapiro–Wilk (Shapiro & Wilk, 1965) test for normality. From the results, we have sufficient evidence to say that all variables, except ln *Primary* and ln *Secondary* are not normally distributed.

The unit root test results, detailed in Table 2.5, provide essential insights into the stationarity properties of the variables in our analysis. Using both the Augmented

Table 2.3 Variance Inflation Factor test for multicollinearity

	Model 1		Model 2		Model 3	
Variable	VIF	1/VIF	VIF	1/VIF	VIF	1/VIF
ln Renewable	2.80	0.358	2.75	0.364	4.93	0.203
ln Fossil	4.57	0.219	6.61	0.151	5.45	0.183
ln HCI	1.68	0.597	3.64	0.275	8.23	0.122
ln Renewable	2.35	0.426	2.52	0.397	2.68	0.372
ln Fossil	1.30	0.770	1.14	0.874	1.14	0.874
Mean VIF	2.54		3.33		4.49	

Note VIF values less than 10 suggest no problem of multicollinearity

Table 2.4 Shapiro–Wilk test

Variable	Obs.	W	V	z	Prob > z
ln Employ	40	0.930**	2.764	2.140	0.016
ln Renewable	40	0.890***	4.344	3.091	0.001
ln Fossil	40	0.940**	2.381	1.826	0.034
ln Primary	40	0.958	1.643	1.045	0.148
ln Secondary	40	0.950*	1.988	1.446	0.074
ln Tertiary	40	0.906***	3.696	2.751	0.003
ln RGDPpc	40	0.851***	5.901	3.736	0.000
ln $MtCO_2pc$	40	0.933**	2.656	2.056	0.020

Note Number of observations is 40. ***, **, * represent statistical significance at the 1%, 5%, and 10% level, respectively

Dickey-Fuller (ADF) test and the Phillip Perron (PP) test, we examined the variables at levels and first differences, considering both constant and constant and trend models. The results consistently indicate that most variables, including employment, CO_2 emissions per capita, primary, secondary, and tertiary school enrolments, GDP per capita, fossil energy consumption, and renewable energy consumption, are non-stationary at their levels. This non-stationarity is evidenced by the non-significant test statistics at levels, implying the presence of unit roots. However, upon differencing, these variables become stationary, as indicated by the highly significant test statistics, thus they are integrated of order one ($I(1)$). This transformation to stationarity upon first differencing ensures that these variables can be reliably used in subsequent econometric models.

Figure 2.1 presents the time series' evolution during the study period, 1980–2019.

Table 2.5 Unit root test

Variable	Levels				First differences				Remark
	Constant		Constant and trend		Constant		Constant and trend		
	t-stat	*p*-value	*t*-stat	*p*-value	*t*-stat	*p*-value	*t*-stat	*p*-value	
Panel A: Augmented Dickey–Fuller test result									
ln Employ	2.823	1.000	− 1.585	0.781	− 3.910***	0.005	− 5.030***	0.001	*I*(1)
ln HCI	− 0.690	0.837	− 3.226	0.095*	− 1.753	0.397	− 1.297	0.874	NS
ln $MtCO_2pc$	− 2.804	0.067*	− 2.748	0.224	− 7.065***	0.000	− 6.989***	0.000	*I*(1)
ln Primary	− 1.750	0.399	− 2.181	0.486	− 4.693***	0.001	− 4.623***	0.004	*I*(1)
ln RGDPpc	− 1.213	0.658	− 2.287	0.430	− 4.297***	0.002	− 3.743**	0.032	*I*(1)
ln Secondary	− 2.593	0.103	− 2.632	0.269	− 4.674***	0.001	− 4.587***	0.004	*I*(1)
ln Fossil	− 1.144	0.689	− 3.072	0.127	− 7.996***	0.000	− 7.896***	0.000	*I*(1)
ln Renewable	− 1.441	0.552	− 2.074	0.543	− 7.014***	0.000	− 6.964***	0.000	*I*(1)
ln Tertiary	− 1.591	0.478	− 1.188	0.899	− 4.782***	0.000	− 4.835***	0.002	*I*(1)
Panel B: Phillip Perron unit root test results									
ln Employ	2.853	1.000	− 1.506	0.810	− 3.972***	0.004	− 5.050***	0.001	*I*(1)
ln HCI	1.255	0.998	− 3.063	0.129	− 1.753	0.397	− 1.297	0.874	NS
ln $MtCO_2pc$	− 2.861	0.059*	− 2.823	0.198	− 7.059***	0.000	− 6.984***	0.000	*I*(1)
ln Primary	− 2.004	0.284	− 2.181	0.486	− 4.570***	0.001	− 4.481***	0.005	*I*(1)
ln RGDPpc	− 0.896	0.779	− 3.233	0.096*	− 4.355***	0.001	− 4.047***	0.015	*I*(1)
ln Secondary	− 2.434	0.139	− 3.031	0.137	− 4.643***	0.001	− 4.516***	0.005	*I*(1)
ln Fossil	− 1.038	0.730	− 3.213	0.097*	− 8.161***	0.000	− 8.067***	0.000	*I*(1)

(continued)

Table 2.5 (continued)

Variable	Levels				First differences				Remark
	Constant		Constant and trend		Constant		Constant and trend		
	t-stat	*p*-value	*t*-stat	*p*-value	*t*-stat	*p*-value	*t*-stat	*p*-value	
ln Renewable	− 1.363	0.590	− 2.032	0.566	− 7.059***	0.000	− 7.058***	0.000	*I*(1)
ln Tertiary	− 3.285	0.023**	− 2.539	0.309	− 4.771***	0.000	− 4.831***	0.002	*I*(1)

Note HCI exhibits non-stationarity (NS) both at levels and first differencing. ***, **, * represent statistical significance at the 1%, 5%, and 10% level, respectively

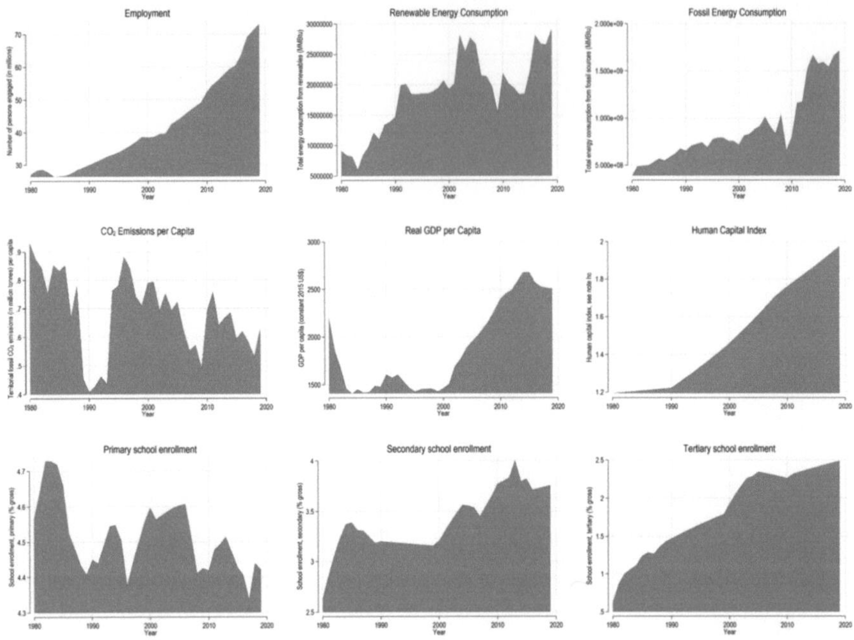

Fig. 2.1 Historical evolution of variables. *Source* Author's compilation

2.3.2 *Empirical Results*

The results of the KRLS analysis, as presented, demonstrate the impact of various factors on employment in Nigeria. The dependent variable in this analysis is employment, and the results table provides estimates for the average marginal effects (Avg.), standard errors (SE), *t*-values (t), *p*-values ($P > t$), and quantiles (P25, P50, P75) for each independent variable. Here's a detailed interpretation of these results:

Renewable Energy Consumption: The average marginal effect of renewable energy consumption on employment is positive and highly significant (*p*-value = 0.00). The results for Models I, II, and III indicate that a 1% increase in renewable energy consumption is associated with a 0.161% (Table 2.6), 0.191% (Table 2.7), and 0.098% (Table 2.8) increase in employment, respectively, holding other factors constant. The low standard errors (0.025–0.026) suggest precise estimates. The quantile results (P25, P50, P75) in Table 2.6 (0.027, 0.170, 0.306) and Table 2.7 (0.039, 0.184, and 0.387) indicate that the positive impact of renewable energy consumption on employment is consistent across different levels, although the effect size varies. Conversely, in Table 2.8 where we control for tertiary enrolment, renewable energy consumption has heterogeneous effect on employment (− 0.036, 0.109, 0.218). The results suggest that at the lower quantile (P25), renewable energy consumption is negatively associated with employment, while at the median or higher quantiles, there is a positive association.

Table 2.6 KRLS pointwise derivatives estimates of model I

Variables	Avg.	SE	t	P > t	P25	P50	P75
ln Renewable	0.161***	0.025	6.408	0.000	0.027	0.170	0.306
ln Fossil	0.156***	0.026	5.973	0.000	0.079	0.174	0.235
ln Primary	0.086	0.084	1.027	0.311	− 0.309	− 0.020	0.528
ln RGDPpc	0.280***	0.038	7.451	0.000	0.068	0.176	0.430
ln $MtCO_2pc$	− 0.030	0.040	− 0.746	0.461	− 0.141	− 0.015	0.091
Model diagnostics							
Lambda = 0.113; Tolerance = 0.04; Sigma = 5; Eff. df = 20; R^2 = 0.988; Looloss = 0.885							

Note Dependent variable is Employment (ln *Employ*). Number of observations is 40. ***, **, * represent statistical significance at the 1%, 5%, and 10% level, respectively

Table 2.7 KRLS pointwise derivatives estimates of model II

Variables	Avg.	SE	t	P > t	P25	P50	P75
ln Renewable	0.191***	0.026	7.243	0.000	0.039	0.184	0.387
ln Fossil	0.138***	0.035	3.966	0.000	0.087	0.166	0.190
ln Secondary	0.103**	0.041	2.506	0.017	− 0.041	0.108	0.195
ln RGDPpc	0.174***	0.054	3.261	0.002	− 0.018	0.148	0.297
ln $MtCO_2pc$	− 0.070	0.043	− 1.631	0.112	− 0.179	− 0.074	0.053
Model diagnostics							
Lambda = 0.064; Tolerance = 0.04; Sigma = 5; Eff. df = 19.56; R^2 = 0.991; Looloss = 0.531							

Note Dependent variable is Employment (ln *Employ*). ***, **, * represent statistical significance at the 1%, 5%, and 10% level, respectively

Table 2.8 KRLS pointwise derivatives estimates of model III

Variables	Avg.	SE	t	P > t	P25	P50	P75
ln Renewable	0.098***	0.025	3.952	0.000	− 0.036	0.109	0.218
ln Fossil	0.093***	0.031	3.010	0.005	0.033	0.077	0.142
ln Tertiary	0.146***	0.020	7.209	0.000	0.105	0.146	0.205
ln RGDPpc	0.228***	0.044	5.190	0.000	0.001	0.163	0.354
ln $MtCO_2pc$	− 0.045	0.036	− 1.225	0.229	− 0.116	− 0.055	0.053
Model diagnostics							
Lambda = 0.059; Tolerance = 0.04; Sigma = 5; Eff. df = 18.92; R^2 = 0.994; Looloss = 0.499							

Dependent variable is Employment (ln *Employ*). ***, **, * represent statistical significance at the 1%, 5%, and 10% level, respectively

Fossil Energy Consumption: The average marginal effect of fossil energy consumption on employment is positive and highly significant (p-value = 0.00). The results for Models I, II, and III indicate that a 1% increase in fossil energy consumption is associated with a 0.156% (Table 2.6), 0.138% (Table 2.7), and 0.093% (Table 2.8) increase in employment, respectively, holding other factors constant. The low standard errors (0.026–0.035) suggest precise estimates. The quantiles (P25, P50, P75) in Table 2.6 (0.079, 0.174, 0.235) and Table 2.7 (0.087, 0.166, 0.190), and Table 2.8 (0.033, 0.077, 0.142). This suggests that increasing fossil energy consumption positively affects employment, albeit to a slightly lesser extent than renewable energy.

Primary School Enrolment (ln *Primary*): In Table 2.6, the coefficient for primary school enrolment is 0.086 with a p-value of 0.311, indicating that this variable is not statistically significant. This suggests that variations in primary school enrolment do not have a clear impact on employment levels in this model. The standard error (0.084) and the quantiles (− 0.309, − 0.020, 0.528) show considerable variability, indicating that the impact of primary education on employment is not consistent and might be influenced by other unobserved factors.

Secondary School Enrolment: In Table 2.7, the average effect of secondary school enrolment on employment is 0.103, with a p-value of 0.017, indicating significance. This suggests that higher secondary education enrolment rates contribute positively to employment levels. The standard error (0.041) and the quantiles (− 0.041, 0.108, 0.195) show that while the relationship is predominantly positive, there is some variability, and in some instances, the effect might be minimal or slightly negative.

Tertiary School Enrolment (ln *Tertiary*): In Table 2.8, The coefficient for tertiary school enrolment is 0.146 with a p-value of 0.000, indicating a highly significant positive relationship with employment. This implies that higher rates of tertiary education enrolment contribute positively to employment levels. The standard error (0.020) and the quantiles (0.105, 0.146, 0.205) show a consistent positive impact, highlighting the importance of higher education for employment.

GDP Per Capita (RGDPpc): The average marginal effect of per capita GDP on employment is positive and highly significant (p-value = 0.00). The results for Models I, II, and III indicate that a 1% increase in fossil energy consumption is associated with a 0.280% (Table 2.6), 0.174% (Table 2.7), and 0.228% (Table 2.8) increase in employment, respectively, holding other factors constant. The quantile results (P25, P50, P75) in Table 2.6 (0.068, 0.176, 0.430) and Table 2.8 (0.001, 0.163, 0.354) suggest that increasing per capita GDP positively affects employment. Conversely, quantile results in Table 2.7 (− 0.018, 0.148, 0.297) suggest that while the central tendency is positive, there are instances where the effect could be negligible or even negative.

CO_2 Emissions Per Capita (MtCO_2pc): The coefficient for CO_2 emissions per capita across models ranges from − 0.070 to − 0.030. However, the coefficients are not statistically significant (p-value > 0.05). Moreover, the results imply a potential negative relationship between CO_2 emissions and employment, suggesting that higher emissions might be associated with lower employment levels, though this relationship is not strongly supported by the data. The quantiles (P25, P50, P75) in Table 2.6 (– 0.141, – 0.015, 0.091), Table 2.7 (– 0.179, – 0.074, 0.053) Table 2.8 (– 0.116, – 0.055, 0.053) reflect the uncertainty and variability in this effect.

Model Summary and Diagnostics: The KRLS model explains a significant portion of the variance in employment, as indicated by the high R-squared value of ~ 0.99 (i.e., 99%) across models. This suggests that the model fits the data very well. The effective degrees of freedom (Eff. df) are ~ 20, indicating the flexibility of the KRLS method in capturing the underlying data structure. The leave-one-out loss (Looloss) ranges from ~ 0.50 to 0.89, which are low, indicating good predictive performance and model stability.

2.4 Discussion

The Kernel-Based Regularized Least Squares (KRLS) analysis provides a comprehensive understanding of the factors affecting employment in Nigeria, focusing on renewable energy consumption, fossil energy consumption, education levels, GDP per capita, and CO_2 emissions. This method, unlike traditional econometric techniques, captures non-linear relationships and offers robust, precise estimates.

The results show that renewable energy consumption positively impacts employment, with a 1% increase leading to a 0.098–0.191% increase in employment. This finding is in line with the ex-post analysis of Musa and Maijama'a (2020), which find that an increase in renewable energy consumption reduces unemployment in Nigeria. Similarly, this finding aligns with Oteng et al. (2024), which demonstrated that renewable energy projects significantly boost employment in the industrial sector in Ghana. Further, the results are in line with ex-ante studies which indicate that increase in renewable energy consumption has positive employment effects in Nigeria (Emenekwe et al., 2022c; Energy Transition Office, 2022; National Climate Change Council, 2023; Okereke et al., 2024; UNDP & ILO, 2021). Overall, these results suggest that increasing renewable energy capacity can drive job creation, particularly in installation, maintenance, and related industries, underscore its role in sustainable development and climate justice.

On the other hand, fossil energy consumption also shows a positive and significant relationship with employment, though the effect is slightly less pronounced than that of renewable energy. A 1% increase in fossil energy consumption results in a 0.093–0.156% increase in employment. This finding is not unexpected, since fossil energy has historically constituted the backbone the modern economy and improvement of living standards globally (Emenekwe et al., 2022a). Similarly, Oteng et al. (2024) find that an increase in fossil fuel consumption increases industrial employment. It is, further argued in the literature, that the unlimited exploitation of natural resources is commonly accepted as the price to pay for fuelling economic growth and providing employment (Dragulanescu & Dragulanescu, 2013; Muazu et al., 2023). Our finding is further supported by Karekezi et al. (2012), who argue that the lack of access to modern, affordable energy services is a significant factor contributing to poverty in a developing region like sub-Saharan Africa, and especially in Nigeria. They emphasize that modern energy access (including natural gas products, which is a fossil fuel) is crucial for alleviating poverty, fostering economic growth and job creation, enhancing social service delivery, and promoting overall sustainable human development.

A key finding in this study is that renewable energy tends to have a significantly larger impact on employment levels compared to fossil energy technologies. Swain et al. (2022) argue that this is primarily due to the labour-intensive nature of renewable energy projects, which create more jobs per unit of energy produced. In contrast, employment in the fossil energy sector is largely tied to fuel supply, a sector with limited domestic job creation in Nigeria. Additionally, Nigeria's status as a net exporter of crude oil and a net importer of refined products, mainly from the Netherlands and Spain, suggests that many potential jobs in the fossil energy sector are generated abroad.

The analysis of educational impacts reveals that tertiary education significantly enhances employment, indicating the importance of higher education in equipping the workforce with necessary skills. This is consistent with findings from Oteng et al. (2024), which emphasized the role of education in supporting the renewable energy transition and job creation. Secondary education also has a positive impact, though with some variability, suggesting that improvements in secondary education could further boost employment. In contrast, primary education does not show a significant impact on employment, highlighting that higher education levels are more critical for job market integration.

GDP per capita emerges as a robust positive determinant of employment, reflecting the general trend that economic growth fosters job creation. This finding aligns with multiple studies, including those by Aigheyisi and Edore (2021), which find a positive relationship between economic growth and employment in Nigeria. Conversely, CO_2 emissions per capita exhibit a negative, though not statistically significant at conventional levels, relationship with employment. The negative relationship in this study is consistent with ILO (2018), indicate that carbon emissions and climate change generally have a net negative effect on employment. This adverse impact is attributed to the detrimental effects of CO_2 on climate change, which disrupts ecosystem services and the jobs that depend on them. Sectors such as farming, fishing, forestry, food and beverage production, wood and paper industries, renewable energy, water management, textiles, chemicals, and environment-related tourism are particularly affected due to their reliance on natural processes (Swain et al., 2022). This suggests potential negative impacts of high emissions on job sustainability, reinforcing the need for environmentally conscious economic policies.

2.5 Conclusion and Policy Recommendations

2.5.1 Conclusion

This study provides a comprehensive analysis of the impact of renewable energy consumption on employment in Nigeria, employing the Kernel-Based Regularized Least Squares (KRLS) method to account for the complex, non-linear relationships between key variables. Our findings underscore the significant positive effects of renewable energy consumption on employment, with a 1% increase in renewable energy leading to a 0.098–0.191% increase in employment. This result aligns with similar studies conducted in other African contexts, where renewable energy projects have been shown to substantially boost employment across various sectors.

In contrast, while fossil energy consumption also positively affects employment, its impact is less pronounced compared to renewable energy. Specifically, a 1% increase in fossil energy consumption is associated with a 0.093–0.156% increase in employment. These findings are consistent with historical trends in Nigeria and other developing countries, where fossil fuels have traditionally driven economic growth and job creation. However, given the global push towards sustainability and climate justice, transitioning to renewable energy sources remains crucial for long-term economic and environmental benefits.

The study also highlights the critical role of higher education in enhancing employment outcomes. Tertiary education, in particular, shows a significant positive impact on employment, emphasizing the importance of advanced education in equipping the workforce with necessary skills for the evolving job market. Secondary education also positively influences employment, suggesting that improvements at this level can further boost job creation. Conversely, primary education does not show a significant impact, indicating that higher education levels are more critical for employment in the Nigerian context.

Furthermore, GDP per capita is identified as a robust positive determinant of employment, reflecting the general trend that economic growth fosters job creation. This finding is consistent with prior research, which also found a positive relationship between economic growth and employment in Nigeria. On the other hand, CO_2 emissions per capita exhibit a negative relationship with employment, although not statistically significant. This suggests potential adverse effects of high emissions on job sustainability, reinforcing the need for environmentally conscious economic policies (ILO, 2018).

2.5.2 Policy Recommendations

Based on the findings, the following policy recommendations are proposed:

Enhance Renewable Energy Adoption: Policymakers should prioritize and incentivize the adoption of renewable energy technologies. This can be achieved through subsidies, tax incentives, and supportive regulatory frameworks that reduce barriers to entry for renewable energy projects. Increasing renewable energy capacity not only contributes to job creation but also aligns with global sustainability goals.

Invest in Education: Strengthening educational systems, particularly at the secondary and tertiary levels, is crucial. Policies aimed at improving access to quality education and aligning curricula with market needs will equip the workforce with the necessary skills to thrive in a transitioning economy. This includes technical and vocational training focused on renewable energy technologies.

Promote Economic Growth: Policies that stimulate economic growth, such as improving infrastructure, fostering innovation, and enhancing trade, will have positive spillover effects on employment. Ensuring a stable macroeconomic environment is essential for attracting investment and fostering job creation.

Adopt Environmentally Conscious Policies: To mitigate the negative impacts of CO_2 emissions on employment, it is essential to adopt policies that encourage sustainable practices. This includes enforcing environmental regulations, promoting energy efficiency, and investing in cleaner technologies.

Support for Transitional Industries: As the economy shifts towards more sustainable energy sources, it is important to support industries and workers transitioning from fossil fuel-based sectors. This can be achieved through retraining programs, financial support for affected workers, and initiatives to attract new industries to regions dependent on fossil fuels.

While this study provides significant insights into the relationship between renewable energy consumption and employment in Nigeria, several limitations must be acknowledged. Firstly, the study focuses on aggregate employment without distinguishing between different sectors or types of employment. This approach overlooks potential variations in the impact of renewable energy consumption across different industries and job categories. Sector-specific analyses could provide a more detailed understanding of how renewable energy adoption affects various parts of the economy. Secondly, while the study includes key variables such as per capita GDP, education levels, and CO_2 emissions, there are other potentially important factors not considered due to data limitations. Variables such as technological advancements, government policies, and international trade dynamics could also play significant roles in shaping the relationship between energy consumption and employment. Lastly, the analysis assumes that the relationships between the variables are stable over time. However, structural changes in the economy, such as shifts in energy policy or major economic reforms, could alter these relationships. Future research could benefit from a dynamic modelling approach to account for such changes.

Despite these limitations, the findings of this study offer valuable insights for policymakers and stakeholders, emphasizing the importance of promoting renewable energy and improving educational outcomes to foster sustainable employment growth in Nigeria. Future research should address these limitations to further refine our understanding of the complex interactions between energy consumption, economic development, and employment.

Acknowledgements Not applicable.

Funding We did not approach for the financial support pertaining to the submitted manuscript.

Data Availability Statement This study used publicly available data.

Ethical Clearance While preparing the manuscript, to the best of our knowledge, we have followed required ethical procedures. We would like to declare that there is no conflict from authors and any other party directly or indirectly.

Conflict of Interest Chukwuemeka Emenekwe is an editor of this handbook and was not involved in the editorial review, or the decision to publish this article. All authors declare that there are no competing interests.

References

Aigheyisi, O., & Edore, J. O. (2021). Economic growth and employment in Nigeria's services sector. *Journal of Economics and Allied Research, 6*(1), 89–102.

Almutairi, K., Thoma, G., & Durand-Morat, A. (2018). Ex-ante analysis of economic, social and environmental impacts of large-scale renewable and nuclear energy targets for global electricity generation by 2030. *Sustainability, 10*(8). https://doi.org/10.3390/su10082884

Arvanitopoulos, T., & Agnolucci, P. (2020). The long-term effect of renewable electricity on employment in the United Kingdom. *Renewable & Sustainable Energy Reviews, 134*. https://doi.org/10.1016/j.rser.2020.110322

Böhringer, C., Keller, A., & van der Werf, E. (2013). Are green hopes too rosy? Employment and welfare impacts of renewable energy promotion. *Energy Economics, 36*, 277–285. https://doi.org/10.1016/j.eneco.2012.08.029

Cai, M., Cusumano, N., Lorenzoni, A., & Pontoni, F. (2017). A comprehensive ex-post assessment of RES deployment in Italy: Jobs, value added and import leakages. *Energy Policy, 110*, 234–245. https://doi.org/10.1016/j.enpol.2017.08.013

Cameron, L., & van der Zwaan, B. (2015). Employment factors for wind and solar energy technologies: A literature review. *Renewable and Sustainable Energy Reviews, 45*, 160–172. https://doi.org/10.1016/j.rser.2015.01.001

Department of Climate Change. (2021). *Nigeria's first nationally determined contribution (NDC): 2021 update*. Department of Climate Change, Federal Ministry of Environment. https://climatechange.gov.ng/wp-content/uploads/2021/08/NDC_File-Amended-_11222.pdf

Dragulanescu, I.-V., & Dragulanescu, N. (2013). Some theories of environmental sustainability. *Romanian Statistical Review, 61*(12), 14–23.

Emenekwe, C. C., Nnamani, U., & Ayodeji, S. (2024). *Green industrialization in Africa: Opportunities and tensions in Africa-Europe relations: Scoping paper on Nigeria*. https://www.iddri.org/sites/default/files/PDF/Publications/CatalogueIddri/Rapport/Ukama_NIGERIA.pdf

Emenekwe, C. C., Okereke, C., Nnamani, U. A., Emodi, N. V., Diemuodeke, O. E., & Anieze, E. E. (2022a). Macroeconomics of decarbonization strategies of selected global south countries: A systematic review. *Frontiers in Environmental Science, 10*. https://doi.org/10.3389/fenvs.2022.938017

Emenekwe, C. C., Onyeneke, R. U., & Nwajiuba, C. U. (2022b). Financial development and carbon emissions in Sub-Saharan Africa. *Environmental Science and Pollution Research, 29*(13), 19624–19641. https://doi.org/10.1007/s11356-021-17161-7

Emenekwe, C. C., Onyeneke, R. U., Nwajiuba, C. U., Anugwa, I. Q., & Emenekwe, O. U. (2023). Determinants of consumption-based and production-based carbon emissions. *Environment, Development and Sustainability*. https://doi.org/10.1007/s10668-023-04311-9

Emenekwe, C. C., Onyeneke, R. U., & Okereke, C. (2022c). *Economic analysis of inclusive green recovery and transition opportunities in Nigeria's energy sector: A policy brief*. Centre for Climate Change and Development (CCCD), Alex Ekwueme Federal University Ndufu-Alike.

Energy Transition Office. (2022). *Nigeria's pathway to achieve carbon neutrality by 2060*. https://energytransition.gov.ng/

Ferwerda, J., Hainmueller, J., & Hazlett, C. (2015). KRLS: A Stata package for kernel-based regularized least squares. *SSRN Electronic Journal, 55*. https://doi.org/10.2139/ssrn.2325523

Friedlingstein, P., Jones, M. W., O'Sullivan, M., Andrew, R. M., Bakker, D. C. E., Hauck, J., Le Quéré, C., Peters, G. P., Peters, W., Pongratz, J., Sitch, S., Canadell, J. G., Ciais, P., Jackson, R. B., Alin, S. R., Anthoni, P., Bates, N. R., Becker, M., Bellouin, N., et al. (2022). Global carbon budget 2021. *Earth System Science Data, 14*(4), 1917–2005. https://doi.org/10.5194/essd-14-1917-2022

Garrett-Peltier, H. (2017). Green versus brown: Comparing the employment impacts of energy efficiency, renewable energy, and fossil fuels using an input-output model. *Economic Modelling, 61*, 439–447. https://doi.org/10.1016/j.econmod.2016.11.012

Hainmueller, J., & Hazlett, C. (2014). Kernel regularized least squares: Reducing misspecification bias with a flexible and interpretable machine learning approach. *Political Analysis, 22*(2), 143–168. http://www.jstor.org/stable/24573219

IEA. (2021). *Net zero by 2050: A roadmap for the global energy sector*. https://www.iea.org/reports/net-zero-by-2050

ILO. (2018). *The employment impact of climate change adaptation: Input document for the G20 climate sustainability working group international labour office*. International Labour Organization (ILO).

Karekezi, S., McDade, S., Boardman, B., Kimani, J., & Lustig, N. (2012). Energy, poverty, and development. In G. E. A. W. Team (Ed.), *Global energy assessment: Toward a sustainable future* (pp. 151–190). Cambridge University Press. https://doi.org/10.1017/CBO9780511793677.008

Lawan, M. W., & Gidigbi, M. O. (2019). Causal link between employment and renewable energy consumption: Evidence from Nigeria. *The European Journal of Applied Economics, 16*(1), 30–40. https://doi.org/10.5937/EJAE15-19730

Le Treut, G., Lefèvre, J., Lallana, F., & Bravo, G. (2021). The multi-level economic impacts of deep decarbonization strategies for the energy system. *Energy Policy, 156*, 112423. https://doi.org/10.1016/j.enpol.2021.112423

Mu, Y., Cai, W., Evans, S., Wang, C., & Roland-Holst, D. (2018). Employment impacts of renewable energy policies in China: A decomposition analysis based on a CGE modeling framework. *Applied Energy, 210*, 256–267. https://doi.org/10.1016/j.apenergy.2017.10.086

Muazu, A., Yu, Q., & Alariqi, M. (2023). The impact of renewable energy consumption and economic growth on environmental quality in Africa: A threshold regression analysis. *Energies, 16*(11). https://doi.org/10.3390/en16114533

Musa, K. S., & Maijama'a, R. (2020). Causal relationship between renewable energy consumption and unemployment in Nigeria: Evidence from Toda and Yamamoto causality technique. *Energy Economics Letters, 7*(1), 46–60. https://doi.org/10.18488/journal.82.2020.71.46.60

National Climate Change Council. (2023). *Nigeria's long-term low-emission development strategy—2060.* National Council on Climate Change. https://natccc.gov.ng/publications/Final_Version_Nigeria's_LT-LEDS.pdf

Okereke, C., Diemuodeke, O. E., Anyanwu, C., Emodi, N. V., Owebor, K., Emenekwe, C., Nnamani, U., Owolabi, A., Ojike, O., Nwachukwu, C. O., Ekwe, E. B., Ghersi, F., & Michoud, B. (2024). *Deep decarbonization pathways (DDP) for Nigeria's low emission development up to 2060.* https://ddpinitiative.org/wp-content/uploads/pdf/nigerias-deep-decarbonization-projec_ddp_-report.pdf

Oteng, C., Iledare, O., Peprah, J. A., & Gamette, P. (2024). Towards just energy transition: Renewable energy transition dynamics and sectorial employment in Ghana. *Sustainability, 16*(9). https://doi.org/10.3390/su16093761

Proença, S., & Fortes, P. (2020). The social face of renewables: Econometric analysis of the relationship between renewables and employment. *Energy Reports, 6*, 581–586. https://doi.org/10.1016/j.egyr.2019.09.029

PWT. (2021). *Penn World Table version 10.0.* https://www.rug.nl/ggdc/productivity/pwt/?lang=en

Ram, M., Osorio-Aravena, J. C., Aghahosseini, A., Bogdanov, D., & Breyer, C. (2022). Job creation during a climate compliant global energy transition across the power, heat, transport, and desalination sectors by 2050. *Energy, 238*, 121690. https://doi.org/10.1016/j.energy.2021.121690

Saboori, B., Gholipour, H. F., Rasoulinezhad, E., & Ranjbar, O. (2022). Renewable energy sources and unemployment rate: Evidence from the US states. *Energy Policy, 168*, 113155. https://doi.org/10.1016/j.enpol.2022.113155

Sarkodie, S. A., & Owusu, P. A. (2020). How to apply the novel dynamic ARDL simulations (dynardl) and kernel-based regularized least squares (krls). *MethodsX, 7*, 101160. https://doi.org/10.1016/j.mex.2020.101160

Shapiro, S. S., & Wilk, M. B. (1965). An analysis of variance test for normality (complete samples). *Biometrika, 52*(3/4), 591–611. https://doi.org/10.2307/2333709

Stavropoulos, S., & Burger, M. J. (2020). Modelling strategy and net employment effects of renewable energy and energy efficiency: A meta-regression. *Energy Policy, 136*. https://doi.org/10.1016/j.enpol.2019.111047

Su, X., Ghersi, F., Teng, F., Le Treut, G., & Liang, M. (2021). The economic impact of a deep decarbonisation pathway for China: A hybrid model analysis through bottom-up and top-down linking. *Mitigation and Adaptation Strategies for Global Change, 27*(1), 11. https://doi.org/10.1007/s11027-021-09979-w

Swain, R. B., Karimu, A., & Gråd, E. (2022). Sustainable development, renewable energy transformation and employment impact in the EU. *International Journal of Sustainable Development & World Ecology, 29*(8), 695–708. https://doi.org/10.1080/13504509.2022.2078902

UNDP & ILO. (2021). *Nigeria green jobs assessment report: Measuring the socioeconomic impacts of climate policies to guide NDC enhancement and a just transition.* United Nations Development Programme (UNDP).

Wang, R., Assenova, V. A., & Hertwich, E. G. (2021). Energy system decarbonization and productivity gains reduced the coupling of CO_2 emissions and economic growth in 73 countries between 1970 and 2016. *One Earth, 4*(11), 1614–1624. https://doi.org/10.1016/j.oneear.2021.10.010

Chapter 3
The Role of Development Assistance, Foreign Direct Investments, Financial Development, and Institutional Quality in Fostering Energy Transition in Africa

Robert Ugochukwu Onyeneke, Daniel Adu Ankrah, Chinenye Judith Onyeneke, and Nkechi Precious Obieche

Abstract Despite the growing importance of foreign direct investment, official development assistance, financial development, and institutional quality in Africa, their plausible combined impact on energy transition has not been explored, creating a void in literature. Our paper contributes to the literature by analysing the effect of foreign direct investment, official development assistance, financial development, and institutional quality on energy transition in Africa between 2009 and 2020. The panel autoregressive distributed lag (PARDL) model was employed for the analysis. The empirical evidence shows that foreign direct investment, official development assistance, and institutional quality significantly increased renewable energy consumption on the continent in the long run. On the other hand, financial development, measured as domestic credit to private sector by banks, tends to reduce renewable energy consumption on the continent both in the short and long runs. This indicates that foreign direct investment, official development assistance, and institutional quality foster just energy transition in Africa, while financial development tends to inhibit energy transition on the continent. These findings underscore the necessity for nuanced policy interventions that integrate foreign direct investment,

R. U. Onyeneke
Department of Agricultural Economics, Alex Ekwueme Federal University Ndufu-Alike, Ikwo, Nigeria
e-mail: robert.onyeneke@funai.edu.ng

D. A. Ankrah
Department of Agricultural Extension, University of Ghana, Legon, Accra, Ghana
e-mail: dankrah@ug.edu.gh

C. J. Onyeneke (✉)
Department of Agricultural Extension, Alex Ekwueme Federal University Ndufu-Alike, Ikwo, Nigeria
e-mail: onyeneke.chinenye@funai.edu.ng

N. P. Obieche
Department of Supply Chain Management, Federal University of Technology Owerri, Owerri, Nigeria

R. U. Onyeneke et al. (eds.), *Energy Transition, Climate Action and Sustainable Agriculture*, https://doi.org/10.1007/978-3-031-83165-2_3

official development assistance with institutional quality objectives, for the promotion energy transition. Policies are also needed to deliberately encourage banks to provide credit facilities for energy transition investments in Africa.

Keywords Energy transition · Financial development · Official development assistance · Institutional quality · Foreign direct investment · Panel autoregressive distributed lag model · Africa

3.1 Introduction

The term "energy transition" has gained grounds in development, political, and academic circles for more than thirty years. It first started off in the global north, then to the global south, and over time, it has attained credence in empirical and theoretical debates around the globe (Akram et al., 2023; Bouyghrissi et al., 2022; Dominkovic et al., 2018; Jangam et al., 2020; Li et al., 2020).

Pursuing an agenda of energy transition cannot be swept under carpet because, it remains important in advancing the sustainable development goals (SDGs), notably the achievement of SDG-7 (affordable and clean energy) and SDG-13 (climate action). The literature (Araújo, 2014; Jiang & O'Neill, 2004) defines energy transition as the process of substituting fossil fuels with energy sources that are low in carbon emissions. Energy transition represents significant structural changes in a country's energy system with regards to the consumption and supply of environment-friendly energy sources. Indeed, the just ended Conference of the Parties (COP)-28 and most development advancement efforts (Zhang et al., 2017), have remained resolute and decisive to replace fossil fuels with more environmentally friendly sources in meeting energy demands for cooking, production, travel and heating.

The quest for a just energy transition in the globe and particularly Africa has indeed gained prominence, but even more striking in the case of Africa, is the challenge of financial constraints that has stifle the pace of development. The question of how Africa's use of foreign direct investment (FDI), official development assistance (ODA), financial development and institutional quality can help advance the attainment of the SDG-7 and SDG-13 remains relevant. Particularly, in the current dispensation, where there is a growing consensus on pursuing an agenda for green growth through the increase consumption of renewable energy, and the cutting down on fossil fuels consumption. This article therefore interrogates how FDI, ODA, financial development and institutional quality influence Africa's attainment of a just energy transition, with an eye to distil important policy ramifications.

Despite the consensus to double efforts towards the attainment of a just energy transition process, its pace in the global south has remained sluggish due to limited attention paid to inclusiveness, and sustainability (Ramzan et al., 2024). The attainment of a just energy transition demands FDI as an important conduit to address energy poverty (Caetano et al., 2022). FDI has been advocated to help transfer new technologies, improves managerial skills (human capital), creates new jobs, and improves economic development (Polat, 2018). And indeed, deliberate efforts have been advanced by the developed countries to developing countries in this regard. For instance, between 2000 and2017, FDI constituted 3.6% source of capital in Africa (Adjei-Mantey & Adams, 2023; World Bank, 2018). Few studies (Adjei-Mantey & Adams, 2023; Caetano et al., 2022; Polat, 2018; Xu et al., 2021) have examined the nexus between FDI, and renewable energy, with mixed results due to regional differences and country context specific factors (Caglar, 2020; Shahbaz et al., 2015). While some literature documents a pollutant effect of FDI, the contrary effect has also been found (Doytch & Narayan, 2016; Mielnik & Goldemberg, 2002; Nepal et al., 2021; Xu et al., 2021). FDI reduces pollution effect through the transfer of energy efficient processes to other countries, especially from the global north to the global south, in what has been defined as the Pollution Halo Hypothesis (PHlH) in the related literature (Balsalobre-Lorente et al., 2019; Polat, 2018). This implies that some developing countries relax their environmental laws and policies to attract FDI. The PHlH states that FDI reduces pollution effect by passing on green technologies (Nepal et al., 2021; Yilanci et al., 2020; Yin et al., 2021). Polat (2018) argued that FDI potentially promotes renewable energy consumption through technological diffusion. Other studies (Hübler & Keller, 2010; Lee, 2013; Zeeb et al., 2015) show no effect between FDI and energy consumption. With others (Aslan et al., 2014; Mohamed & Mamat, 2016; Paramati et al., 2016) showing a positive effect. Generally, few studies have examined the nexus between FDI and renewable energy consumption (Er et al., 2018; Ghazouani, 2018; Khandker et al., 2018; Kilicarslan, 2019).

Official development assistance (ODA) remains paramount in developing countries quest to explore renewable energy consumption and decrease dependence on fossil fuels. Like FDI, ODI serves as a purpose to assist developing countries promote economic development and achieve improved welfare outcomes (Carfora et al., 2021). Here again the literature documents mixed findings. Some literature (Gualberti et al., 2014; Hübler & Keller, 2010; Ince et al., 2016; Nakhooda et al., 2014; Zhang & Zhou, 2016) finds a positive relationship between ODA and renewable energy consumption, while others (Bae et al., 2016) find a negative relationship. Some literature (Murshed, 2020) shows no relationship. In order, to maximize the efficiency of ODA and increase the understanding of its impact and the logic of attribution of ODA to renewable energy consumption, some researchers (Bourguignon & Platteau, 2015; Li et al., 2021; Pickering et al., 2017) have called for further studies to shed light on the issue. Particularly on how green ODA influences renewable energy consumption. Our study therefore pursues this recommendation to deepen understanding in this regard.

Overall, institutions and proper governance has been found to promote environmental quality. In this regard, various economists have used different proxies (political rights, corruption, right to information, government effectiveness, regulatory quality, freedom of speech, law and order) for institutional quality (Inglesi-Lotz, 2024; Rafiq et al., 2024). Esty and Porter (2005) argued that the nexus between energy consumption and environmental quality has been less explored in sub-Saharan Africa (SSA). Azam et al. (2021) recently echoed the limited account on the nexus between renewable energy and institutional quality in Africa. The few studies (Amoah et al., 2020; Chang & Wang, 2017; Cifor et al., 2015; Wirth, 2014) which have examined this nexus, provides mixed accounts, thereby leading to limited solutions. While a strand of literature (Ali et al., 2020; Amoah et al., 2020; Barbier, 2015; Bernauer et al., 2012; Dasgupta & Mäler, 2004; Güngör et al., 2021; Udeagha & Ngepah, 2022) supports a positive impact between institutional quality and renewable energy consumption, others have established a negative relationship (Gallagher & Thacker, 2008), with others (Scruggs & Rivera, 2008) showing no relationship. Arguments have been made to show that democracy positively impacts environmental quality through freedom of information and political rights (Barbier, 2015; Bernauer et al., 2012).

Sadorsky (2011) argues that financial development increases energy consumption. This happens through three routes. First, financial development directly influences consumers to borrow money to purchase luxury items that increases energy consumption. Second, there is a business effect where enterprises borrow money to finance production, thereby increasing energy demand. Finally, the wealth effect improves participation on the stock market that raises business and consumer confidence. Overall, this improves economic activities and increases energy consumption. The assertion that financial development increases energy consumption is given credence by other literatures (Charfeddine & Kahia, 2019; Fung, 2009; Kahouli, 2017; Liu et al., 2018; Ntow-Gyamfi et al., 2020; Sadorsky, 2010, 2011; Shahbaz & Lean, 2012; Wu & Broadstock, 2015). On the contrary, other literatures (Assi et al., 2020; Kahouli, 2017) finds a negative impact of financial development on energy consumption. Denisova (2020) finds no effect of financial development on renewable energy consumption. Chiu and Lee (2020) argues that so far, limited studies have examined the nexus between financial development and energy consumption. Our study attempts to deepen understanding.

The summary of the related literature shows studies that consider foreign direct investment, official development assistance, financial development, and institutional quality in Africa, in silos, no attempt has been advanced in the extant literature to assess the plausible combined impacts of our covariates on energy transition in Africa. The objective of our paper is to assess the potential impact of FDI, ODA, financial development, and institutional quality on renewable energy transition. To the best of our knowledge, no related study has assessed the joint impacts of our study's covariates on renewable energy transition in Africa. Our empirical findings indicate that FDI, ODA, financial development and institutional quality significantly increases renewable energy consumption on the continent in the long run. On the other hand, financial development, measured as domestic credit to private sector by banks, tends to reduce renewable energy consumption on the continent both in the short and long run. This indicates that foreign direct investment, official development assistance, and institutional quality foster just energy transition in Africa, while financial development tends to inhibit energy transition on the continent.

In summary, our study makes a threefold contribution to the extant literature. First, distinct from prior studies, our study jointly accounts for the impact FDI, ODA, financial development and institutional quality from a single study. This will help governments and policy makers in African countries to examine the joint impacts and rightly put in place policies that potentially addresses challenges found and foster green growth. Second, our study incorporates six (6) institutional quality indicators and created an index of institutional quality using principal component analysis. This remains relevant, given that most African countries have challenges with institutional quality, thus renewable energy transition cannot take place, without considering institutional quality indicators. Finally, our study employs a panel autoregressive distributed lag (PARDL) model, to account for short and long run impacts. Thereby contributing to the estimation rigour and bringing into perspectives policy drives that can be put in place in the short and long run.

3.2 Methodology

3.2.1 Data Source

This study evaluates the role of foreign direct investments, financial development, institutional quality, and official development assistance in fostering energy transition in Africa. Africa was chosen because the continent is rich in renewable energy resources, which have not been adequately developed and utilized. Also, the continent is uniquely positioned at the crossroads of growth and significant environmental impact (Wang et al., 2023). Africa relies heavily on fossil fuels and faces distinct challenges and opportunities in the energy transition due to development pressures (Destek et al., 2022). The final forty-four sample countries were chosen based on data availability, which also ensured the reliability and the representativeness of the findings. Studying these African countries provides valuable insights into the energy transition process and the findings have far-reaching implications for attainment of the sustainable development goals and achieving the ambitions of the Paris Climate Agreement. Our study provides crucial knowledge that may guide policy and investment decisions critical for achieving global sustainable development goals. Table 3.1 provides the list of sample countries while Table 3.2 provides the description and source of data of the variables.

3.2.2 Analytical Technique

Our dependent variable, energy transition, is proxied by renewable energy consumption (ratio of renewable energy consumption to total final energy consumption). Our independent variables include institutional quality, official development assistance, foreign direct investments, and financial development proxied by domestic credit to private sector by banks. Institutional quality was computed using principal component analysis. There were six indicators of institutional quality identified in the

Table 3.1 List of countries studied

Algeria	Cote d'Ivoire	Malawi	Sierra Leone
Angola	Egypt	Mali	South Africa
Benin	Equatorial Guinea	Mauritania	Tanzania
Botswana	Eswatini	Mauritius	Togo
Burkina Faso	Gabon	Morocco	Tunisia
Burundi	Ghana	Mozambique	Uganda
Cape Verde	Guinea	Namibia	Zambia
Cameroon	Guinea Bissau	Niger	Zimbabwe
Chad	Kenya	Nigeria	
Comoros	Lesotho	Rwanda	
Congo Democratic Republic	Liberia	Sao Tome and Principe	
Congo Republic	Madagascar	Senegal	

Table 3.2 Description of the variables

Acronym	Variables	Definition	Data source
RENEW	Renewable energy consumption	This is the ratio of renewable energy consumption to total final energy consumption in a country measured by % of total final energy consumption	WDI
IQ	Institutional quality	Institutional quality calculated using six indicators of institutional quality measured as the principal component score	WDI
ODA	Net ODA received per capita	This is the average net official development assistance received by a country and measured by current US $ per capita	WDI
FDI	Foreign direct investments	This is the net inflows of foreign direct investment measured as percentage of GDP	WDI
FinD	Financial development	This is the development of financial systems in a country by providing domestic credit to the private sector by banks measured as percentage of GDP	WDI

Source World Development Indicators (WDI) (World Bank Group, 2024)

literature. They include control of corruption (number of sources), political stability and absence of violence/terrorism (number of sources), government effectiveness (number of sources), rule of law (number of sources), regulatory quality (number of sources), and voice and accountability (number of sources). These indicators/proxies were subjected to factor analysis and the data reduced to one principal component, which we used as the index of institutional quality in this paper. The index was not transformed to natural logarithm for further analysis since it has already been subjected to factor analysis. The implicit model of our panel autoregressive distributed lag (PARDL) is stated as follows:

$$\text{RENEW} = f(\text{ODA}, \text{FinD}, \text{INSQ}, \text{FDI}) \tag{3.1}$$

where:

RENEW	Energy transition (renewable energy consumption) (% of total final energy consumption)
ODA	Net ODA received per capita (current US $)
FinD	Financial development (domestic credit to private sector by banks) (% of GDP)
FDI	Foreign direct investment, net inflows (% of GDP)
INSQ	Institutional quality (principal component score).

To control for possible heteroskedasticity in our dataset, we converted the real values of other variables to their natural logarithmic values. The panel autoregressive distributed lag model with the logarithms is presented thus:

$$\ln \text{RENEW}_{it} = \beta_0 + \beta_1 \ln \text{ODA}_{it} + \beta_2 \ln \text{FinD}_{it} + \beta_3 \text{INSQ}_{it} + \beta_4 \ln \text{FDI}_{it} + \varepsilon_{it} \tag{3.2}$$

where; i: 1, 2, 3, …, 44 countries; t: 2009, 2010, 2011, …, 2020 year; ln denotes natural logarithm; ε is the error term. Furthermore, β_1, β_2, β_3, and β_4 define the estimated percentage change in renewable energy consumption caused by a one percent change in overseas development assistance, financial development (domestic credit to private sector by banks), institutional quality, and foreign direct investments, respectively, while all other factors remained constant. The annual data from 2009 to 2020 provides sufficient temporal coverage for the comprehensive analysis of patterns and trends for the study variables. Besides this, a log transformation of all data series except IQ (which is a principal component score) was done to unify the scale and unite the variables to facilitate comparison (Fu et al., 2022).

3.3 Results and Discussion

3.3.1 Descriptive Statistics

Table 3.3 shows the summary of the descriptive statistics of the dependent and independent variables of the study. The average renewable energy consumption was 58.98% of the total energy consumption while the median was 71.96%. The average

Table 3.3 Descriptive statistics

Variables/ statistics	Renewable energy consumption (% of total final energy consumption)	Net ODA received per capita (current US $)	Domestic credit to private sector by banks (% of GDP)	Institutional quality (principal component score)	Foreign direct investment, net inflows (% of GDP)
Number of observations	528	528	528	528	528
Mean	58.98	69.96	22.81	0.40	4.50
Median	71.96	56.09	16.52	0.48	2.52
Std. deviation	29.21	69.02	18.60	0.79	9.05
Skewness	− 0.67	3.63	1.86	− 0.74	6.27
Kurtosis	− 0.97	18.00	3.469	− 0.06	53.86
Minimum	0.06	0.33	0.00	− 1.83	− 18.92
Maximum	97.03	627.35	104.85	1.61	103.34

and median values of net official development assistance were 69.96 current US $ and 56.09 current US $, respectively. The minimum value was 0.33 current US $. Domestic credit to the private sector by banks—a proxy for financial development—recorded an average value of 22.81% of GDP and a median value of 16.52% of GDP. Institutional quality was derived from the principal component score of a factor analysis of six indicators of institutional quality. The mean score was 0.40 while the median value was 0.48. The mean and median values of foreign direct investments were 4.50% of GDP and 2.52% of GDP, respectively. The mean value of most of the variables—official development assistance, financial development (domestic credit to private sector by banks), and foreign direct investments—is greater than the median value implying that the distribution of these variables are symmetric while that of renewable energy consumption and institutional are asymmetric (Gyamfi et al., 2021).

3.3.2 Multicollinearity Test

Table 3.4 presents the result of collinearity diagnostics conducted using the variance inflation factor (VIF). The values of the VIF were far below 5, which is a generally accepted cut-off point for multicollinearity. Many studies have used 5 as an acceptable limit of multicollinearity (see Chidiebere-Mark et al., 2022; Elom et al., 2024; Emenekwe et al., 2022a; Onyeneke et al., 2023a, 2023b, 2023c, 2024a). This implies that VIF values of less or equal to 5 indicate absence of multicollinearity. Our study demonstrates absence of multicollinearity among the independent variables.

Table 3.4 Multicollinearity statistics based on variance inflation factor (VIF)

Variable	Tolerance	VIF
ln ODA	0.92	1.09
ln FinD	0.98	1.02
INSQ	0.98	1.02
ln FDI	0.91	1.10

Table 3.5 Cross section dependence test

Variable	CD-test	*p*-value
ln RENEW	24.98	0.000
ln ODA	11.08	0.000
ln FinD	25.15	0.000
INSQ	59.86	0.000
ln FDI	9.39	0.000

3.3.3 Cross-Section Dependence Test

We conducted cross-section dependence test to eliminate the data series' time-dependence properties to confirm our data's validity and reliability (Munonye et al., 2024; Onyeneke et al., 2024c). The cross-section dependence test result is presented in Table 3.5. The result showed there is cross-section dependence in our dataset. We adopted second-generation unit root tests of Pesaran's CADF and Cross-sectionally Augmented Im–Pesaran–Shin (CIPS) because of the rejection of the null hypothesis of cross-sectional independence.

3.3.4 Unit Root Test

This sub-section applied the Pesaran's CADF test, Cross-sectionally Augmented Im–Pesaran–Shin (CIPS) test, and Cross-sectionally Augmented Im–Pesaran–Shin (CIPS) test with trend to check the unit root properties and order of integration of the model variables. Table 3.6 shows the results. The result indicates that renewable energy consumption, official development assistance, and financial development were stationary at level under the Pesaran's CADF test, while official development assistance, financial development, institutional quality, and foreign direct investments were stationary at level under Cross-sectionally Augmented Im–Pesaran–Shin (CIPS) test. For the Cross-sectionally Augmented Im–Pesaran–Shin (CIPS) test with trend, official development assistance, financial development, and foreign direct investments were stationary at level. All the model variables were stationary at first difference under the three-unit root tests conducted. This implies that there is cointegrating relationships among our variables (Emenekwe et al., 2022b, 2023;

Table 3.6 Unit root test

H_0 = All panels contain unit roots
H_0 = series have unit roots

Variable	At level I(0)	At first difference I(1)	Decision: H_0	Result
	t-statistic	*t*-statistic		
Pesaran's CADF test				
ln RENEW	– 2.42***	– 2.36***	Reject	I(0) at 1%
ln ODA	– 2.53***	– 2.99***	Reject	I(0) at 1%
ln FinD	– 2.34***	– 3.01***	Reject	I(0) at 1%
INSQ	– 1.97	– 2.73***	Reject	I(1) at 1%
ln FDI	– 1.77	– 2.04**	Reject	I(1) at 5%
Cross-sectionally Augmented Im–Pesaran–Shin (CIPS) test				
ln RENEW	– 1.92	– 2.92***	Reject	I(1) at 1%
ln ODA	– 2.74***	– 3.75***	Reject	I(0) at 1%
ln FinD	– 2.26***	– 3.08***	Reject	I(0) at 5%
INSQ	– 2.15**	– 3.32***	Reject	I(0) at 5%
ln FDI	– 2.57***	– 3.87***	Reject	I(0) at 1%
Cross-sectionally Augmented Im–Pesaran–Shin (CIPS) test with trend				
ln RENEW	– 2.28	– 2.74**	Reject	I(1) at 5%
ln ODA	– 2.99***	– 3.73***	Reject	I(0) at 1%
ln FinD	– 2.82***	– 3.07***	Reject	I(0) at 1%
INSQ	– 2.53	– 3.48***	Reject	I(1) at 1%
ln FDI	– 2.80**	– 3.76***	Reject	I(0) at 5%

Note **:and *** indicate significance at 5% and 1% levels, respectively

Onyeneke et al., 2023d, 2024b, 2024c). We therefore conducted cointegration tests to determine the existence of long-run and short-run impacts between our dependent and independent variables.

3.3.5 Cointegration Tests

The co-integration tests result is presented in Table 3.7. The table shows that all the cointegration tests—Kao, Pedroni and Westerlund test—were statistically significant indicating presence of short-run and long-run association between the independent and dependent variables. This justifies our choice of panel autoregressive distributed lag model which models long-run and short-run impacts of independent variables on dependent variable from panel data.

Table 3.7 Cointegration test

H_0: No cointegration
H_a: All panels are cointegrated

Test	Statistic	*p*-value
Kao test for cointegration		
Modified Dickey–Fuller t	– 2.34	0.010
Dickey–Fuller t	– 3.58	0.000
Augmented Dickey–Fuller t	– 5.72	0.000
Unadjusted modified Dickey–Fuller t	– 1.34	0.091
Unadjusted Dickey–Fuller t	– 3.05	0.001
Pedroni test for cointegration		
Modified Phillips–Perron t	8.42	0.000
Phillips–Perron t	– 6.31	0.000
Augmented Dickey–Fuller t	– 7.70	0.000
Westerlund test for cointegration		
Variance ratio	6.98	0.000

3.3.6 Impact of Development Assistance, Financial Development, Institutional Quality, and Foreign Direct Investments on Just Energy Transition

We applied the pooled mean group and dynamic fixed effect frameworks of the panel autoregressive distributed lag approach to model the long-run and short-run impacts of development assistance, financial development, institutional quality, and foreign direct investments on energy transition. The results are presented in Table 3.8. Based on the Hausman test of poolability result, we chose the pooled mean group for further interpretation and discussion.

Official development assistance recorded a significant long-run positive impact on energy transition. A one percent increase in official development assistance yielded a 2.10% increase in renewable energy consumption in Africa in the long run. This finding suggests that official development assistance enhances energy transition in Africa.

Financial development recorded a significant negative impact on energy transition both in the long-run and short-run. A one percent increase in credit to private sector by banks yielded a 1.45% and 0.07% decrease in renewable energy consumption in Africa in the long-run and short-run, respectively. This finding suggests that financial development opposes energy transition in Africa. Banks in Africa may not have been supporting energy transition, green development and climate action on the continent.

Institutional quality recorded a significant long-run positive impact on energy transition. A one percent increase in institutional quality resulted to a 0.76% increase

Table 3.8 PARDL results from pooled mean group (PMG) and dynamic fixed effect (DFE) estimators

Variables	PMG	DFE
Panel A: Long-run estimates		
ln ODA	2.10***	0.15***
	(3.41)	(3.22)
ln FinD	– 1.45***	– 0.20**
	(– 2.69)	(– 2.37)
INSQ	0.76**	0.05
	(2.44)	(0.77)
ln FDI	0.18**	– 0.01
	(2.47)	(– 0.60)
Panel B: Short-run estimates		
ECT	0.10**	0.22***
	(2.12)	(6.21)
Δln ODA	0.02	0.03***
	(1.07)	(3.36)
Δln FinD	– 0.07*	– 0.01
	(– 1.93)	(– 0.28)
ΔINSQ	0.01	– 0.02
	(0.17)	(– 1.13)
Δln FDI	– 0.02*	– 0.002
	(– 1.67)	(– 1.11)
Constant	0.02	– 0.83***
	(1.48)	(– 5.63)
Hausman test of poolability (H_o: difference in coefficients not systematic)		PMG and DFE
$\chi^2(4)$		0.00
p-value		1.000
Decision		The PMG is preferred over the DFE

in renewable energy consumption in Africa in the long run. This finding suggests that institutional quality promotes just energy transition in Africa.

Foreign direct investments recorded a significant long-run positive impact on energy transition. A one percent increase in foreign direct investments resulted to a 0.18% increase in renewable energy consumption in Africa in the long run. This finding suggests that foreign direct investments support just energy transition in Africa.

3.4 Conclusions, Policy Implications and Limitations

This paper explored the impacts of foreign direct investment, financial development, institutional quality, and official development assistance on renewable energy consumption (used as a proxy for energy transition) in Africa. We used panel data spanning a period of 12 years (2009–2020) and from 44 African countries obtained from the World Development Indicators database. The paper leverages panel data econometrics to analyse the link between the variables studied. We infer from the results that official development, institutional quality, and foreign direct investment foster energy transition in Africa, while financial development tend to slow energy transition on the continent. The findings of our study have implications for policymaking and program targeting. Policymakers and institutions, international development partners and foreign investors should prioritize increasing investment and support to sustainable energy practices in Africa, while banks should be encouraged to redirect their credit facilities/interventions to energy transition investments on the continent.

Despite the interesting outcomes of our paper, the study is not without limitations. One limitation needs to be stated. We used data spanning a period of twelve years and this may not be a long enough period to sufficiently explore the determinants of energy transition. Data availability informed the choice of the number of countries and period covered. We acknowledge these potential limitations. We therefore suggest that effort should be intensified in data collection of all important variables that may affect energy transition in Africa. When such data become available, future studies should focus on incorporating more variables and countries and covering a longer period to allow better understanding of the determinants of just energy transition in Africa. Such developments are needed to broaden the understanding of the intricacies involved in the energy transition within Africa. These limitations notwithstanding, our study provided nuanced understanding of the role of official development assistance, financial development, institutional quality, and foreign direct investment in fostering energy transition in Africa and other regions with similar contexts.

Funding The study received no funding.

Data Availability Statement The data used in this study is publicly available and can be downloaded at https://databank.worldbank.org/source/world-development-indicators#.

Ethical Clearance Not applicable. Competing Interests: Robert Onyeneke is an editor of this handbook and was not involved in the editorial review, or the decision to publish this article. All authors declare that there are no competing interests.

References

Adjei-Mantey, K., & Adams, S. (2023). Renewable energy, foreign direct investment and carbon dioxide emissions: Do sectoral value additions and policy uncertainty matter? *Energy Nexus, 10*, 100193.

Akram, V., Rath, B. N., & Sahoo, P. K. (2023). Club convergence in per capita carbon dioxide emissions across Indian states. *Environment, Development and Sustainability,* 1–28. https://doi.org/10.1007/s10668-023-03443-2

Ali, S., Yusop, Z., Kaliappan, S. R., & Chin, L. (2020). Dynamic common correlated effects of trade openness, FDI, and institutional performance on environmental quality: Evidence from OIC countries. *Environmental Science and Pollution Research, 27*(11), 11671–11682. https://doi.org/10.1007/s11356-020-07768-7

Amoah, A., Kwablah, E., Korle, K., & Offei, D. (2020). Renewable energy consumption in Africa: The role of economic well-being and economic freedom. *Energy and Sustainable Society, 10*(1), 1–7. https://doi.org/10.1186/s13705-020-00264-3

Araújo, K. (2014). The emerging field of energy transitions: Progress, challenges, and opportunities. *Energy Research and Social Science, 1*, 112–121.

Aslan, A., Apergis, N., & Topcu, M. (2014). Banking development and energy consumption: Evidence from a panel of middle eastern countries. *Energy, 72*, 427–433.

Assi, A. F., Isiksal, A. Z., & Tursoy, T. (2020). Highlighting the connection between financial development and consumption of energy in countries with the highest economic freedom. *Energy Policy, 147*, 111897.

Azam, M., Liu, L., & Ahmad, N. (2021). Impact of institutional quality on environment and energy consumption: Evidence from developing world. *Environment, Development and Sustainability, 23*, 1646–1667.

Bae, J. H., Kang, H., & Park, T. (2016). Can green aid help mitigating climate change? *International Journal of Ecological Economics and Statistics, 37*, 40–58.

Balsalobre-Lorente, D., Gokmenoglu, K. K., Taspinar, N., & Cantos-Cantos, J. M. (2019). An approach to the pollution haven and pollution halo hypotheses in MINT countries. *Environmental Science and Pollution Research, 26*, 23010–23026.

Barbier, E. B. (2015). Climate change impacts on rural poverty in low-elevation coastal zones. *Estuarine, Coastal and Shelf Science, 165*, A1–A13.

Bernauer, T., Böhmelt, T., & Koubi, V. (2012). Environmental changes and violent conflict. *Environmental Research Letters, 7*(1), 015601.

Bourguignon, F., & Platteau, J. P. (2015). The hard challenge of aid coordination. *World Development, 69*, 86–97. https://doi.org/10.1016/j.worlddev.2013.12.011

Bouyghrissi, S., Murshed, M., Jindal, A., Berjaoui, A., Mahmood, H., & Khanniba, M. (2022). The importance of facilitating renewable energy transition for abating CO_2 emissions in Morocco. *Environmental Science and Pollution Research, 29*(14), 20752–20767.

Caetano, R. V., Marques, A. C., & Afonso, T. L. (2022). How can foreign direct investment trigger green growth? The mediating and moderating role of the energy transition. *Economies, 10*(8), 199.

Caglar, A. E. (2020). The importance of renewable energy consumption and FDI inflows in reducing environmental degradation: Bootstrap ARDL bound test in selected 9 countries. *Journal of Cleaner Production, 264*, 121663.

Carfora, A., Scandurra, G., & Thomas, A. (2021). Factors affecting official development assistance distribution. A panel investigation. *Journal of Cleaner Production, 304*, 126970.

Chang, Y. C., & Wang, N. (2017). Legal system for the development of marine renewable energy in China. *Renewable and Sustainable Energy Reviews, 75*, 192–196.

Charfeddine, L., & Kahia, M. (2019). Impact of renewable energy consumption and financial development on CO_2 emissions and economic growth in the MENA region: A panel vector autoregressive (PVAR) analysis. *Renewable Energy, 139*, 198–213.

Chidiebere-Mark, N. M., Onyeneke, R. U., Uhuegbulem, I. J., Ankrah, D. A., Onyeneke, L. U., Anukam, B. N., & Chijioke-Okere, M. O. (2022). Agricultural production, renewable energy consumption, foreign direct investment, and carbon emissions: New evidence from Africa. *Atmosphere, 13*(12), 1981. https://doi.org/10.3390/atmos13121981

Chiu, Y. B., & Lee, C. C. (2020). Effects of financial development on energy consumption: The role of country risks. *Energy Economics, 90*, 104833.

Cifor, A., Denholm, P., Ela, E., Bri-Mathias, H., & Reed, A. (2015). The policy and institutional challenges of grid integration of renewable energy in the western United States. *Utility Policy, 33*, 34–41.

Dasgupta, P., & Mäler, K. G. (2004). The economics of non-convex ecosystems: Introduction. In P. Dasgupta & K. G. Mäler (Eds.), *The economics of non-convex ecosystems. The economics of non-market goods and resources* (Vol. 4). Springer. https://doi.org/10.1007/1-4020-2515-7_1

Denisova, V. (2020). Financial development and energy consumption: Evidence from Germany. *International Journal of Energy Economics and Policy, 10*(2), 35–39.

Destek, M. A., Manga, M., Cengiz, O., & Destek, G. (2022). Investigating the potential of renewable energy in establishing global peace: Fresh evidence from top energy consumer countries. *Renewable Energy, 197*, 170–177. https://doi.org/10.1016/j.renene.2022.07.097

Dominkovic, D. F., Bacekovic, I., Pedersen, A. S., & Krajacic, G. (2018). The future of transportation in sustainable energy systems: Opportunities and barriers in a clean energy transition. *Renewable and Sustainable Energy Reviews, 82*, 1823–1838.

Doytch, N., & Narayan, S. (2016). Does FDI influence renewable energy consumption? An analysis of sectoral FDI impact on renewable and nonrenewable industrial energy consumption. *Energy Economics, 54*, 291–301.

Elom, C. O., Onyeneke, R. U., Ayerakwa, H. M., Atta-Ankomah, R., Deffor, E. W., & Uwaleke, C. C. (2024). The role of information and communication technologies and access to electricity on education in Africa. *Education and Information Technologies, 29*, 15501–15532. https://doi.org/10.1007/s10639-024-12504-6

Emenekwe, C. C., Onyeneke, R. U., & Nwajiuba, C. U. (2022a). Financial development and carbon emissions in sub-Saharan Africa. *Environmental Science and Pollution Research, 29*, 19624–19641. https://doi.org/10.1007/s11356-021-17161-7

Emenekwe, C. C., Onyeneke, R. U., & Nwajiuba, C. U. (2022b). Assessing the combined effects of temperature, precipitation, total ecological footprint, and carbon footprint on rice production in Nigeria: A dynamic ARDL simulations approach. *Environmental Science and Pollution Research, 29*, 85005–85025. https://doi.org/10.1007/s11356-022-21656-2

Emenekwe, C. C., Onyeneke, R. U., Nwajiuba, C. U., Anugwa, I. Q., & Emenekwe, O. U. (2023). Determinants of consumption-based and production-based carbon emissions. *Environment, Development and Sustainability*. https://doi.org/10.1007/s10668-023-04311-9

Er, B., Güneysu, Y., & Ünal, H. (2018). Financing renewable energy projects: An empirical analysis for Turkey. *International Journal of Energy Economics and Policy, 8*(6), 180–185.

Esty, D., & Porter, M. (2005). National environmental performance: An empirical analysis of policy results and determinants. *Environment and Development Economics, 10*(4), 391–434.

Fu, Z., Chen, Z., Sharif, A., & Razi, U. (2022). The role of financial stress, oil, gold and natural gas prices on clean energy stocks: Global evidence from extreme quantile approach. *Resources Policy, 78*, 102860. https://doi.org/10.1016/j.resourpol.2022.102860

Fung, H. K. (2009). Financial development and economic growth: Convergence or divergence? *Journal of International Money and Finance, 28*, 56–78.

Gallagher, K. P., & Thacker, S. C. (2008). *Democracy, income, and environmental quality* (PERI Working Paper 164). University of Massachusetts Amherst.

Ghazouani, T. (2018). *Re-examining the foreign direct investment, renewable energy consumption and economic growth nexus: Evidence from a new bootstrap ARDL test for cointegration* (MPRA Paper No. 89975).

Gualberti, G., Martins, L. F., & Bazilian, M. (2014). An econometric analysis of the effectiveness of development finance for the energy sector. *Energy for Sustainable Development, 18*, 16–27.

Güngör, H., Olanipekun, I. O., & Usman, O. (2021). Testing the environmental Kuznets curve hypothesis: The role of energy consumption and democratic accountability. *Environmental Science and Pollution Research, 28*, 1464–1478.

Gyamfi, B. A., Bein, M. A., Udemba, E. N., & Bekun, F. V. (2021). Investigating the pollution haven hypothesis in oil and non-oil sub-Saharan Africa countries. Evidence from quantile regression technique. *Resources Policy, 73*, 102119. https://doi.org/10.1016/j.resourpol.2021.102119

Hübler, M., & Keller, A. (2010). Energy savings via FDI? Empirical evidence from developing countries. *Environment and Development Economics, 15*(1), 59–80.

Ince, D., Vredenburg, H. & Liu, X. (2016). Drivers and inhibitors of renewable energy: A qualitative and quantitative study of the Caribbean. *Energy Policy, 98*, 700–771.

Inglesi-Lotz, R. (2024). Exploring the impact of institutional quality to South Africa's transition to renewables. *Energy Sources, Part B: Economics, Planning, and Policy, 19*(1), 2291433.

Jangam, B. P., Sahoo, P. K., & Akram, V. (2020). Convergence in electricity consumption across Indian states: A disaggregated analysis. *International Journal of Energy Sector Management, 14*(3), 624–637. https://doi.org/10.1108/IJESM-03-2019-0009

Jiang, L., & O'Neill, B. C. (2004). The energy transition in rural China. *International Journal of Global Energy Issues, 21*(1/2), 2–26.

Kahouli, B. (2017). The short and long run causality relationship among economic growth, energy consumption and financial development: Evidence from South Mediterranean Countries (SMCs). *Energy Economics, 68*, 19–30.

Khandker, L. L., Amin, S. B., & Khan, F. (2018). Renewable energy consumption and foreign direct investment: Reports from Bangladesh. *Journal of Accounting, Finance and Economics, 8*(3), 72–87.

Kilicarslan, Z. (2019). The relationship between foreign direct investment and renewable energy production: Evidence from Brazil, Russia, India, China, South Africa and Turkey. *International Journal of Energy Economics and Policy, 9*(4), 291–297.

Lee, J. W. (2013). The contribution of foreign direct investment to clean energy use, carbon emissions and economic growth. *Energy Policy, 55*, 483–489.

Li, D. D., Rishi, M., & Bae, J. H. (2021). Green official development aid and carbon emissions: Do institutions matter? *Environment and Development Economics, 26*(1), 88–107.

Li, H. X., Edwards, D. J., Hosseini, M. R., & Costin, G. P. (2020). A review on renewable energy transition in Australia: An updated depiction. *Journal of Cleaner Production, 242*, 118475.

Liu, L., Zhou, C., Huang, J., & Hao, Y. (2018). The impact of financial development on energy demand: Evidence from China. *Emerging Markets Finance and Trade, 54*, 269–287.

Mielnik, O., & Goldemberg, J. (2002). Foreign direct investment and decoupling between energy and gross domestic product in developing countries. *Energy Policy, 30*, 87–89.

Mohamed, K. A., & Mamat, M. N. (2016). Examining the relationship between FDI, economic growth, energy consumption and exports in Yemen. *Journal of Advanced Social Research, 6*.

Munonye, J. O., Onyeneke, R. U., Ankrah, D. A., Agyarko, F. F., Onyeneke, C. J., Nejad, J. G., & Chukwu, C. (2024). Do climate change, access to electricity and renewable energy consumption matter in aquaculture production in Africa? *Natural Resources Forum.* https://doi.org/10.1111/1477-8947.12427

Murshed, M. (2020). Are trade liberalization policies aligned with renewable energy transition in low- and middle-income countries? An instrumental variable approach. *Renewable Energy, 151*, 1110–1123.

Nakhooda, S., Norman, M., Barnard, S., Watson, C., Greenhill, R., Caravani, A., Canales Trujillo, N., & Banton, G. (2014). *Climate finance: Is it making a difference? A review of the effectiveness of multilateral climate funds.* Overseas Development Institute.

Nepal, R., Paija, N., Tyagi, B., & Harvie, C. (2021). Energy security, economic growth and environmental sustainability in India: Does FDI and trade openness play a role? *Journal of Environmental Management, 281*, 111886. https://doi.org/10.1016/j.jenvman.2020.111886

Ntow-Gyamfi, M., Bokpin, G. A., Aboagye, A. Q. Q., & Ackah, C. G. (2020). Environmental sustainability and financial development in Africa; does institutional quality play any role? *Development Studies Research, 7*, 93–118. https://doi.org/10.1080/21665095.2020.1798261

Onyeneke, R. U., Agyarko, F. F., Onyeneke, C. J., Osuji, E. E., Ibeneme, P. A., & Esfahani, I. J. (2023a). How does climate change affect tomato and okra production? Evidence from Nigeria. *Plants, 12*(19), 3477. https://doi.org/10.3390/plants12193477

Onyeneke, R. U., Chidiebere-Mark, N. M., Ankrah, D. A., & Onyeneke, L. U. (2023b). Determinants of access to clean fuels and technologies for cooking in Africa: A panel autoregressive distributed lag approach. *Environmental Progress & Sustainable Energy, 42*(3), e14147. https://doi.org/10.1002/ep.14147

Onyeneke, R. U., Ankrah, D. A., Atta-Ankomah, R., Agyarko, F. F., Onyeneke, C. J., & Nejad, J. G. (2023c). Information and communication technologies and agricultural production: New evidence from Africa. *Applied Sciences, 13*(6), 3918. https://doi.org/10.3390/app13063918

Onyeneke, R. U., Emenekwe, C. C., Adeolu, A. I., & Ihebuzor, U. A. (2023d). Climate change and cattle production in Nigeria: Any role for ecological and carbon footprints? *International Journal of Environmental Science and Technology, 20*, 11121–11134. https://doi.org/10.1007/s13762-022-04721-8

Onyeneke, R. U., Osuji, E. E., Anugwa, I. Q., & Chidiebere-Mark, N. M. (2024a). Impacts of biocapacity, climate change, food vulnerability, readiness and adaptive capacity on cereal crops yield: Evidence from Africa. *Environmental Development and Sustainability, 26*, 11979–12003. https://doi.org/10.1007/s10668-023-03615-0

Onyeneke, R. U., Chidiebere-Mark, N. M., & Ayerakwa, H. M. (2024b). Impact of information and communication technologies and renewable energy consumption on carbon emissions in Africa. *Carbon Research, 3*, 53. https://doi.org/10.1007/s44246-024-00130-3

Onyeneke, R. U., Ejike, R. D., Osuji, E. E., & Chidiebere-Mark, N. M. (2024c). Does climate change affect crops differently? New evidence from Nigeria. *Environment Development and Sustainability, 26*(1), 393–419. https://doi.org/10.1007/s10668-022-02714-8

Paramati, S. R., Ummalla, M., & Apergis, N. (2016). The effect of foreign direct investment and stock market growth on clean energy use across a panel of emerging market economies. *Energy Economics, 56*, 29–41.

Pickering, J., Betzold, C., & Skovgaard, J. (2017). Special issue: Managing fragmentation and complexity in the emerging system of international climate finance. *International Environmental Agreements: Politics, Law and Economics, 17*(1). https://doi.org/10.1007/s10784-016-9349-2

Polat, B. (2018). The influence of FDI on energy consumption in developing and developed countries: A dynamic panel data approach. *Journal of Yasar University, 13*(49), 33–42.

Rafiq, S., Paramati, S. R., Alam, M. S., Hafeez, K., & Shafiullah, M. (2024). Does institutional quality matter for renewable energy promotion in OECD economies? *International Journal of Finance & Economics*. https://doi.org/10.1002/ijfe.2926

Ramzan, M., Razi, U., Usman, M., Sarwar, S., Talan, A., & Mundi, H. S. (2024). Role of nuclear energy, geothermal energy, agriculture, and urbanization in environmental stewardship. *Gondwana Research, 125*, 150–167. https://doi.org/10.1016/j.gr.2023.08.006

Sadorsky, P. (2010). The impact of financial development on energy consumption in emerging economies. *Energy Policy, 38*, 2528–2535.

Sadorsky, P. (2011). Financial development and energy consumption in Central and Eastern European frontier economies. *Energy Policy, 39*, 999–1006.

Scruggs, L., & Rivera, C. (2008). Political regimes, democratic institutions, and environmental sustainability: A cross-national analysis. In *Proceedings of the Midwest Political Science Association Meeting in Chicago* (Vol. 36), Chicago, IL, USA. http://www.sp.uconn.edu/~scruggs/mpsa08.pdf

Shahbaz, M., & Lean, H. H. (2012). Does financial development increase energy consumption? The role of industrialization and urbanization in Tunisia. *Energy Policy, 40*, 473–479.

Shahbaz, M., Nasreen, S., Abbas, F., & Anis, O. (2015). Does foreign direct investment impede environmental quality in high-, middle-, and low-income countries? *Energy Economics, 51*, 275–287.

Udeagha, M. C., & Ngepah, N. (2022). Dynamic ARDL simulations effects of fiscal decentralization, green technological innovation, trade openness, and institutional quality on environmental sustainability: Evidence from South Africa. *Sustainability, 14*(16), 10268. https://doi.org/10.3390/su141610268

Wang, Q., Guo, J., Li, R., Mikhaylov, A., & Moiseev, N. (2023). Does technical assistance alleviate energy poverty in sub-Saharan African countries? A new perspective on spatial spillover effects of technical assistance. *Energy Strategy Reviews, 45*, 101047.

Wirth, S. (2014). Communities matter: Institutional preconditions for community renewable energy. *Energy Policy, 70*, 236–246.

World Bank. (2018). *The World Bank annual report 2018*. World Bank. https://openknowledge.worldbank.org/handle/10986/30326. License: CC BY-NC-ND 3.0.

World Bank Group. (2024). *World development indicators (WDI)*. Retrieved February 15, 2024, from https://databank.worldbank.org/source/world-development-indicators

Wu, L., & Broadstock, D. C. (2015). Does economic, financial and institutional development matter for renewable energy consumption? Evidence from emerging economies. *International Journal of Economic Policy in Emerging Economies, 8*(1), 20–39.

Xu, C., Zhao, W., Zhang, M., & Cheng, B. (2021). Pollution haven or halo? The role of the energy transition in the impact of FDI on SO_2 emissions. *The Science of the Total Environment, 763*, 143002. https://doi.org/10.1016/j.scitotenv.2020.143002

Yilanci, V., Bozoklu, S., & Gorus, M. S. (2020). Are BRICS countries pollution havens? Evidence from a bootstrap ARDL bounds testing approach with a Fourier function. *Sustainable Cities and Society, 55*, 102035.

Yin, Y., Xiong, X., & Hussain, J. (2021). The role of physical and human capital in FDI-pollution-growth nexus in countries with different income groups: A simultaneity modeling analysis. *Environmental Impact Assessment Review, 91*, 106664. https://doi.org/10.1016/j.eiar.2021.106664

Zeeb, A., Maqsood, F., & Munir, F. (2015). Impact of foreign direct investment on energy saving in South Asian countries. *Journal of Asian Development Studies, 4*(3), 14–26.

Zhang, C., & Zhou, X. (2016). Does foreign direct investment lead to lower CO_2 emissions? Evidence from a regional analysis in China. *Renewable and Sustainable Energy Reviews, 58*, 943–951.

Zhang, N., Yu, K., & Chen, Z. (2017). How does urbanization affect carbon dioxide emissions? A cross-country panel data analysis. *Energy Policy, 107*, 678–687.

Chapter 4
Assessing Energy Flexibility: A Comparative Review of Nigeria and the United Kingdom

Gbubemi Kevin Akporhonor and Pegah Mirzania

Abstract This paper provides a comparative analysis of the energy systems in Nigeria and the United Kingdom (UK), focusing on market instruments for energy flexibility to achieve NetZero emissions under the Paris Agreement. Through a systematic review using Scopus database, the study identifies key energy flexibility instruments in Nigeria, including energy efficiency initiatives, smart metering, peer-to-peer energy trading, decentralised energy systems, and independent power plants. Supporting policies include Renewable Energy Master Plan, National Energy Efficiency Action Plan, and National Renewable Energy Action Plan. However, Nigeria faces challenges like low electrification rates, underutilised grid capacity, and fragile infrastructure, resulting in power outages. Conversely, the UK's energy system features advanced Demand-Side Response Programmes, Smart Grid Technologies, Energy Storage Solutions, and Virtual Power Plants, supported by policies and initiatives like Agile Octopus Smart Meters. Learning from the UK's successes evidenced by its high ranking on the World Energy Council Energy Trilemma Index, this study recommends that Nigeria improve its power infrastructure and incorporate energy flexibility instruments. By expanding power capacity, upgrading the grid, enhancing flexibility mechanisms, creating roadmaps, and transparent allocation and use of financial resources. These actions are essential for enhancing the reliability and sustainability of Nigeria's energy system to achieve a Just Transition.

Keywords Energy trilemma · Renewable energy · Just transition · Energy flexibility · Developing economy

---

G. K. Akporhonor (✉)
School of Water, Energy and Environment, Cranfield University, Cranfield, Bedfordshire, UK
e-mail: gbubemi.akporhonor.788@cranfield.ac.uk

P. Mirzania
School of Water, Energy and Environment, Centre for Energy Systems and Strategy, Cranfield University, Cranfield, Bedfordshire, UK

R. U. Onyeneke et al. (eds.), *Energy Transition, Climate Action and Sustainable Agriculture*, https://doi.org/10.1007/978-3-031-83165-2_4

4.1 Introduction

The global energy transition has accelerated as nations strive to limit global temperature rise and meet the Paris Agreement's NetZero goals. This transition emphasises the critical need for integrating renewable energy sources, reducing dependence on fossil fuels, and enhancing energy efficiency. To achieve a balance between supply and demand, there is an increasing need for energy flexibility instruments globally (Cruz et al., 2024; Dannheim et al., 2020; Jung & Schindler, 2020; Malaekeh et al., 2022; Neupane et al., 2022; Ofgem, 2021; Paraschiv et al., 2022; Patidar et al., 2022; Vallejo-Díaz et al., 2022; Virtanen et al., 2022; Wang et al., 2024).

Energy flexibility instruments are crucial for ensuring the reliability of power systems, particularly with the growing integration of variable renewable energy sources (RES). Traditionally, supply-side flexibility has played a significant role in stabilising power systems. However, as RES become more prevalent, they displace traditional supply-side flexibility providers and increase generation variability. This shift has intensified discussions on the need for flexibility to support a just energy transition in both developed and developing nations (Ahmadiahangar et al., 2020).

The stochastic nature of RES presents operational and security challenges, particularly when their share in the generation mix is significant. The unpredictability of power generation from RES necessitates a greater emphasis on flexibility to maintain power balance. In the ongoing transition towards a decarbonised society, driven by intermittent and non-deterministic renewable sources, flexibility becomes essential for creating resilient, efficient, and sustainable energy systems (Abbà et al., 2024). This need for flexibility extends across global regions, including both the Global North and South (Reiter & Jakob, 2021).

Energy flexibility tools are essential for maintaining the reliability of power systems, especially as variable renewable energy sources, such as solar, wind, hydropower, geothermal, biomass, tidal, and wave energy, are increasingly integrated. Historically, supply-side flexibility has been key to stabilising power systems. However, as the share of these renewable energy sources grows, they start to replace conventional supply-side flexibility options, leading to more variability in power generation. This change has heightened the conversation around the need for flexibility to facilitate a fair energy transition in both developed and developing countries (Ahmadiahangar et al., 2020). The unpredictable nature of renewable energy sources, particularly when they constitute a large portion of the energy mix, poses operational and security challenges. The irregularity of power generation from these sources underscores the importance of flexibility in maintaining a stable power balance. As the world transitions to a decarbonised future, driven by renewable sources that are inherently intermittent and unpredictable, the role of flexibility becomes crucial in developing resilient, efficient, and sustainable energy systems (Abbà et al., 2024). The necessity for such flexibility spans across the globe, affecting both the Global North and South.

While regions in the Global North have made progress in developing and advancing both supply and demand-side flexibility (Tziovani et al., 2020), many

Global South nations, including Nigeria, continue to face ongoing challenges such as unreliable power supply and unstable grids. For instance, Nigeria's energy sector is characterised by a significant gap between available energy and system capacity. Despite an installed power generation capacity of 15 GW, only one-third operates at peak levels, leaving over 20 million homes without adequate electricity. The country's electrification rate is 56.5%, indicating widespread issues with reliable energy access. Frequent power outages, insufficient capacity, and low levels of renewable energy integration further complicate Nigeria's energy landscape. Socioeconomic factors, such as a 5.75% unemployment rate, a 2.6% population growth rate, and high fossil fuel emissions, further contribute to these challenges (Attabo et al., 2023; Ben et al., 2021; Eweka et al., 2022; IRENA, 2023c; Ndukwu et al., 2021; Olaofe, 2017; Ritchie et al., 2020; Shigenobu et al., 2017; Statista, 2023).

Despite Nigeria's abundant non-renewable and renewable energy resources—including solar PV, hydroelectric, offshore and onshore wind, tidal, and biomass—clean energy sources only contribute about 26% of the country's annual energy production (Akporhonor et al., 2023; Ben et al., 2021; Climatescope, 2022; Eweka et al., 2022; IRENA, 2023a; Somoye, 2023). Addressing these challenges is crucial for ensuring a just energy transition as Nigeria increases its integration of renewable energy sources. This transition must be accompanied by corresponding advancements in flexibility instruments to improve the country's energy landscape.

In Nigeria, as in many developing countries, load shedding is commonly used to balance energy demand and supply, albeit inconsistently. This approach aims to distribute electricity equitably among users, despite its limitations. However, there is a lack of sufficient knowledge and literature regarding other energy flexibility instruments utilised in Nigeria. A literature review indicated that limited research has focused on this topic, highlighting the need for further study and exploration in this area.

This presents an opportunity to assess and enhance energy flexibility instruments in Nigeria, thereby identifying areas for improvement and better integration.

4.1.1 Aim and Objectives

The primary aim of this study is to investigate and evaluate the existing energy flexibility instruments in Nigeria. The primary aim of this study is to assess the factors affecting the success of energy flexibility instruments in Nigeria. This will involve a comparative review with similar instruments in the United Kingdom to evaluate their effectiveness and identify areas for improvement. This will be done by addressing two main questions:

(i) What are the existing energy flexibility instruments in Nigeria? and
(ii) How effective are the existing energy flexibility programs, and how can they be improved, based on lessons learned from the United Kingdom?

To answer these questions, the following objectives have been set:

- To evaluate Nigeria's power system and key policies that support flexibility.
- To evaluate existing market instruments for flexibility within Nigeria's energy system.
- To explore the opportunities and uncertainties for energy flexibility in Nigeria.
- To explore best practices and insights from the UK energy flexibility market that can be applied to the Nigerian context.

4.2 Methodology

To understand the energy flexibility instruments in Nigeria, a critical evaluation of the country's energy system and supporting policies was conducted. Following this, a systematic review of academic publications was undertaken to address the research questions and objectives outlined above. Following the Collaboration for Environmental Evidence (CEE) Guidelines, a systematic review involves a meticulous data collection process that emphasises thoroughness, transparency, reproducibility, and minimising bias (Gusenbauer & Haddaway, 2020).

Consequently, the following approach was employed.

4.2.1 Database Selection

The Scopus database was selected as the primary source due to its wide acceptance among researchers and frequent use in systematic and comparative studies in the energy sector.

4.2.2 Search String Development and Search

The search string was developed in two parts. Primary keywords derived from the study's title were searched under the "Article title" category. Secondary keywords likely to appear in the study's abstract were searched under the "Article title, Abstract, Keywords" category. The initial search string (TITLE (energy AND flexibility AND (mechanism OR instrument*)) AND TITLE-ABS-KEY (energy AND trilemma OR renewable AND energy OR energy AND transition OR energy AND flexibility OR just AND transition)). The initial data search yielded two publications even after search results were limited to feature only publications focused on "energy" and "engineering". Thereafter, to narrow search resulted to focused on Nigeria, search results was subjected to exclusion criteria A. This resulted in the inclusion of zero publication, indicating a narrow search focus as seen in Fig. 4.1 PRISMA flow chart for the initial set of search strings.

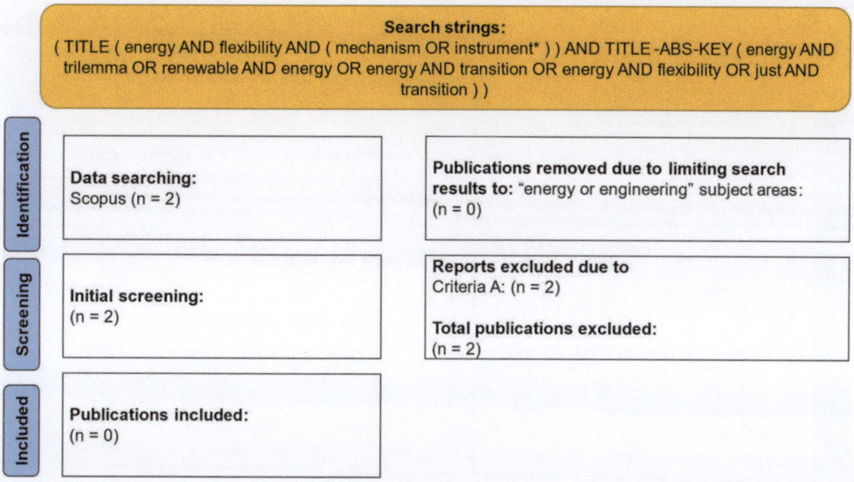

Fig. 4.1 PRISMA flow chart for the initial search string

- **Criteria A**: Does the publication focus on the Nigeria Energy sector?

To expand the search focus, the terms "AND (mechanism OR instrument*)" were removed, refining the search string to (TITLE (energy AND flexibility) AND TITLE-ABS-KEY (energy AND trilemma OR renewable AND energy OR energy AND transition OR energy AND flexibility OR just AND transition)). Given the inclusion of zero publications from the initial search, this adjustment aimed to broaden the scope and increase the relevance of the search results.

As indicated in the PRISMA flow chart captioned Fig. 4.2, this revised search yielded 108 publications at first search. Thereafter the search resulted were limited to focus on "energy and engineering", resulting in the exclusion of 31 publications. This study is focused on assessing Nigeria energy flexibility mechanism, consequently, the search results were further subject to exclusion criteria A, "focused on Nigeria". This led to the exclusion of all remaining 77 publications, leaving the total included publication as zero.

The indication of both search procedures suggested a limitation in study across Nigeria, regarding Energy flexibility mechanisms, further buttressing the need for this study.

4.2.3 Backward Snowballing

To move forward, the expert recommendation was considered to understand better the Nigeria energy sector and a key paper study from recommendations, "Systematic Review of Demand-Side Management Strategies in Power Systems of Developed and Developing Countries", was reviewed. Although this study did not deal directly

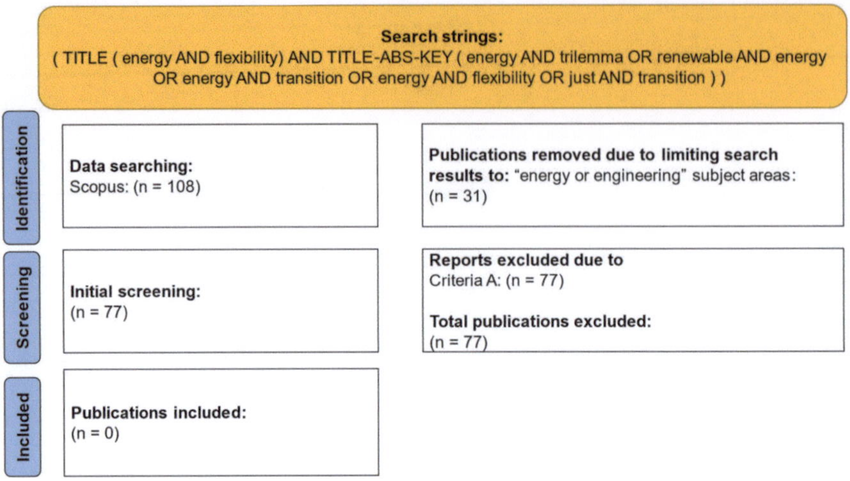

Fig. 4.2 PRISMA flow chart for the second set of search strings

with energy flexibility instruments in Nigeria, it focused on demand side management, which is an energy flexibility instrument. Consequently, this study was adopted as a set paper to conduct a backward snowballing process to include more publications. Backward snowballing is a widely used method for review papers particularly in studies with limited publication availability. This method involves the identification, review and adoption of references from selected set papers (Wohlin, 2014). By subjecting the references to the following criteria:

- **Criteria B**: Publications must make substantial contributions in the initial set papers.
- **Criteria C**: The contribution in the Initial set paper must largely be related to "Nigeria Energy flexibility Mechanism" either as an instrument, policy or statement highlighting energy flexibility in Nigeria.

4.2.4 Comparative Review

There are no universally established criteria for evaluating the success of energy flexibility mechanisms. Therefore, this study employed a comparative review approach, focusing on Nigeria and a leading nation in the implementation of energy flexibility instruments, particularly demand-side management, such as the United Kingdom. The analysis centred on three key documents: Ofgem's "Transitioning to a Net Zero Energy System: Smart Systems and Flexibility Plan 2021," "Guidance: The Change Process for the Capacity Market Rules" (2023), and the Department of Energy and Climate Change's "Electricity Market Reform Delivery Plan" (2013). These documents were selected for their detailed insights into the UK's strategic, regulatory, and operational frameworks for energy flexibility, providing a basis for comparison and highlighting potential improvements for Nigeria's energy sector.

4.3 Results and Discussion

4.3.1 Relevant Publication

Following the decision to adopt a start-set paper and proceed to employ a backward snowballing process. A total of nine publications were included in this process as indicated in Table 4.1. The inclusion criteria required that the retrieved papers significantly contribute to the initial set and reference the Nigerian energy sector. Among the selected works, were three academic publications including studies by Adewole et al. (2023) on peer-to-peer energy trading, Ikpe and Torriti (2018) on demand-side management, and Nitte and Salahudeen (2023) on energy transitions. Four of the included papers were Organisational publications including Agusto & Co Store (2023) on Nigerian independent electric power producers, the Energy Commission of Nigeria (2022) with the National Energy Policy, and the Nigerian Electricity Regulatory Commission (2018) discussing MAP regulations. National Energy Policy Revised 2022 by the Energy Commission of Nigeria was adopted as a policy document. Additionally, articles from three websites, including The Cable (2023) on NERC licensing and the IEA (2022) with the World Energy Outlook, were the final included papers.

These include nine publications that were adopted to understand Nigeria's entire energy system and key policies that support flexibility, evaluate market instruments for flexibility within Nigeria's energy system, assess the success of the energy flexibility instruments in Nigeria, and explore the opportunities and uncertainties for energy flexibility in Nigeria.

4.3.2 Energy Flexibility and Nigerian Whole Energy System

4.3.2.1 Energy Policy Targets

Nigeria's energy policy targets aim to enhance flexibility and sustainability while reducing dependence on non-renewable resources (Usman et al., 2022). The National Energy Policy emphasises conservation and efficient technology adoption, with the Renewable Energy Master Plan (REMP) targeting a 36% increase in renewable energy production by 2030, focusing on solar, wind, and hydropower (Energy Commission of Nigeria, 2022; Usman et al., 2022). Initiatives like rooftop solar systems and off-grid connections as energy flexibility instruments for five million communities post-COVID-19 are underway. Challenges include transitioning from

Table 4.1 Relevant publications

Publication title	Publication type	References
2023 The Nigerian Independent Electric Power Producers—Agusto & Co Store	Report	Agusto & Co Store (2023)
A Means to An Industrialisation End? Demand Side Management in Nigeria	Journal article	Ikpe and Torriti (2018)
Demand Response Strategy Management with Active and Reactive Power Incentive in the Smart Grid: A Two-Level Optimization Approach	Journal article	Shigenobu et al. (2017)
MAP Regulations Will Fast-Track Closure of Metering Gap	Report	Nigerian Electricity Regulatory Commission (2018)
NERC Licenses 17 Power Distribution Networks, Seeks Unbundling of System Operations	News article	The Cable (2023)
National Energy Policy Revised 2022	Policy document	Energy Commission of Nigeria (2022)
Peer-to-Peer Energy Trading, Independence Aspirations and Financial Benefits Among Nigerian Households	Journal article	Adewole et al. (2023)
Transmission Company of Nigeria	Organisation website	Transmission Company of Nigeria (2023)
World Energy Outlook 2022	Report	IEA (2022)

non-renewables, rural electrification, and policy implementation inconsistencies (Energy Commission of Nigeria, 2022; Shigenobu et al., 2017).

Nigeria has also introduced various policies like the National Energy Efficiency Action Plan (NEEAP) and the National Renewable Action Plan (NREAP) to advance energy efficiency and increase electricity access. The Nigerian Electricity Regulatory Commission (NERC) has played a role in promoting energy efficiency through initiatives such as energy-efficient lighting and addressing the metering gap. However, challenges remain, including gas supply limitations, distribution network constraints, and infrastructure issues, highlighting the need for comprehensive reforms (Nitte & Salahudeen, 2023).

4.3.2.2 Nigerian Whole Energy System

The Nigerian energy system is categorised by different stakeholders, as indicated in Table 4.2. The Federal Government is responsible for policy and oversight, while the Ministry of Power implements energy policies. The Energy Commission of Nigeria serves as the regulatory and planning authority. The Electric Power Sector Reform Act (ESPRA) 2005 initiated privatisation, leading to Generation Companies

Table 4.2 Key stakeholders in Nigeria's energy system

Stakeholders	Roles and responsibilities
Federal Government	Oversight and policymaking for the energy sector
Ministry of Power	Implementation of government energy policies and regulations
Energy Commission of Nigeria (ECN)	Regulatory and planning authority for the energy sector
Electric Power Sector Reform Act (ESPRA) 2005	Legislation leading to energy system privatisation
Generation Companies (GenCos)	Responsible for bulk power generation
Transmission Company of Nigeria (TCN)	Ensures the transmission of high voltage electricity
Distribution Companies (DisCos)	Responsible for distributing electricity to consumers
Nigeria Bulk Electricity Trading (NBET)	Manages electricity trading across the Nigerian Electricity Supply Industry (NESI)
Rural Electrification Agency (REA)	Focuses on electrification of underserved communities

Source IRENA (2023b), Nigerian Bulk Electricity Trading (2023), Transmission Company of Nigeria (2023) and Usman et al. (2022)

(GenCos) for power generation, Transmission Company of Nigeria (TCN) for transmission, and Distribution Companies (DisCos) for electricity distribution. Nigeria Bulk Electricity Trading (NBET) manages electricity trading, and the Rural Electrification Agency (REA) focuses on electrifying underserved communities (IRENA, 2023b; Nigerian Bulk Electricity Trading, 2023; Transmission Company of Nigeria, 2023; Usman et al., 2022).

4.3.3 Existing Energy Flexibility Instruments in Nigeria and Effectiveness of Market Instruments

In pursuit of enhancing supply and demand equilibrium within the nation, Nigeria has implemented several energy flexibility instruments within its energy sector. The energy flexibility landscape in Nigeria is characterised by a variety of market dynamics as discussed below.

4.3.3.1 Demand Side Management

A key initiative is Demand Side Management, which aims to reduce energy demand by promoting the adoption of energy-efficient technologies such as light-emitting diodes (LEDs) as replacements for incandescent and fluorescent lighting in households. This aligns with the objectives outlined in the National Energy Efficiency Action Plan (NEEAP) 2015–2030 and poses the potential to reduce peak demand by 3–4 GW with NEEAP targeting a 40% improvement in lighting efficiency by 2020 and 100% by 2030 (IRENA, 2023b; Nitte & Salahudeen, 2023; Usman et al., 2022).

4.3.3.2 Peer-to-Peer (P2P) Energy Trading

Nigeria has also embraced peer-to-peer (P2P) energy trading as a means to facilitate energy flexibility. This short-term strategy, in line with the National Energy Policy, encourages the sale of excess electricity generated by industries to other users, contributing to improved energy supply (IRENA, 2023b). P2P trading, while conceptually promising, faces significant challenges due to the absence of a grid infrastructure capable of facilitating such transactions (Adewole et al., 2023). This highlights the crucial role that grid development plays in enabling advanced energy trading mechanisms.

4.3.3.3 Decentralised Energy Systems

The integration of decentralised energy systems using renewable energy sources, including isolated microgrids and grid-connected microgrids, serves as a pivotal energy flexibility instrument, particularly in rural and peri-urban areas. These systems, supported by Solar Home Systems, align with the objectives of the National Energy Policy (NEP) to enhance renewable energy integration (Energy Commission of Nigeria, 2022). The adoption of solar home systems (SHS) has seen substantial growth to meet Nigeria's energy demand. REN21's report noted a 5% increase in off-grid solar markets in 2019, and Nigeria stands out with one of the largest mini-grid support programs. This includes notable installations, such as two solar hybrid mini grids in 2020, facilitated by Nigeria's Rural Electrification Authority (REA) (Ren21, 2021). Presently, over 45 million people in sub-Saharan Africa utilise SHS with a capacity exceeding 11 Watts peak rating (IEA, 2022).

4.3.3.4 Independent Power Plants (IPPs)

Approved by NERC, there are currently 12 IPPs operating in Nigeria, with nine of them supplying power to the grid and three serving off-grid regions without access to the grid (Agusto & Co Store, 2023; The Cable, 2023). These power plants primarily rely on gas as their energy source, enabling them to provide ancillary services due to their black start capability and short response time. Independent Power Plants (IPPs) have gained a strong foothold in Nigeria's energy market, contributing a substantial 31.2% to the total generated power capacity in 2022 (Agusto & Co Store, 2023). This underscores their effectiveness in enhancing the energy supply landscape.

4.3.3.5 Smart Metering

Another notable energy flexibility instrument is the introduction of electricity smart meters under the Meter Provider framework, a strategy aimed at improving demand-side response and providing valuable information to suppliers for optimising energy distribution (Ikpe & Torriti, 2018; Nigerian Electricity Regulatory Commission, 2018). This initiative was part of the efforts of NERC's inter-ministerial committees on renewable energy and energy efficiency strategies (Usman et al., 2022). The Nigerian Electricity Regulatory Commission (NERC) has made significant strides in metering consumers, achieving levels ranging from 80 to 100% for maximum-demand consumers, and considerable coverage for non-maximum-demand consumers as of December 2022 (The Cable, 2023) This progress indicates an improvement in monitoring and management capabilities within the energy market.

4.3.3.6 Energy Flexibility in the UK

A world leader in energy systems, the United Kingdom has numerous energy flexibility instruments, like Demand Side Response Programs, Smart Grid Technologies, Energy Storage Solutions, Virtual Power Plants (VPPs), and Market-Based Mechanisms supported by policies like Smart Systems and Flexibility Plan, Electricity Market Reform (EMR), Flexible Power Auctions and Energy Storage R&D Funding etc. as indicated in Table 4.3. These instruments are supported by policies like the Smart Systems and Flexibility Plan, Electricity Market Reform (EMR), Flexible Power Auctions, and Energy Storage R&D Funding.

4.3.4 Comparing Energy Flexibility in the UK to Nigeria

4.3.4.1 Case Study: UK Agile Octopus Smart Meters

The strength of the UK flexibility instruments was displayed in a case study in June 2018, when Octopus Energy introduced a pioneering half-hourly time-of-use tariff known as Agile Octopus in Britain. This innovative tariff utilises data from smart meters to reflect the changing price of electricity to customers throughout the day. As a result, customers can benefit from lower electricity costs during periods of low demand or when available electricity generation is abundant. This approach not only allows customers to access affordable electricity but also supports balance and optimisation of the grid for cleaner and more sustainable power sources (Ofgem, 2021).

The introduction of Agile Octopus brought to light several key lessons regarding time-of-use tariffs. Firstly, it underscored the critical need for raising consumer awareness about these tariffs, as many customers were unfamiliar with the concept

Table 4.3 Flexibility instruments in the United Kingdom

Energy flexibility instrument	Description	Corresponding policies	Policy relation to instrument	References
Demand response programs	Initiatives incentivising consumers or businesses to adjust energy consumption based on grid needs or price signals	Smart Systems and Flexibility Plan	Supports the implementation of demand response initiatives by fostering technological advancements and regulatory frameworks for flexible energy usage	Ofgem (2021)
		Capacity Market	Encourages participation in demand-side response by providing financial incentives to reliable energy sources during peak demand periods	Ofgem (2023)
Smart grid technologies	Utilises digital communication and control to manage electricity supply and demand efficiently	Smart Systems and Flexibility Plan	Promotes the adoption of smart technologies to enhance grid management and flexibility in handling energy supply and demand	Ofgem (2021)
		Electricity Market Reform (EMR)	Introduces policies to drive investment in low-carbon electricity generation and grid stability through smart grid technologies	Department of Energy and Climate Change (2013)
Energy storage solutions	Includes battery storage systems to store excess energy for later use, helping to balance supply and demand fluctuations	Smart Systems and Flexibility Plan	Encourages advancements in energy storage technologies for effective utilisation and storage of excess energy	Ofgem (2021)
		Electricity Market Reform (EMR)	Facilitates the integration of storage solutions into the energy market, ensuring better management of supply fluctuations	Department of Energy and Climate Change (2013)

(continued)

Table 4.3 (continued)

Energy flexibility instrument	Description	Corresponding policies	Policy relation to instrument	References
Virtual Power Plants (VPPs)	Aggregated systems using distributed energy resources to function as a single power plant, enhancing flexibility	Smart Systems and Flexibility Plan	Supports the development and integration of virtual power plants by enabling coordination among diverse energy resources	Ofgem (2021)
		Electricity Market Reform (EMR)	Encourages the deployment of aggregated resources to act as a single power source, improving system flexibility	Department of Energy and Climate Change (2013)
Market-based mechanisms	Systems facilitating trading or exchange of energy between suppliers, consumers, and market players to optimise energy usage	Capacity Market	Ensures a competitive market by incentivising the exchange of energy resources and demand response participation	Ofgem (2023)
		Electricity Market Reform (EMR)	Establishes frameworks for market-based mechanisms to enhance flexibility and optimise energy usage	Department of Energy and Climate Change (2013)

of fluctuating electricity prices throughout the day. Secondly, the inconvenience of strictly adhering to off-peak usage was evident, highlighting the challenges consumers might face in limiting electricity-intensive activities to specific periods. Additionally, the tariff's suitability for families with children, who have varying electricity needs, was called into question, emphasising the importance of tailored tariff options. Moreover, the observation of high tariffs during winter peak times emphasised the significance of budgeting for seasonal variations in energy costs. Finally, the implementation of Agile Octopus revealed that lifestyle changes might be necessary for customers to maximise cost savings, emphasising the need for clear communication and support. In conclusion, Agile Octopus provided valuable insights, emphasising consumer awareness, flexibility, and tailored solutions as crucial considerations for effective time-of-use tariffs (Smart Home Charge, 2021).

4.3.4.2 Case Study: Nigeria 2022 Power Outage

The rapid growth of population across the most populous and largest economy in Africa, Nigeria has resulted in an exponential increment in energy demand predicted to unlock the economic development even further (IRENA, 2023b). Consequently, the World Bank approved a $500 million aid to enable the improvement of electricity distribution and energy flexibility across the Nation through large-scale projects (Africa–EU Renewable Energy Cooperation Programme, 2021). Nonetheless today Nigeria is still plagued with insufficient energy generation to meet demand, thus energy flexibility instruments although existing can only contribute little to guaranteeing an energy system equilibrium. For example, the lack of a smart grid system resulted in a significant power outage, across Nigeria in 2022. Due to an explosion and fire at a transmission line connecting power plants in Niger state. The total grid collapse would have been averted if the grid system had been decentralised and flexible (BBC News, 2022). The employment of a smart grid system creates pathways to overcome single faults.

4.3.5 Measuring Success: Energy Flexibility in the UK Versus Nigeria

The criteria for measuring the success of energy flexibility and balancing mechanisms within global energy systems have not been universally defined. However, an essential indicator in today's energy sector is the ability of an energy system to dynamically adjust the timing and location of energy generation and consumption in real-time. This adaptability should align with the goals of ensuring a secure, sustainable, and equitable energy supply, often referred to as the energy trilemma (Ofgem, 2021; World Energy Council, 2022c).

In the 2022 World Energy Council Energy Trilemma Index, the United Kingdom ranked impressively. It secured the 10th position in energy security with a score of 70.8, just 6 points below the top-ranked Canada, highlighting its capacity to manage its energy system effectively. Sustainability ranked 7th, with a score of 83.2, indicating its commitment to minimising environmental impact. Additionally, in providing reliable and affordable energy, it stood 12th with an index score of 96.4. However, concerns emerged regarding energy equity due to inflation and geopolitical tensions. Overall, according to the World Energy Trilemma Index, the UK's energy system demonstrated effectiveness, focusing on balance and environmental sustainability, improving its 2022 ranking to 4th place in 2023 with an index score of 82.4 (World Energy Council, 2022b).

In contrast, Nigeria's energy system faces challenges in all three areas of the energy trilemma. Nigeria ranked 84th in the 2022 energy trilemma index with a score of 39.1 (grade DDD) across all categories. While energy security improved slightly, sustainability remained stagnant. Energy equity showed some improvement, with a

3.6% increase since previous assessments. However, the trilemma scores for energy security, sustainability, and equity were 46.2, 47.9, and 24.9, respectively, indicating room for enhancement (World Energy Council, 2022a).

More specifically, the execution of the agile octopus, although plagued by behavioural limitations, provides reliable and affordable electricity tariffs when used during off-peak hours, encourages the use of cleaner and more sustainable energy by reducing the tariff for greener energy, and provides secure energy. Consequently, when underpinned below the world energy trilemma criteria for success, the case study gets an A in all three grades.

Considering the grid failure in 2022, energy security would get an F, due to the lack of secure power supply, however, the grade for energy sustainability and equity would maintain the general grade for Nigeria "D" by the World Energy Council as indicated in because it would be difficult to measure the sustainability and equity of energy that was not supplied.

4.3.6 Discussion: Assessing Uncertainties and Opportunities for Flexibility in Nigeria

4.3.6.1 Expanding Power Capacity

Nigeria's current power generation capacity falls short of meeting the needs of its growing population, which exceeds 206 million people. Estimates suggest that around 24 GW of capacity is required to address this demand. To achieve energy security and enhance energy flexibility, substantial investments, and the integration of renewable energy sources, particularly solar and wind, are imperative. The addition of alternative energy sources would help Nigeria move away from load shedding.

4.3.6.2 Upgrading the Grid (Smart Grid)

Despite having an installed power capacity ranging from 12 to 15 GW, Nigeria faces challenges with its grid infrastructure. The average power generation of approximately 5 GW falls short of the potential 8 GW due to distribution and transmission losses stemming from an ageing grid. Initiatives like the partnership between Nigeria and Siemens aim to comprehensively enhance the grid in a three-phase approach, increasing the grid capacity from 7 to 25 GW (Omonfoman, 2019; Siemens, 2019).

4.3.6.3 Poor Policies Implementation

According to the authors, Governmental policies and initiatives have played a central role in promoting energy efficiency. However, the success of these policies has been mixed, highlighting the need for more targeted strategies and robust implementation (Adewole et al., 2023). Nigeria has devised an array of energy policies, strategies, and plans, yet their actualisation is marred by delays and often unrealised potential. For instance, the Energy Transition Plan, which aspires to achieve universal energy access by 2030 and establish a carbon–neutral energy system by 2060 (Ashurt, 2022). Although great, the plan remains ambitious without a clear roadmap, effective measuring and monitoring to achieving (World Economic Forum, 2023).

4.3.6.4 Corruption

Corruption poses a significant and formidable obstacle to the attainment of energy flexibility, vividly demonstrated by Nigeria's smart metering initiative. Despite securing a substantial $500 million from the World Bank to advance smart metering, Nigeria's progress remains slow, with smart meter installations around 40% (Africa–EU Renewable Energy Cooperation Programme, 2021).

4.3.6.5 Insecurity

Insecurity, particularly in Nigeria's northern and eastern regions, has hindered the reach of energy flexibility instruments into these areas. Prolonged conflicts and insurgencies have created an environment of risks, making the deployment of energy infrastructure difficult. This situation severely curtails access to reliable energy services.

4.3.6.6 Expanding Energy Flexibility Mechanisms

To promote energy flexibility, it is essential to extend mechanisms such as improved ancillary services, demand-side response, and energy efficiency to a broader population. Currently, a significant portion of Nigeria's population lacks access to adequate energy flexibility options and remains unaware of their benefits. While efforts are underway to enhance energy supply, exemplified by the construction of power plants in Abuja, further education is necessary to increase awareness and adoption.

Boosting power capacity and modernising the grid are undeniably pivotal endeavours in fortifying energy flexibility within Nigeria. With advancing technology, these goals not only become increasingly achievable but also more imperative in the face of the nation's escalating energy needs. Furthermore, the synchronisation of government policies with international net-zero objectives, placing a significant emphasis on decentralisation, decarbonisation, and digitisation, offers a hopeful path for the enhancement of Nigeria's energy sector. Nevertheless, it is crucial to emphasise that maintaining this trajectory and proactively adapting to forthcoming shifts in energy demand, technology, and policy are imperative prerequisites. This adaptability is vital to establishing a secure, equitable, and sustainable energy landscape that can effectively meet Nigeria's evolving requirements.

4.4 Conclusion, Recommendations and Limitations

4.4.1 Conclusion

The imperative of limiting global temperature rise and achieving the Paris Agreement's NetZero goal has spurred the global energy transition, emphasising the need to integrate renewable energy sources, reduce fossil fuel dependence, and enhance energy efficiency. This transition demands robust engagement with energy flexibility instruments to achieve supply and demand equilibrium worldwide. Nigeria, with its large population and significant landmass, faces unique energy challenges, including a low electrification rate, underutilised power grid capacity, and a fragile grid. This paper critically evaluates Nigeria's energy system, focusing on market instruments for flexibility by comparing it with the United Kingdom using Agile Octopus Smart Meters as a case study and the energy trilemma as comparison measures.

The UK has effectively implemented various energy flexibility mechanisms such as Demand Side Response Programmes, Smart Grid Technologies, Energy Storage Solutions, Virtual Power Plants (VPPs), and Market-Based Mechanisms, supported by strong policies and innovative projects such as the Agile Octopus Smart Meters. The case study of Agile Octopus Smart Meters showcases the effectiveness of these instruments in promoting time-of-use tariffs and optimising energy usage. It highlights the importance of consumer awareness, flexibility, tailored solutions, and seasonal variations in energy costs as key lessons learned.

Energy flexibility instruments identified in Nigeria include demand-side management, smart metering, peer-to-peer energy trading, decentralised energy systems, and the operation of independent power plants (IPPs). Despite its vast energy potential and investment support from organisations like the World Bank, Nigeria faces a significant energy flexibility shortfall. The country's rapid population growth has led to a substantial increase in energy demand, but insufficient energy generation and infrastructure challenges persist. For instance, Nigeria's recent power outage in 2022, caused by a transmission line explosion, could have been avoided with a decentralised and flexible grid system. Advanced economies like the United Kingdom use grid systems with multiple pathways to ensure continuous electricity supply, reducing dependency on a single point of failure and enhancing overall system resilience. The absence of a smart grid system, coupled with the centralised nature of the energy infrastructure, leaves Nigeria vulnerable to such disruptions.

The UK's efforts in employing energy flexibility instruments have contributed to its high ranking in the World Energy Council Energy Trilemma Index, ranking 4th globally. The UK energy system demonstrates a balanced approach to energy security, sustainability, and equity, ensuring a just energy transition across the nation.

In contrast, Nigeria's energy system remains challenged by insufficient infrastructure and centralised energy generation, resulting in frequent power outages and limited energy access. Nigeria ranks 84th with percentage indices of 46.2, 47.9, and 24.9 for energy security, sustainability, and equity, respectively.

Although the UK has successfully implemented several policies to support energy flexibility, such as demand-side response initiatives and advanced grid management technologies, these policies cannot be directly applied to Nigeria without significant adaptation. The primary reasons for this are rooted in the distinct differences between developing and developed economies. The UK benefits from a mature infrastructure, stable political climate, and robust economic frameworks that facilitate the implementation of advanced energy policies. In contrast, Nigeria faces numerous challenges, including underdeveloped infrastructure, political instability, and economic volatility. These factors present substantial barriers to the direct adoption of UK policies.

Implementing similar processes in Nigeria would require significant political and economic changes. This involves not only infrastructural upgrades but also regulatory reforms and substantial investment in technological advancements. Given the current political and economic landscape in Nigeria, such changes may not be feasible shortly. The socio-political complexities and financial constraints create an environment where replicating the UK's energy flexibility measures becomes highly challenging, if not impossible.

In conclusion, while lessons can be drawn from the UK's experience with energy flexibility, any attempt to apply these in Nigeria must be carefully tailored to address the unique challenges of a developing economy. This highlights the need for a customized approach that considers Nigeria's specific socio-economic and political context.

4.4.2 Recommendations

Consequently, this study proposes the following recommendations to improve Nigeria's energy system, integrate flexibility instruments and aid a just transition

- **Expanding power capacity and upgrading the grid**: Substantial investments in renewable energy sources and modernising grid infrastructure with smart technologies are essential. These actions will reduce transmission losses, enhance reliability, and support the transition to a more sustainable energy system.
- **Enhancing flexibility mechanisms**: Expanding the integration and implementation of energy flexibility instruments, such as energy efficiency measures, demand-side responses, and ancillary services, can help prevent issues like grid collapse. This approach enhances system flexibility, making the energy infrastructure more resilient and responsive.
- **Creating strategic roadmaps and implementing policies**: There is a need to create strategic roadmaps and implement policies to achieve targets, such as the

Energy Transition Plan, which aims for universal energy access by 2030 and a carbon–neutral energy system by 2060. Achieving these goals requires clear roadmaps and effective implementation strategies.

- **Financial allocation and transparency in energy development**: Financial resources should be allocated to enhance the reliability of energy systems by expanding power capacity, improving and expanding the grid, and integrating energy flexibility instruments. It is also essential to establish transparent processes for allocating and using these financial resources, including regular audits and public reporting. As well as creating systems to monitor these projects from conception to completion and during their operation life cycle to ensure the effective use of resources. This transparency will help, prevent misuse and foster accountability.

4.4.3 Limitations

A key limitation of this study was the high dependence on expert recommendations and the backward snowballing method to gather and review relevant articles. This approach was primarily adopted due to the limited availability of studies when employing the specific search string used for this research. The limitation could have been mitigated by expanding the search string to include a wider range of terms, potentially capturing articles that, while not directly related to the topic, might have provided a more holistic view of the Nigerian energy sector. Nonetheless, this research serves as a strong foundation, offering valuable insights into the use of energy flexibility mechanisms within the Nigerian energy sector to improve the balance between energy supply and demand across the Nigerian whole energy sector.

Acknowledgements The authors would like to express my gratitude to everyone at Cranfield University for providing the necessary resources for this research. The corresponding author would also like to acknowledge Delta State University and the Tertiary Education Trust Fund for their educational sponsorship, which enabled him to undertake the MSc program. The modules taught and assignments completed during the program formed the foundation of this publication.

Disclaimer This paper was curated from an assignment submission made by the Author for the course module "Energy Policy and Economics" delivered at Cranfield University, resulting in a high similarity index. The original submission to Cranfield University received a Turnitin similarity score of 10%, comprising individual components, each contributing 2% or less.

Funding No funding was received for conducting this study.

Availability of Data and Materials No primary or secondary data were used in this research. The research focused on a literature-based analysis.

Ethical Clearance Ethical clearance was obtained as required.

Competing Interest The authors declare that they have no known competing financial interests or personal relationships that could have appeared to influence the work reported in this paper.

References

Abbà, I., La Bella, A., Corgnati, S. P., & Corsetti, E. (2024). Assessing flexibility in networked multi-energy systems: A modelling and simulation-based approach. *Energy Reports, 11*, 384–393. https://doi.org/10.1016/J.EGYR.2023.11.049

Adewole, A., Shipworth, M., Lemaire, X., & Sanderson, D. (2023). Peer-to-peer energy trading, independence aspirations and financial benefits among Nigerian households. *Energy Policy, 174*, 113442. https://doi.org/10.1016/J.ENPOL.2023.113442

Africa–EU Renewable Energy Cooperation Programme. (2021). *How does the energy sector work in Nigeria?* https://www.africa-eu-renewables.org/how-does-the-energy-sector-work-in-nigeria/

Agusto & Co Store. (2023). *2023 the Nigerian independent electric power producers*. Agusto & Co Store. https://www.agustoresearch.com/report/2023-the-nigerian-independent-electric-power-producers/

Ahmadiahangar, R., Rosin, A., Palu, I., & Azizi, A. (2020). Investigating different sources of flexibility in power system. In R. Ahmadiahangar, A. Rosin, I. Palu, & A. Azizi (Eds.), *Demand-side flexibility in smart grid* (pp. 27–37). Springer Singapore. https://doi.org/10.1007/978-981-15-4627-3_3

Akporhonor, G. K., Otuagoma, S. O., & Akporhonor, T. A. (2023). Nigerian wind energy status. *Wind Engineering, 48*(2), 310–322. https://doi.org/10.1177/0309524X231206559

Ashurt. (2022). *Nigeria's energy transition*. https://www.ashurst.com/en/insights/nigerias-energy-transition/

Attabo, A. A., Ajayi, O. O., Oyedepo, S. O., & Afolalu, S. A. (2023). Assessment of the wind energy potential and economic viability of selected sites along Nigeria's coastal and offshore locations. *Frontiers in Energy Research, 11*, 1186095. https://doi.org/10.3389/fenrg.2023.1186095

BBC News. (2022). *Nigeria hit by widespread blackout in "total system collapse"*. https://www.bbc.co.uk/news/world-africa-66810202?_hsenc=p2ANqtz--qfkWlPSD-3VlohLb0Jg3yFiF-DEroi1mng5FRWBzur5_gpVMM_3-66mMyhWUWIoDxjAle

Ben, U. C., Akpan, A. E., Mbonu, C. C., & Ufuafuonye, C. H. (2021). Integrated technical analysis of wind speed data for wind energy potential assessment in parts of southern and central Nigeria. *Cleaner Engineering and Technology, 2*, 100049. https://doi.org/10.1016/j.clet.2021.100049

Climatescope. (2022). *Nigeria*. https://www.global-climatescope.org/markets/ng/

Cruz, R. O., Duarte, G. R., Pires de Lima, B. S. L., & Jacob, B. P. (2024). Optimization of steel jackets to support offshore wind turbines using evolutionary algorithms. *Journal of Offshore Mechanics and Arctic Engineering, 146*(2), 022001. https://doi.org/10.1115/1.4062625

Dannheim, J., Bergström, L., Birchenough, S. N. R., Brzana, R., Boon, A. R., Coolen, J. W. P., et al. (2020). Benthic effects of offshore renewables: Identification of knowledge gaps and urgently needed research. *ICES Journal of Marine Science, 77*(3), 1092–1108. https://doi.org/10.1093/icesjms/fsz018

Department of Energy and Climate Change. (2013). *Electricity market reform delivery plan*. www.nationalarchives.gov.uk/doc/open-government-licence/

Energy Commission of Nigeria. (2022). *National energy policy revised 2022*. https://www.energy.gov.ng/NEP.php

Eweka, E. E., Lopez-Arroyo, E., Medupin, C. O., Oladipo, A., & Campos, L. C. (2022). Energy landscape and renewable energy resources in Nigeria: A review. *Energies, 15*(15), 5514. https://doi.org/10.3390/en15155514

Gusenbauer, M., & Haddaway, N. R. (2020). Which academic search systems are suitable for systematic reviews or meta-analyses? Evaluating retrieval qualities of Google Scholar, PubMed, and 26 other resources. *Research Synthesis Methods, 11*(2), 181–217. https://doi.org/10.1002/jrsm.1378

IEA. (2022). *World energy outlook 2022*. https://www.iea.org/reports/world-energy-outlook-2022

Ikpe, E., & Torriti, J. (2018). A means to an industrialisation end? Demand side management in Nigeria. *Energy Policy, 115*, 207–215. https://doi.org/10.1016/J.ENPOL.2018.01.011

IRENA. (2023a). *Energy profile (United Kingdom).* www.irena.org

IRENA. (2023b). *Renewable energy roadmap: Nigeria.* International Renewable Energy Agency. www.irena.org

IRENA. (2023c). *World energy transitions outlook 2023.* https://www.irena.org/Digital-Report/World-Energy-Transitions-Outlook-2023

Jung, C., & Schindler, D. (2020). Introducing a new approach for wind energy potential assessment under climate change at the wind turbine scale. *Energy Conversion and Management, 225*, 113425. https://doi.org/10.1016/j.enconman.2020.113425

Malaekeh, S. M., Safaie, A., Shiva, L., & Tabari, H. (2022). Spatio-temporal variation of hydro-climatic variables and extreme indices over Iran based on reanalysis data. *Stochastic Environmental Research and Risk Assessment, 36*(11), 3725–3752. https://doi.org/10.1007/S00477-022-02223-0

Ndukwu, M. C., Onwude, D. I., Bennamoun, L., Abam, F. I., Simo-Tagne, M., Horsfall, I. T., & Briggs, T. A. (2021). Nigeria's energy deficit: The challenges and eco-friendly approach in reducing the energy gap. *International Journal of Sustainable Engineering, 14*(3), 442–459. https://doi.org/10.1080/19397038.2020.1842546

Neupane, D., Kafle, S., Karki, K. R., Kim, D. H., & Pradhan, P. (2022). Solar and wind energy potential assessment at provincial level in Nepal: Geospatial and economic analysis. *Renewable Energy, 181*, 278–291. https://doi.org/10.1016/J.RENENE.2021.09.027

Nigerian Bulk Electricity Trading. (2023). *NBET*. https://nbet.com.ng/epsra.html

Nigerian Electricity Regulatory Commission. (2018). *MAP regulations will fast-track closure of metering gap.* https://nerc.gov.ng/index.php/media-library/press-releases/551-map-regulations-will-fast-track-closure-of-metering-gap

Nitte, I. S., & Salahudeen, T. M. (2023). Energy transitions in Nigeria: The role of policies for the adoption of low-carbon technologies and system integration. *EPRA International Journal of Research & Development, 8*(3), 128–139. https://doi.org/10.36713/EPRA12666

Ofgem. (2021). *Transitioning to a net zero energy system: Smart systems and flexibility plan 2021.* https://assets.publishing.service.gov.uk/government/uploads/system/uploads/attachment_data/file/1003778/smart-systems-and-flexibility-plan-2021.pdf

Ofgem. (2023). *Guidance the change process for the capacity market rules.* https://www.ofgem.gov.uk/publications/final-guidance-capacity-market-cm-rules

Olaofe, Z. O. (2017). Assessment of the offshore wind speed distributions at selected stations in the south-west coast, Nigeria. *International Journal of Renewable Energy Research, 7*(2), 565–577. https://doi.org/10.20508/ijrer.v7i2.5439.g7031

Omonfoman, O. (2019). *Analyzing the Siemens electricity roadmap proposal.* Business day NG. https://businessday.ng/opinion/article/analyzing-the-siemens-electricity-roadmap-proposal/

Paraschiv, S., Paraschiv, L. S., Serban, A., & Cristea, A. G. (2022). Assessment of onshore wind energy potential under temperate continental climate conditions. *Energy Reports, 8*, 251–258. https://doi.org/10.1016/J.EGYR.2022.02.118

Patidar, H., Shende, V., Baredar, P., & Soni, A. (2022). Comparative study of offshore wind energy potential assessment using different Weibull parameters estimation methods. *Environmental Science and Pollution Research, 29*(30), 46341–46356. https://doi.org/10.1007/s11356-022-19109-x

Reiter, U., & Jakob, M. (2021). What is the flexibility potential in the tertiary sector? In D. Möst, S. Schreiber, A. Herbst, M. Jakob, A. Martino, & W. R. Poganietz (Eds.), *The future European energy system* (pp. 137–153). Springer. https://doi.org/10.1007/978-3-030-60914-6_8

Ren21. (2021). *Trends in Nigeria facts from the renewables 2021 global status report Nigeria's top highlights in 2020.* https://www.ren21.net/gsr

Ritchie, H., Roser, M., & Rosado, P. (2020). *CO_2 and greenhouse gas emissions.* Our World in Data. https://ourworldindata.org/co2/country/nigeria

Shigenobu, R., Adewuyi, B., Yona, A., & Senjyu, T. (2017). Demand response strategy management with active and reactive power incentive in the smart grid: A two-level optimization approach. *AIMS Energy, 5*(3), 482–505. https://doi.org/10.3934/energy.2017.3.482

Siemens. (2019). *Siemens and Nigerian Government signed implementation agreement for electrification roadmap.* Siemens. https://press.siemens.com/global/en/pressrelease/siemens-and-nigerian-government-signed-implementation-agreement-electrification

Smart Home Charge. (2021). *Octopus agile review.* https://www.smarthomecharge.co.uk/reviews/octopus-agile-review-would-it-work-for-you/

Somoye, O. A. (2023). Energy crisis and renewable energy potentials in Nigeria: A review. *Renewable and Sustainable Energy Reviews, 188*, 113794. https://doi.org/10.1016/J.RSER.2023.113794

Statista. (2023). *Nigeria: Unemployment rate from 2003 to 2022.* https://www.statista.com/statistics/382366/unemployment-rate-in-nigeria/

The Cable. (2023). *NERC licenses 17 power distribution networks, seeks unbundling of system operations.* https://www.thecable.ng/nerc-licenses-17-power-distribution-networks-seeks-unbundling-of-system-operations#google_vignette

Transmission Company of Nigeria. (2023). *Transmission Company of Nigeria.* https://tcn.org.ng/index.php

Tziovani, L., Savva, M., Asprou, M., Kolios, P., Kyriakides, E., Tapakis, R., et al. (2020). Assessing the operational flexibility in power systems with energy storage integration. In B. Németh & L. Ekonomou (Eds.), *Flexitranstore. Lecture Notes in Electrical Engineering* (Vol. 610). Springer. https://doi.org/10.1007/978-3-030-37818-9_1

Usman, R., Mirzania, P., Alnaser, S. W., Hart, P., & Long, C. (2022). Systematic review of demand-side management strategies in power systems of developed and developing countries. *Energies, 15*(21), 7858. https://doi.org/10.3390/EN15217858

Vallejo-Díaz, A., Herrera-Moya, I., Fernández-Bonilla, A., & Pereyra-Mariñez, C. (2022). Wind energy potential assessment of selected locations at two major cities in the Dominican Republic, toward energy matrix decarbonization, with resilience approach. *Thermal Science and Engineering Progress, 32*, 101313. https://doi.org/10.1016/j.tsep.2022.101313

Virtanen, E. A., Lappalainen, J., Nurmi, M., Viitasalo, M., Tikanmäki, M., Heinonen, J., et al. (2022). Balancing profitability of energy production, societal impacts and biodiversity in offshore wind farm design. *Renewable and Sustainable Energy Reviews, 158*, 112087. https://doi.org/10.1016/J.RSER.2022.112087

Wang, L., Wang, B., Cen, W., Xu, R., Huang, Y., Zhang, X., et al. (2024). Ecological impacts of the expansion of offshore wind farms on trophic level species of marine food chain. *Journal of Environmental Sciences, 139*, 226–244. https://doi.org/10.1016/j.jes.2023.05.002

Wohlin, C. (2014). Guidelines for snowballing in systematic literature studies and a replication in software engineering. *ACM International Conference Proceeding Series, 38*, 1–10. https://doi.org/10.1145/2601248.2601268

World Economic Forum. (2023). *Nigeria is tackling barriers to the green energy transition.* https://www.weforum.org/agenda/2023/05/how-nigeria-is-tackling-barriers-to-its-green-energy-transition/

World Energy Council. (2022a). *WEC trilemma: Country profile Nigeria.* https://trilemma.worldenergy.org/#!/country-profile?country=Nigeria&year=2022

World Energy Council. (2022b). *WEC trilemma: Country profile United Kingdom.* https://trilemma.worldenergy.org/#!/country-profile?country=United%20Kingdom&year=2022

World Energy Council. (2022c). *World energy trilemma index.* https://www.worldenergy.org/assets/downloads/World_Energy_Trilemma_Index_2022.pdf?v=1669842216

Chapter 5
Fuel Subsidy Removal: Implications on the Nigerian Economy

Stella Ngozi Okoroafor, Michael Chidiebere Okoroafor, and Mark Umunna Amadi

Abstract Energy subsidies and indeed fossil-fuel subsidies is at the front burner discussions for most developing economies. Nigeria is not an exception, whereby the managers of the economy recently in 2023 announced the removal of premium motor spirit subsidy, leading into the era of removal of all forms of fuel subsidy in the country. This study investigates the implications of fuel subsidy reforms and the state of the Nigerian economy with complete removal of fossil-fuel subsidies. Given the nature of this topic of debate over the decades, this study adopts a systematic review (SR) of the scientific literature to assess the effect of fuel subsidy removal on the Nigerian economy. Using the standard Preferred Reporting Items for Systematic Reviews and Meta-Analyses (PRISMA) guidelines, a total of ten scientific papers within years 2012–2024 on the research topic were thoroughly reviewed. A mix of bibliometric and narrative analyses was used to synthesize, analyse, and interpret the findings of the review. The findings of the review reveal the consequences of eliminating fuel subsidies by examining the intricate and complex interaction between economic, political, environmental, and social factors. The study showed negative effects of fuel subsidy removal on household and business sector in terms of their income, welfare and poverty indices, with the more vulnerable groups disproportionately affected. On the wider economy scale, the subsidy removal increased the inflation rate, reduced consumer demand and decreased output especially in the industrial sector, whereas it increases national income through government savings and revenue while reducing environmental costs in form of carbon emissions. The socio-political effects involve social unrest due to loss of means of livelihood and issues

S. N. Okoroafor (✉)
Department of Accountancy, Alvan Ikoku Federal University of Education, Owerri, Imo State, Nigeria
e-mail: ngozichi98@yahoo.com

M. C. Okoroafor
Eberelinks International Nigeria Limited, Owerri, Imo State, Nigeria
e-mail: eberelinks99@gmail.com

M. U. Amadi
Department of Agricultural Economics, Alex Ekwueme Federal University Ndufu-Alike, Ikwo, Ebonyi State, Nigeria
e-mail: amadi.mark@funai.edu.ng

R. U. Onyeneke et al. (eds.), *Energy Transition, Climate Action and Sustainable Agriculture*, https://doi.org/10.1007/978-3-031-83165-2_5

of addressing corruption and transparency in the petroleum sector. It was recommended that the government should be rather proactive in assessing the impacts of eliminating fuel subsidies on individuals and businesses and offer targeted assistance in areas like public transportation and energy infrastructure that will immediately lessen the adverse effects. Then also government must work towards mitigating the intensity of the adverse effects from global rise in petroleum prices by reviving the existing refineries/building modern ones and establishing social safety nets and social reinvestment pathways for societal welfare.

Keywords Fuel · Subsidy removal · Impact · Economic effect · Socio-political effect · Systematic review

5.1 Introduction

Nigeria is one of the leading oil-producing nations and the largest oil producer in Africa, ranking 6th among the Organization of Petroleum Exporting Countries (OPEC, 2022). The oil and gas industry has grown to be the dominant sector of the Nigerian economy, making up most government income and earnings in foreign currency and contributing to the largest portion of the country's GDP. This has been at the expense of solid minerals, manufacturing, and agriculture since the Gowon era. Over time, the country's economy was centred on the provision and sale of inexpensive petroleum products. In Nigeria, fuel subsidies have been in place since the 1970s and an average household depends on subsidized oil for domestic and commercial use, transportation, and power generation. According to Ovaga and Okechukwu (2012), a fuel subsidy is a government policy that reduces the market price of fossil fuels, making customers pay less than the actual fuel price. When subsidies are in place, consumers would pay less per litre of the petroleum product. Since fuel subsidies have a significant impact on residents' well-being and can sustain a nation's budget, they are an important issue of discussion nationwide.

According to international agency projections, global fossil fuel subsidies would increase from $325 billion in 2018 to $1 trillion in 2022. This amount is significantly more than the projected $204 billion in global aid in 2022 and more than the combined government revenue of all developing nations (Ozili & Ozen, 2021). In Nigeria, fuel subsidies significantly drain the government's finances, accounting for a large portion of the country's budget. This has led to proposals for the elimination of the fossil fuel subsidy to use the saved funds to assist the impoverished, disadvantaged and poor people in the country (Mouhoud & Couharde, 2020; Ozili & Ozen, 2021). Fuel subsidy removal is controversial, though, as some argue before its removal that it qualifies as help because it makes fuel more affordable for the poor. In 2023, the newly elected president announced the exclusion of fuel subsidies; the after-effect was untold hardship for the people. The removal made fuel too expensive for the poor. However, literature documented the negative effects of fuel subsidies, such as increased greenhouse gas emissions, air pollution, traffic jams and early deaths, lost tax revenue and a rise in wealth and poverty inequality (McCulloch et al., 2021). Other negative effects include fiscal burden (where the government diverts resources from other critical public programs), inefficient resource allocation, environmental degradation, distortion of market prices, reduced investment in alternative energy, dependence on fossil fuel, and reduced government revenue. Removing fuel subsidies and reforming fuel subsidies are challenging policies for legislators in many nations because they could lead to significant increases in electricity and fuel prices that low-income earners would find difficult to afford. It might spark massive demonstrations and increase the likelihood of a revolution or the removal of the current administration.

In the 1970s, Nigeria implemented fuel subsidies for the first time in reaction to the 1973 oil price shock. Fuel subsidies were partially eliminated in 1986. Since then, the fuel subsidies have remained in place. In 2012, the government abruptly discontinued its fuel subsidy policy. After the removal, there were massive protests aimed at persuading the government to reinstate the fuel subsidy that it had eliminated. In 2012, the government reinstated fuel subsidies in response to the massive protests. Since then, Nigeria has made significant increases in fuel subsidy payments. The government's 2022 budget of ₦17.126 trillion was split by fuel subsidies, which came to ₦4 trillion. As a result, Nigeria could not keep providing fuel subsidies in 2023, and the government announced that the subsidies would stop (Ozili & Obiora, 2024).

Current literature demonstrates that fuel subsidies have diverse consequences. While some studies highlight the advantages of fuel subsidies amidst the call for their elimination as well as their management, other studies draw attention to the drawbacks and call for their elimination. For instance, Omitogun et al. (2021) contend that the removal of fuel subsidies in Nigeria could potentially lead to a significant reduction in the country's carbon emissions, a crucial environmental benefit. Similarly, Adekunle and Oseni (2021) contend that eliminating fuel subsidies might restrict the growth in carbon emissions through low energy usage, even though implementing it would result in higher energy prices. Ovaga and Okechukwu (2012) maintain that fuel subsidy encourages dishonesty exploitation in the country. A dishonest sectional has been interfering with the operation of existing refineries, and they undermined efforts to construct new refineries in the country to maintain fuel importation and preserve fuel subsidies for their self-centered interest. Scholars emphasize the impact of removing fuel subsidies. Umeji and Eleanya (2021) argued that the removal of fuel subsidies could have serious effects, which could be abated by government precision regarding the use of the money recovered from removing fuel subsidies for infrastructure development. According to Ozili and Obiora (2024), the elimination of fuel subsidies could lead to more macroeconomic instability in Nigeria due to rising energy prices and inflation. Additionally, McCulloch et al.'s (2021) research demonstrates that a significant portion of Nigerians are against the removal or reform of fuel subsidies due to their perception that the government is dishonest and incapable of implementing clear and clean reforms.

Notwithstanding the evidence in the literature, there is little discussion about the effect of Nigeria's recent removal of fuel subsidies in 2023. There was debate concerning the impact of eliminating fuel subsidies on the Nigerian economy and the citizens since this was done without initial provision for some palliatives. As a result, it is critical to recognize and comprehend both the macroeconomic and microeconomic effects of the removal of fuel subsidies in 2023 in Nigeria. This study offers a thorough analysis of the implications of removing fuel subsidies, focusing on the macroeconomic and microeconomic implications of the recent removal of fuel subsidies in Nigeria, the largest economy in Africa. This study adds to the corpus of knowledge that examines the benefits and harmful effects of fuel subsidies as studied by some scholars (Adekunle & Oseni, 2021; Asare et al., 2020; Omitogun et al., 2021; Umeji & Eleanya, 2021). The study contributes to the ongoing discussion concerning the implication of fuel subsidies on the Nigerian economy, hence it is designed to provide a comprehensive analysis of the implications of fuel subsidy removal on the Nigerian economy. Using a discourse technique and systematic review (SR) of the scientific literature, the effect of fuel subsidy removal on the Nigerian economy was conducted based on PRISMA (Preferred Reporting Items for Systematic Reviews and Meta-Analyses) guidelines suggested by Moher et al. (2009).

According to the Agency of International Energy, worldwide fuel subsidies are expected to reach over $1 trillion by 2022, increasing from about $325 billion in 2018. Due to these costs, proposals have been made to end the world's fuel subsidies so that the money saved can be used for other productive projects in developing nations. The main economic sectors in Nigeria are negatively impacted by low earnings that are already being used up by inflated subsidies. To increase the capacity of local production, remove reliance on fuel imports, and increase the trade balance, the $20 million trade deficit that resulted from low crude oil export receipts in November 2022 necessitated the prompt removal of fuel subsidies (Abayomi, 2023). Mismanagement, exploitation, inconsistency, and improper use of public funds undermined the fuel subsidies. Between January 2020 and June 2022, N3.92 trillion was set aside for fuel subsidies; this amount is more than the total federal budget for agriculture, healthcare, education, and capital infrastructure and defence cumulative over 30 months. Nigeria spent over 10 trillion Naira on petroleum subsidies between 2006 and 2018, whereas N5.82 trillion was consumed between 2021–2022, and N3.36 trillion suggested for the first half of 2023 (Aniemeke, 2024). These numbers point to a substantial financial burden on the government, limiting its capacity to invest in vital sectors that could support economic expansion and improve the lives of citizens. Nigeria has not benefited from the increase in oil prices because of its poor oil production and rising fuel subsidy costs. Fuel subsidy removal is expected to boost economic growth and development. This expectation appears to be entirely academic and is not likely to materialize anytime soon, given that the international market price of crude oil continues to rise, which is a major factor in determining fuel prices in Nigeria. It follows that the current policy on eliminating fuel subsidies has both positive and negative implications for the populace and the nation at large. Therefore, this study combined discourse technique analysis and systematic review methodology to thoroughly investigate the implications of fuel subsidy removal on the Nigerian economy.

5.2 Conceptual Review

According to Ovaga and Okechukwu (2012), a fuel subsidy is a government policy that reduces the market price of fossil fuels, making customers pay less than the actual fuel price. It is a program that reduces consumer fuel costs by providing financial support to fuel producers or retailers. The subsidy is usually paid to the fuel suppliers or retailers. There are different types of subsidies, including consumer subsidies, producer subsidies, and retailer subsidies. Fuel subsidies are often used to reduce consumer fuel prices, increase energy access and affordability and support economic development. Nigeria's economy since the 1960s has been subsidized in various ways, most notably in the areas of fuel, education, power, and foreign exchange.

Fuel subsidies first appeared in the 1970s and were made official in 1977 with the passage of the Price Control Act, which outlawed the sale of petrol and other products for more than the suggested retail price. Although the concept of subsidies was admirable, the Nigerian gasoline subsidy program has been accused of several instances of corruption and inadequate management. This led to the deregulation of diesel in 2003 and the removal of the kerosene subsidy in 2016 (Manjo, 2024). A subsidy is a type of support or financial aid given to an industry, usually with the intention of advancing social and economic policy. The word "subsidy" refers to any support and assistance that can take different forms, including interest-free loans and cash grants, as well as indirect ones like insurance, tax advantages, low-interest loans, and rent rebates, price discounts, depreciation. In addition, consumption subsidies lower the cost of products and services for consumers, and production subsidies guarantee producers benefits by providing payments to production components or market price support (Manjo, 2024). Since the fourth republic was inaugurated in 1999, the fuel subsidy has grown to be the largest threat to succeeding administrations.

Despite several allegations of corruption in the program, a significant amount of the country's financial resources is allocated to paying fuel subsidies each year. Additionally, the budgetary vote on subsidies increased dramatically each year, even though the amount of fuel consumed nationally remained controversial. In January 2012, President Jonathan announced the elimination of fuel subsidies for the first time, which led to a price increase of N65 to N140 per litre, consequently fifteen days of widespread protests known as "Occupy Nigeria" led to its review. Before the Tinubu administration, fuel subsidies provided by Nigerian governments made the cost of gasoline cheaper. Literally, subsidies are economic interventions that are contrary to the market's demand. It is also government assistance given directly to individuals or private businesses that are believed to benefit the public to alleviate negative drawbacks and market failures and increase efficiency in the economy.

On May 29th, 2023, President Bola Tinubu, in his inaugural address, declared that "Nigeria's fuel subsidies would be scrapped, citing budgetary concerns." Anticipating this, President Bola Ahmed Tinubu said that the fuel subsidy would no longer be provided; Premium Motor Spirit (PMS) increased by nearly 200%, from N190/litre ($0.25/litre) in May 2023 to N600/litre ($0.8/litre). The removal of subsidies for Automotive Gas Oil (AGO), often known as "Diesel," had earlier caused a price increase of over 200% as at 2022. This negatively affected firms that primarily rely on diesel generators to supplement the intermittent electrical grid supply. Nigerians have experienced both positive and negative effects from eliminating fuel subsidies. The highly notable positive effect is the apparent savings of over N1 trillion on a bimonthly basis following the removal of fuel subsidies, while the most immediate negative effect is a sharp increase in inflation caused by the sharp rise in transportation costs (Ojeabulu, 2024). Additionally, Micro, Small, and Medium-Sized Enterprises (MSMEs) are currently having trouble finding reasonably priced power. Fuel subsidy elimination presents both challenges and opportunities.

The current gasoline price increase may encourage investments in reasonably priced renewable energy infrastructure, resulting in a more environmentally friendly economy and sustainable transportation systems, while helping the United Nations reach its goal of having no net emissions by 2050 (Ojeabulu, 2024). For example, the government acknowledged that eliminating fuel subsidies will increase government revenue, which increased from approximately N786 billion in May 2023 to approximately N1.9 trillion in June 2023. President Bola Ahmed Tinubu has stated that part of the reserves from the elimination of gasoline subsidies will go to boosting the economy. These include giving loans to small-scale enterprises at low or single-digit interest and purchasing buses for mass transit to encourage cleaner transportation (Ojeabulu, 2024). Fuel subsidy removal has both positive and negative macroeconomic and microeconomic implications.

5.3 Methodology

5.3.1 Study Approach

Systematic review (SR) of scientific literature on the effect of fuel subsidy removal on the Nigerian economy was conducted. This was based on Preferred Reporting Items for Systematic Reviews and Meta-Analyses (PRISMA) guidelines suggested by Moher et al. (2009). PRISMA framework supports a transparent approach for SR and ensures a replicable review procedure (including search strategy, inclusion and exclusion criteria, article selection criteria, data extraction, etc.).

A total of ten scientific papers on the research topic were reviewed. One electronic database (ScienceDirect) was used. Also, the literature search was limited to scientific articles published between 2012 and 2024, to increase the accuracy of the analysis, considering the popular nature of fuel subsidies topic such that it is fraught with opinions all over the print and electronic media in response to the rapid changing energy transitions.

A mix of bibliometric and narrative analyses was used in analysing, interpreting, and synthesising the findings of the review. Bibliometric analysis enables the identification of relationships among different aspects of the scientific literature through the analysis of publications and documents according to specific characteristics, such as authors, journals, institutions, and countries (Esen et al., 2018). On the other hand, the narrative analysis involved summarising and explaining the findings of multiple studies primarily with words and text.

5.3.2 *Epistemological Stance of This Study*

This paper adopts a positivist philosophical stance, within an epistemology that posits that knowledge emerges from objective and dispassionate review, and interpretation of the evidence. This is evident in the goals of identifying, evaluating, describing and synthesising the results of the investigated literature through strictly specified and impartially applied methods (Ferreira González et al., 2011; Mendoza De Los Santos et al., 2022). This approach guided the entire review process which is based on the principles of objectivity, empirical evidence, and the application of the scientific method.

5.3.3 *Search String Strategy*

It is widely regarded in published literature on SR that an appropriate search strategy should be developed to ensure that the relevant literature is identified for the study (Brunton et al., 2017). Developing this search strategy is an important first step to ensure that only literature relevant to the topic of the SR process is identified and utilized. One of the initial efforts of this study was to create a search string that would appropriately capture the broad research objectives to enable the researchers to assess the effect of fuel subsidy removal on the Nigerian economy. The initial challenge was the task of creating a search strategy that would appropriately capture all the strings embedded in "effect," "fuel subsidy removal," and "Nigerian economy."

Boolean operators were used to combine relevant keywords that facilitated the finding of published articles relevant to the topic. Furthermore, the use of parentheses helped in the grouping of related terms, while the operator OR expanded the search to include synonyms or related terms. For instance, the search string used synonymous terms like "fuel subsidy removal," "fossil fuel subsidy removal," "fuel subsidy reforms," and "energy subsidy removal" to specifically capture published studies on all types of fossil fuel subsidy reforms.

Details of the Boolean search strategy and combination of search terms are presented in Table 5.1.

A combination of the Boolean strings was used to search the Science Direct database. Science direct is a well-known database which houses Elsevier journals (top-tier peer-reviewed journals). The terms "impact" and "fuel subsidy" received the most hits given its broad views, understanding and interpretation. Nigeria does not actually have synonymous uses in the published literature. However, not all literature

Table 5.1 Boolean search strings used in the search for relevant literature

Topic and cluster	Search terms
Effect	"Impact" OR "Effect" OR "Effects" OR "Influence" OR "Implications"
Fuel subsidy removal	"Fuel subsidy removal" OR "Fuel subsidy reform" OR "Fossil fuel subsidy removal" OR "Fossil fuel subsidy reform" OR "Fossil energy subsidy removal" OR "Energy subsidy removal" OR "Fossil energy subsidy reform" OR "Energy subsidy reform" OR "Subsidy reforms" OR "Subsidy removal"
Nigerian economy	"Nigerian economy" OR "Nigeria economy" OR "Nigeria" OR "Nigerian society"

Source Author's composition

on Nigeria mentions it in the title. The search yielded a lot of rich materials (open and closed access) that are relevant to the research topic. This enabled the research to proceed to the next step in the systematic review.

5.3.4 *Inclusion and Exclusion Criteria*

Strict inclusion and exclusion criteria were implemented to keep the research nimble, slim and focused and to weed out ambivalences that are likely to compromise the integrity of the research. Table 5.2 presents the criteria used in including or excluding materials from the review process, eight filters were established.

5.3.5 *Study Selection and Screening Process*

The process of identifying, screening and selecting the final materials included in the SR followed the stepwise PRISMA flowchart framework described by Moher et al. (2009) and presented in Fig. 5.1. The first Excel codebook was created to guide the process of screening and selecting from relevant articles from the pool of 537 articles identified through the search strategy. This initial effort was strictly guided by the inclusion and exclusion criteria described in Table 5.2.

The first screening process was broad and involved an initial pool of all the source materials identified from ScienceDirect database (n = 537). Majority of the articles were empirical research articles (n = 229). This initial screening involved title and abstract screening which resulted in the exclusion of all articles that did not meet the selection criteria. Following this initial screening, a total of 146 records were removed based on studies published earlier than 2012.

The second step in the selection process involved checking all the papers and classifying them in terms of the article type within the valid years of publication. Thus, selecting the ones approved for further content analysis. This second screening

Table 5.2 Inclusion and exclusion criteria in the study

S. No.	Inclusion criteria	Exclusion criteria
1	Studies published between 2012 and 2024 to increase the accuracy of the current analysis	Studies published before 2012 were excluded. Also excluded are books, book chapters, book reviews, abstracts, conference papers, theses, dissertations, reports, videos, podcasts and blogs on the subject matter
2	Full-text, studies published in peer-reviewed journals	Journals that are not peer-reviewed
3	Empirical studies and literature reviews	Conceptual studies and theoretical studies
4	Quantitative and qualitative studies that specifically examined the link between fuel subsidy removal and the Nigeria economy	
5	Studies conducted in Nigeria	Studies conducted in settings other than the Nigerian economy
6	Studies assessing the impact of fuel subsidy removal on Nigeria economy	Studies assessing the impact of fuel subsidy removal on other economies of the world or across continents
7	Studies which discuss influence of the fuel subsidy removal in Nigeria at least one dimension of the economy	Studies which do not discuss the influence of the fuel subsidy removal in Nigeria at least one dimension of the economy
8	Studies published in the English language	Studies not published in the English language

Source Author's composition

led to the removal of 130 records. These records were excluded for reasons of being neither research articles nor review articles.

The final stage of the selection and screening process yielded 25 eligible articles for preliminary bibliometric analysis based on inclusion criteria. The following six characteristics were mainly examined: (i) year of study, (ii) country of study, (iii) analysis on fossil fuel subsidy, (iv) discussion on the impact of fossil fuel subsidy removal, (v) discussion on the impact of fossil fuel subsidy reform on Nigeria economy and (vi) reference to a fuel subsidy reform regime in Nigeria.

Following the implementation of the above bibliometric parameters, ten (10) articles were identified for the SR in the final stage of the screening and selection process (see Fig. 5.1).

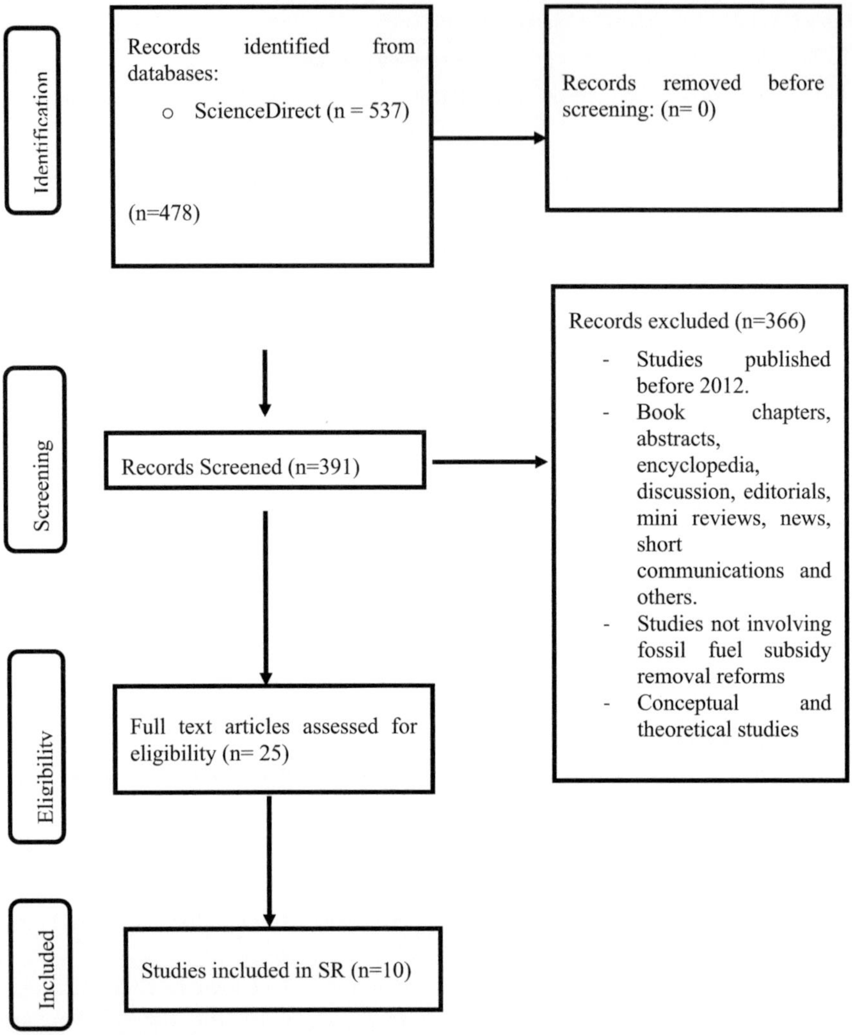

Fig. 5.1 PRISMA flowchart

5.3.6 *Data Extraction and Critical Appraisal*

A second Excel codebook (Data Extraction Codebook) was created to capture the following categories of information extracted from each of the ten (10) articles:

(1) Author(s)
(2) Year of Publication
(3) Country of Study
(4) Purpose/Objective of the Study

(5) Study Design/Method
(6) Instrument/Data Collection Method
(7) Data Analysis Technique
(8) Findings/Result related to impact of fuel subsidy removal on the Nigerian economy.

Each study was independently checked, coded and selected for inclusion based on the above eight bibliometric categories and characteristics. All inconsistencies and discrepancies were removed. A critical appraisal of the 10 studies that met all the inclusion/exclusion criteria described in Table 5.2 was conducted. Special attention was paid to methodological strengths and weaknesses to ensure accurate interpretation of the data and identification of biases. A summary of the study characteristics for each of the reviewed studies is then presented in Fig. 5.1.

5.4 Results and Discussion

This section focuses on the findings from the discourse technique analysis and systematic reviews with respect to the extent of the widespread interest on the impact of fuel subsidy removal on the Nigerian economy; the micro-economic and macro-economic implications; the socio-political and environmental perspectives; and categorisations of the interventions projected.

5.4.1 Description of Selected Studies

The description of the selected articles is provided to deepen our understanding of the criteria used in the selection. This description is also intended to show objectivity in keeping with the original inclusion and exclusion criteria of the systematic review as described in Table 5.2.

a. Year of Publication and Country of Research: The available information shows that current journal articles (published between 2012 and 2024) were included in the selection. The use of these journal articles published within this period is appropriate since the debate on fuel subsidy and major policy decisions taken in the removal of dual-purpose kerosene subsidy (2016) and premium motor spirit subsidy (2023) happened within the period. For the study to ensure specificity, only articles which independently analyzed the Nigerian situation with respect to fuel subsidy were further reviewed.
b. The purpose and context of each study played a role in the selection. To ensure consistency with the original design of the study, only studies with a stated purpose of assessing the impact, effect or implication of fuel subsidy removal on the Nigerian economy were selected.

c. All the studies included in the selection employed both quantitative and qualitative research paradigm or approach. This is in conformity with the inclusion and exclusion criteria of the present study.
d. The various studies adopted different methods of data collection which include semi-structured interviews (Agbonifo, 2023; Bazilian & Onyeji, 2012; Houeland, 2020; McCulloch et al., 2021; Osunmuyiwa & Kalfagianni, 2017), questionnaire (McCulloch et al., 2021), and secondary data (Okorie & Wesseh, 2024; Osunmuyiwa & Kalfagianni, 2017; Rentschler, 2016; Shittu et al., 2024; Siddig et al., 2014). Some studies (Agbonifo, 2023; Bazilian & Onyeji, 2012; Gamette & Oteng, 2024; Houeland, 2020; Osunmuyiwa & Kalfagianni, 2017) also used other secondary sources of information like journals, articles, periodicals, news tabloids, web news, agency information, online media space, etc. Conducting a primary data survey for the issues investigated in this study seems an expensive and daunting task owing to the quality of data needed, hence, only one identified study undertook the rigours of carrying-out a structured survey.
e. The studies employed various statistical analytical procedures ranging from descriptive statistical analysis to inferential analysis and simulations. Most studies used computable general equilibrium models (Okorie & Wesseh, 2024; Shittu et al., 2024; Siddig et al., 2014) and leveraged on available social accounting matrix data (Okorie & Wesseh, 2024; Shittu et al., 2024; Siddig et al., 2014). Other analytical methods used include logistic regression, literature reviews, and theory-based narratives.
f. Findings and results from all the 10 articles are reported with respect to the revealed impact or effects of fuel subsidy removal reforms on the Nigeria economy.

5.4.2 Fuel Subsidy Reform Regimes

Originally, subsidies were economic instruments used in the public domain in the form of government transfer payments to provide some level of support to segments of the population. In like manner, fuel subsidy in Nigeria was historically adopted in recognition of the oil endowment as a universal benefit of its derivatives to her citizenry. Unfortunately, the means to this derived benefit were not adequately internalized, which ironically resulted in regimes of fuel subsidy removal and reforms, with economic reasons at the forefront.

Figure 5.2 shows different fuel subsidy removal/reform regimes attributed to those reported by the articles within the review period. Five fuel subsidy removal/reform regimes were referenced or discussed by the ten (10) selected articles. These include the fuel subsidy reform of 2009 (Rentschler, 2016); fuel subsidy removal of 2012 (Agbonifo, 2023; Bazilian & Onyeji, 2012; Gamette & Oteng, 2024; Houeland, 2020; Siddig et al., 2014); fuel subsidy removal of 2016 (Gamette & Oteng, 2024; McCulloch et al., 2021; Osunmuyiwa & Kalfagianni, 2017); fuel subsidy reform

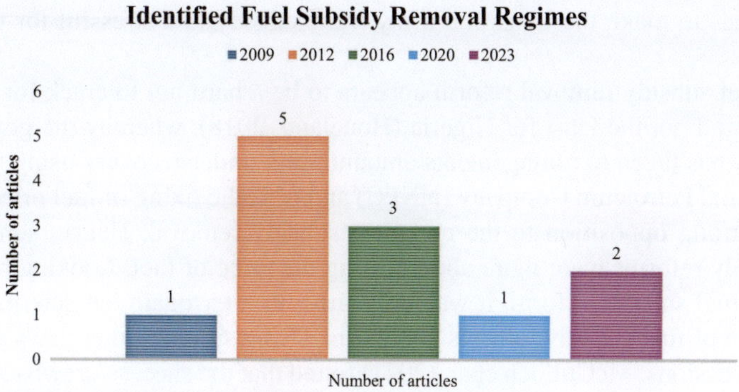

Fig. 5.2 Fuel subsidy removal regimes discussed in the articles

of 2020 (Okorie & Wesseh, 2024); and fuel subsidy removal of 2023 (Gamette & Oteng, 2024; Shittu et al., 2024).

The fuel subsidy removal of 2012 is the most discussed in the articles, perhaps for certain reasons. The subsidy removal affected both household kerosene (HHK) and premium motor spirit (PMS) products, and it was seriously protested by various components of the Nigerian society. Whereas the government of the day presented economic reasons (subsidy expenditure was too high with international oil prices rising so much) for its decision to remove the fuel subsidies, the masses alluded to various other narratives for their protests against the subsidy removal. The protests were borne out of the grievances of the masses that the dividends of democracy in terms of economic distribution of wealth and political inclusion were still lacking, years since the latest (1999) civilian rule (Houeland, 2020). The civil society and labour unions championed the anti-subsidy removal protest, pointing to the lack of transparency in the petroleum sector while also questioning the reality of fuel subsidies in the first place (Agbonifo, 2023). It can be said that the issue of trust was at the center stage of the government and citizenry disagreements on fuel subsidy removal or reforms, hence, allegations of corrupt practices in the government modalities of collection of revenue in the implementation of fuel subsidy (Gamette & Oteng, 2024).

Siddig et al. (2014) argued in this respect, highlighting the unsustainable and misplaced nature of the subsidy in terms of not fulfilling its articulated purpose of benefitting the less privileged groups in the economy. Emphasizing how the fuel subsidy rather works in the opposite direction of PMS subsidizing the rich, encouraging rentier and corrupt tendencies, enriching the few petroleum importing companies and draining the national vault and reserves. However, recognizing the inefficiencies that come with the perennial subsidization of fuels, which fails to promote efficient use of energy (Bazilian & Onyeji, 2012) draws attention to the more damning nature of such subsidy withdrawals in halting local production (therefore impeding economic progress), especially for developing economies. Again, they are focused on the crucial role of the State in ensuring that appropriate infrastructure (especially for the energy sector) is in place, supported by good governance in the utilization of resources to make the transition away from subsidy less stressful for the ailing economy.

The fuel subsidy removal reform appears to be a hard nut to crack for different countries and not the least for Nigeria (Houeland, 2018), whereby the government of Nigeria has taken to arrangements amounting to under-recovery using the Nigerian National Petroleum Company (NNPC) and periodic fixing of fuel prices. There is often strong opposition to the complete subsidy removal. Hence, some of the fuel subsidy reforms were more about raising the price of fuel. Looking back after the 2016 fuel subsidy reform, it was imperative to interrogate the determinants of acceptance of fuel subsidy reforms in Nigeria. Using surveys, interviews and focus group discussions, McCulloch et al. (2021) found that the success/drawbacks of fuel subsidy reforms depend on some economic variables (the fuel price and its availability) and other socio-political variables (level of government service delivery, government capacity and perceived corruption levels). Fuel (e.g. PMS) is almost necessary for Nigerians' day-to-day activities. Hence, its scarcity is so challenging that people are often willing to pay higher prices to access it. The observed or perceived competence level of the government of the day in discharging its responsibilities (which includes obligated services to the citizenry) influences the acceptance level of the government's new policies (e.g. fuel subsidy reforms). Conversely, the level of transparency and accountability communicated to the public via the actions of the government determines how the reforms are accepted. The alleged corruption in the petroleum and energy sector energizes the public outcry against fuel subsidy reforms. These findings of McCulloch et al. (2021) resonate with the events which heralded the 2016 fuel subsidy reform; whereby amidst the economic downturn and protracted petrol product scarcity, the reform was implemented without resistance from the masses. Also, this process might have scaled through because there was a new government in place that ousted the previous government in the 2015 presidential election on the mantra of the focused fight against corruption. Hence, the masses were still positively expectant of what the new regime would provide.

In response to the usual public resistance to fuel subsidy removal reforms, Osunmuyiwa and Kalfagianni (2017) posits that the government also learnt to strategize in such a manner as to push through their policy thrust amidst the economic pressure and intended goals by employing redistributive measures, discursive framings, and institutional strategies. In their narrative (Osunmuyiwa & Kalfagianni, 2017) believe these three-prong strategies characterized the different government actions of redistribution of rent (increments in wage, the Petroleum Development Trust Fund, the Youth Enterprise with Innovation in Nigeria, Youth Entrepreneurship Support Program, the Subsidy Reinvestment and Empowerment Program, N-Power, school children feeding program and conditional cash transfers); discursive framings (use of the news agencies and press briefings, town hall meetings to express subsidy pros and cons, campaigns); institutional strategies (National Economic Empowerment Development Strategy, privatizations of public enterprises, inclusion of renewables in the energy master plan and expansion of the energy sector, etc.).

Agbonifo (2023) analysis of the fuel subsidy regimes, shows how power asymmetries or differing layers of decision-making and authority control the narrative.

The findings reveal how the citizenry resists being taken for granted and demands some level of consideration in the decision-making processes of the energy sector, which affects their welfare. Gamette and Oteng (2024) highlight the critical role of well-planned and targeted palliatives for/from the subsidy reforms to guarantee its political feasibility and social acceptance. This highlights the need for subsidy reforms that can catalyze a transition in public perception regarding fossil fuels by engaging with the public and raising awareness about the benefits of the reform (Osunmuyiwa & Kalfagianni, 2017). This is a critical aspect for the government in building the 'social contract' because while the masses may not have a detailed understanding of essential issues warranting the fuel subsidy reforms, they usually develop strong opinions about it (McCulloch et al., 2021).

5.4.3 Effects of Fuel Subsidy Removal Reforms on the Nigeria Economy

Like various oil-producing countries of the world, fossil-fuel subsidy reforms (removal or price reviews), though popularly conceived as an economic decision, are fraught with varying consequences and impacts on national development objectives. The selected articles in this study reveal the diverse nature of the effects of the fuel subsidy reforms in Nigeria, with implications in the micro-economic, macro-economic, environmental and socio-political dimensions deduced.

5.4.3.1 Microeconomic Effects

Negative microeconomic effects. Consumer-based subsidies, supposedly targeted at the poor segment of society, are generally conceived to aid the end-users in obtaining lower prices and improving their living standards, leading to more consumption of the good. Such subsidies are said to create dependency or a "lock-in" effect in the economy (Sovacool, 2017), thereby making its withdrawal complicated, given the possible regressive effects it may result in.

Siddig et al. (2014) found negative effects of fuel subsidy removal on household income. Their simulations show that the urban households are affected more than the rural ones due to their larger direct purchase of petroleum products. Similarly, Shittu et al. (2024) and Okorie and Wesseh (2024), focusing on the 2023 fuel subsidy removal, found that outright withdrawal of PMS subsidy for Nigeria leads to a decrease in the household real consumption expenditure predicated by the general rise in the price level. (CPI). These studies (Gamette & Oteng, 2024; Okorie & Wesseh, 2024; Shittu et al., 2024) suggest that this situation subsequently creates adverse effects on household welfare. However, on household income level, Okorie and Wesseh (2024) found that poor households (urban or rural) are affected more, whereas Shittu et al. (2024) found that vulnerable urban households suffer the most from this removal.

Given the above effects on the household, it is appropriate that the poverty implications are analyzed. Gamette and Oteng (2024) suggested that fuel subsidy removal indirectly pushes more vulnerable households into poverty, whereas Rentschler (2016) analysis shows that with outright fuel subsidy removal, the poverty rate is increased for States in the southern region of Nigeria. Bazilian and Onyeji (2012) found that despite the calls for more efficient use of energy through the removal of fuel subsidies, the resulting effect is the increase in the production costs (decrease in the cost-competitiveness) of the business sector (firms) due to the lack of efficient power infrastructure in the economy.

Fuel subsidy removal has led to an increase in the price of petroleum products, leading to low demand for petroleum products and a decrease in the quantity of fuel bought. Small enterprises that depend on fuel will see a reduction in earnings due to the decrease in demand. The impact will be felt by all, but as usual, the poor and low-income earners will bear the brunt of the consequences. While there is a claim that the elimination of fuel subsidies in 2023 will increase competition among petroleum marketers and drive down prices, this prediction is theoretical and is not likely to come true anytime soon since the price of Nigeria's petroleum products primarily depends on the crude oil price in the global market. The implication is that the recent rising petroleum price will not abate for some time (Tughhemba & Akputu, 2024).

Fuel subsidy removal has the unfavourable microeconomic implication of temporarily increasing poverty. As a result, families will face immediate hardship and hunger. The removal of fuel subsidies and the absence of palliatives may lead to decreased disposable income, reduced food production, restricted access to healthcare, and inability to pay for basic education. There will be an increase in the number of hungry families. The purchasing power of middle-class and lower-class citizens will decline, and small enterprises will experience pressure on their profit margins due to increased expenses and lower sales volumes. In addition, the removal of fuel subsidies may have a more disproportionate adverse effect on low-income and vulnerable populations in the absence of social assistance or safety net programs to mitigate the financial difficulties brought on by the removal of fuel subsidies.

Fuel subsidy removal may result in job losses in the unorganized sector, which mostly depends on Premium Motor Spirit (PMS) (Houeland, 2020). The formal sector uses mainly diesel for operations, while the informal sector mainly uses petrol. Small-scale enterprises whose earnings or profit margins have been affected negatively by the elimination of fuel subsidies in the formal sector and who are unable to pay the rising fuel cost may close due to the increase in petrol prices. Some may try to pass on the cost to the consumer. Attempts to pass on the higher cost to consumers have resulted in low business patronage and sales volumes.

The primary effect of fuel subsidy removal is that the cost of living will rise because rising fuel prices will probably result in higher pricing for goods and services. Eliminating fuel subsidies will directly result in high transportation costs, while the indirect effect is higher prices for goods and services. It will also lead to many Nigerians being malnourished and families going hungry.

Public unrest has already resulted from the removal of fuel subsidies, and if the administration cannot persuade the populace of the policy's long-term benefits, this unrest may intensify (Aniemeke, 2024). Furthermore, removing fuel subsidy can make individuals think the government doesn't care about their needs. It has led to the agitation for an increase in the minimum wage to meet the current cost of goods and services, making the Nigerian labour force embark on a national strike on June 3rd, 2024. Furthermore, removing the fuel subsidy could increase fuel smuggling and other illicit activity.

Positive microeconomic effects. Eliminating fuel subsidies could have the positive microeconomic impact of putting an end to the corruption associated with fuel subsidy payments. According to Itumo and Onyejiuba (2019), there is a view that the fuel subsidy is an attempt to continue transferring Nigeria's hard-earned foreign currency into private accounts located outside the nation. Ozili and Obiora (2024) claims that although Nigeria's supply of crude oil has increased and is now producing almost two million barrels per day, the price of oil has not increased significantly globally. However, Nigeria's foreign reserve has been steadily dropping. Even when an oil marketer only imports fifteen metric tons of gasoline, he reports importing seventy-five metric tons to the Petroleum Product Pricing Regulatory Agency (PPPRA) in Abuja. To include the inflated sixty metric tons, the importers conspire with some PPPRA personnel. This is another kind of corruption that existed under the fuel subsidy program. However, such corruption will cease when fuel subsidies are eliminated. The only fuel that importers will be paid for is the actual petroleum that they bring into the country.

Revitalizing domestic refineries is a potential microeconomic benefit of the removal of fuel subsidies. Nigeria's domestic refineries have been in dilapidation since the start of the fuel subsidy scheme (Okongwu & Imoisi, 2022). Fuel subsidies have not led to the revival of domestic refineries because they are associated with extensive corruption. Following the removal of gasoline subsidies, the government might make some adjustments and bring Nigeria's regional refineries back to life. This might lead to the resuscitation of domestic oil-producing companies, leading to a rise in the country's crude oil output.

Removing fuel subsidies has the good microeconomic effect of allowing supply and demand dynamics to set the price of Premium Motor Spirit (PMS), as opposed to government regulation or subsidies (Su et al., 2020). This will prevent fuel from being underpriced and stop corruption that results from overstating the amount of PMS imported under the fuel subsidy program. Also, fuel subsidy removal will result in correct pricing that accurately reflects the state of the global crude oil market.

Fuel subsidy removal will bring about market-determined pricing, which may eventually lead to lower product prices because of healthy competition, which is another significant microeconomic benefit of fuel subsidy removal (Bagirov & Mateus, 2019). Complete oil industry deregulation will increase market competitiveness and boost downstream sector effectiveness. Additionally, flexible pricing methods that are completely based on market forces are made possible by deregulation. The removal of fuel subsidies will promote competition and create opportunities for new players; in addition, it encouraged investors to withdraw from the Nigerian National Petroleum Corporation Limited (NNPCL) and join the downstream industry monopoly on fuel importation into Nigeria. The market sets the price of fuel as competition arises, and as the level of competition increases, the price of fuel will decrease.

According to Tughhemba and Akputu (2024), the current economic challenges in Nigeria will make Nigerians more disciplined and prudent. Many Nigerians will learn to cut down on costs to cope with the current challenge of high costs. They will learn to restrain the culture of wasteful spending. A lot of wisdom will have to be applied in the management of income and revenue earned.

5.4.3.2 Macroeconomic Effects

Positive macroeconomic effects. Nigerian fuel subsidy removal has a beneficial macroeconomic impact since the money that would have been used to pay for the subsidies might now be channelled to building vital public infrastructure in the country. Economic scholars argue that public infrastructure spending might be funded by the money utilized for subsidy payments (Bazilian & Onyeji, 2012; Majekodunmi, 2013). Nigeria lacked the resources to finance the construction of vital public infrastructure before the petroleum subsidy was removed. The government had to take on massive debt to finance the budget because of a lack of finance. However, after the fuel subsidy was eliminated in 2023, the government might be in the position to use this money to build vital infrastructure in Nigeria. The result is only possible if the government can be open, truthful, and accountable in ensuring that the funds saved by removing fuel subsidies are utilized to build vital public infrastructure.

Fuel subsidy removal would allow the existing budget deficit to be funded. Removing fuel subsidies can improve fiscal sustainability by reducing the budget deficit. A budget deficit is a situation where government expenses exceed its revenues, resulting in a shortfall or deficit. Subsidies often contribute to budget shortfalls, and its removal can help the government better manage finances. Prior research indicates that fuel subsidies contribute to Nigeria's increasing fiscal deficit, and their removal is necessary (Adagunodo, 2022; Harun et al., 2018; Ozili & Obiora, 2024). More recently, the approved budget 2023 was just ₦21.83 trillion, while the fuel subsidies were estimated to utilize 4 trillion naira in 2022 and a staggering 17 trillion naira in 2023. This suggests that fuel subsidy was expected to account for around 77% of the budget, pushing the country closer to bankruptcy and creating a persistent budget deficit (Ozili & Obiora, 2024). Furthermore, Nigeria's external debt service accounts for 90% of its earnings, which made the country's financial circumstances much more precarious throughout the petroleum subsidy program. The current elimination of the fuel subsidy is, in fact, a good thing for the country's funds since it would lower the recent budget deficit and strengthen the national budget.

The detrimental impact of fuel subsidy payments on government borrowing has been debated (Okongwu & Imoisi, 2022). Subsidies removal can help the government reduce its reliance on borrowing and overdependence on imported petrol. This ultimately may result in reduced debt levels and interest payments. The Nigerian government has been in debt since the fuel subsidy program began, and during the recession of 2016 and the COVID-19 epidemic of 2020, the debt became more challenging (Ozili & Obiora, 2024). In 2022, the government has been borrowing continuously from the Central Bank of Nigeria (CBN) to pay subsidies and settle loans or debt. The only option available to the administration was to raise central bank borrowing. The Federal Government recently securitized the ₦22.7 trillion that the government owes the Central Bank, with the approval of the national assembly in 2023. The government's recent decision to discontinue the fuel subsidy raises the possibility that it may stop borrowing funds from the Central Bank as it is assumed that the saved money will now be available to meet public spending.

Negative macroeconomic effects. The popular dimension of the discourse on subsidy reforms is often about the fiscal irresponsibility and wastefulness it portends. Subsidies create larger budget deficits and misallocate funds, which would have better served the public in terms of social amenities while curtailing public expenditure (El-Katiri & Fattouh, 2017; Sovacool, 2017). Governments expenditure on fossil-fuel subsidies is projected to have surpassed the amount of development aid provided by some international organizations for these governments (Merrill & Chung, 2015). Hence, estimations of these effects must be carefully analyzed, bearing in mind the complexities of such reforms and its intricacies with the economy.

Given the nature of most fossil-fuel subsidies, creating an underpricing mechanism for energy leading to higher consumption, it often becomes a critical decision for withdrawal. This is because fuel subsidies eventually subsidize the different economic units (both the consumers and producers) in society, creating a trap. Hence, fuel subsidy reforms in Nigeria led to hikes in the prices of goods and services, resulting in a general rise in the price level (inflation) (Gamette & Oteng, 2024; Shittu et al., 2024; Siddig et al., 2014). Okorie and Wesseh's (2024) simulation model found that the complete fuel subsidy removal leads to an increase in the national output (measured as GDP) through the channel of price increases in the various production activities and industries in the economy. Similarly, Siddig et al. (2014) found an increase in the real GDP (national output) for a total fuel subsidy removal, but the level of increase can go higher when domestic refined oil production is subsidized. Additionally, there is a fall in petroleum products imports (improved balance of trade) and a rise in government disposable income. Hence, government expenditure rises. Conversely, Shittu et al. (2024) found that the real GDP declines because of the fuel subsidy removal as a result of high product prices (pushed by a rise in price levels), creating less consumer demand, constricting industrial output, which results in the fall of the national production (GDP).

In a slightly different dimension, Osunmuyiwa and Kalfagianni (2017) articulated the effect for renewable energy development, and found that fuel subsidy reform encouraged interest in alternative energy source, hence the increase in renewable energy investment which drives increase in the number of renewable energy companies and subsequent increase in renewable energy jobs.

Removing fuel subsidies slows down economic growth, which has certain detrimental macroeconomic effects (Houeland, 2020). The removal of fuel subsidies has resulted in higher costs for necessities. As a result of growing costs, there will be insufficient disposable income in the hands of the citizens as the national minimum wage will no longer meet the needs of people and small businesses. This will lead to a decline in consumption spending, leading to an aggregate decline in demand. If consumption declines, there will be less demand from customers for the goods and services that companies offer. As a result, economic output and GDP may fall, and the rate of economic growth may decline.

According to Mohammed et al. (2020), removing fuel subsidies will also have a negative macroeconomic effect of rising inflation. Fuel subsidy removal often leads to an instant increase in fuel prices. The price of fuel increased from ₦190 in May 2023 to ₦537 in June 2023 and ₦617 in July 2023, ₦780 in May 2024, and ₦1200 in October 2024. This price increase has triggered a chain reaction, caused higher transportation costs and ultimately contributed to inflation. In the meantime, because of the high cost of transportation, the prices of consumable goods increased as most consumer and industrial goods became much more expensive. Both the price of essential commodities and local transit rises, making it more difficult for the poor and those with low incomes to pay. The poor and the rich are affected, but the vulnerable are usually the most hit because their purchasing power is significantly reduced. The Federal Government's delayed implementation of palliatives to support the poor and households affected by the rise in the price of goods and services after the removal of fuel subsidies will worsen the inflationary impact. The elimination of fuel subsidies may result in a rise in the unemployment rate, which is a major drawback. Additionally, it will widen the poverty gap in Nigeria, especially among the impoverished and vulnerable residents, including the youth, hence leading to a high crime rate among the youths (Aniemeke, 2024).

5.4.3.3 Environmental Effects

The challenge with the fossil-fuel subsidies is the hidden costs on the environment which researchers and policymakers often forget or disregard. In other words, there are environmental implications for fuel subsidy removal. Fuel subsidies have been in place for donkey years, encouraging economic activity dependent on fossil fuels that raise carbon emissions and air pollution in Nigeria. Fuel subsidies are partially to blame for the increase in carbon emission damage in Nigeria, which increased from US $1.5 billion in 1998 to US $5.23 billion in 2021 (Ozili & Obiora, 2024). The assumption is that by 2030, eliminating fuel subsidies in Nigeria would lessen the country's contribution to emissions of greenhouse gases worldwide and help continuing efforts to mitigate climate change. Eliminating fuel subsidies would also reduce Nigeria's supply and demand for fossil fuels, lowering the country's carbon dioxide output (Omitogun et al., 2021).

The fuel subsidy removal produces a positive effect on the environment by reducing the level of CO_2 emissions through the channel of increased prices of petroleum products, resulting in less import and use of these fossil fuels (Okorie & Wesseh, 2024). The effects are estimated as abatement cost, hence, the cost is reduced as more carbon is abated or removed from the atmosphere. Their finding is in line with Gamette and Oteng (2024) and Gelan (2018) that the removal of fossil fuel subsidies improves the environmental quality by dissuading its use. Sovacool (2017) opined that fossil fuel subsidies are known to generate carbon footprints. Shittu et al. (2024) analyzed the fuel subsidy reform under three basic frameworks for society (total removal, removal with compensation and removal with reinvestment), and these scenarios all revealed a decrease in CO_2 emission in the environment.

Osunmuyiwa and Kalfagianni (2017) uniquely assessed the environmental implications of fuel subsidy removal indirectly by estimating its effects on the renewable energy transition. This analysis provided useful insights into environmental concerns in terms of emissions because the processes involved in the fuel subsidy reform increased the use of renewable energy technologies, which will, in turn, reduce the consumption of fossil fuels as intended.

5.4.3.4 Socio-political Effects

Energy subsidies are common with energy exporting economies as they could create avenues for investment in the energy sector and give access to energy supply for its population. However, despite documented inefficiencies of energy subsidies, some governments still need help to eliminate them due to the political economy behind the policy (Lockwood, 2015). Governments have adopted energy subsidies as a means of redistributing income (Cheon et al., 2013), and once the citizens start enjoying the benefits of the subsidy, they fiercely protest against the removal because they often consider the subsidy a social right (Houeland, 2018; Koplow, 2014). Subsidies can be wasteful and unsustainable, and this applies to the case of petroleum subsidy in Nigeria (Bazilian & Onyeji, 2012), which then warranted the efforts of governments to remove it.

It has almost become a norm that fuel subsidy removal reforms in Nigeria are met by opposition from different groups in the society (labour unions, students associations, civil society organizations, etc.), to the extent that the labour union leaders have assumed the role of "a price-setting mechanism" in their negotiations with the government on fuel subsidy reforms. These events resonate with Usman (2020), who adduced that the government and managers of the economy should consider social actors in society in making decisions on extractive resources (e.g. petroleum).

Houeland (2020) analysis of the effects of fuel subsidy removal (2012 reform), reveals the following outcomes: civil mobilization to achieve institutionalized political inclusion; presented a platform to reassess the social and economic failures in governance; engagement of the youth in civic action; mobilized action of civil society organizations to demand transparency; deepened political cohesion among like minds. According to Agbonifo (2023), the fuel subsidy removal reforms inspired the institutionalization of accountability in the energy sector by creating Nigeria's Extractive Industry and Transparency Initiative (NEITI). It brought up the debate on corruption in the petroleum sector. McCulloch et al. (2021) suggested that the fuel subsidy removal reforms awakened the notion of inclusivity between the government and citizens in decisions that affect all. It revealed the masses' trust deficit in the government's delivering on their responsibilities or promises. This is related to the public lack of confidence in governance due to the alleged corruption in the petroleum sector (Gamette & Oteng, 2024).

Central to the outcomes of the subsidy reforms were protests and group mobilizations for the perceived common good of the citizenry. Spinoffs from the fuel subsidy removal regimes revealed the need for the government (political elites) and the people to maintain a progressive 'social contract' kind of niche to advance inclusive development.

5.4.3.5 Other Implications

Higher-income groups gain disproportionately from fuel subsidies since they use more fuel. Since the removal of the subsidy, fuel costs have increased by more than 200% (though they are predicted to gradually decline over time due to market competition). This increase in the cost of living has undoubtedly had a disproportionate impact on lower-income households. The increase in fuel prices because of fuel subsidy removal has put additional strain on public infrastructure. Owing to the elevated expenses of fuel, a portion of the population has chosen to utilize public transit, resulting in elevated demand and extended wait times. The general well-being and productivity of the populace have suffered as a result. Nigeria has experienced economic difficulties recently, with concern about the middle class disappearing and the population shifting into the lower economic strata. Most of the population's economic power is predicted to be further impacted by the removal of fuel subsidies, which could, in the short term, lead to a bigger percentage of social inequalities than upward social advancement. The nation's cost of doing business has increased dramatically because of the elimination of fuel subsidies, driving up the price of goods and services as well as transportation. To keep up with these growing costs, businesses need more encouragement to boost their profits and sales. As business costs continue to rise, aspiring entrepreneurs who want to launch a new venture may encounter difficulties. Fuel subsidy withdrawal has negative social implications now since it takes time for businesses and the public to fully adjust to the new reality while they wait for the economy to stabilize (Kreston, 2023).

5.5 Conclusions and Recommendations

Removal of fuel subsidies in Nigeria can alleviate the government's financial burden, hence providing investment in another sector of the economy. It is important to emphasize that to properly counteract the related socioeconomic pressures, fuel subsidy reform calls for calculated steps and activities. Fuel subsidy removal has disproportionately impacted the poor and the vulnerable. The removal of fuel subsidies has led to an increase in the price of petroleum products and a decline in the demand for the product. The decreasing consumer demand has narrowed the profit margin of small businesses that rely on fuel products. The federal and state governments are not effectively managing the social support programs or financial safety nets that are supposed to mitigate the financial difficulty resulting from the removal of fuel subsidies. Hence, everyone is feeling the effects. The argument that the removal of fuel subsidies in 2023 would increase competition among petroleum marketers and drive down prices has been made, but this expectation appears to be mainly academic and is not likely to materialize anytime soon, given that the crude oil prices on the global market continue to rise, which is a major factor in determining petroleum product prices in Nigeria. It follows that the current extreme petroleum prices will not change for a very long time.

To mitigate these impacts, the government should implement policies that address the adverse implications of removing subsidies. To lessen this, the government should consider implementing a targeted social assistance program to shield disadvantaged groups from the consequences of fuel subsidy removal.

The government should carefully assess the impacts of eliminating fuel subsidies on individuals and businesses and offer palliative assistance and various financial assistance to lessen the adverse effects. The government should provide financial assistance by introducing financial assistance programs to offset the immediate impact on vulnerable populations. These programmes can provide direct cash subsidies to low-income households to provide temporary relief while individuals adjust their budgets and seek alternative solutions. Infrastructure that benefits households, well-funded primary health care and education programs, increased access to food subsidy programs, cash benefits for the impoverished, and increased investment in public transportation to mitigate the high cost of transportation should be provided. The government must be sincere in its commitment to social considerations and reforms. Reviving the existing refineries with obsolete equipment and building new ones is crucial to significantly reduce the nation's reliance on imported petroleum products.

The government should invest in renewable energy as a long-term solution. The impact of fuel subsidy removal can be lessened by increasing clean energy efforts (energy derived from clean, renewable sources with no emissions that, when used, do not harm the environment), diversifying the energy mix, and lowering reliance on fossil fuels. In addition to promoting a clean future, improved investment in green energy will boost employment and the economy.

Acknowledgements We are thankful especially for the helpful suggestions and contributions of Dr. Robert Onyeneke and his team of editors/organizers of the maiden edition of the conference on Climate Justice and Just Energy Transition in Developing Countries.

Funding No financial assistance in form of funding aids or grants were received for the purpose of conducting this study.

Data Availability Statement The data used for this study are available from the corresponding authors upon reasonable request.

Ethical Clearance The authors upheld the values of integrity, honesty, and fairness in the research process for this study. Hence, declare that any opinions, findings, and conclusions or recommendations expressed in this material are solely those of the authors.

Conflict of Interest The authors declare that they have no known conflicting interests of either financial or personal relationships that could have appeared to influence the work reported in this study.

References

Abayomi, O. (2023, June 18). Consequences of fuel subsidy removal on Nigeria's balance of trade. *Vanguard News Paper*.

Adagunodo, M. (2022). The effect of oil receipts and fuel subsidy payment on the current account deficit in Nigeria and Venezuela. *Annals of Spiru Haret University. Economic Series, 22*(1), 137–152.

Adekunle, I. A., & Oseni, I. O. (2021). Fuel subsidies and carbon emission: Evidence from asymmetric modelling. *Environmental Science and Pollution Research, 28*, 22729–22741.

Agbonifo, J. (2023). Fuel subsidy protests in Nigeria: The promise and mirage of empowerment. *Extractive Industries and Society*. Retrieved September 20, 2024, from https://doi.org/10.1016/j.exis.2023.101333

Aniemeke, E. H. (2024, May 15). The microeconomic and macroeconomic implications of fuel subsidy removal in Nigeria. Retrieved June 4, 2024, from https://rsisinternational.org/journals/ijriss/articles/the-microeconomic-and-macroeconomic-implications-of-fuel-subsidy-removal-in-nigeria/

Asare, J., Reguant, M., Saab, M., & Sacchetto, C. (2020, August). *Low oil prices during COVID-19 and the case for removing fuel subsidies* (pp. 1–15). International Growth Centre.

Bagirov, M., & Mateus, C. (2019). Oil prices, stock markets and firm performance: Evidence from Europe. *International Review of Economics & Finance, 61*, 270–288.

Bazilian, M., & Onyeji, I. (2012). Fossil fuel subsidy removal and inadequate public power supply: Implications for businesses. *Energy Policy, 45*, 1–5. https://doi.org/10.1016/j.enpol.2012.02.057

Brunton, V. J., Stansfield, C., Caird, J., & Thomas, J. (2017). *Finding relevant studies*. Sage.

Cheon, A., Urpelainen, J., & Lackner, M. (2013). Why do governments subsidize gasoline consumption? An empirical analysis of global gasoline prices, 2002–2009. *Energy Policy, 56*, 382–390.

El-Katiri, L., & Fattouh, B. (2017). A brief political economy of energy subsidies in the Middle East and North Africa. In *Combining economic and political development* (pp. 58–87). Brill Nijhoff.

Esen, M., Bellibaş, M., & Gümüş, S. (2018). The evolution of leadership research in higher education for two decades (1995–2014): A bibliometric and content analysis. *International Journal of Leadership in Education*, 1–15.

Ferreira González, I., Urrútia, G., & Alonso-Coello, P. (2011). Systematic reviews and meta-analysis: Scientific rationale and interpretation. *Revista Espanola de Cardiologia, 64*(8), 688–696.

Gamette, P., & Oteng, C. (2024, June 1). Fuel subsidy removal in global south oil-producing economies: A review of literature. *Extractive Industries and Society, 18*. https://doi.org/10.1016/j.exis.2024.101468

Gelan, A. (2018). Economic and environmental impacts of electricity subsidy reform in Kuwait: A general equilibrium analysis. *Energy Policy, 112*, 381–398.

Harun, M., Mat, S. H. C., Fadzim, W. R., Khan, S. J. M., & Noor, M. S. Z. (2018). The effects of fuel subsidy removal on input costs of productions: Leontief input-output price model. *International Journal of Supply Chain Management, 7*(5), 529–534.

Houeland, C. (2018). Between the street and Aso Rock: The role of Nigerian trade unions in popular protests. *Journal of Contemporary African Studies, 36*(1), 103–120.

Houeland, C. (2020). Contentious and institutional politics in a petro-state: Nigeria's 2012 fuel subsidy protests. *The Extractive Industries and Society, 7*(4), 1230–1237.

Itumo, A., & Onyejiuba, E. I. (2019). Oil subsidy and development of local refineries in Nigeria: A critical analysis. *African Journal of Politics and Administrative Studies, 12*(1).

Koplow, D. (2014). *5 global energy subsidies scale, opportunity costs, and barriers to reform*. Retrieved September 20, 2024, from www.earthtrack.net

Kreston, P. (2023, June 21). *Assessing the implication of fuel subsidy removal in Nigeria: Economic, social and political considerations*. Retrieved May 22, 2024, from https://krestonpedabo.com/assessing-the-implication-of-fuel-subsidy-removal-in-nigeria-economic-social-and-political-considerations/

Lockwood, M. (2015). Fossil fuel subsidy reform, rent management and political fragmentation in developing countries. *New Political Economy, 20*(4), 475–494. https://doi.org/10.1080/13563467.2014.923826

Majekodunmi, A. (2013). The political economy of fuel subsidy removal in Nigeria. *International Journal of Management and Social Sciences Research, 2*(7), 76–81.

Manjo, Y. G. (2024, January). Effects of fuel subsidy removal on economy and citizen's welfare in Nigeria. *Lapai International Journal of Politics, 9*(1), 203–220. https://www.researchgate.net/publication/377701283_EFFECTS_OF_FUEL_SUBSIDY_REMOVAL_ON_ECONOMY_AND_CITIZEN'S_WELFARE_IN_NIGERIA

McCulloch, N., Moerenhout, T., & Yang, J. (2021). Fuel subsidy reform and the social contract in Nigeria: A micro-economic analysis. *Energy Policy, 156*.

Mendoza De Los Santos, O. E., María, G., Riquelme Alcantar, L., & De Hoyos Bermea, A. (2022). Systematic reviews and their epistemological foundations: A narrative review of the literature. *Journal of Positive School Psychology* (8), 10298–10312.

Merrill, L., & Chung, V. (2015). Financing the sustainable development goals through fossil-fuel subsidy reform: Opportunities in Southeast Asia, India and China.

Mohammed, A. B., Ahmed, F. F., & Adedeji, A. N. (2020). Assessment of impact of fuel subsidy removal on socio-economic characteristics: A survey of households in Maiduguri, Borno State, Nigeria. *Journal of Business and Economic Development, 5*(1), 10.

Moher, D., Liberati, A., Tetzlaff, J., Altman, D. G., Altman, D., Antes, G., & Tugwell, P. (2009, July 1). Preferred reporting items for systematic reviews and meta-analyses: The PRISMA statement. *PLoS Medicine, 6*. Retrieved September 21, 2024, from https://doi.org/10.1371/journal.pme

Mouhoud, S., & Couharde, C. (2020). Fossil fuel subsidies, income inequality, and poverty: Evidence from developing countries. *Journal of Economic Surveys, 34*(5), 981–1006.

Ojeabulu, P. (2024). *Impact of fuel subsidy removal in Nigeria*. Retrieved May 18, 2024, from Sun-Connect Sub-Saharan Africa News: https://sun-connect.org/impact-of-fuel-subsidy-removal-in-nigeria/

Okongwu, C. J., & Imoisi, S. E. (2022). Removal of petrol subsidy: Legal implications for the Nigerian economy. *Nnamdi Azikiwe University Journal of International Law and Jurisprudence, 13*(2), 135–139.

Okorie, D. I., & Wesseh, P. K. (2024). Fossil fuel subsidy removal, economic welfare, and environmental quality under alternative policy schemes. *Journal of Cleaner Production, 450*. https://doi.org/10.1016/j.jclepro.2024.141991

Omitogun, O., Longe, A. E., Muhammad, S., & Adekomi, I. J. (2021). Environmental impact of economic growth and fuel subsidy in Nigeria. *Economic Insights—Trends & Challenges* (1).

OPEC. (2022). *Historical production data*. Organization of the Petroleum Exporting Countries. Retrieved May 22, 2024, from https://opec.org/opec_web/en/data_graphs/335.htm

Osunmuyiwa, O., & Kalfagianni, A. (2017). The oil climax: Can Nigeria's fuel subsidy reforms propel energy transitions? *Energy Research and Social Science, 27*, 96–105. https://doi.org/10.1016/j.erss.2017.03.003

Ovaga, O. H., & Okechukwu, M. E. (2012). Subsidy in the downstream oil sector and the fate of the masses in Nigeria. *Kuwait Chapter of Arabian Journal of Business and Management Review, 1*(16), 1–20.

Ozili, P. K., & Obiora, K. (2024, March). *Implications of fuel subsidy removal on Nigerian economy*. Retrieved May 18, 2024, from https://mpra.ub.uni-muenchen.de/120509/

Ozili, P. K., & Ozen, E. (2021). Global energy crisis: impact on the global economy. In *Proceedings of IAC in Budapest 2021* (Vol. 1, pp. 85–89). Czech Institute of Academic Education.

Rentschler, J. (2016). Incidence and impact: The regional variation of poverty effects due to fossil fuel subsidy reform. *Energy Policy, 96*, 491–503. https://doi.org/10.1016/j.enpol.2016

Shittu, I., Abdul Latiff, A. R., & Baharudin, S. 'A. (2024). Assessing the compensation and reinvestment plans for fuel subsidy rationalization in Nigeria: A dynamic computable general equilibrium approach. *Energy, 293*. Retrieved September 23, 2024, from https://doi.org/10.1016/j.ener

Siddig, K., Aguiar, A., Grethe, H., Minor, P., & Walmsley, T. (2014). Impacts of removing fuel import subsidies in Nigeria on poverty. *Energy Policy, 69*, 165–178. https://doi.org/10.1016/j.enpol.2014.02.006

Sovacool, B. K. (2017, May 1). Reviewing, reforming, and rethinking global energy subsidies: Towards a political economy research agenda. *Ecological Economics, 135*, 150–163. https://doi.org/10.1016/j.ecolecon.2016.12.009

Su, C. W., Qin, M., Tao, R., Moldovan, N. C., & Lobonţ, O. R. (2020). Factors driving oil price—From the perspective of United States. *Energy, 197*, 117219.

Tughhemba, T. N., & Akputu, I. W. (2024, June). Examining economic hardship resulting from fuel subsidy removal in Nigeria: A critical reflection through the lens of Karl Jaspers' boundary situations. *Journal of Contemporary Dialectrical Sociology, 12*(1).

Umeji, G., & Eleanya, E. (2021). Assessing the impact of fuel subsidy removal in Nigeria on the poor in the COVID-19 era. *SERBD-International Journal of Multidisciplinary Sciences*.

Usman, Z. (2020). The successes and failures of economic reform in Nigeria's post-military political settlement. *African Affairs, 119*(474), 1–38. https://doi.org/10.1093/afraf/adz026

Chapter 6
Checking the Validity of Environmental Kuznets Curve Hypothesis for Nigeria

Iyke Uwazie Uwazie, Kelechi Clara Anyanwu, and Valentine Lotaya Ikeokwu

Abstract This study investigates the validity of the Environmental Kuznets Curve (EKC) hypothesis for Nigeria over the period 1980–2022, using the Autoregressive Distributed Lag (ARDL) econometric model. The EKC hypothesis suggests that environmental degradation initially worsens with economic growth but improves after a certain income threshold is reached. The analysis incorporates key variables such as CO_2 emissions, per capita income, oil rents, and urbanization. The findings reveal statistically significant evidence supporting the EKC hypothesis in Nigeria, with a positive coefficient for per capita income and a negative coefficient for its square, indicating that economic growth initially increases CO_2 emissions but reduces them after reaching a turning point. Additionally, oil rents are found to exacerbate environmental pollution, while urbanization contributes to environmental improvements. Based on these results, the study recommends stricter environmental regulations in key polluting industries, the promotion of advanced low-emission technologies, and urban planning policies aimed at developing eco-friendly cities. Limitations include the focus on CO_2 emissions and the exclusion of other environmental factors such as deforestation and water pollution. Future research should broaden the range of environmental indicators and account for external shocks, such as oil price fluctuations and technological changes, to provide a more comprehensive understanding of Nigeria's environmental dynamics.

Keywords CO_2 emissions · ARDL · EKC · Economic growth · Urbanization · Environmental pollution

I. U. Uwazie (✉) · V. L. Ikeokwu
Department of Economics, College of Management Sciences, Michael Okpara University of Agriculture, Umudike, Umuahia, Abia State, Nigeria
e-mail: ui.uwazie@yahoo.com

K. C. Anyanwu
Department of Economics and Development Studies, Alex Ekwueme Federal University Ndufu Alike, Ikwo, Ebonyi State, Nigeria

R. U. Onyeneke et al. (eds.), *Energy Transition, Climate Action and Sustainable Agriculture*, https://doi.org/10.1007/978-3-031-83165-2_6

6.1 Introduction

The growing concern of global warming and alterations in climatic circumstances are consequences of heightened energy consumption in the industrial sector and robust economic expansion, which have resulted in environmental variations worldwide (Waqih et al., 2019). The worldwide community aims to restrict the average global temperature to 2 °C above pre-industrial levels in response to this problem. Consequently, the United Nations ratified the '2030 Agenda for Sustainable Development' and the 'Paris Climate Change Agreement' in 2015. The Intergovernmental Panel on Climate Change has projected that global temperatures may increase by up to 5.8 °C by 2100.

According to Pao and Chen (2019), this underscores the immense challenge of mitigating climate change, necessitating the formulation of policies to address climate change and avert its catastrophic impacts on humanity and the environment (Halliru et al., 2020). A primary concern impacting the global ecosystem is the continual rise in carbon emissions. Carbon emissions have precipitated climate change, resulting in environmental deterioration and biodiversity loss, among other impacts (Kole et al., 2023). Human-related variables, including energy consumption, transportation, and the import and export of products and services across international borders, significantly contribute to environmental deterioration (Zamir & Mujahid, 2022). Human activities such as agriculture, industrialization, deforestation, and mining significantly impact the natural ecosystem, adversely affecting the globe (Nathaniel et al., 2021).

The EKC hypothesis is valid for numerous developed nations, particularly in Europe and North America, which have embraced environmentally sustainable technologies in their economic practices and advocated for a pollution-free world, thereby improving climate quality and fostering an environmentally friendly environment for all (Martin & Vaitilingam, 2014). Conversely, numerous developing economies, particularly in Latin America, Asia, and Africa, persist in exacerbating environmental degradation with each increment in the production of products and services (Bell & Russell, 2002). These countries predominantly operate their production processes in a manner that generates significant pollution alongside value creation, as they heavily depend on fossil fuel consumption, utilize firewood for domestic cooking, rely on fuel-intensive transportation methods, and engage in horizontal housing expansion, resulting in deforestation and increased carbon dioxide emissions (Falade & Adeyemi, 2022). Nigeria has consequently become heavily reliant on non-renewable energy sources. Recently, there has been a minor shift in the utilization of renewable energy, particularly in solar power (Kole et al., 2023).

The substantial expense associated with the switch to renewable energy technologies is a significant worry for policymakers in Nigeria. Nigeria primarily relies on fossil fuels for energy, with revenue from crude oil constituting a significant portion of the nation's foreign exchange revenues (Kole et al., 2023). Nonetheless, the fluctuations in crude oil prices in the global market, together with the ongoing movement towards renewable energy, are significant factors. Urbanization significantly influences economic progress and environmental sustainability. The extant literature presents varying conclusions regarding the influence of urbanization on environmental quality. Empirical research indicates that the irrational development and exploitation of natural resources during urbanization result in environmental degradation and contribute to air, noise, and water contamination (Rahman & Alam, 2021). This study's primary contribution to the existing literature is its re-evaluation of the Environmental Kuznets Curve (EKC) and identification of the principal factor contributing to CO_2 emissions in Nigeria. This is crucial as it aids in comprehending Nigeria's standing in the United Nations 2030 Sustainable Development Goals, thereby informing policymakers on areas requiring enhancement. The study also contributes to the current knowledge on urbanization and oil rent about the Environmental Kuznets Curve (EKC). To address two basic research questions:

a. What is the impact of CO_2 emissions and per capita income in Nigeria?
b. Does the Environmental Kuznets Curve hypothesis apply in Nigeria?

The Autoregressive Distributed Lag model is utilized to identify these gaps and address issues of endogeneity and heterogeneity. The initial segment of the study contains the introduction, while the literature review is in section two. The method is delineated in section three, whereas the results and discussions are articulated in section four. Section five addressed the summary, conclusion, and recommendations.

6.2 Literature Review

Bekhet et al.'s (2020) study, which used data from 1971 to 2013 on ARDL technique to examine the relationship between the EKC hypothesis and the urban environment transition hypothesis, reveals that Malaysia has an inverted U-shaped EKC hypothesis. The study identified the negative impact of urbanization on carbon emissions, attributing it to technological innovation and sustainable development. Similarly, Oluwaseyi et al. (2020) investigate whether Gross Domestic Income, trade openness, GDP, and capital mitigate CO_2 emissions in Nigeria. The study utilized ARDL and Vector Autoregressive (VAR) methodologies, analysing data from 1980 to 2018. The results indicate that a growth in GDP and FDI decreases CO_2 emissions, hence corroborating the validity of the EKC hypothesis in Nigeria over the examined period. The analysis also demonstrates the existence of cointegration among the variables examined.

In a similar vein, Sreenu (2022) employs ARDL and NARDL methodologies on data from 1990 to 2020 to examine the effects of foreign direct investment, crude oil prices, and economic growth on CO_2 emissions in India. The study concluded that economic expansion decreases CO_2 emissions, hence corroborating the Environmental Kuznets Curve hypothesis regarding economic growth and CO_2 emissions in India. Naz et al. (2019) investigates the moderating and mediating effects of renewable energy consumption, foreign direct investment inflows, and economic growth on

CO_2 emissions in Pakistan, utilizing annual data from 1975 to 2016. The study's findings refute the EKC hypothesis regarding the relationship between per capita income and CO_2 emissions. Nonetheless, the data corroborates the pollution haven hypothesis, demonstrating that foreign direct investment adversely affected Pakistan's natural flora. Setyari and Kusuma investigate the EKC theory across 62 countries utilizing annual data from 1992 to 2017. The Correction Model (ECM) yielded equivocal results for the four sub-samples, demonstrating a neutral (N-shaped) relationship between GDP per capita and CO_2 emissions.

Mitsis (2021) examines the income-pollution relationship within the framework of the EKC hypothesis across 35 nations, revealing an N-shaped correlation between income and CO_2 emissions, indicating the inconclusiveness of the EKC model. The correlation between economic growth and environmental degradation is elucidated by the environmental Kuznets curve hypothesis, which delineates three mechanisms through which economic growth influences the environment: the scale effect (augmented economic activity and corresponding energy consumption), the composition effect (alteration of economic structure favouring either polluting or sustainable goods), and the technique effect (transition to more environmentally benign technologies). Alshehry and Belloumi (2016) documented a consistently growing correlation between economic growth and environmental degradation in their analysis of Saudi Arabia. The study by Wang et al. (2018), done in China, confirmed the Environmental Kuznets Curve concept, indicating that economic expansion leads to environmental improvement after beyond a certain income threshold. The research conducted by Koçak and Şarkgüneşi (2018) in Turkey identified an inverted U-shaped correlation between economic growth and environmental factors, thereby corroborating the Environmental Kuznets Curve (EKC) hypothesis. Unexpectedly, in Turkey, Ozcan et al. (2018) failed to verify the EKC hypothesis, instead reporting a monotonous pattern, indicating that findings may vary within a single nation. A recent study by Destek and Sarkodie (2019) in Singapore revealed an inverted U-shaped relationship between economic growth and the environment, thereby validating the Environmental Kuznets Curve (EKC) theory.

The examined literature clearly indicates that the validity of the EKC hypothesis remains empirically contentious, hence perpetuating an ongoing research void for exploration. The EKC framework is supported by the majority of the literature reviewed for this study on Nigeria but given the lack of clarity surrounding the empirical findings across nations especially to determinant as oil rent and urbanization, it is necessary to reevaluate the EKC hypothesis in Nigeria by taking other factors that contribute to environmental degradation into account in order to determine whether it will still be valid.

6.3 Methodology

6.3.1 Theoretical Framework

Based on the Environmental Kuznets Curve (EKC) hypothesis, which posits an inverted U-shaped relationship between environmental degradation and economic progress, indicating that environmental degradation initially escalates with economic growth but subsequently diminishes after a specific income threshold is attained (Grossman & Krueger, 1995). This concept has been extensively utilized in both rich and emerging economies; nevertheless, its validity in low- and middle-income nations, including Nigeria, continues to be a subject of ongoing research and discussion. Nigeria, the predominant economy in Africa, has undergone swift economic expansion in recent decades, propelled by sectors including oil, agriculture, and industries. This growth has resulted in considerable environmental difficulties, including as deforestation, polluted air, and oil-related deterioration, prompting inquiries over the relevance of the EKC theory in the nation. The Environmental Kuznets Curve (EKC) hypothesis is expressed as follows:

$$ENV_t = f\left(PCI_t, PCI_t^2\right) \tag{6.1}$$

where Env indicates environmental sustainability, PCI indicates per capital income.

6.3.2 Model Specification

Following the studies of Jahanger et al. (2023). This study utilizes the conventional EKC model in its quadratic, the baseline econometric model is expressed as follows:

$$CO_{2t} = \beta_0 + \beta_1 PCI_t + \beta_2 PCI_t^2 + \beta_3 OIL_t + \beta_4 URB_t + \varepsilon_t \tag{6.2}$$

Equation (6.2) is further expressed in it log form to give

$$LnCO_{2t} = \beta_0 + \beta_1 LnPCI_t + \beta_2 PCI_t^2 + \beta_3 LnOIL_t + \beta_4 LnURB_t + \varepsilon_t \tag{6.3}$$

where CO_2 is Carbon Dioxide, PCI is Per capita GDP, PCI^2 is Square of Per capita GDP, OIL is Oil rent, URB is Urbanization, ε_t is the error term at time t.

6.3.3 Econometric Technique

6.3.3.1 Auto-Regressive Distributed Lag (ARDL)

The Autoregressive Distributed Lag (ARDL) model is a widely used econometric technique for analyzing long-term and short-term dynamics in time series data, especially when the variables exhibit different orders of integration. Introduced by Pesaran and Shin (1999), the ARDL approach has become popular because it allows for the analysis of both I(0) and I(1) variables, without requiring pre-testing for unit roots, which is necessary in traditional cointegration methods such as the Engle–Granger approach or the Johansen method. The ARDL model is particularly effective in small sample sizes, offering flexibility in dealing with variables that are not necessarily integrated of the same order.

The ARDL model incorporates both the autoregressive part (where the dependent variable is regressed on its own past values) and the distributed lag part (where the current and past values of the independent variables are included) (Pesaran et al., 2001). One of the key advantages of the ARDL model is that it can simultaneously estimate the long-run equilibrium relationship between variables and the short-run dynamics of the system, making it a suitable tool for economic and financial data analysis, such as testing the Environmental Kuznets Curve (EKC) hypothesis or assessing the relationship between energy consumption and economic growth (Nkoro & Uko, 2016).

The ARDL model also allows for the bounds testing approach to cointegration, where the existence of a long-run relationship can be tested even when the order of integration of the variables is mixed. This flexibility is particularly useful in cases where it is unclear whether the variables are I(0), I(1), or a combination of both. If the computed F-statistic in the bounds test exceeds the upper bound critical value, cointegration is confirmed, suggesting a long-run relationship among the variables (Pesaran et al., 2001). The ARDL mode is here expressed as follows:

The ARDL equation is given as:

$$\begin{aligned} CO_2 = \beta o + \sum_{i+1}^{p+1} \beta \Delta CO_2 t + 1 \sum_{i+0}^{p+1} \beta 2 \Delta InPCLt - I + \sum_{i+0}^{p+1} \beta 3 \Delta InPCI2t - I \\ + \sum_{i+0}^{p+1} \beta 4 \Delta InOILt - I + \sum_{i+0}^{p+1} \beta 5 \Delta InURBt - I + \beta 1 lnPCLt \\ + \beta 3 lnPCI2t + \beta 4 OILt + \beta 5 lnURB + \varepsilon_t \end{aligned} \tag{6.4}$$

The ECM to be used in this analysis is the ECM for ARDL because; it estimates the model while taking the dependent variable at lag 1 as an independent variable. The ECM equation is given as follows:

Table 6.1 Description of variables

Variables	Symbol	Unit of measurement	Sources
Carbon emissions	CO_2	Metric tons	WDI 2023
GDP per capita	GDP	In constant 2010 USD	WDI 2023
GDP^2	GDP^2	GDP squared	Author's computation
Oil rent	OIL	% of GDP	WDI 2023
Urbanization	URB	% of population	WDI 2023

Source Author's compilation

$$CO_2 = \beta o + \sum_{i+0}^{p+1} \beta 1 \Delta CO_2 t - I + \sum_{i+0}^{p+1} \beta 2i \Delta InPCIt - I$$
$$+ \sum_{i+0}^{p+1} \beta 3i \Delta InPCI2t - I + \sum_{i+0}^{p+1} \beta 5 \Delta InURBt - I + \varepsilon_t \tag{6.5}$$

where, Δ = change in the variable over time, ε_t = Error term of previous year, and $\beta_1, \beta_2, \beta_3$, and β_4, are parameters while CO_2, PCI, PCI^2, OIL and URB are as defined.

6.3.4 Data Sources

See Table 6.1.

6.4 Results and Discussion

6.4.1 Unit Root Test

Based on the data in Table 6.2, we can see that the variables CO_2, InPCI, $InPCI^2$, and OIL are stationary when their differences are taken, indicating an order of integration of 1. On the other hand, the variable InURB is stationary at its original level, indicating an order of integration of 0. This is because the test-statistic for the variables exceeds the critical values in absolute terms at a significance level of 5%. Thus, we can deduce that the variables are in a state of stationarity.

Table 6.2 ADF unit root test result

Level			First difference			Order of integ.
Variables	ADF stat	5% critical value	Variables	ADF stat	5% critical value	
CO_2	– 2.256	– 2.928	CO_2	– 7.964	– 2.930	I(1)
InPCI	– 1.740	– 2.930	InPCI	– 4.968	– 2.930	**I(1)**
$InPCI^2$	– 1.638	– 2.930	$InPCI^2$	– 4.954	– 2.930	**I(1)**
OIL	– 1.546	– 2.946	OIL	– 4.963	– 2.946	**I(1)**
InURB	– 3.929	– 2.946	InURB	–	–	**I(0)**

Source Authors' compilation
The level of significance is defined as the best accepted level in 1%, 5% and 10% levels

Table 6.3 Result of bound test approach to ARDL

Test statistic	Value	K
F-statistic	5.503	4
Critical value bounds		
Significance	10 bound	11 bound
10%	2.45	3.52
5%	2.86	4.01
2.5%	3.25	4.49
1%	3.74	5.06

Source Authors' compilation

6.4.2 ARDL Bounds Test

Based on the information provided in Table 6.3, we can conclude that there is co-integration and a long run link among the variables because the computed F-statistic (5.503) is higher than the upper and lower bounds at a 5% significant level. Thus, from the ARDL bound test there is evidence of long run relationship amongst the variables of interest.

6.4.3 Autoregressive Distributed Lag (ARDL) Model

According to Table 6.4, the short-run coefficients are estimated using the error correction model (ECM). Table 6.4 presents the ECM version of the autoregressive distributed lag (ARDL) model, based on the Akaike information criterion (4, 1, 3, 0, 1). The results show that the error correction term is negative and statistically significant, confirming the existence of a stable long-term relationship among the variables. Specifically, the error correction coefficient (– 0.846) indicates that approximately 84.6% of the short-term deviation from the equilibrium is corrected each period, implying a rapid adjustment towards long-term equilibrium.

The short-run coefficients reveal a positive relationship between per capita income and CO_2 emissions. A 1.632 percentage point increase in per capita income leads to a corresponding rise in CO_2 emissions. This relationship is statistically significant at the 5% level, as indicated by the *p*-value of 0.034, which is below the 0.05 threshold. Additionally, the association between oil rents and CO_2 emissions is also significantly positive, with a *p*-value of 0.015, further reinforcing the impact of oil rents on environmental degradation.

On the other hand, an inverse relationship is observed between the square of per capita income and CO_2 emissions, suggesting that as income increases beyond a certain point, CO_2 emissions start to decrease. Specifically, a 0.111 decrease in CO_2 emissions occurs for every one percentage point increase in the square of per capita income. This relationship is also statistically significant, as indicated by the *p*-value of 0.047, which is below the 0.05 significance level. These results align with the Environmental Kuznets Curve (EKC) hypothesis, indicating that as economic growth progresses, environmental degradation initially rises but eventually declines.

An increase of one percentage point in per capita income leads to a 3.517% rise in CO_2 emissions. This relationship is statistically significant at the 5% level, as evidenced by the absolute t-value of 3.403, which exceeds the critical value of 1.96, and the probability value of 0.002, which is less than the threshold of 0.05. Economically, this finding suggests that as Nigeria's economy grows and individuals' income increases, the environmental impact worsens in the short term. The positive correlation between rising income and emissions can be attributed to increased consumption, industrialization, and energy demand, primarily driven by fossil fuel use, which exacerbates environmental degradation. This underscores the challenge for policymakers

Table 6.4 Estimated coefficients of the short run dynamic error correction model (ECM) dependent variable: CO_2

Variable	Coefficient	Std. error	t-statistic	Prob.
D(CO_2(− 1))	0.209	0.165	1.262	0.218
D(CO_2(− 2))	0.514	0.162	3.180	0.004
D(CO_2(− 3))	0.389	0.134	2.903	0.007
D(ln PCI)	1.632	0.731	2.231	0.034
D(ln PCI^2)	− 0.111	0.054	− 2.080	0.047
D(ln PCI^2(− 1))	0.002	0.007	0.288	0.776
D(ln PCI^2(− 2))	− 0.012	0.005	− 2.555	0.016
D(OIL)	0.006	0.002	2.521	0.018
D(ln URB)	− 7.018	4.969	− 1.412	0.169
CointEq(− 1)	− 0.846	0.181	− 4.684	0.000

Source Authors' compilation

to balance economic growth with environmental sustainability during early stages of development (Table 6.5).

An increase in oil rents by one unit results in a statistically significant rise in CO_2 emissions by 0.007 units, with an absolute t-value of 2.579 and a probability value of 0.016, both below the 5% significance level. This indicates that Nigeria's dependence on oil revenues directly contributes to higher carbon emissions. Economically, this result reflects the environmental costs associated with the country's oil-driven growth. As oil production intensifies to generate revenue, it leads to increased extraction activities, pollution, and CO_2 emissions. This emphasizes the need for diversification of Nigeria's economy away from fossil fuels and towards cleaner energy sources to mitigate the adverse environmental effects of oil dependence.

In contrast, a one percentage point increase in urbanization leads to a 0.532% decrease in CO_2 emissions. This relationship is highly statistically significant, with an absolute t-value of 5.223 and a probability value of 0.000, well below the 5% threshold. The economic implication of this result suggests that as more people move to urban areas, there may be a shift toward more efficient and environmentally friendly practices, such as the adoption of cleaner energy technologies, better infrastructure, and enhanced energy efficiency. Urbanization could thus contribute to a reduction in CO_2 emissions through improved access to cleaner energy sources and more sustainable living conditions, providing an opportunity for Nigeria to leverage urban growth for environmental improvements.

Lastly, the ARDL model, with an F-statistic probability of 0.007 (below the 0.05 significance level) and a coefficient of 3.044 (above 1.96), confirms the overall statistical significance of the model. This suggests that the variables in the model adequately explain the dynamics of CO_2 emissions in Nigeria, offering robust insights into the relationship between economic factors and environmental outcomes.

Table 6.5 Estimated long-run coefficients of ARDL model dependent variable: CO_2

Variable	Coefficient	Std. error	t-statistic	Prob.
InPCI	3.517	1.034	3.403	0.002
$InPCI^2$	− 0.233	0.073	− 3.171	0.004
OIL	0.007	0.003	2.579	0.016
InURB	− 0.532	0.102	− 5.223	0.000
C	− 10.463	3.429	− 3.052	0.005
R-squared	0.586	Durbin–Watson stat		2.136
Adjusted R-squared	0.393			
F-statistics	3.044			
F-prob.	0.007			

Source Authors compilation

6.4.4 ARDL Diagnostic Test

The Breusch–Godfrey Serial Correlation LM Test for autocorrelation, conducted using data from Table 6.6, indicates that the probability values of 0.416 and 0.309 are both over the significance level of 0.05. Consequently, there is no association between the series.

Based on the data presented in Table 6.6, the F-probability values of 0.928 and chi square probability of 0.879 are both greater than 0.05. Therefore, we can conclude that there is no heteroscedasticity in the model, indicating that the model is homoskedastic.

The Jarque–Bera probability value of 0.694, as shown in Fig. 6.1, above the significance threshold of 0.05. Therefore, we may reject the null hypothesis that there is no normal distribution and conclude that the error components of the model are normally distributed at a 5% level of significance.

Table 6.6 Diagnostic test summary table

Test performed (Nigeria)	Nature of test	*P* value (a = 0.05)	
		Prob. F	Prob. chi-square
Breusch–Godfrey Serial Correlation LM test	Auto correlation	0.416	0.309
Breusch–Pagan–Godfrey heteroskedasticity test	Heteroskedasticity	0.928	0.879
Jarque–Bera test	Normality test	0.694	
CUSUM	Stability test	Falls within	

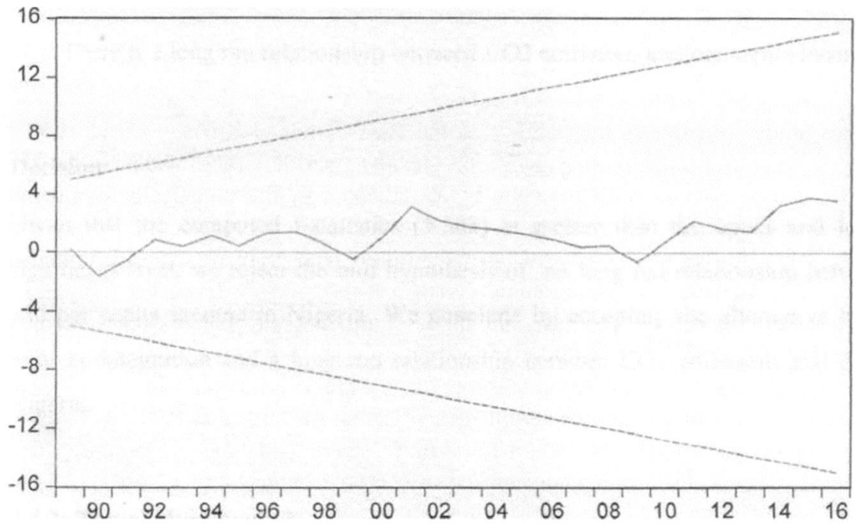

Fig. 6.1 CUSUM graph

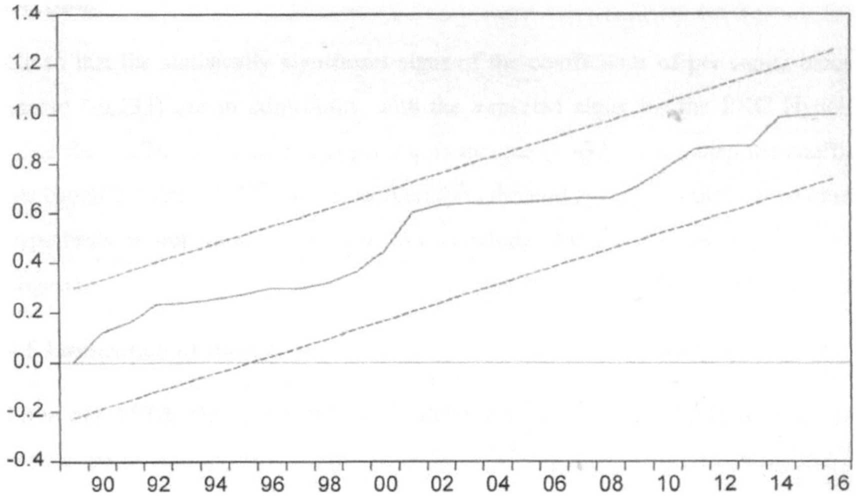

Fig. 6.2 Cumulative sum of squares of recursive residuals. *Note* CUSUM of squares 5% significance

Stability Test

Figure 6.2 illustrates that the plots of the CUSUM and CUSUMSQ statistics fall below the critical limitations of a 5% significance level. Hence, we can deduce that the model is indeed stable.

6.5 Conclusions, Policy Recommendations and Limitation

This study investigates the validity of the Environmental Kuznets Curve (EKC) hypothesis as it applies to Nigeria. The results provide strong evidence that the EKC hypothesis holds for the country, as indicated by the statistically significant positive coefficient of per capita income and the negative coefficient of its square. This suggests that while economic growth initially leads to increased environmental degradation, there is a turning point after which further growth reduces CO_2 emissions. Furthermore, the findings highlight the significant role of oil rents in worsening environmental pollution, while urbanization appears to positively impact environmental quality. These insights underscore the need for careful policy interventions to balance economic development with environmental sustainability in Nigeria. Based on the study's findings, the following policy recommendations are proposed that the Nigerian government should enforce stricter environmental regulations, particularly in industries that contribute significantly to pollution, such as oil extraction and manufacturing. The government should promote the use of advanced, low-emission technologies in key sectors, particularly the extractive industries. This would help reduce emissions and waste discharge. Policies that encourage the development of eco-friendly cities and communities should be prioritized. A balanced distribution of the population across regions would reduce pressure on urban centres and contribute to better environmental management. Legislation should be introduced to regulate key sources of emissions, such as the use of old vehicles and machinery, controlled burning of vegetation, grazing practices, fertilizer use, and extractive activities.

This study has several limitations. First, it focuses primarily on CO_2 emissions, potentially overlooking other critical environmental indicators such as water and soil pollution, deforestation, and biodiversity loss. Second, the data used in the study may not fully capture informal economic activities and other non-measurable factors that could influence the environmental-economic relationship in Nigeria. Additionally, the ARDL model, while suitable for analysing long-run and short-run dynamics, may not account for structural breaks or external shocks, such as the impact of global economic crises or major technological changes.

Future research should aim to address these limitations by incorporating a wider range of environmental indicators to provide a more comprehensive analysis of the EKC hypothesis. Additionally, future studies could explore the role of external shocks, such as global oil price fluctuations or technological innovations, in shaping the environmental-economic relationship in Nigeria.

Acknowledgements The authors wish to acknowledge the organizers of the maiden climate justice and just energy transition in developing countries conference in Honour of Prof. Chinedum Nwajiuba @60 and to the editor and anonymous peer reviewers of this paper.

Funding This study did not receive any specific fundings.

Data Availability Statement The data for this study will be available on request.

Conflict of Interest The Authors declares that there is no conflict of interest.

Ethical Clearance All ethical clearance was followed.

References

Alshehry, A. S., & Belloumi, M. (2016). Energy consumption, carbon dioxide emissions and economic growth: The case of Saudi Arabia. *Renewable and Sustainable Energy Reviews, 41*, 237–247.

Bekhet, H. A., Othman, N. S., & Yasmin, T. (2020). Interaction between environmental Kuznets curve hypothesis and urban environment transition hypothesis in Malaysia. *International Journal of Energy Economics and Policy, 10*(1), 384–402. https://doi.org/10.32479/ijeel.8389

Bell, R. G., & Russell, C. (2002). Environmental policy for developing countries. *Issues in Science and Technology, 18*(3), 38–51.

Destek, M. A., & Sarkodie, S. A. (2019). Investigation of environmental Kuznets curve for ecological footprint: The role of energy and financial development. *Science of the Total Environment, 650*, 2483–2489.

Falade, O., & Adeyemi, W. (2022). The effect of sectorial contributions to GDP on environmental degradation: A verification of the environmental Kuznets curve hypothesis in Niger. *Journal of Economics and Behavioral Studies, 14*(4), 1–10.

Grossman, G. M., & Krueger, A. B. (1995). Economic growth and the environment. *The Quarterly Journal of Economics, 110*(2), 353–377. https://doi.org/10.2307/2118443

Halliru, A., Loganathan, N., Hassan, A., Mardani, A., & Kamyab, H. (2020). Re-examining the environmental Kuznets curve hypothesis in the Economic Community of West African States: A panel quantile regression approach. *Journal of Cleaner Production.* https://doi.org/10.1016/j.jclepro.2020.124247

Jahanger, A., Ozturk, I., Onwe, J., Ogwu, S., Hossain, M., & Abdallah, A. (2023). Do pro-environmental interventions matters in restoring environmental sustainability? *Gondwana Research.* https://doi.org/10.1016/j.gr.2023.07.010

Koçak, E., & Şarkgüneşi, A. (2018). The impact of foreign direct investment on CO_2 emissions in Turkey: New evidence from cointegration and bootstrap causality analysis. *Environmental Science and Pollution Research, 25*(1), 790–804.

Kole, A., Umar, S., Muktar, A., & Geidam, A. (2023). Re-examining the environmental Kuznets curve hypothesis in Nigeria: Recent evidence from ARDL model. *Fane-Fane International Multidisciplinary Journal, 7*(2), 94–101.

Martin, L., & Vaitilingam, R. (2014). The impact of 'clean innovation' on economic growth: Evidence from the transport and energy industries. www.cep.lse.ac.uk

Mitsis, P. (2021). Examining the environmental Kuznets curve hypothesis using Bayesian model averaging. *Regional Statistics, 11*(1).

Nathaniel, S. P., Bekun, F. V., & Fizulayev, A. (2021). Modelling the impact of energy consumption, natural resources, and urbanization on ecological footprint in South Africa: Assessing the moderating role of human capital. *International Journal of Energy Economics and Policy, 11*(3), 130–139.

Naz, S., Sultan, R., Zaman, K., Aldakhil, A. M., Nassani, A. A., & Abro, M. M. Q. (2019). Moderating and mediating role of renewable energy consumption, FDI inflows, and economic growth on carbon dioxide emissions: Evidence from robust least square estimator. *Environmental Science Pollution Research Institute, 26*(3), 2806–2819. https://doi.org/10.1007/s11356-018-3837-6

Nkoro, E., & Uko, A. K. (2016). Autoregressive Distributed Lag (ARDL) cointegration technique: Application and interpretation. *Journal of Statistical and Econometric Methods, 5*(4), 63–91. https://doi.org/10.47260/jsem/v5-i4

Oluwaseyi, A. Z., Abdul-Rahim, S. A., & Dankumo, A. M. (2020). Does gross domestic income, trade integration, foreign direct investment inflows, gross domestic product, and capital reduce CO_2 emissions in Nigeria? *Current Research in Environmental Sustainability, 2*, 100009.

Ozcan, B., Apergis, N. & Shahbaz, M. (2018). A revisit of the environmental Kuznets curve hypothesis for Turkey: New evidence from bootstrap rolling window causality. *Environmental Science and Pollution Research, 25*, 32381–32394. https://doi.org/10.1007/s11356-018-3165-x

Pao, H. T., & Chen, C. C. (2019). Decoupling strategies: CO_2 emissions, energy resources, and economic growth in the Group of Twenty. *Journal of Cleaner Production, 206*, 907–919.

Pesaran, M. H., & Shin, Y. (1999). An autoregressive distributed-lag modelling approach to cointegration analysis. In S. Strom (Ed.), *Econometrics and Economic Theory in the 20th Century: The Ragnar Frisch Centennial Symposium* (pp. 371–413). Cambridge University Press.

Pesaran, M. H., Shin, Y., & Smith, R. J. (2001). Bounds testing approaches to the analysis of level relationships. *Journal of Applied Econometrics, 16*(3), 289–326. https://doi.org/10.1002/jae.616

Rahman, M. M., & Alam, K. (2021). Clean energy, population density, urbanization and environmental pollution nexus: Evidence from Bangladesh. *Renewable Energy, 172*, 1063–1072.

Sreenu, N. (2022). Impact of FDI, crude oil price and economic growth on CO_2 emission in India. Symmetric and asymmetric analysis through ARDL and non-linear ARDL approach. *Environmental Science Pollution Research Institute, 29*(28), 42452–42465. https://doi.org/10.1007/s11356-022-19597

Wang, N., Zhu, H., Guo, Y., & Peng, C. (2018). The heterogeneous effect of democracy, political globalisation, and urbanisation on $PM_{2.5}$ concentrations in G20 countries: Evidence from panel quantile regression. *Journal of Cleaner Production, 194*, 54–68.

Waqih, M. A. U., Bhutto, N. A., Ghumro, N. H., Kumar, S., & Salam, M. A. (2019). Rising environmental degradation and impact of foreign direct investment: Empirical evidence from SAARC region. *Journal of Environmental Management, 243*, 472–480.

Zamir, A., & Mujahid, N. (2022). Nexus among green energy consumption, foreign direct investment, green innovation technology, and environmental pollution on economic growth. *Environment Science Pollution Research Institute, 29*(51), 76501–76513. https://doi.org/10.1007/s11356-022-23184-5

Chapter 7
Determinants of Energy Choices for Cooking and Lighting Among Rural Households in Imo State, Nigeria

Nneka Maris Chidiebere-Mark and Paschal Chukwunonso Adikaibe

Abstract Energy consumption patterns in rural households have significant implications for socio-economic development, environmental sustainability, and public health of rural communities. Promoting clean energy transitions in line with global sustainability goals underscores the need to investigate the determinants of energy choices for cooking and lighting among rural households in Imo State. The study examined the demographic characteristics of the households in the study area; identified the sources of energy available to the rural households for cooking and lighting; ascertained the share of clean energy in the energy mix consumed by the households; ascertained the households' knowledge of the benefits of clean energy consumption; examined the determinants of the choice of energy used by the households in Imo State; and identified the constraints to clean energy consumption by the households. A multi-stage sampling technique was employed to select 185 rural households, and data were collected through a structured questionnaire. Data collected was analysed using descriptive statistics and binary logistic regression. The findings reveal that 77% of households used dirty fuels for cooking, while 52% relied on dirty fuels for lighting. Critical determinants of energy consumption include income, education, and membership in social organizations. Barriers to clean energy consumption include the high cost of clean fuels, limited access to clean energy technologies, and socio-cultural factors. The study concludes that many rural households still consume dirty fuels for cooking and lighting. Promoting clean energy use in rural communities in Imo State requires targeted interventions and implementing awareness campaigns on the benefits of clean energy.

Keywords Energy · Clean fuels · Dirty fuels · Determinants · Energy poverty

N. M. Chidiebere-Mark (✉)
Department of Agricultural Economics, Extension and Rural Development, Faculty of Agriculture, Imo State University, Owerri, Imo State, Nigeria
e-mail: nnekamark@gmail.com; nnekamark@imsuonline.edu.ng

P. C. Adikaibe
Sahel Consulting Agriculture and Nutrition Limited, Abuja, Nigeria

R. U. Onyeneke et al. (eds.), *Energy Transition, Climate Action and Sustainable Agriculture*, https://doi.org/10.1007/978-3-031-83165-2_7

7.1 Introduction

Energy is crucial in driving economic development, improving living standards, and serving various essential purposes globally. The significance of energy has gained increasing recognition in recent decades (Onyeneke et al., 2023; Yi-Bin & Chien-Chiang, 2020). In Nigeria, energy is central to economic growth as it facilitates production, income generation, and investments across critical sectors. Household energy consumption, which accounts for one-third of global primary energy demand, is vital because it impacts the environment and public health (Han & Wei, 2021).

Most rural households in developing countries rely on energy sourced from both renewable and non-renewable resources for cooking, lighting, and space heating (Chidiebere-Mark et al., 2018; Dioha, 2018), with a heavy dependence on solid biomass and fossil fuels.

According to the International Energy Agency (IEA, 2023), approximately 2.3 billion people globally still lack access to clean cooking facilities. Many rural households rely on solid biomass (e.g., wood, crop residues) and fossil fuels (e.g., kerosene, coal) for their energy needs (Ben-Iwo et al., 2016; Olanrewaju et al., 2019). Parikh et al. (2009) note that households resort to alternative sources of energy such as petrol and diesel for lighting homes—through the use of generators, while charcoal, kerosene, and liquefied petroleum gas (LPG) are their alternative sources of energy for cooking.

Solid biomass and fossil fuels are classified as dirty fuels, as they emit harmful pollutants, which are associated with severe health risks, including respiratory and cardiovascular diseases, particularly among women and children, who are exposed to household air pollution (Martey et al., 2024; Rahut et al., 2016; Smith et al., 2024; World Health Organization, 2023). The incomplete combustion of dirty energy releases many greenhouse gases and hazardous air pollutants, such as carbon gases, sulfur dioxide, and ambient particulate matter (Malla & Timilsina, 2014). Additionally, collecting and purchasing these fuels impose a significant time and financial burden on rural households. Women and children, often responsible for gathering fuel, spend time that could otherwise be used for education or income-generating activities (Choudhuri & Desai, 2021).

Despite the efforts of the Nigerian government to expand energy access through the promotion of renewable energy and rural electrification programs (Olaniyan et al., 2024), many rural households continue to rely heavily on dirty energy sources. This situation highlights the persistent issue of energy poverty, characterized by limited access to clean, affordable and reliable energy. The transition to clean energy sources is essential for improving energy security, enhancing public health, and promoting environmental sustainability.

Clean energy are fuel sources that create less negative externality (pollution) by emitting less amount of greenhouse gases [GHG], such as CO_2, to the atmosphere (Maina et al., 2019). Clean energy technologies like solar power, biogas, and improved cook stoves provide significant health, environmental, and economic benefits over traditional energy sources. They help reduce indoor air pollution, lowering the risk of respiratory diseases, particularly for women and children. Furthermore, the shift to clean energy aligns with global sustainable development goals (SDG 7), which aim to ensure universal access to affordable, reliable, and sustainable energy. However, the slow pace of transitioning to sustainable energy solutions has been attributed to barriers such as high initial costs for clean energy technologies and fuels, inadequate infrastructure, limited awareness, and socio-economic, cultural, and institutional factors (Onyeneke et al., 2023; Smith et al., 2024).

Despite the growing global emphasis on transitioning to clean energy, significant research gaps exist, particularly regarding the energy choices of rural households in developing countries like Nigeria. Previous research has documented the health risks (Rahut et al., 2016; WHO, 2020, 2023) and environmental impacts of using dirty fuels (Dioha, 2018; Olaniyan et al., 2024; Onyeneke et al., 2023). However, much of this research focuses on broad regional or national trends and urban settings, with limited attention to specific rural contexts where energy poverty is severe. The study by Maina et al. (2019) focused on rural and urban households' demand for clean energy. However, they failed to study the factors influencing the market. The survey data was in 2016, and the economic factors prevailing in Nigeria (e.g., prices, inflation) have changed. Also, Osunkwo et al. (2020) used a systematic review approach in their study on rural and urban households' choice of energy use. Though systematic reviews are essential in social research, they are limited in using direct responses from the respondents. The potential to miss or ignore relevant studies compromises the validity of the result. More so, there is heterogeneity in the prevailing economic and cultural settings, which may need to be captured in the reporting.

It is essential to understand the factors influencing energy choices to support the development of policies and interventions to promote clean energy adoption in rural communities. The lack of localized studies that explore the factors affecting energy choices at a community level informs the research. Hence, the study analysed the determinants of energy use in rural households in Imo State. The specific objectives were to examine the demographic characteristics of households in the study area, identify the available energy sources for cooking and lighting; assess the share of clean energy in the household energy mix, evaluate household knowledge of the benefits of clean energy consumption, examine the determinants of household energy choices; and identify the constraints to clean energy adoption.

7.2 Theoretical Literature

This study on the determinants of energy choices among rural households in Imo State, Nigeria, is anchored on three fundamental theories: the Energy Ladder Theory, the Energy Stacking Model, and the Social Cognitive Theory. These frameworks comprehensively examine household energy consumption, particularly in low-income rural regions.

7.2.1 Energy Ladder Theory

The Energy Ladder Theory outlines a progressive pattern of household energy use that correlates with income. Generally, households move through three stages of energy consumption: traditional, transitional, and modern fuels. At the lower end of the ladder, households rely on traditional, polluting sources like firewood, crop residues, and charcoal. These fuels are considered dirty due to their environmental and health impacts. As household income rises, they transition to intermediate energy sources such as kerosene or coal, which are cleaner than biomass but still pose environmental risks. Ultimately, wealthier households adopt modern fuels like liquefied petroleum gas (LPG), electricity, or solar energy, which are both cleaner and more efficient.

Thus, the theoretical assumption underlying the energy ladder hypothesis is that low living standards induce greater dependence on firewood and other biomass fuels owing to a combination of income and substitution effects (Baland et al., 2006). This model, notable mainly for its ability to explain the income dependency of fuel choices, has been criticized as insufficient to represent actual energy consumption dynamics (Foster et al., 2000). It assumes that households will ascend the imaginary ladder in a somewhat linear progression pattern by switching entirely to higher-level fuels as their income improves (Osiolo, 2009).

Within this framework, income is the primary factor determining where a household falls on the energy ladder. Lower-income households are often restricted by financial constraints, relying on cheaper, dirtier energy. As incomes and education improve, there is a greater likelihood of moving up the ladder and adopting cleaner technologies. However, in rural settings, this progression is only sometimes straightforward. High costs, lack of access, and cultural preferences can prevent households from fully transitioning, even when they have the means to do so.

7.2.2 Energy Stacking Model

While the Energy Ladder Theory suggests a straightforward progression, the Energy Stacking or Multiple Fuels Model introduces more complexity. This model was posited by Masera et al. (2000), and it assumes that household energy use patterns depend on several factors not limited to income but could be social, economic, cultural, or even personal preferences. Therefore, rather than transiting linearly to cleaner fuels, households use multiple energy sources simultaneously rather than fully transitioning from one to another (Risseeuw, 2012). This phenomenon, known as "energy stacking," reflects a flexible, layered approach to energy use based on availability, cost, and convenience.

For instance, a rural household might cook with firewood while using kerosene for lighting, or they might have solar panels for lighting but still rely on biomass for heating. This pattern of mixed energy use highlights the complicated nature of energy decisions in rural areas, where households juggle economic limitations, cultural habits, and energy resource availability. This model is particularly relevant to the study's findings in Imo State, where many households continue using dirty and clean fuels due to persistent socio-economic and infrastructure barriers.

By integrating the Energy Ladder Theory and the Energy Stacking Model, this framework provides a well-rounded understanding of the determinants of energy choices among rural households. While economic factors, such as income and fuel costs, remain central to energy decisions (as suggested by the Energy Ladder Theory), the Energy Stacking Model captures the multifaceted energy use patterns typical of rural areas.

7.3 Methodology

7.3.1 Study Area

The study was carried out in Imo State, Nigeria. Imo State is one of the 36 States that make up Nigeria. It lies in the rainforest agroecological region of Nigeria. The State is bounded on the east and northeast by Abia State, South by Rivers State, and West and northwest by Anambra State (SEREDEC, n.d.). Imo State covers an area of about 5,067.20 km^2, with an estimated population of 5,167,722 (National Bureau of Statistics, 2020).

7.3.2 Sampling Technique

We conducted a cross-sectional survey of rural households in Imo State, Nigeria, between May 2022 and July 2022. Imo State, Nigeria's rural households are predominantly smallholder farmers, petty traders, and pensioners. Sixteen rural communities were chosen across the 3 agricultural zones of the State. Data on demographic characteristics, awareness of the benefits of clean energy use, the share of clean energy in the energy mix for cooking and lighting, and constraints to the use of clean energy were collected with a structured questionnaire. The authors recruited and trained enumerators for the data collection. Rural household heads and/or their spouses were interviewed.

A multi-stage sampling technique was employed to select the households for the study. The three agricultural zones (Owerri, Orlu and Okigwe) that make up the State were all included in the study. The first stage involved the purposive sampling of 8 Local Government Areas (LGA) based on the low rate of urbanization compared to other LGAs, and the high volume of agricultural activities. They include Njaba and Oru East (Orlu agricultural Zone), Isiala Mbano and Ehime Mbano (Okigwe agricultural zone) and Ikeduru, Aboh-Mbaise, Ahiazu Mbaise, and Ngor Okpala (Owerri agricultural zone). In the second stage, a proportionate sampling technique of two (2) communities was purposively selected from each of the selected LGAs based on the proximity and level of agricultural activities in the community. In the third stage, one Village was selected from each community, giving a total of 16 villages for the study. Finally, 12 households were randomly selected from each of the villages, to give a sample size of 192 households.

A structured questionnaire was administered to the 192 selected households. However, only 185 questionnaires were correctly filled and returned, and this was used for analysis. The questionnaire collected primary data on household demographics, income levels, education, energy sources, and energy consumption patterns. The questionnaire collected quantitative data on household demographics, income levels, education, energy sources, and energy consumption patterns.

7.3.3 Data Analysis

The data was analysed using mean, percentages, and Binary Logistic regression. Descriptive statistics such as mean, frequency, and percentage were used to analyse the objectives (i).

The objective (iii) was analysed using a Binary Logistic regression. The binary logistic model was specified as follows:

$$\Pr ob(Y = 1/x) = \frac{\exp(x\beta)}{1 + \exp(x\beta)} \tag{7.1}$$

$$\frac{LogPi}{1 - Pi} = Zi = a + \beta Xi + \varepsilon i \tag{7.2}$$

where,

P_i	is the probability that the event occurs to an individual with a given set of characteristics, X_i
a	is the intercept or constant
β_i	is the vector of coefficients of the vector of covariates or coefficients, X_i
$P_i/[1 - P_i]$	is the odds ratio of households with a given set of characteristics using clean energy or dirty energy
Ln(Y)	Probability that the households use clean energy = 1 and otherwise = 0.

The dependent variables were:

$$Y = f(X)$$

Y	Energy choice (Dummy, 1 = if a household uses clean energy for cooking, 0 = dirty energy for cooking)

X_1 Household size (No. of persons)
X_2 Income (Naira)
X_3 Gender (1 = Female, 0 = Male)
X_4 Housing Tenure (1 = Self owned; 2 = Rented)
X_5 Age (Years)
X_6 Education (Years)
X_7 Price of energy (Naira)
X_8 Membership of social organization (Member = 1, 0 = Non-member).

The a priori expectations of the independent variables regarding the likelihood of a household choosing clean energy are based on theoretical and empirical insights from energy consumption literature.

- Household Size (X_1). This is expected to be negative a priori. Larger households prefer cheaper, dirty fuels due to higher energy needs.
- Income (X_2). It is expected a priori to be positive. Higher income allows households to afford cleaner energy.
- Gender (X_3). A mixed relationship is expected a priori. Female-headed households prefer clean energy but face financial constraints.
- Housing Tenure (X_4) is expected to be positive a priori. Homeowners are more likely to invest in clean energy technologies, as they stand to benefit from long-term energy cost savings and better living conditions.
- Age (X_5). A priori is expected to be negative. Older individuals are more likely to stick to traditional energy sources.
- Education (X_6). A priori is expected to be positive. Educated households are more aware of the benefits of clean energy.
- Price of Energy (X_7). It is expected to be negative a priori. Higher energy prices discourage the use of clean fuels.
- Membership in Social Organizations (X_8). A priori is expected to be positive. Social networks can promote access to clean energy technologies.

7.4 Results and Discussion

7.4.1 Demographic Characteristics of Rural Households in Imo State

The distribution of the demographic characteristics of the rural households is presented in Table 7.1. The result shows that 72.8% of the rural households' heads were between the ages of 36 and 65 years, with a mean age of 52 years. Older individuals tend to have more conservative energy consumption patterns, which can be attributed to their lifestyle, income, and environmental attitudes. They are also more likely to use traditional energy sources such as firewood and kerosene for cooking and lighting (Oladejo et al., 2020). In contrast, younger individuals are more open to using modern clean energy sources such as solar, wind, and biomass. They are also more likely to adopt energy-efficient technologies and practices, such as using energy-saving light bulbs and appliances, and insulating their homes Bae et al. (2024). The findings are consistent with a study conducted by Adewumi et al. (2022), which found that the majority of households in rural Nigeria are headed by individuals over 50 years of age.

The findings show that 55.14% of the household heads were female, while 44.86% were male. The gender of the household head may play a role in clean energy consumption patterns in the study area, as women primarily make the cooking choices. More so, the household head is responsible for making decisions on household energy expenditure. Females play a pivotal role in the transition towards sustainable energy practices in households, as they are primary energy users (Shrestha et al., 2021).

The results show the marital status of the household heads in Imo State, Nigeria. The results reveal that 69.72% were married, 11.35% were single. The majority of the household heads were married. This may be because of cultural norms that place a high value on marriage and family life. Married household heads may have higher energy needs because of the presence of children or other family members, which could affect their energy consumption. The findings further revealed that the majority (67%) of the rural households in Imo State, Nigeria, had a range of 5–8 household sizes. The mean household size was 6 persons. This is consistent with the report of Egwuonwu and Nweke (2018), which states that the average number of persons in rural households in Imo State is six persons.

The majority (97.84%) of the household heads had attained some level of formal education. 40.54% of the respondents had spent between 7 and 12 years which can be categorized as secondary education based on the education system in present Nigeria. While 20.54 and 36.76 had attained only primary and tertiary education respectively. This implies that a greater percentage of the respondents were literate. According to Guta (2018) found that higher levels of education among rural households were significantly associated with increased adoption of clean energy technologies, such as solar panels and clean cook stoves. This is because education can increase awareness of the benefits of clean energy, as well as provide the knowledge and skills necessary to use and maintain these technologies.

The findings on the source of livelihood revealed that 56.0% of the respondents identified farming as their major source of livelihood. Traders constitute 32.6%, while civil servants were 27.7%. The source of livelihood of the rural households may determine the choice of energy, as clean energy usage saves time needed to carry out other productive activities.

The findings show that the majority (84.3%) of the rural households' respondents are members of social clubs, while 15.7% identified as non-members of any social club in Imo State. Social club membership is beneficial to rural and community development as it contributes greatly to the adoption or rejection of social changes in society. Karlan et al. (2017) posited that membership in social clubs such as savings groups and farmer groups were associated with access to information, resources, and social support that enabled them to make more informed decisions. Housing

Table 7.1 Distribution of the rural households by socioeconomic characteristics

Demographic characteristics	Frequency	Percentage	Mean
Age (years)			
21–35	22	12.00	
36–50	62	32.60	
51–65	74	40.20	
66–80	29	15.20	52 years
Gender			
Male	83	44.86	
Female	102	55.14	
Marital status			
Single	21	11.35	
Married	129	69.72	
Household size (persons)			
1–4	43	23.30	
5–8	124	67.00	
9–12	18	9.70	6 persons
Educational attainment			
None	4	2.16	
1–6 years	38	20.54	
7–12 years	75	40.54	
13–18 years	68	36.76	12 years
Source of livelihood			
Farmer	103	56.00	
Civil servant	51	27.70	
Pensioner	18	9.80	
Remittance	13	7.10	
Trader	60	32.60	
Membership of social organization			
Member	156	84.30	
Non-member	29	15.70	
Housing tenure			
Self-owned	146	78.90	
Rented	39	21.10	
Monthly income (₦)			
₦1000–₦30,000	52	28.10	
₦31,000–₦60,000	81	43.80	
₦61,000–₦90,000	52	28.10	₦54,607
Employment			

(continued)

Table 7.1 (continued)

Demographic characteristics	Frequency	Percentage	Mean
Gainfully employed	81	43.78	
Not gainfully employed	104	56.22	

is one of the necessities of man for sustainability. The housing tenure refers to the binding ownership of shelter in the study area. The findings reveal that 78.9% of the households owned their houses, while 21.1% of the respondents lived in rented houses. The rural households that owned their houses had greater control over their homes and had greater incentives to invest in clean energy consumption as they could benefit from lower energy costs and improved indoor air quality in the long term. According to Bisu et al. (2016) personal ownership of a dwelling gives the owner the freedom to use any fuel for cooking.

The average monthly income of the households was estimated at ₦54,607.02, which is higher than the prevailing minimum wage in Nigeria, pegged at ₦30,000 per month at the time. The finding implies that rural households are financially stable with the energy of their choice. The results reveal that 56.22% of the respondents posited that they are not gainfully employed while 43.78% identified that they are gainfully employed. Agricultural production in most rural areas in Imo State is mainly for subsistence, with little sold in the market. The livelihood of these families is supported by pensions and remittances from migrant family members who may live in cities or abroad. Remittances have a significant positive effect on rural households, particularly in terms of income and food consumption. Studies have shown that remittances enable households, especially in rural areas, to adopt modern renewable energy sources and improve their access to electricity and clean cooking systems (Acheampong et al., 2024; Barkat et al., 2023). This financial support can alleviate energy poverty by increasing household income, which allows for more significant investment in adequate energy solutions.

7.4.2 Different Sources of Energy Available to Rural Households

The different sources of energy available to rural households for cooking and lighting are presented in Tables 7.2 and 7.3.

7.4.2.1 Different Sources of Energy Available for Cooking

The different energy sources used for cooking by the rural households is presented in Table 7.2. The findings show that Kerosene stove was used by 90.8% of the rural households as their primary cooking energy source, while 82.7% indicated

Table 7.2 Distribution of different sources of energy available for cooking

Cooking energy source	Frequency[a]	Percentage
Kerosene stove	168	90.80
LPG cooker	61	32.97
Electric stove	27	14.60
Firewood	153	82.70
Plant debris	7	3.70
Charcoal	74	39.40
Biofuel	7	3.70

[a] Multiple responses were recorded

Table 7.3 Distribution of different sources of energy available for lighting

Lighting energy source	Frequency	Percentage
Kerosene lantern	149	80.54
Oil lamp	60	32.40
Generator	119	63.30
Rechargeable	121	65.40
Dry battery	133	71.89
Inverter	3	1.62
PHCN (Power Holding Company of Nigeria)	89	48.10

[a] Multiple responses were recorded
Source Field survey data, 2022

that firewood was their cooking energy source in Imo State. This is because of the ease in sourcing these energy types and the cost implication associated with the choice of their cooking energy. The finding is in line with Adekoya et al. (2023) that firewood and kerosene are the most dominant source of cooking energy among rural households. LPG stoves, electric stove and biofuel are only used by 32.97, 14.6 and 3.7% of the rural households. This implies that the rural households use inefficient and dirty fuels for their cooking needs in the study area.

7.4.2.2 The Different Sources of Energy Available for Lighting Among the Rural Households

Table 7.3 shows the different energy sources used for lighting by the rural households. The result in Table 7.3 shows that Kerosene lantern, Oil Lamp, Generator, Rechargeable Lantern, Dry battery torch, Inverter batteries, Solar panels and Electricity as the sources of lighting for rural households in Imo State. The results also show that 80.54 and 71.89% of the respondents use kerosene lantern and dry battery torch as lighting source. In a similar way, 65.4%, 63.3% and 48.1% indicated their lighting source as Rechargeable lantern, Generator and PHCN respectively. Furthermore, 32.4% posited oil lamp as their lighting source while 1.62% of the rural population had inverter/solar panel system as a lighting source for their households.

This finding aligns with Chukwuedo and Ofuyekpone (2019) who observed that greater percentages of the source of energy for domestic lighting are fossil fuel. This implies that kerosene lantern is the prevalent lighting source for rural household in Imo State resulting to energy insecurity. The inverter system is the least used lighting source because of the financial implication associated with its installation, hence, the reason for low consumption of clean energy among the respondents.

7.4.3 *Households' Knowledge of the Benefits of Clean Energy Consumption*

The rural households' knowledge on the benefits of clean energy consumption in Imo State is shown in Table 7.4. The result showed that majority of respondents are aware of the social benefits and cost associated with clean energy consumption. The result showed that majority of the respondents identified that they are aware that the consumption of clean energy is beneficiary to their wellbeing based on the findings in Table 7.4. However, many people in rural areas of Imo State are not aware that the consumption of clean energy is climate friendly, reduces child and maternal mortality rate, agronomical cost during crop production and flooding that have experienced in the previous years. According to Ogwumike et al. (2014), the non-sustainable consumption of dirty energy in Nigeria has resulted to environmental hazards. The implication of these findings is the vulnerability of rural households to human environmental hazards like air pollutions which affects their health and livelihood to the broader economy.

Table 7.4 Knowledge of the benefits of clean energy consumption

Benefits	Frequency[a]	Percentage
Reduction in cardiovascular disease	128	69.20
Reduces exposure from household air pollution (HAP)	152	87.60
Saves time	152	87.60
Reduces climate change impact	76	41.10
Reduces child and maternal mortality from exposure from fossil fuels	73	39.50
Less hospital visits from household air pollution exposure	106	57.30
Reduces house renovation cost	153	82.70

[a] Multiple responses were recorded

7.4.4 *Share of Clean Energy in the Energy Mix of the Rural Households*

Share of clean energy in the energy mix by the households for lighting and cooking were presented in Figs. 7.1 and 7.2.

7.4.4.1 The Share of Clean Energy in the Energy Mix for Cooking by the Rural Households

In Fig. 7.1, the results indicate that 23% of the rural households use clean energy like LPG, Electric cooker and Biofuel for cooking in Imo State, while 77% of the households use dirty fuels from biomass and fossil fuels for cooking. This implies that the use of dirty energy like firewood, charcoal, and kerosene stove for cooking

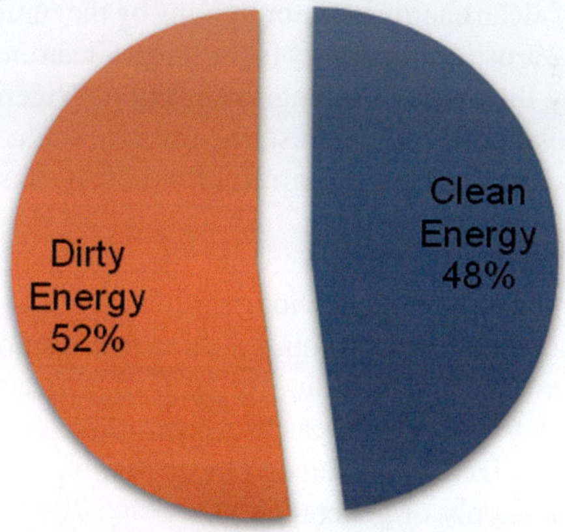

Fig. 7.1 Share of clean energy for cooking among rural households in Imo State

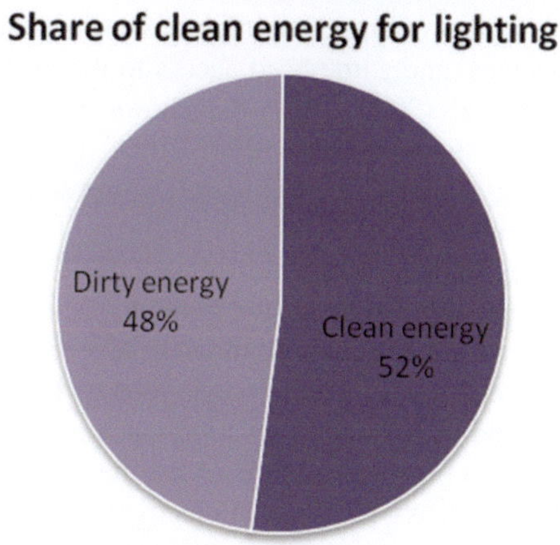

Fig. 7.2 Share of clean energy for lighting among rural households

is prevalent among the rural population. This may result in a high mortality rate common in women and children who are exposed to household air pollution from the use of dirty energy, loss of economic time as they spend more time cooking for their family, and an increase in adverse effects of climate because of the deforestation associated with the energy source for cooking and also diseases like cardiovascular, as a result of steady inhalation of carbon emission gases and fumes as a result of the cooking energy choice. The findings are consistent with the findings of Roche et al. (2024) that two-thirds of the population of Nigeria use solid biomass for cooking, mainly in the rural areas.

7.4.4.2 The Share of Clean Energy in the Energy Mix for Lighting by Rural Households

Rural households use multiple energy sources depending on the year's season and the energy source's availability. The result presented in Fig. 7.2 shows that the share of clean energy used for lighting by the rural households in Imo State was 48%, while 52% used dirty energy for lighting. Clean energy sources include electricity generated by the Power Holding Company of Nigeria, rechargeable lamps and torches, and solar bulbs. In contrast, the primary source of dirty energy for lighting is kerosene lamps, which emit harmful fumes that cause indoor air pollution and various health challenges.

In Nigeria, the National Electric Power Authority (NEPA) until 2005 had the monopoly of generation, transmission, and distribution of electricity. Power generation in Nigeria is from gas-fired thermal power plants and hydro dams, with natural gas plants accounting for 80% of power generation. Natural gas used to produce electricity, which is a clean energy, is a pollutant as it emits CO_2. In 2005, NEPA was privatized and renamed Power Holding Company of Nigeria (PHCN), still with the monopoly of generating electricity. 2013, the Company was unbundled, and some investors were given ownership rights to generate and distribute electricity.

Nevertheless, electricity distribution is still unstable and unreliable in Nigeria, particularly in rural communities, as most households are exposed to frequent power outages, and some need access to the grid (Arowolo & Perez, 2020). A report by IEA et al. (2023) shows that shows only 27% of the rural population in Nigeria had access to electricity in 2022. This situation results in households using alternative energy sources like Generators powered by fossil fuels. More so, though most rural households rely heavily on electricity from the Power Holding Company and other energy distribution companies (DISCOs) due to competitive pricing compared to other energy sources for lighting, the frequent epileptic power supply constrains the rural households from using dirty energy sources like kerosene lanterns. Some households use clean energy sources like rechargeable batteries, but they need to depend on an electricity supply to charge the batteries. Also, some households use solar and inverter batteries, classified as clean energy.

The heavy reliance on electricity for lighting explains the contrast between the share of clean energy for cooking (48%) and the share for lighting (52%). Though the share of clean energy for lighting is relatively higher, rural households have fewer options than those available for cooking.

7.4.5 Determinants of Energy Choices by Rural Households

The determinants of the energy choices of the rural households were estimated using a Binary Logistic Model, and the results are presented in Table 7.5. The results show that income, education, and price coefficients were statistically significant at the 1 and 5% significance levels. This indicates that these factors determine the energy choices of the rural household for cooking and lighting.

The coefficient for income is positive (0.000) and statistically significant (p-value = 0.036) at 5%, indicating that higher income is associated with a higher likelihood of using cleaner energy choices. This indicates that an increase in the income of the household will result in a reduction in the use of dirty energy. The result is consistent with the energy ladder hypothesis, which posits that as a household's economic status improves, it will progressively transition from using traditional and polluting fuels to cleaner and more efficient energy sources. Nevertheless, the households may use different fuels for different purposes, supporting the energy stacking model. This result supports the assertion of Ogwumike et al. (2014) that energy consumption in Nigeria supports the energy stacking model, which assumes that households in developing countries use a mix of traditional and modern fuels, even as their income increases. Instead of transitioning to the exclusive use of cleaner, more efficient fuels, households often use a combination of fuel sources.

Table 7.5 Determinants of rural households' energy choice

Variables	B	S.E.	Wald	Df	Sig.	Exp(B)
Household size	– 0.132	0.087	2.303	1	0.129	0.876
Income	0.000	0.000	4.378	1	0.036**	1.000
Gender	0.382	0.351	1.182	1	0.277	1.465
Housing tenure	– 0.017	0.488	0.001	1	0.972	0.983
Age	0.022	0.018	1.471	1	0.225	1.022
Education	– 0.122	0.052	5.562	1	0.018**	0.885
Price	– 0.000	0.000	15.702	1	0.000*	1.000
Membership of social organization	0.341	0.504	0.458	1	0.499	1.406
Constant	3.254	1.425	5.214	1	0.022	25.882
Number of observations: 185						
– 2 log likelihood = 210.838						
Prob. > Chi^2 = 0.0000						

The coefficient for education is positive (− 0.122) and statistically significant (p-value = 0.018), indicating that higher education levels of the household heads increase the likelihood of using cleaner energy and reduce household consumption of dirty energy. Education may enable individuals to make more informed choices that benefit them, as they are more aware of the consequences of using dirty fuels. The result corroborates the findings of Rahut et al. (2020), who posit that the number of households using clean cooking fuels will increase with the rise in educational levels.

The coefficient for price is positive (− 0.000) and statistically significant (p-value = 0.000), indicating that higher energy prices increase the probability of selecting cleaner energy over dirty energy. Individuals select the best bundle of goods that provides maximum utility. At higher energy prices, households will choose cleaner fuels that are more efficient than dirty fuels that consume more time when used for cooking.

7.4.6 Constraints Affecting the Energy Choices of Rural Households

The distribution of respondents by constraints to energy choices of the rural households is presented in Table 7.6. The results showed that 86.5% of the rural households were constrained by the high costs associated with clean energy technologies such as solar panels, LPG stoves, and other clean cooking solutions and lighting technologies. The initial investment required for clean energy technologies is often beyond the reach of many rural households, who typically have low incomes and are mostly subsistence farmers and petty traders. Households opt for cheaper, less efficient, and polluting options like firewood, kerosene, and charcoal. These findings align with the energy ladder hypothesis, suggesting that transitioning from traditional polluting fuels to cleaner alternatives is likely as households' incomes rise. Low income was reported by 78.9% of the households as a constraint to clean energy consumption. Most rural households have limited disposable income, most of which is allocated to food. The rising cost of inflation further reduces their ability to afford clean energy technologies. The findings align with Maina et al. (2019), who state that income determines household energy choices. The result further corroborates the findings of Abubakar et al. (2024), who assert that poor households prefer traditional energy sources.

About 69.7% of rural households indicate that inadequate access to clean energy technologies and fuels is the significant factor underlying their choice of energy used for cooking and lighting. This can be attributed to inadequate infrastructure in rural areas, where distribution networks for clean energy solutions, such as LPG, are poorly developed. Also, the distance from markets and vendors providing these clean energy solutions may need to be improved so that rural households can adopt clean

Table 7.6 Constraints affecting the energy choices of households

Constraints	Frequency	Percentage
Inadequate access to clean energy technologies and fuels	128	69.2
Technology complexity	44	23.8
Cultural preferences	65	31.5
High cost of clean fuels and technologies	160	86.5
Government policy	55	29.7
Low income	146	78.9
Inflation	107	57.8

Multiple responses recorded

fuels. As a result, households rely on easily accessible traditional energy sources despite being aware of the health and environmental risks associated with their use.

Similarly, 31.5% identified that cultural preferences play a significant role in influencing energy choices. Culture can shape people's motivations and incentives to adopt different energy practices. Many rural households have long relied on firewood and kerosene for cooking and lighting, and the transition to clean energy technologies may appear to them to be complicated and unnecessary. Additionally, 23.8% of rural households cited the complexity of clean energy technologies as a barrier to adopting clean alternatives. Education and exposure of individuals may limit their adoption of new technologies, as less educated people may perceive modern cooking and lighting solutions as complex. They may be unwilling to adapt to changes. The result reiterates the argument of Goggins et al. (2022) that culture plays a role in advancing sustainable energy practices.

7.5 Conclusions, Recommendations and Limitations

This study explored the energy security situation of rural households in Imo State, Nigeria, providing essential insights into socio-economic characteristics and the factors influencing energy choices in the state. A significant portion of the rural population continues to rely on traditional, polluting energy sources, which raises concerns about environmental sustainability and the effect it will pose on agriculture. Factors such as household size, income, gender of the household head, housing tenure, age of the household head, and social club membership significantly influence the choice of energy sources that rural households use. Rural households are constrained by the high cost of clean fuels and technologies, income size, and inadequate access to clean energy technologies and fuels when making clean energy choices. These findings highlight the need for targeted policies and interventions to overcome these barriers and promote the use of cleaner energy in rural areas.

The study recommends that the Government subsidizes clean energy technologies for rural households as the households are constrained by income and price of clean energy technologies and fuels. This will foster the drive towards the actualization of SDG 7 by 2030 by promoting reliable, modern and affordable access to modern energy for all. This can be achieved by partnering with local financial institutions and renewable energy companies to finance clean energy products. The government should expand rural electrification networks to provide reliable electricity in communities dependent on dirty fuels.

Efforts should be made to increase the availability of clean energy technologies in rural areas. This could involve government programs, NGOs, and private sector involvement in distributing and promoting clean energy solutions like solar lamps, efficient cook stoves, and biogas systems. The Government should strengthen policy frameworks at the State and community levels to promote clean energy use. This can be achieved by enforcing regulations that limit the use of highly polluting fuels in rural areas.

NGOs, Governments, and public health groups should implement community-based awareness campaigns on clean energy benefits. This should be done in collaboration with community leaders and social organizations, using various mediums to raise awareness about the health and environmental and economic benefits of clean energy use.

Despite the valuable insights provided by this study, the study is geographically limited to rural communities in Imo State, Nigeria, which may restrict the generalization of the findings to other rural areas. There is a need for further research to involve urban households both in the State of Nigeria and in sub-Saharan Africa, as it is necessary to achieve SDG 7 by 2030. The study relied on primary data collected through structured questionnaires and interviews. There is a risk of response bias where participants may need to accurately recall or underreport their energy consumption patterns or socioeconomic conditions, and it may affect the accuracy of the findings, particularly in terms of the determinants of energy choices.

Funding The authors did not receive support from any organization for the submitted work.

Data Availability Statement The datasets used or analysed during the current study are available from the corresponding author upon reasonable request.

Conflict of Interest The authors declare no conflict of interest.

References

Abubakar, I. R., Alola, A. A., Bekun, F. V., & Onifade, S. T. (2024). Investigating the determinants of household energy consumption in Nigeria: Insights and implications. *Energy Sustainability and Society, 14*(1). https://doi.org/10.1186/s13705-024-00451-6

Acheampong, A. O., Brahim, M., & Dzator, J. (2024). Addressing rural energy poverty and rural-urban energy access gap in developing countries: Does international remittances matter? *Applied Economics,* 1–18. https://doi.org/10.1080/00036846.2024.2331971

Adekoya, A. E., Adenikinju, A. F., Olubusoye, O. E., Oyeranti, O. A., Otekunrin, O. A., Ogunbayo, I. E., et al. (2023). Household food insecurity and cooking energy access in Nigeria: A panel data approach. *Energy Nexus, 12*, 100242. https://doi.org/10.1016/j.nexus.2023.100242

Adewumi, A., Ahmed, Y., Lawal, S., & Shiyaki, U. A. (2022). Livelihood activities and food security status of farming households in Niger State, Nigeria. *Journal of Agripreneurship and Sustainable Development, 5*(3), 129–140. https://doi.org/10.59331/jasd.v5i3.341

Arowolo, W., & Perez, Y. (2020). Market reform in the Nigeria power sector: A review of the issues and potential solutions. *Energy Policy, 144*, 111580. https://doi.org/10.1016/j.enpol.2020.111580

Bae, C., Lee, J. Y., Kim, D., & Chun, C. (2024). Understanding residents' energy consumption profiles: Behavioral segmentation analysis of heat energy consumption in Korean apartment residences. *Energy and Buildings, 319*, 114482. https://doi.org/10.1016/j.enbuild.2024.114482

Baland, J., Bardhan, P., Das, S., Mookherjee, D., & Sarkar, R. (2006). Managing the environmental consequences of growth forest degradation in the Indian Mid-Himalayas. *India Policy Forum, 3*(1), 125–277. https://eml.berkeley.edu/~webfac/bardhan/papers/inpolforrevfin.pdf

Barkat, K., Alsamara, M., & Mimouni, K. (2023). Can remittances alleviate energy poverty in developing countries? New evidence from panel data. *Energy Economics, 119*, 106527. https://doi.org/10.1016/j.eneco.2023.106527

Ben-Iwo, J., Manovic, V., & Longhurst, P. (2016). Biomass resources and biofuels potential for the production of transportation fuels in Nigeria. *Renewable and Sustainable Energy Reviews, 63*, 172–192. https://doi.org/10.1016/j.rser.2016.05.050

Bisu, D. Y., Kuhe, A., & Iortyer, H. A. (2016). Urban household cooking energy choice: An example of Bauchi metropolis, Nigeria. *Energy, Sustainability and Society, 6*(1), 1–12. https://doi.org/10.1186/s13705-016-0080-1

Chidiebere-Mark, N. M., Agunanne, U. T., & Anyanwu, U. G. (2018). Determinants of cooking energy consumption among farming households in Owerri Agricultural Zone, Imo State. *International Journal of Energy and Environmental Research, 6*(2), 55–62.

Choudhuri, P., & Desai, S. (2021). Lack of access to clean fuel and piped water and children's educational outcomes in rural India. *World Development, 145*, 105535. https://doi.org/10.1016/j.worlddev.2021.105535

Chukwuedo, J. A., & Ofuyekpone, O. D. (2019). Statistical analysis of domestic lighting energy consumption in Nigeria: A case of Owerri, Imo State, Nigeria. *International Journal of Applied Research and Technology, 8*(3), 79–93.

Dioha, M. (2018). Modelling the impact of Nigeria household energy policies on energy consumption and CO_2 emissions. *Engineering Journal, 22*, 1–19. https://doi.org/10.4186/ej.2018.22.6.1

Egwuonwu, H. A., & Nweke, A. P. (2018). Influence of socio economic factors of rural households on fuelwood consumption in Orlu agricultural zone of Imo State, Nigeria. *Journal of Energy Research and Reviews*, 1–6. https://doi.org/10.9734/jenrr/2019/v2i129725

Foster, V., Tre, J., Wodon, Q. (2000). Energy prices, energy efficiency and fuel poverty. *World Bank Mimeo.*

Goggins, G., Rau, H., Moran, P., Fahy, F., & Goggins, J. (2022). The role of culture in advancing sustainable energy policy and practice. *Energy Policy, 167*, 113055. https://doi.org/10.1016/j.enpol.2022.113055

Guta, D. D. (2018). Determinants of household adoption of solar energy technology in rural Ethiopia. *Journal of Cleaner Production, 204*, 193–204. https://doi.org/10.1016/j.jclepro.2018.09.016

Han, X., & Wei, C. (2021). Household energy consumption: State of the art, research gaps, and future prospects. *Environment Development and Sustainability, 23*(8), 12479–12504. https://doi.org/10.1007/s10668-020-01179-x

IEA. (2023). *SDG 7: Data and projections.* IEA. https://www.iea.org/reports/sdg7-data-and-projections

IEA, IRENA, UNSD, World Bank, & WHO. (2023). *Tracking SDG 7: The energy progress report*. World Bank. https://www.irena.org/-/media/Files/IRENA/Agency/Publication/2023/Jun/Tracking_SDG7_energy_progress_2023.pdf

Karlan, D., Savonitto, B., Thuysbaert, B., & Udry, C. (2017). Impact of savings groups on the lives of the poor. *Proceedings of the National Academy of Sciences, 114*(12), 3079–3084. https://doi.org/10.1073/pnas.1611520114

Maina, Y. B., Kyari, B. G., & Maina, M. B. (2019). Households' clean energy demand in Nigeria and its implication on the environment. *Journal of Agricultural Economics, Environment and Social Sciences, 5*(1 and 2), 35–55. https://jaeess.com.ng/index.php/jaeess/article/view/15

Malla, S., & Timilsina, G. R. (2014). *Household cooking fuel choice and adoption of improved cookstoves in developing countries: A review* (World Bank Policy Research Working Paper No. 6903). https://ssrn.com/abstract=2445749

Martey, E., Etwire, P. M., Armah, R., & Nordjo, R. E. (2024). Empirical analysis of solid biomass fuel and ill-health. *Environmental and Sustainability Indicators, 23*, 100437. https://doi.org/10.1016/j.indic.2024.100437

Masera, O. R., Saatkamp, B. D., & Kammen, D. M. (2000). From linear fuel switching to multiple cooking strategies: A critique and alternative to the energy ladder model. *World Development, 28*(12), 2083–2103. https://doi.org/10.1016/s0305-750x(00)00076-0

National Bureau of Statistics NBS. (2020). *Demographic statistics bulletin*. https://nigerianstat.gov.ng/download/1241121

Ogwumike, F. O., Ozughalu, U. M., & Abiona, G. A. (2014). Household energy use and determinants: Evidence from Nigeria. *International Journal of Energy Economics and Policy, 4*(2), 248–262.

Oladejo, O. S., Dahunsi, S. O., Adesulu-Dahunsi, A. T., Ojo, S. O., Lawal, A. I., Idowu, E. O., & Evbuomwan, I. (2020). Energy generation from anaerobic co-digestion of food waste, cow dung and piggery dung. *Bioresource Technology, 313*, 123694. https://doi.org/10.1016/j.biortech.2020.123694

Olaniyan, A., Caux, S., & Maussion, P. (2024). Rural electrification in Nigeria: A review of impacts and effects of frugal energy generation based on some of e-waste components. *Heliyon, 10*(11), e31300. https://doi.org/10.1016/j.heliyon.2024.e31300

Olanrewaju, F., Andrews, G., Li, H., & Phylaktou, H. (2019). Bioenergy potential in Nigeria. *Chemical Engineering Transactions, 74*, 61–66. https://doi.org/10.3303/cet1974011

Onyeneke, R. U., Chidiebere-Mark, N. M., Ankrah, D. A., & Onyeneke, L. U. (2023). Determinants of access to clean fuels and technologies for cooking in Africa: A panel autoregressive distributed lag approach. *Environmental Progress & Sustainable Energy, 42*(3), e14147. https://doi.org/10.1002/ep.14147

Osiolo, H. (2009). Enhancing Household Fuel Choice and Substitution in Kenya, Kippra Discussion Paper, No. 102, 1–55. The Kenya Institute for Public Policy Research and Analysis.

Osunkwo, F. O. C., Onwusogbulu, R. D., & Uma, K. E. (2020). Rural and urban households choice of energy use in Nigeria: Problems and prospects. (08), 25–33.

Parikh, J., Panda, M., Ganesh-Kumar, A., & Singh, V. (2009). CO2 emissions structure of Indian economy. *Energy, 34*(8), 1024–1031. https://doi.org/10.1016/j.energy.2009.02.014

Rahut, D. B., Ali, A., & Behera, B. (2016). Domestic use of dirty energy and its effects on human health: Empirical evidence from Bhutan. *International Journal of Sustainable Energy, 36*(10), 983–993. https://doi.org/10.1080/14786451.2016.1154855

Rahut, D. B., Ali, A., Mottaleb, K. A., & Aryal, J. P. (2020). Understanding households' choice of cooking fuels: Evidence from urban households in Pakistan. *Asian Development Review, 37*(1), 185–212. https://doi.org/10.1162/adev_a_00146

Risseeuw, N. (2012). *Household energy in Mozambique: A study on the socio-economic and cultural determinants of stove and fuel transitions* (Master's thesis). Vrije Universiteit, IVM Institute for Environmental Studies.

Roche, M. Y., Slater, J., Malley, C., Sesan, T., & Eleri, E. O. (2024). Towards clean cooking energy for all in Nigeria: Pathways and impacts. *Energy Strategy Reviews, 53*, 101366. https://doi.org/10.1016/j.esr.2024.101366

SEREDEC (n.d.) Imo State. South East Region Economic Development Company. https://southeast.ng/the-region/imo-state/

Shrestha, B., Tiwari, S. R., Bajracharya, S. B., Keitsch, M. M., & Rijal, H. B. (2021). Review on the importance of gender perspective in household energy-saving behaviour and energy transition for sustainability. *Energies, 14*(22), 7571. https://doi.org/10.3390/en14227571

Smith, J. D., Al-Rubaye, H., Zangana, M. H. S., Nagapurkar, P., Zhou, Y., & Gelles, G. (2024). Optimum operation of microgrid systems that employ PV solar and battery systems. In *Science and technology—Recent updates and future prospects* (pp. 13–30). https://doi.org/10.9734/bpi/strufp/v3/7829a

World Health Organization. (2023). *Household air pollution.* https://www.who.int/news-room/fact-sheets/detail/household-air-pollution-and-health

World Health Organization (WHO). (2020). *Household air pollution.* https://www.who.nt/airpollution/household/en

Yi-Bin, C., & Chien-Chiang, L. (2020). Effects of financial development on energy consumption: The role of country risks. *Energy Economics, 90*(C). https://doi.org/10.1016/j.eneco.2020.104833

Part II
Climate

Chapter 8
Green Transition: Exploring the Clean Energy and Public Health Outcome Nexus in Nigeria (1990–2022)

Daniel Chinedu Nwaogu, Raphael Ndubuisi Echebiri, Joseph Chimere Onwumere, Igwe Oscar Obasi, Blessing Chidimma Nwaogu, and Clinton Nnamdi Usulor

Abstract The combustion of solid fuels generates certain effluents such as black carbon and carbon dioxide that are harmful to human health. Substantial evidence suggest that these substances are among the leading cause of non-communicable diseases such as chronic obstructive pulmonary disease, acute lower respiratory infections in children, strokes, lung cancer and heart attacks, particularly, in countries that are termed "low-income". By employing the ARDL model, this study investigated the suspected link between consuming clean energy and public health outcomes in Nigeria from 1990 to 2022. Empirical findings revealed that renewable energy utilization and access to clean cooking energy technologies increased public health outcomes by reducing air pollution related mortality, while exposure to CO_2 and black carbon led to a decreased public health outcome by increasing mortality resulting from air pollution. Considering the Nigerian perspective, these results underscore the impact of green transition on public health outcomes in addition to the influence on environmental quality. Global attempts towards ensuring a just transition to and increasing the share of more renewable energy will in addition to drastically reducing carbon emissions and its consequent damage on the climate, lead to a robust reduction in health costs incurred due to air pollution.

D. C. Nwaogu (✉) · C. N. Usulor
Department of Agribusiness and Management, Alex Ekwueme Federal University, Ndufu-Alike, Ikwo, Ebonyi State, Nigeria
e-mail: nwaogu.chinedu@funai.edu.ng

R. N. Echebiri · I. O. Obasi
Department of Agricultural Economics, Michael Okpara University of Agriculture, Umudike, Abia State, Nigeria

J. C. Onwumere
Department of Agribusiness and Management, Michael Okpara University of Agriculture, Umudike, Abia State, Nigeria

B. C. Nwaogu
Department of Psychology, Alex Ekwueme Federal University, Ndufu-Alike, Ikwo, Ebonyi State, Nigeria

R. U. Onyeneke et al. (eds.), *Energy Transition, Climate Action and Sustainable Agriculture*, https://doi.org/10.1007/978-3-031-83165-2_8

Keywords Public health · Just transition · Carbon emissions · Renewable energy · ARDL

8.1 Introduction

Nigeria currently has over 200 billion kg biomass capacity with the combination of charcoal and wood fuel accounting for over 80% of her households' energy consumption for cooking and heating (IEA, 2021). The report also has it that wood fuel as the traditional biomass in Nigeria alone accounts for about 94% of cooking fuel. The report further showed that biofuels and wastes accounted for about 79% of the final energy consumption with coal, natural gas, oil products, and electricity accounting for 1.1%, 3.1%, 15.1% and 1.7% respectively.

With high consumption of biofuels, carbon dioxide (CO_2) emissions are expected to be high too. The Greenhouse Gas Emissions from Energy Data Explorer (2021) reported that CO_2 emissions from fuel combustion in Nigeria was about 101.461 Mt which accounted for about 0.3% global share of CO_2 emissions from combustible fuels. The report further stated that about 63% of these emissions came from oil combustion. This high consumption of biofuels has a number of well documented consequences ranging from public health concerns, reduced economic opportunities, especially those affecting women and girls responsible for biomass collection, environmental consequences such as deforestation and forest degradation which significantly contribute to global warming and climate change, and triggering a cascade of events (Agan, 2024; Magazzino et al., 2023; Mele et al., 2022; Musah et al., 2021; Raggad, 2018; Sola et al., 2016).

The combustion of solid fuels generates numerous harmful air pollutants and toxins, which degrade and damage both indoor and outdoor air quality (Liu et al., 2016; Shen et al., 2019; Zhang et al., 2012; Zhao, 2018). Exposure to these pollutants undeniably impacts public health, leading to premature mortality (Archer-Nicholls et al., 2016; Ebenstein et al., 2017; Lelieveld et al., 2015). Global statistics indicate that in 2019, indoor and outdoor air pollution contributed to the premature deaths of about 5.4 million people (IEA, 2020), with associated healthcare costs reaching an astronomical \$2.9 trillion in 2018 (CREA, 2018).

To effectively mitigate these human and economic costs that will result to enhanced indoor and outdoor air quality, transitions to clean energy remains the way out (Aunan et al., 2018; Barrington-Leigh et al., 2019; Cohen et al., 2017; Wang, 2017), giving that the consequent health benefits are extensive and direct, echoed by marked reductions in morbidity and premature mortality (Zhao, 2018). Despite substantial health benefits (WHO, 2018), clean energy remains unacceptably expensive and less accessible, particularly in developing countries like Nigeria.

About three billion people are still heavily dependent on polluting energy sources such as coal, charcoal, firewood, and crop residues rather than cleaner alternatives and sources (Cameron, 2016; WHO, 2014). According to Smith et al. (2013), about two-fifths of the global population are exposed to air pollution as a result of inefficient and incomplete combustion of solid fuels used for cooking and heating. Gallagher and Holloway (2020) highlight that this damaging exposure to air pollution contributes to approximately 4.2 million deaths annually, constituting the largest environmental impact on human health. In addition, outdoor air pollution contributed to about 3.3 million deaths in 2010 (Wang & Orris, 2015). The notable health burden from exposure to these poorly combusted fuels, notably black carbon and carbon dioxide, underscore their severe human health impacts (Bridges et al., 2015; Hirschberg et al., 2016; Kim et al., 2020). In the energy sector, renewable energy sources (RESs) and energy efficiency (EE) represent promising avenues for achieving emissions reduction goals (Wang & Orris, 2015). These sources do not emit harmful pollutants or, when considering their full life cycle, their emissions are significantly fewer than traditional fossil fuels. Therefore, integrating RESs into the energy mix can lead to a substantial decrease in life-threatening emissions and subsequently reduce health damages linked to pollution. Silva et al. (2021) opined that promoting access to non-polluting and sustainable energy has considerable potential to enhancing public health outcomes. This effort is particularly critical in addressing health issues related to poor indoor air quality prevalent in developing countries due to the combustion of traditional fuels for cooking and heating. Thus, replacing these energy traditional fuels with clean and reliable alternatives is essential for curbing these associated health challenges.

The central objective of climate and energy policy should be to foster innovation that ensures energy is both environmentally friendly and economically competitive. Moreover, clean energy innovations substantially contribute to economic development by improving productivity (Gyedu et al., 2021; Mughal et al., 2022), fostering innovation and creativity, creating a competitive advantage in the global market (Ahmad et al., 2021; Li & Solaymani, 2021), generating new job opportunities, driving infrastructure development, leading to cross-sectoral spillover effects, expanding global trade opportunities and ultimately improving the standard of living through advancements in various sectors such as education, communication and healthiness (Zhang et al., 2023).

This research significantly contributes to the burgeoning fields of energy, health, and environmental economics by highlighting on their interplay, notably, the implementation of green energy policies while robustly emphasizing the role of greenhouse gas emissions in sustainable development pursuit of developing countries, with Nigeria in view. By utilizing the research outcomes, the study provides justifiable guidance on how national energy policies could be effectively aligned with the goals of the Paris agreement and Sustainable Development Goals (SDGs). The aim of this paper is therefore to synthesize documented evidence of the impact of clean and reliable energy on mortality and emphasize the mechanisms by which these impacts manifest, as documented in the literature. This research is strongly driven by the existing knowledge gap in understanding the interaction between these two critical development concerns—clean energy transition and public health outcomes. The author strongly believes that this research is a leading attempt to synthesize available efforts in this area. This study is viewed from the Nigeria perspective, which has the

Table 8.1 Choice variables, definitions, measurements and sources

Variables	Definition	Measurement	Source
DTH	Deaths from air pollution	Total deaths caused by air pollution from solid fuels' combustion	WDI
ACE	Electricity access	Percentage of population with electricity access	WDI
BC	Black carbon emissions	Metric tons per capita	ourworldindata.org
CO_2	CO_2 emissions	Metric tons per capita	ourworldindata.org
RE	Consumption of renewable energy	Percentage of total final energy consumption	ourworldindata.org
CE	Access to clean fuels and technologies for cooking	(% of population)	ourworldindata.org

Source The author's compilation

third highest emission of CO_2 in Africa and a very low consumption of renewable energy (FAO, 2015; IEA, 2020).

8.2 Description of Data and Estimation Procedure

8.2.1 Description of Data

This study evaluates the nexus between clean energy utilization and mortality (measured by number of deaths caused by air pollution). Data used for the study included access to electricity, black carbon emissions, CO_2, consumption of renewable and access to clean cooking fuels and tech technologies. These data were sourced from various databases for the period of 1990–2022 and they are presented in Table 8.1.

8.2.2 Estimation Procedure

8.2.2.1 The Autoregressive Distributed Lag

The ARDL model specified by Pesaran et al. (2001) was adopted in this study to assess the long-term interrelationship between the variables and stated as:

$$\Delta \text{DTH}_t = \delta_0 + \sum_{i=1}^{p} \delta_1 \Delta \text{DTH}_{t-1} + \sum_{i=1}^{p} \delta_2 \Delta \text{ACE}_{t-1} + \sum_{i=1}^{p} \delta_3 \Delta \text{BC}_{t-1}$$

$$+\sum_{i=1}^{p}\delta_4\Delta CO_{2t-1}+\sum_{i=1}^{p}\delta_5\Delta RE_{t-1}+\sum_{i=1}^{p}\delta_6\Delta CE_{t-1}$$
$$+\sum_{i=1}^{p}\omega_1\Delta DTH_{t-1}+\sum_{i=1}^{p}\omega_2\Delta ACE_{t-1}+\sum_{i=1}^{p}\omega_3\Delta BC_{t-1}$$
$$+\sum_{i=1}^{p}\omega_4\Delta CO_{2t-1}+\sum_{i=1}^{p}\omega_5\Delta RE_{t-1}+\sum_{i=1}^{p}\omega_6\Delta CE_{t-1}+U_t \tag{8.1}$$

where

Δ	the first-difference operator,
δ's and ω's	long run and short run coefficients, respectively.

The null hypothesis of no cointegration is that H_0: $\delta_1=\delta_2=\delta_3=\delta_4=\delta_5=\delta_6=0$ against the alternative hypothesis H_1: $\delta_1\neq\delta_2\neq\delta_3\neq\delta_4\neq\delta_5\neq\delta_6\neq 0$. To establish a cointegrating relationship among the variables, the F-statistic is used. Following the tabulation of the critical bounds by Pesaran et al. (2001), two bounds exist—the upper critical bounds (UCB) and low critical bounds (LCB) following the assumption that all variables are I(1) and I(0) respectively. There are three possible outcomes; (i) accept the null hypothesis of no cointegration if the UCB is lower than the F-statistic, (ii) reject the null hypothesis of no cointegration if the LCB is higher than the F-statistic and (iii) an inconclusive outcome exists if the F-statistic lies between UCB and LCB.

8.2.2.2 Error Correction Model (ECM)

The ECM version of ARDL model, following the short run dynamics form stipulates:

$$\Delta DTH_t=\delta_0+\sum_{i=1}^{p}\delta_1\Delta DTH_{t-1}+\sum_{i=1}^{p}\delta_2\Delta ACE_{t-1}+\sum_{i=1}^{p}\delta_3\Delta BC_{t-1}$$
$$+\sum_{i=1}^{p}\delta_4\Delta CO_{2t-1}+\sum_{i=1}^{p}\delta_5\Delta RE_{t-1}+\sum_{i=1}^{p}\delta_6\Delta CE_{t-1}$$
$$+\eta ECT_{t-1}+U_t \tag{8.2}$$

ECT_{t-1}	EC term lagged by one period,
η	coefficient of the EC term.

The Error Correction Term (ECT) depicts short-term adjustments and accounts for the rate of return to equilibrium. Typically, the ECT value as a general rule is expected to range between − 1 and 0 and should be negatively significant. The goodness of fit of the ARDL model's fit was assessed using stability tests of OLS cumulative sum.

8.3 Empirical Results and Discussion

8.3.1 Summary Statistics of Research Data

Table 8.2 is a summary of the variables that are hypothesized to influence the mortality as a result of air pollution (combustion of solid fuels) in Nigeria from 1990 to 2022.

Findings showed that the value observed for renewable energy, RE is 2,466,068, ranging between 2,225,275 and 4,320,767. Death from air pollution (DAP) has a mean value of 154,279.8, minimum and maximum values of 128,259.3 and 197,166.5 respectively. Access to clean electricity (ACE) ranges from 34.619 to 59.3, with a mean of 37.236. Black carbon (BC) falls between 110,631.8 and 201,742.6, with a mean of 158,513.4. Clean energy ranges from 0.9 to 16.8, with a mean of 3.070 while carbon dioxide (CO_2) has a minimum and maximum values of 0.537 and 0.916 respectively and a mean value of 0.644.

Skewness measures the degree of asymmetry or departure of the variables from a normal distribution centered around zero while Kurtosis measures the degree of flatness or peakness of a series. The results for skewness showed that death from air pollution, access to electricity and carbon dioxide were negative, implying a leftward skew while renewable energy, black carbon and clean energy were all positively skewed. However, the value of black carbon was close to zero (0), suggesting a relatively symmetric distribution. Similarly, the Kurtosis result showed that all the variables were positive with renewable energy, access to electricity and black carbon all platykurtic. Death from air pollution, clean energy and carbon dioxide were all leptokurtic, since their values were above 3.

8.3.2 Unit Root Test of the Variables

To evaluate the stochasticity of the data, unit root tests were conducted. The conventional tests, namely Augmented Dickey–Fuller (ADF) and the Phillips–Perron (PP) were employed to examine the stationarity of the series. The ADF and PP test statistics were compared with their respective critical values at 5% level, thus, variables that were significant at the 5% level region were selected for further analyses as shown in Table 8.3.

The ADF and PP results are similar with insignificant differences. The results showed that ACE, BC, CO_2, DTH and RE were all stationary at first difference while CE was stationary at level. Given that the series were either stationary at level or the first difference, the data are suitable for ARDL estimations. In addition, Table 8.4 shows the VAR Lag Selection Criteria, necessary for the selection of correct lag orders for the ARDL analyses.

Table 8.2 Summary Statistics for the study variables (1990–2022)

Variable	N	Mean	Minimum	Maximum	Std. dev.	Skewness	Kurtosis
Death from pollution	33	154,279.800	128,259.300	197,166.500	53,352.910	− 2.180	6.853
Renewable energy	33	2,466,068.000	2,225,275.00	4,320,767.000	1,436,242.000	− 0.759	2.367
Access to electricity	33	37.236	34.619	59.300	19.275	− 1.133	2.916
Black carbon	33	158,513.400	110,631.800	201,742.600	29,353.530	0.081	1.631
Clean energy	33	3.070	0.900	16.800	4.639	1.731	4.868
Carbon dioxide	33	0.644	0.538	0.916	0.204	− 1.727	6.757

Source Authors' computation

Table 8.3 Unit root test of variables

Variable	Augmented Dickey–Fuller (ADF)				Phillips–Perron (PP)				Decision
	Level	5%	1st diff	5%	Level	5%	1st diff	5%	
ACE	− 1.095	− 2.957	− 6.251	− 2.960	− 1.110	− 2.957	− 6.251	− 2.960	I(1)
BC	− 1.097	− 2.957	− 4.841	− 2.960	− 1.077	− 2.957	− 4.830	− 2.960	I(1)
CE	− 5.850	− 2.964	− 4.617	− 2.972	− 1.956	− 2.957	− 0.803	− 2.960	I(0)
CO_2	0.165	− 2.957	− 5.347	− 2.960	0.620	− 2.957	− 5.369	− 2.960	I(1)
DTH	0.713	− 2.957	− 5.363	− 2.960	1.147	− 2.957	− 5.363	− 2.960	I(1)
RE	− 1.047	− 2.957	− 6.137	− 2.998	− 1.146	− 2.957	− 5.518	− 2.960	I(1)

Source Authors' computation

Table 8.4 VAR lag order selection criteria

Lag	Log L	LR	FPE	AIC	SC	HQ
0	− 342.7226	NA	7.35e+08	23.24817	23.52841	23.33783
1	− 326.9873	7.397741*	3.19e+08*	22.39915*	22.81951*	22.53363*

* represents the selected lag order. *Source* Authors' computation

8.3.3 Bounds Test

Following the determination of the integration order of the model series, the bound testing approach was estimated to test for the presence of a co-integrating (long run) relationship among the series. Ordinary least squares (OLS) regression was employed to ascertain the presence or not of co-integration. Subsequently, an F-test was carried out to examine the hypothesis of no co-integration between the series. The F-statistic was also employed to test for a long-run relationship among the variables. If the computed F-statistic exceeds the UCV, the null hypothesis of no co-integration is rejected and vice-versa. Conversely, if the F-statistic falls within the range of the critical value, the results are inconclusive. The bounds test result is presented in Table 8.5.

Findings show that the calculated F-statistic (5.052) is greater than the upper and lower bond values of 3.41 and 4.68 respectively, statistically significant at 1%, which suggests the existence of a long run relationship among the series. In essence, the ECM and ARDL models can be estimated.

8.3.4 ARDL Estimations

To examine the effect of ACE, BC, CE, CO_2 and RE on public health outcomes (measured by mortality accounted for by air pollution) in Nigeria, ARDL model was employed. The result is presented in Table 8.6.

Table 8.5 Co-integration bounds test

Test statistic	Value	Significance (%)	I(0)	I(1)
			Asymptotic: n = 1000	
F-statistic	5.051	10	2.26	3.35
k	5	5	2.62	3.79
		2.5	2.96	4.18
		1	3.41	4.68
t-bounds test		Null hypothesis: no levels relationship		
t-statistic	– 4.981	10	– 2.57	– 3.86
		5	– 2.86	– 4.19
		2.5	– 3.13	– 4.46
		1	– 3.43	– 4.79

Source Author's computation

Table 8.6 ARDL estimates

Variables	Long run			Short run		
	Coeff.	t-stat	*p* value	Coeff.	t-stat	*p* value
Constant				12.321	3.46***	0.02
Ln Dth	– 0.393	– 2.98***	0.006			
Ln RE	– 0.601	– 0.97	0.340	0.296	1.17	0.254
Ln ACE	1.188	2.34**	0.028	0.468	2.16**	0.041
Ln BC	1.237	1.80*	0.085	– 0.487	– 1.72*	0.099
Ln CE	– 0.084	– 2.33**	0.015	– 0.033	– 1.57	0.131
Ln CO_2	0.178	1.72	0.061*	0.070	0.62	0.541

Note Dependent variable is DTH. The symbols ***, **, and * 0.01, 0.05, and 0.1 respectively
Source Authors' computation

Findings showed a 1% negative long run relationship between mortality accounted for by air pollution (DTH) and the first lags DTH. There was also a 5% positive relationship between ACE and DTH at both LR and SR. Black carbon (BC) was positively related to death through air pollution (DTH) at 10%, implying that exposure to black carbon is directly linked to deaths from air pollution. In addition, the coefficient of clean energy (CE) was negative in the LR (suggesting that households with access to clean cooking energy are less likely to die from air pollution) while CO_2 was positive in the same period, at 5% and 10% levels respectively. The positive coefficient of CO_2 suggests that people who are exposed to inhaling CO_2 are likely to die from it. This long run effect entails that carbon dioxide is a gradual killer. The use of renewable energy, RE, led to a reduction in the number of deaths due to air pollution in the long run.

According to the Global Burden of Disease (GBD) analysis conducted in 2019, outdoor air pollution, particularly in the form of black carbon, poses a much greater public health concern than previously understood, with an annual contribution to over 4.3 million premature deaths globally, an unacceptably chunk of years of healthy life lost (Chowdhury et al., 2022; Murray et al., 2020). Documented evidence showed that long-term exposure to black carbon has been strongly linked to a wide range of damaging effects on human health, such as increased mortality from cardiovascular diseases, cerebrovascular diseases, acute lower respiratory tract infections, diabetes, lung cancer, adverse birth outcomes, and neonatal diseases (Cohen et al., 2017; Lelieveld et al., 2015; Murray et al., 2020). As a component of air pollution, black carbon significantly contributes to more than 8 million premature deaths per year, including over 500,000 deaths of children under five each year (Clean Air Fund, 2023). It also contributes to an estimated $8.1 trillion in annual global health costs, a 6.1% reduction in global GDP and 1.2 billion workdays lost globally each year (Clean Air Fund, 2023). Economically and socially disadvantaged groups, including ethnic and racial minorities, are most likely to be exposed to black carbon and suffer its health impacts. For example, women and children within low-income regions and households are more exposed to pollution from kerosene lamps and stoves (Curto, 2019). Low-income communities are often more likely to live close to busy roads with air pollution from diesel trucks and buses and suffer health impacts as a result (Khan, 2024).

At higher levels, extended exposure to carbon dioxide concentrations exceeding 5000 ppm can cause symptoms such as headaches, dizziness, and nausea. At even elevated levels, CO_2 can lead to a drastic decline of the level of oxygen in the bloodstream, posing a risk of asphyxiation. Documented evidence suggests that very high exposure to environmentally relevant increases in atmospheric CO_2 is hazardous and could be responsible for terminal health challenges like inflammation, reduced cognitive function, bone demineralization, kidney calcification, oxidative stress, and endothelial dysfunction (Jacobson et al., 2019). However, chronic exposure to elevated atmospheric CO_2 could therefore represent an underappreciated stressor in today's and future environments.

Households with access to clean cooking energy and technologies have less exposure to air pollutants and are less likely to die from it. The use of clean cookstoves technologies significantly reduces exposure to black carbon and other harmful air pollutants in households, thereby enhancing both health outcomes and women's well-being globally, since they are mostly involved in preparation of meals (CCAC, 2020). Each year, about 3.8 million premature deaths globally are related to exposure to smoke from traditional cooking fuels (WHO, 2018). The use of inefficient cooking stoves or burning fuels like dung, wood, and coal in open hearths exposes primary cooks and household members to these air pollutants which have been linked to serious health conditions such as lung cancer, stroke, heart attacks, chronic obstructive pulmonary disease, pneumonia, and asthma (WHO, 2021). Thus, to reduce maternal mortality and under-five mortality rates in affected regions, increasing access to clean fuels and improved cooking technologies is inevitable.

It is pertinent to highlight that clean fuels have a low carbon intensity, that is, they emit minimal to no greenhouse gases or pollutants into the atmosphere. Consequently, the utilization of clean fuels, along with access to improved cooking technologies and electricity, plays a pivotal role in reducing all forms of mortality relating to air pollution. Previous research have consistently demonstrated that the use of clean energy improves public health outcomes by reducing mortality due to air pollution, which aligns with the findings of this study (Shobande, 2020; Wang, 2019; Zhu et al., 2023).

The findings further support the narrative that increased consumption of renewable energy enhances life expectancy and this corroborates the efforts of Hanif (2018) which similarly stresses the beneficial impacts of renewable energy on public health outcomes, such as reduced mortality rates and tuberculosis incidences. Wang (2019) also agrees with these benefits, highlighting a notable decrease in maternal mortality due to the utilization of clean technologies in China. Renewable energy therefore contributes to enhanced life expectancy by reducing the amount of carbon emitted into the atmosphere while ensuring environmental quality. Conversely, environmental pollution, as identified by Majeed and Luni (2020) and Byaro et al. (2024), diminishes healthy and quality living.

8.4 Diagnostic Tests

8.4.1 Cumulative Sum Test for Parameter Stability

The CUSUM test is was used to detect structural breaks or changes in the parameters of the ARDL regression model over time and presented in Table 8.7.

The result showed that the cumulative sum (test statistic) was less than the critical values at all levels, thus, the null hypothesis of parameter stability (no structural break) is accepted, and we conclude that the relationship between the dependent and independent variables are stable over time. This confirms that the ARDL model's parameters are stable over time, and the model can be relied upon for predictions or policy conclusions.

Table 8.7 Cumulative sum test for parameter stability

Type	Test statistic	Critical value		
		1%	5%	10%
OLS	0.7757	1.6276	1.3581	1.2238

Table 8.8 Heteroskedasticity and omitted variables test

Test	Value
Breusch–Pagan–Godfrey test for heteroskedasticity	F-statistic = 0.017; *p*-value = 0.8964
Durbin–Watson stat (8, 32)	1.938604
Ramsey RESET test for omitted variables (1, 10)	0.5720
White's test of homoskedasticity	($p = 0.4167$)

8.4.2 *Heteroskedasticity, Autocorrelation and Omitted Variables Test*

The tests are presented in Table 8.8.

The null hypothesis of the heteroskedasticity is that there is no heteroskedasticity among the variables. The decision rule is to accept this hypothesis if the Chi-square value is significant and vice-versa. The result gave a Chi-square probability value of 0.8964 which is higher than the 0.05 probability level, thus the chi-square is considered insignificant and the null hypothesis of no heteroskedasticity, skewness and kurtosis is accepted. Similarly, the Ramsey RESET test for omitted variables showed an insignificant *p*-value of 0.5720, thus the need to accept the null hypothesis of no omitted variables. The Durbin–Watson (DW) statistic is used to detect the presence of autocorrelation (serial correlation) in the residuals of a regression analysis. Specifically, it tests for first-order autocorrelation, which occurs when the residuals from one time period are correlated with the residuals from the previous period. A Durbin–Watson statistic of 1.939 in the context of an Autoregressive Distributed Lag (ARDL) model suggests that there is no autocorrelation in the residuals. For the heteroskedasticity test, the *p*-value associated with the test statistic (0.4167) is **greater than** the 5% significance level (0.05). Thus, we **do not reject the null hypothesis of no homoskedasticity**, and therefore conclude that h**omoskedasticity is present**.

8.5 Conclusion, Recommendations and Limitations

This study examined the interrelationship between clean energy utilization and public health outcomes (measured by mortality attributed air pollution) in Nigeria. By conducting some time-series analyses from 1990 to 2022, the study employed the conventional unit root tests (PP and ADF) to ascertain the series' order of integration. To determine the short and long-term effects of these exogenous variables on the public health outcomes, the study utilized the ARDL estimator. It was revealed that the adoption of renewable, clean energy and access to clean cooking energy technologies increased public health outcomes by reducing the mortality attributed to air pollution while exposure to CO_2 and black carbon led to a decreased public health outcome by increasing mortality as a result of air pollution. From a Nigerian point of view, these findings provide novel insights into the impacts of green transition on public health outcomes in addition to its economic importance on environmental quality. Based on these study outputs, policy shifts that advocate for a proactive transitioning to a low-carbon economy and utilization of renewable energy sources are recommended. In the event of underutilization of clean fuels and cooking technologies in Nigeria, efforts towards prioritizing the utilization of clean fuels could assist the Nigerian government in wooing development financiers to support implementation. In addition, by increasing the proportion of renewables in the energy mix, emission reduction as well as reduction in pollution related healthcare costs would be simultaneously achieved. Therefore, the Nigerian government should initiate and expedite internal mechanisms that would subsidize the costs involved in accessing renewable energy which would serve as a contributory effort to establishing a low-carbon economy, globally.

A major issue challenging the determination of the exact health status of most developing countries is the unavailability of credible data. When they are available, researchers are expected to pay some huge amounts.

Acknowledgements The authors wish to acknowledge their individual and collective efforts towards the success of this manuscript. Special appreciation to Dr. Nwaogu who painstakingly developed the concept of the research and guided other authors' contributions to a great end.

Funding This work was solely funded by the authors.

Data Availability Statement The authors are willing to provide the data for this study to the editors and the publisher on request.

Conflict of Interest There are no conflicts of interests.

Ethical Clearance There are no ethical issues surrounding this study.

References

Agan, B. (2024). Assessing the dynamic impacts of climate change adaptation and clean energy innovations on economic growth in OECD countries. *Natural Resources Forum,* 1–24. https://doi.org/10.1111/1477-8947.12428

Ahmad, M., Jiang, P., Murshed, M., Shehzad, K., Akram, R., Cui, L., & Khan, Z. (2021). Modelling the dynamic linkages between eco-innovation, urbanization, economic growth and ecological footprints for G7 countries: Does financial globalization matter? *Sustainable Cities and Society, 70*(3), 102881.

Archer-Nicholls, S., et al. (2016). The regional impacts of cooking and heating emissions on ambient air quality and disease burden in China. *Environmental Science and Technology, 50*, 9416.

Aunan, K., Hansen, M. H., & Wang, S. (2018). Introduction: air pollution in China. *The China Quarterly*. https://doi.org/10.1017/S0305741017001369

Barrington-Leigh, C., Baumgartner, J., Carter, E., Robinson, B. E., Tao, S., & Zhang, Y. (2019). An evaluation of air quality, home heating and well-being under Beijing's programme to eliminate household coal use. *Nature Energy, 4*, 416–423.

Bridges, A., Felder, F. A., McKelvey, K., & Niyogi, I. (2015). Uncertainty in energy planning: Estimating the health impacts of air pollution from fossil fuel electricity generation. *Energy Research & Social Science, 6*, 74–77.

Byaro, M., Mmbaga, N. F., & Mafwolo, G. (2024). Tackling energy poverty: Do clean fuels for cooking and access to electricity improve or worsen health outcomes in Sub-Saharan Africa? *World Development Sustainability, 4*, 100125. ISSN 2772-655X. https://doi.org/10.1016/j.wds.2024.100125

Cameron, C. (2016). Policy trade-offs between climate mitigation and clean cook-stove access in South Asia. *Nature Energy, 1*, 15010.

CCAC. (2020). *Opportunities for 1.5°C consistent black carbon mitigation.*

Chowdhury, S., Pozzer, A., Haines, A., Klingmüller, K., Münzel, T., Paasonen, P., Sharma, A., Venkataraman, C., & Lelieveld, J. (2022). Global health burden of ambient $PM_{2.5}$ and the contribution of anthropogenic black carbon and organic aerosols. *Environment International, 159*. https://doi.org/10.1016/j.envint.2021.107020

Clean Air Fund. (2023). *The state of global air quality funding 2023*. Clean Air Fund.

Cohen, A. J., et al. (2017). Estimates and 25-year trends of the global burden of disease attributable to ambient air pollution: An analysis of data from the Global Burden of Diseases Study 2015. *Lancet, 389*, 1907–1918.

CREA. (2018). *Quantifying the economic costs of air pollution from fossil fuels.*

Curto, A. (2019). Predictors of personal exposure to black carbon among women in southern semirural Mozambique. *Environment International, 131*, 104962.

Ebenstein, A., Fan, M., Greenstone, M., He, G., & Zhou, M. (2017). New evidence on the impact of sustained exposure to air pollution on life expectancy from China's Huai River Policy. *Proceedings of the National Academy of Sciences of the United States of America, 114*, 10384–10389.

FAO. (2015). *Regional overview of food insecurity: African food insecurity prospects brighter than ever*. FAO.

Gallagher, C. L., & Holloway, T. (2020). Integrating air quality and public health benefits in U.S. decarbonization strategies. *Frontiers in Public Health, 8*.

Gyedu, S., Heng, T., Ntarmah, A. H., He, Y., & Frimppong, E. (2021). The impact of innovation on economic growth among G7 and BRICS countries: A GMM style panel vector autoregressive approach. *Technological Forecasting and Social Change, 173*(3), 121169.

Hanif, I. (2018). Energy consumption habits and human health nexus in Sub-Saharan Africa. *Environmental Science and Pollution Research, 25*(22), 21701–21712.

Hirschberg, S., Bauer, C., Burgherr, P., Cazzoli, E., Heck, T., Spada, M., & Treyer, K. (2016). Health effects of technologies of power generation: Contributions from normal operation, severe accidents and terrorist threat. *Reliability Engineering and System Safety, 145*, 373–387.

IEA. (2020). *World energy outlook 2020.*

IEA. (2021). *World energy balance highlights.*

Jacobson, T. A., Kler, J. S., & Hernke, M. T. (2019). Direct human health risks of increased atmospheric carbon dioxide. *Nature Sustainability, 2*, 691–701. https://doi.org/10.1038/s41893-019-0323-1

Khan, R. N. (2024). Traffic-related air pollution in marginalized neighborhoods: A community perspective. *Inhalation Toxicology, 15*, 1–12. https://doi.org/10.1080/08958378.2024.2331259

Kim, C., Henneman, L. R. F., Choirat, C., & Zigler, C. M. (2020). Health effects of power plant emissions through ambient air quality. *Journal of the Royal Statistical Society Series A: Statistics in Society, 183*, 1677–1703.

Lelieveld, L., Evans, J. S., Fnais, M., Giannadaki, D., & Pozzer, A. (2015). The contribution of outdoor air pollution sources to premature mortality on a global scale. *Nature, 525*, 367–371.

Li, Y., & Solaymani, S. (2021). Energy consumption, technology innovation and economic growth nexuses in Malaysian. *Energy, 232*, 121040.

Liu, J., Mauzerall, D. L., Qi, C., Qiang, Z., & Zhu, T. (2016). Air pollutant emissions from Chinese households: A major and underappreciated ambient pollution source. *Proceedings of the National Academy of Sciences of the United States of America, 113*, 7756–7761.

Magazzino, C., Mele, M., Drago, C., Kuşkaya, S., Pozzi, C., & Monarca, U. (2023). The trilemma among CO_2 emissions, energy use, and economic growth in Russia. *Scientific Reports, 13*(1), 10225.

Majeed, M. T., & Luni, T. (2020). Renewable energy, circular economy indicators and environmental quality: A global evidence of 131 countries with heterogeneous income groups. *Pakistan Journal of Commerce and Social Sciences, 14*(4), 866–912.

Mele, M., Magazzino, C., Schneider, N., Gurrieri, A. R., & Golpira, H. (2022). Innovation, income, and waste disposal operations in Korea: Evidence from a spectral granger causality analysis and artificial neural networks experiments. *Economia Politica, 39*(2), 427–459.

Mughal, N., Arif, A., Jain, V., Chupradit, S., Shahzad, M., Ramos-Meza, C. S., & Zhanbayev, R. (2022). The role of technological innovation in environmental pollution, energy consumption and sustainable economic growth: Evidence from South Asian economies. *Energy Strategy Reviews, 39*, 100745.

Murray, C. J. L., Aravkin, A. Y., Zheng, P., Abbafati, C., Abbas, K. M., Abbasi-Kangevari, M., Abd-Allah, F., Abdelalim, A., Abdollahi, M., Abdollahpour, I., Abegaz, K. H., Abolhassani, H., Aboyans, V., Abreu, Z.-J., Zhao, J. T., Zhao, X.-J.G., Zhao, Y., Zhou, M., Ziapour, A., … Lim, S. S. (2020). Global burden of 87 risk factors in 204 countries and territories, 1990–2019: A systematic analysis for the Global Burden of Disease Study 2019. *Lancet, 396*, 1223–1249. https://doi.org/10.1016/S0140-6736(20)30752-2

Musah, M., Kong, Y., Mensah, I. A., Antwi, S. K., Osei, A. A., & Donkor, M. (2021). Modelling the connection between energy consumption and carbon emissions in North Africa: Evidence from panel models robust to cross-sectional dependence and slope heterogeneity. *Environment, Development and Sustainability, 45*(1), 0123456789.

Pesaran, M. H., Shin, Y., & Smith, R. J. (2001). Bounds testing approaches to the analysis of level relationships. *Journal of Applied Econometrics, 16*(3), 289–326. https://doi.org/10.1002/jae.616

Raggad, B. (2018). Carbon dioxide emissions, economic growth, energy use, and urbanization in Saudi Arabia: Evidence from the ARDL approach and impulse saturation break tests. *Environmental Science and Pollution Research, 25*(15), 14882–14898.

Shen, G., Ru, M., Du, W., Zhu, X., & Tao, S. (2019). Impacts of air pollutants from rural Chinese households under the rapid residential energy transition. *Nature Communications, 10*, 1–8.

Shobande, O. A. (2020). The effects of energy use on infant mortality rates in Africa. *Environmental and Sustainability Indicators, 5*, 100045. https://doi.org/10.1016/j.indic.2019.100015

Silva, S., Laranjeira, E., & Soares, I. (2021). Health benefits from renewable electricity sources: A review. *Energies, 14*, 6678. https://doi.org/10.3390/en14206678

Smith, K., Frumkin, H., Balakrishnan, K., Butler, C., Chafe, Z., Fairlie, I., Kinney, P., Kjellstrom, T., Mauzerall, D. L., & McKone, T. E. (2013). Energy and human health. *Annual Review of Public Health, 34*, 159–188.

Sola, P., Ochieng, C., Yila, J., & Iiyama, M. (2016). Links between energy access and food security in Sub-Saharan Africa: An exploratory review. *Food Security, 8*, 635–642. https://doi.org/10.1007/s12571-016-0570-1

Wang, J., & Orris, P. (2015). *Healthy energy initiative. The health impacts of energy choices: A briefing paper for the health community; Health care without harm*. Brussels, Belgium.

Wang, S. (2017). Natural gas shortages during the "coal-to-gas" transition in China have caused a large redistribution of air pollution in winter 2017. *Proceedings of the National Academy of Sciences of the United States of America, 117*, 31018–31025.

Wang, S. (2019). Residential energy consumption and its linkages with life expectancy in mainland China: A geographically weighted regression approach and energy-ladder-based perspective. *Energy, 177*, 347–357. https://doi.org/10.1016/j.energy.2019.04.099

WHO. (2014). *Burden of disease from household air pollution for 2012: Summary of results.* Public Health, Social and Environmental Determinants of Health Department, World Health Organization.

WHO. (2018). *Household air pollution and health: Key facts.*

Zhang, Q., He, K., & Huo, H. (2012). Cleaning China's air. *Nature, 484*, 161–162.

Zhang, T., Ma, Z., & Shang, Y. (2023). Higher education, technological innovation, and green development—Analysis based on China's provincial panel data. *Sustainability, 15*(5), 1–18.

Zhao, B. (2018). Change in household fuels dominates the decrease in $PM_{2.5}$ exposure and premature mortality in China in 2005–2015. *Proceedings of the National Academy of Sciences of the United States of America, 115.*

Zhu, L., Liao, H., & Burke, P. J. (2023). Household fuel transitions have substantially contributed to child mortality reductions in China. *World Development, 164*, 106174. https://doi.org/10.1016/j.worlddev.2022.106174

Chapter 9
Climate Smart Water Conservation Strategy for Sustainable Development in Nigeria: A Systematic Review

Sunday Olufemi Adesogan and Blessing Funmbi Sasanya

Abstract Water scarcity is a growing concern in Nigeria, threatening sustainable development and human well-being. Effective water conservation strategies are crucial to tackle this challenge. This systematic review examines the existing literature on the present and future directions of climate-smart water conservation strategies in Nigeria by addressing their strengths, weaknesses, and potential for sustainable development. The review aims to identify effective approaches, challenges, and opportunities for improving water management in the face of climate change. A comprehensive search of peer-reviewed articles, government reports, and grey literature was conducted. Themes and sub-themes addressed in the study were determined and an average of ten relevant literatures were included to address each sub-theme. The findings highlight the need for integrated water resources management, efficient irrigation systems, use of water saving technology, design and construction of climate resilient water infrastructure as well as community-led initiatives. The review also emphasizes the importance of addressing institutional, financial, and technical barriers to successful implementation. The need for a coordinated approach to ensure sustainable development using the climate-smart water conservation framework in Nigeria is of utmost importance.

Keywords Water conservation · Climate resilience · Water saving technologies · Water resources management · Climate change

S. O. Adesogan
Department of Civil Engineering, University of Ibadan, Ibadan, Nigeria
e-mail: sunday.adesogan@ui.edu.ng

B. F. Sasanya (✉)
Department of Crop and Soil Science, University of Port Harcourt, Choba, Rivers State, Nigeria
e-mail: blessing.sasanya@uniport.edu.ng

Department of Civil and Environmental Engineering, Morgan State University, Baltimore, Maryland, USA

R. U. Onyeneke et al. (eds.), *Energy Transition, Climate Action and Sustainable Agriculture*, https://doi.org/10.1007/978-3-031-83165-2_9

9.1 Introduction

Water is a vital natural resource that underpins human existence. Despite significant advancements in human development, no alternative to water has been discovered thus making it an indispensable resource. However, abundant water resources are becoming increasingly scarce, impacting sectors like health, food production, and natural ecosystems (Mishra, 2023). Although 71% of the Earth's surface is water-covered, a staggering 96% of this is found in oceans, leaving only 2.4% in rivers, ponds, and lakes, and approximately 1.6% in groundwater aquifers. A mere 0.001% exists as vapour clouds and precipitation (Adesogan, 2014). Water plays crucial roles in agriculture and irrigation, heat removal, sanitation, hygiene, and most importantly, human and animal consumption, maintaining good health balances and supporting numerous other uses (Vinci et al., 2021). Despite its immense importance, a minimal amount of freshwater exists on the Earth's surface, highlighting the need for responsible management and conservation (Ahmed et al., 2022).

The importance of sustainable water supply, with both sufficient quantity and good quality, cannot be overstated, particularly in rural communities of developing countries where water availability is dwindling. Access to fresh and potable water for domestic and industrial uses is essential. Among the myriad challenges facing developing countries, including poverty, energy shortages, food insecurity (hunger and malnutrition), and climate change, water scarcity stands out as a critical issue, because it intersects with and exacerbates all other challenges (He et al., 2021; Molden, 2020). Water scarcity or unavailability often has severe consequences, including health complications such as cholera, dysentery, malaria, fever, and gastroenteritis (Moslen et al., 2024). Furthermore, it has significant socio-economic impacts, including loss of productivity, violence against women and children, delayed school attendance, and poor sanitation (Algur et al., 2021; Sasanya & Adesogan, 2021). The far-reaching effects of water scarcity, particularly during dry seasons in rural communities, necessitate immediate attention and action. The situation demands urgent intervention to mitigate suffering and ensure access to this essential resource.

Water scarcity poses a significant threat globally, especially in rural areas of developing nations such as Nigeria during annual dry seasons, typically spanning from November to March and recently more exacerbated by climate change (Ayejoto et al., 2023). During these dry periods, community members are forced to walk long distances to queue for limited water, often of poor quality (Sarkar, 2020; Sasanya & Adesogan, 2021). Some also await the replenishment of dry hand-dug wells before gaining access to water, especially during dry seasons (Sasanya & Adesogan, 2021; Walker, 2022). Furthermore, surplus water during wet seasons is often compromised due to overland and sub-surface flow into both protected and unprotected water wells, thus highlighting the need for effective water management strategies (Nanekely, 2020; Walker, 2022).

Water availability fluctuates with seasons, geology, and geographical locations, affecting rural populations who rely heavily on surface and groundwater for their daily needs (Mishra, 2023). Various methods, including groundwater abstraction, stream damming, and rainwater harvesting, have been explored to address these issues. However, economic constraints and unfavourable geological formations can limit the adoption of these methods in rural communities of developing countries (Wang et al., 2020). In Nigeria, for instance, 60% of the population rely on ponds, streams, and rivers for drinking water (Isukuru et al., 2024; Omotosho & Sangodoyin, 2013). Unfortunately, untreated industrial wastewater often flows unchecked into watercourses, leading to surface and groundwater pollution, increased Biochemical Oxygen Demand (BOD) in freshwater habitats, and high mortality rates among aquatic life, among other detrimental effects (Jan et al., 2022).

This review therefore addresses the following research questions: (i) What is the current state of water insecurity in Nigeria? (ii) Which water sources in Nigeria are overexploited, underutilised, or currently in use? (iii) What measures are currently in place in Nigeria to combat water scarcity amidst worsening climate change? (iv) How is climate change affecting Nigeria's water resources, and how can these impacts be mitigated? (v) What water conservation techniques are available to address climate change? The review aims to identify effective approaches, challenges, and opportunities for improving water management in the face of climate change. This review document serves as a valuable resource for assessing Nigeria's adaptation to climate change in terms of water resource management and distribution.

9.2 Methodology

This systematic review was conducted in accordance with the guidelines set out by the Preferred Reporting Items for Systematic Reviews and Meta-Analyses (PRISMA) (Parums, 2021). To avoid duplication and reinventing the wheel, a search of the review title was carried out on reputable databases, including Google Scholar and ResearchGate. Mendeley, a free reference management software, was utilised to screen, cite, and manage all references included in this study.

9.3 Search Strategy, Inclusion and Exclusion

As the searches were conducted across selected databases, numerous articles were yielded. However, the following criteria were applied to determine the inclusion or exclusion of studies for this research: (i) Studies focused on and addressed issues related to water scarcity, risks and conservation practices. (ii) Only studies published after 2020, that is, not more than four years ago, were included. (iii) Only studies published in peer-reviewed journals managed by reputable publishers were considered. (iv) Articles were required to be fully published in the English language.

Literature searches were conducted on online platforms and repositories between May and August 2024. The online platforms utilised included Google Scholar and ResearchGate. The following search phrases were used, enclosed in inverted commas: "Water Conservation in Africa", "Water Conservation for Sustainable Development", "Climate Smart Water Conservation Techniques OR Strategy" "Water Scarcity" OR "Water Insecurity" AND "Water Conservation for Sustainable Development", "Integrated Water Resources Management", "Climate Resilience", "Ecosystem Services", "Efficient irrigation System". As a standard criterion, the publication year of selected peer-reviewed articles was restricted to between 2020 and 2024. The search was geographically limited, as this study focused specifically on Nigeria and by extension, Africa.

The selected references from all databases were downloaded and imported into Mendeley, a reference management software, for efficient organisation. To ensure accuracy, the citing, listing, and management processes were also manually cross-checked to prevent duplication and formatting errors. The selected papers underwent further screening by two independent researchers, who reviewed the title and abstract of each research article. The researchers worked independently, but their selections aligned, as they had previously agreed on the review objectives to be addressed in this study. Consequently, there were no discrepancies in the selected articles. The systematic selection process utilized specific inclusion and exclusion criteria to refine the initial pool of articles, ultimately yielding the final set of 73 studies that informed the analysis, as illustrated in Fig. 9.1.

9.4 Data Analysis

Data analysis was conducted in two phases. In the first phase, two researchers independently extracted information from the selected articles, including the study's objective, focus, citation, and methodology. The second phase involved coding and thematic analysis, primarily focusing on theme development from peer-reviewed articles. The agreed-upon themes formed the basis for subtitles and subheadings in this review. Theme generation employed inductive models, contrasting to deductive approaches. The thematic analysis followed Thomas and Harden's (2008) three-stage framework: (i) Identifying phrases related to climate smart water conservation strategy for sustainable development within the selected articles' main findings, which were then grouped into themes and subthemes. (ii) Categorising related findings into themes and subthemes. (iii) Interpreting and discussing the themes and subthemes in detail through analysis and inferences drawn from other related studies.

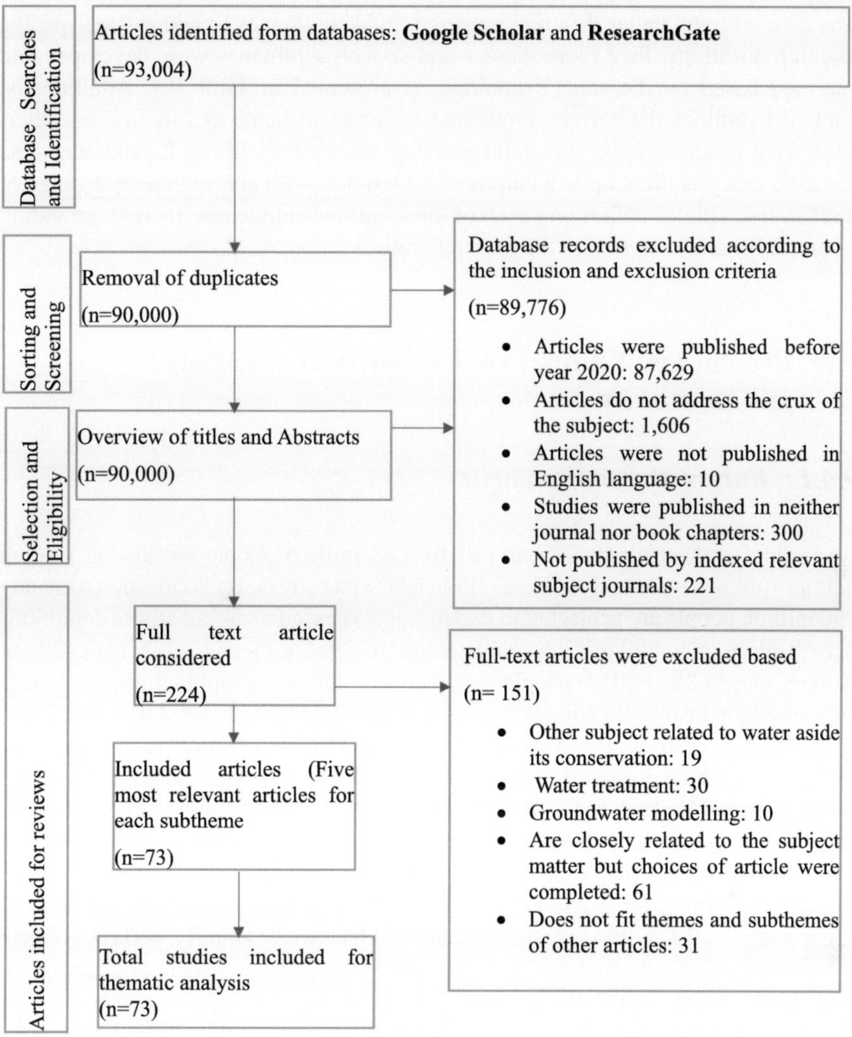

Fig. 9.1 Flow chart of article inclusion and exclusion criteria

9.5 Results and Discussion

9.5.1 Article Search and Selection

Searches conducted on Google Scholar and ResearchGate yielded 81,697 and 10,307 articles, respectively. After eliminating duplicates and irrelevant studies, the articles were narrowed down to 73. The titles, abstracts, and findings of the remaining 73 articles (following exclusions) were thoroughly reviewed and categorised. Each paper was then grouped under different predetermined subthemes to address the specific research questions. Five main themes and sixteen subthemes were developed and analysed based on the study's findings, as presented in Table 9.1. Additionally, Table 9.1 outlines the various subthemes under each theme and the relevant literature used to address the research questions associated with each subtheme. The thematic analysis drew upon a corpus of 73 articles, with approximately a range of three to five articles addressing each of the identified subthemes, thereby providing a comprehensive insight into the research topic.

9.6 Present and Future Problems of Water Scarcity and Insecurity

9.6.1 Future Water Insecurity

According to the United Nations (2016), 663 million people worldwide rely on unimproved water sources, while 2 billion face water stress. By 2030, approximately 250 million people are projected to live in water-stressed areas, and these conditions may displace up to 700 million people across Africa (du Plessis, 2023; Naik, 2017). Approximately 884 million people lack access to safe water supplies, and 2.4 billion are exposed to health risks due to inadequate sanitation (Sasanya & Adesogan, 2021; United Nations, 2016). By 2025, 2.8 billion people in 48 countries will experience water stress, which may worsen due to population growth (Lukenga, 2015; Marques, 2020; Neto & Camkin, 2022). By 2050, 54 countries and 4 billion people (9.4% of the projected global population) may face water stress (Zhao et al., 2024). African and Asian countries like Kenya, Egypt, Ethiopia, and Nigeria already experience annual water scarcity (Elhassnaoui et al., 2023; Peña-Ramos et al., 2022). Projections suggest that by 2025, 230 million Africans will face water scarcity, 460 million will live in water-stressed countries, and 31 countries (8% of the global population) will face chronic freshwater shortages, potentially resorting to energy-intensive seawater desalination (Shan et al., 2020). The global water crisis is largely attributed to population growth and increased water consumption, which has risen significantly from 383 km^3/year in the twentieth century to 1500 km^3/year in the twenty-first century (Gelfan et al., 2023; Kummu et al., 2016; Tzanakakis et al., 2020). These circumstances foreshadow a future of high-water insecurity for developing nations if left unattended to. Water security, defined as access to acceptable water quality and quantity for living, socio-economic development, sanitation, and ecosystem preservation in a peaceful and politically stable environment (Poudel, 2020), is crucial.

Table 9.1 Theme and subthemes with references reviewed in this study

S. No.	References (authors and year)	Theme	Subthemes
1.	du Plessis (2023) Neto and Camkin (2022) Poudel (2020) Sasanya and Adesogan (2021) Tzanakakis et al. (2020)	Present and future problems of water scarcity and insecurity	Future water insecurity
	Andrés et al. (2021) Baggio et al. (2021) Leal Filho et al. (2022) Lee et al. (2020) Shi et al. (2021)		Problems associated with water insecurity
2.	Danert and Healy (2021) Martínez-Santos et al. (2020) Rey-Romero et al. (2022) Suganthi and Elango (2020) Weatherl et al. (2021)	Water sources harnessed in Nigeria	Surface and groundwater resources
	Abdulla et al. (2021) Huang et al. (2021) Mazurkiewicz et al. (2022) Saeedi and Goodarzi (2020) Teston et al. (2022)		Rainwater harvesting and storage
	Dunkerley (2021) Huo et al. (2021) Rammal and Berthier (2020) Sasanya and Adesogan (2023) Simpson et al. (2023)		Surface runoff as a water source
3.	Aikowe and Mazancová (2021) Mishra (2023) Mushtaq et al. (2020) Ngene et al. (2021) Sasanya et al. (2024)	Integrated water management	Water management problems in Nigeria

(continued)

Table 9.1 (continued)

S. No.	References (authors and year)	Theme	Subthemes
	Adeoti et al. (2024) Amanambu et al. (2020) Olabode and Comte (2024) Shiru et al. (2020)		Water insufficiency problems and amelioration efforts in Nigeria
	Adeniran et al. (2021) Ngene et al. (2021) Olaniyi et al. (2023) Olise et al. (2023)		Nigerian government past efforts to combat water insecurity
4.	Mir et al. (2021) Ogunrinde et al. (2021a) Ogunrinde et al. (2021b) Rignot (2022) Srivastava and Chinnasamy (2021)	The effects of climate change on Nigeria's water resources: enhancing adaptation and resilience	Climate change impacts on water resources
	Bozorgzadeh and Mousavi (2023) Kumar and Thakur (2024) Leal Filho et al. (2021) Olatunde et al. (2024) Seigerman et al. (2023)		Sustainable water management and climate change mitigation
	Almulhim and Abubakar (2023) Amin et al. (2023) Leal Filho et al. (2022) Ndimele et al. (2024) Olatunde et al. (2024)		Adaptation strategies for water resources management
5.	Fayed (2020) Jamroen et al. (2020) Lakhiar et al. (2024) Liang et al. (2020) Ray and Majumder (2024)	Water conservation techniques to combat climate change	Efficient irrigation systems

(continued)

Table 9.1 (continued)

S. No.	References (authors and year)	Theme	Subthemes
	Craswell (2020) El-Beltagi et al. (2022) Kumawat et al. (2020) Mgolozeli et al. (2020) Niziolomski et al. (2020)		Mulching and soil conservation techniques
	Gautam and Dahiya (2022) Gupta et al. (2020) Tzanakakis et al. (2020) Wells and Vermeer (2020) Yusuf et al. (2020)		Water-saving technologies and innovations
	Davids et al. (2021) Julio et al. (2021) Shahangian et al. (2022)		Community-based water management projects
	Ariom et al. (2022) Cui et al. (2021) Ehret et al. (2021) Ngene et al. (2021) Vushe (2021)		Future directions for water conservation in Nigeria

9.6.2 Problems Associated with Water Insecurity

Africa, the driest continent after Australia, is severely impacted by water scarcity, particularly in arid regions. Contributing factors include climate change, inadequate water infrastructure, industrial pollution, salinization, acidification, eutrophication, poor environmental policy implementation, and urbanization (Besada & Werner, 2015; Bwapwa, 2018). Africans experience water scarcity in various forms, including water stress, demand deficits, economic scarcity, physical scarcity, and water crises (Leal Filho et al., 2022; Naik, 2017).

Water-related diseases have a more significant effect on the world population than diseases attributed to other causes combined (Lee et al., 2020). This is largely due to inadequate access to improved sanitation, with only 62% of the global population benefiting from such facilities, while 1.2 billion people lack access to basic sanitation infrastructure (Andrés et al., 2021; Lukenga, 2015). Furthermore, less than 1% of the world's freshwater is readily available for direct human consumption (Baggio et al., 2021). According to Lukenga (2015) and Gürsu (2024), water-related diseases result in 3.575 million annual fatalities, with 43% of these deaths attributed to diarrhoea, and a staggering 98% occurring in developing countries. A study by Adekitan et al. (2017) revealed alarming prevalence rates of waterborne illnesses, including Cholera (3%), dysentery (7%), typhoid (38%), diarrhoea (11%), malaria (40%), and nausea (1%) in the Oluyole Local Government Area of Oyo state, southern Nigeria. Notably, 32% of the sampled population frequently experienced these ailments, and a significant proportion were unaware of water treatment methods before or after use. Moreover, impoverished individuals living in slums or rural areas of developing countries are forced to pay five to ten times more for water than their urban counterparts, exacerbating the existing water insecurity (Shi et al., 2021).

In 2016, approximately 384 million people worldwide were affected by drought, resulting in significant financial losses, particularly in agricultural production, in regions such as the USA, China, and others globally (Fahad et al., 2019). More than half of the global population experiences water scarcity for at least 30 days per year (Mekonnen & Hoekstra, 2016). Several densely populated countries in Asia, the Middle East, Africa, and North Africa face severe groundwater depletion, especially during dry seasons, leading to decreased freshwater tables and increased risk of saltwater intrusion (Mishra, 2023; Rakib et al., 2020).

9.7 Water Sources Harnessed in Nigeria

9.7.1 Surface and Groundwater Resources

Currently, many Nigerians rely on various water sources, including surface water (springs, streams, rivers), hand-dug wells, rainwater harvesting, piped water from water corporations, deep wells, and water vendors (Ajibade et al., 2015). Surface water, which is exposed to the atmosphere, is replenished and sustained by surface runoff and groundwater flow during dry periods (Weatherl et al., 2021). However, the quality of surface water is often compromised by the presence of silts, organic matter, and microorganisms, while its quantity is influenced by factors such as rainfall intensity, climate, vegetation, geological formation, and watershed topography (Rey-Romero et al., 2022; Tay, 2017).

Groundwater resources is the primary drinking water source in most areas of Nigeria and across Africa (Danert & Healy, 2021). Groundwater comprises of the portion of percolated rainfall into underground aquifers, and they generally exhibit better quality than surface water (Islam, 2023). Groundwater extraction can occur through various methods, including springs, deep wells, shallow wells, and infiltration galleries (Lachassagne et al., 2021; Naves et al., 2021). However, in Nigeria, the cost of drilling for groundwater extraction is often prohibitively expensive, limiting access to a select few who can afford it (Martínez-Santos et al., 2020). These individuals often extract groundwater indiscriminately, without conducting necessary geophysical surveys or obtaining required permissions (Ajibade et al., 2015; Saha et al., 2021).

Changes in precipitation patterns and higher evaporation rates have led to reduced groundwater recharge, particularly in Nigeria's arid and semi-arid regions (Amanambu et al., 2020; Yenehun et al., 2020). Despite these challenges, groundwater storage in many African countries has also been subjected to over-extraction, over-reliance, and depletion due to declining surface water availability and its deteriorating quality (Olalekan et al., 2023). Over-abstraction of groundwater has, in some areas, resulted in land subsidence (Suganthi & Elango, 2020).

9.7.2 Rainwater Harvesting and Storage

Rainwater harvesting involves collecting and storing rainwater from impermeable surfaces like rooftops or permeable surfaces like soil, directing it towards a collection point for either groundwater recharge (passive harvesting) or storage in a tank (active harvesting) (Saeedi & Goodarzi, 2020). Active rainwater harvesting systems often include pumps, water level sensors, system controllers, and water treatment facilities for efficient operation (Palermo et al., 2022). According to Huang et al. (2021), rainwater harvesting also includes runoff from land surfaces flowing into surface waters like streams, rivers, and oceans. The process is relatively inexpensive in terms of construction and maintenance, increasing access to water in public places (Richards et al., 2021). The components of rainwater harvesting can be broadly divided into three categories: supply (rainfall), demand (point of usage), and distribution system (Teston et al., 2022). Harvested rainwater can be directly pumped to the point of use, allowed to flow by gravity, or collected in an underground storage tank, pumped to an overhead tank, and then distributed to the point of use (Abdulla et al., 2021).

Unfortunately, many harvested rainwaters are either biologically or chemically polluted, failing to meet set standards for potable water. Harvested rainwater is susceptible to physical, chemical, and microbiological contaminations, which can be pathogenic (often due to the presence of faecal wastes) or non-pathogenic (non-disease-causing organisms, including Protozoa, Algae, Bacteria, and Viruses) (Mazurkiewicz et al., 2022). Some factors can influence the type and level of contaminants in harvested rainwater. Such factors include catchment area conditions, storage facilities, atmospheric conditions, roof type, and roof age (Abegunrin et al., 2014; Tengan & Akoto, 2022). Contaminants like debris, volatile organic chemicals, synthetic organic chemicals, minerals, and metals may be present in harvested rainwater (Marszałek et al., 2023). Airborne emissions deposited on the collection surface significantly impact the quality of the rainwater collected (Zdeb et al., 2020). Rainwater is typically free from microbiological pollutants until harvested and stored (Mazurkiewicz et al., 2022). Consequently, most harvested and stored rainwater is only suitable for irrigation purposes and non-drinking household uses, such as sanitation and domestic washing (Rahman, 2021).

9.7.3 Surface Runoff as a Water Source

Surface runoff serves as a primary pathway for the transport of terrestrial pollutants to wetlands, streams, lakes, and coastal waters, with urban areas identified as a significant source of water pollution in many countries, including Nigeria (Datta et al., 2022). The characteristics of surface runoff are strongly influenced by factors such as precipitation amount, infiltration rate, percolation amount, vegetative interceptions, flood waves, soil erosion, increased soil water level, groundwater storage, and changes in surface roughness (Kaletova & Nemetova, 2017; Rammal & Berthier, 2020). Surface runoff volumes play crucial roles in the design of hydraulic structures, sediment transport, and the fate of agrochemicals on the soil surface (Verlicchi, & Ghirardini, 2022). Surface runoff significantly influences water table levels and groundwater recharge, leading to overflow and pollution in shallow wells during and after rainfall events (Yenehun et al., 2020). Furthermore, surface and subsurface runoff have impact on the available water levels in ponds, streams and rivers and also determines the landscape characteristics and flood size and occurrence (Cantonati et al., 2020).

The volume, timing, and duration of overland flow are essential factors affecting water supply, flood propagation, and various hydrological processes (Jiang & Yu, 2022). The rainfall-runoff process is governed by several factors, including rainfall intensity, terrain slope, land use, vegetation, and soil properties, among others (He et al., 2020; Sitterson et al., 2018). Raghunath (2006) and Huo et al. (2021) categorized the factors influencing runoff into four groups: storm characteristics, meteorological factors, watershed and storage characteristics of the basin, and additional factors. Storm characteristics comprise storm distribution, type, nature, intensity, duration, frequency, and direction. Meteorological factors include temperature, humidity, wind velocity, antecedent moisture content, and atmospheric pressure. The basin's watershed and storage characteristics encompass factors such as size, shape, slope, topography, soil, geological characteristics, proximity to water bodies, and prominent vegetation types, as well as storage characteristics like depression storages, ponds, lakes, channels, reservoirs, gullies, and flood plains. However, increased rainfall intensity can lead to increased runoff due to the formation of crust on soil surfaces and thus reducing infiltration rates (Dunkerley, 2021). Additionally, as terrain slope increases, infiltration decreases, resulting in increased runoff (Chen et al., 2020; Dunkerley, 2021). He et al. (2020) also observed that factors like vegetative cover and initial soil moisture content significantly affect runoff, although their impact diminishes with increased rainfall intensity.

The effective management of runoff can convert it into a valuable water resource, suitable for flood plain agricultural practices (Antolini et al., 2020; Sasanya & Adesogan, 2023). However, surface runoff collection and usage pose several challenges, including increased flow, sedimentation in channels and basins, high Biochemical Oxygen Demand (BOD), deterioration of aquatic habitats, fish losses, and decreased water quality due to increased solid waste flow along water channels (Md Anawar et al., 2020; Shukla et al., 2021). Additionally, surface runoff can introduce excess nutrients, metals, hydrocarbons, and bacteria into water bodies (Sharma et al., 2021). Remediation of polluted water through biological, chemical, physical, ecological and/or engineering processes is thus recommended. Surface runoff or overland flow significantly contributes to ground and surface water pollution and recharge (Weatherl et al., 2021). Pollution typically occurs during and after storms or rainfall events, characterized by debris, pathogens, insoluble suspended solids, and high turbidity. The content of surface runoff is influenced by factors such as flow velocity and roughness coefficient (Kaletova & Nemetova, 2017). Factors including precipitation characteristics runoff and percolation rates, pollutant runoff and transport rates, sediment deposition and resuspension rates, time interval between rainfall events, size of municipal drainage, land use type and intensity, degree of imperviousness, tree cover, soil type, slope and drainage the level of pollutants in runoff water (Jiang et al., 2020; Simpson et al., 2023).

9.8 Integrated Water Management

9.8.1 Water Management Problems in Nigeria

Water is a vital and indispensable resource for life, but its conservation is of significant concern due to excessive pollution, climate change, and improper waste disposal (Datta et al., 2025; Mishra, 2023). Although, Nigeria boasts an abundance of water resources, albeit with spatial variability in their distribution (Sasanya et al., 2024). The country's southern and north central geopolitical zones receive high annual precipitation amounts, averaging around 3000 mm/year, whereas the far northern regions receive significantly less, approximately 500 mm/year (Ibebuchi & Abu, 2023). Despite the country's abundant water resources, a substantial population faces water scarcity and inadequate access to safe drinking water and sanitation (Ngene et al., 2021).

A significant proportion of rural communities in Nigeria lack access to improved water supply systems, such as piped water or boreholes (Aikowe & Mazancová, 2021). Consequently, these communities suffer the socio-economic and health consequences of water scarcity. In rural communities, water scarcity has been addressed through stream water utilization, reservoir construction, and shallow well digging (Ritchie et al., 2021). However, these solutions are not without challenges: streams are often polluted, stored water in reservoirs can develop unpleasant odours, and dug wells may be situated near septic tanks, compromising water quality (Adesogan, 2014; Mushtaq et al., 2020). Nevertheless, in regions where conventional water supply systems are ineffective, and groundwater is inaccessible due to geological formations, or streams are unavailable, rainwater harvesting from roof catchments becomes the only viable option (Adesogan & Sasanya, 2021). Therefore, it is essential to research and develop alternative frameworks for water supply, particularly in areas where non-existent, minimal, or failing catchments hinder rainfall harvesting

(Adesogan & Sasanya, 2021). In general, goal of every water supply scheme should be to provide adequate, wholesome, and sufficient water to combat water insecurity (Belhassan, 2021).

9.8.2 Water Insufficiency Problems and Amelioration Efforts in Nigeria

Although Nigeria's combined surface and groundwater resources can meet the water demands of its growing population, several critical constraints hinder effective water supply and management, leading to water scarcity (Ajibade et al., 2021; Olabode & Comte, 2024). Notwithstanding the abundance of water resources and substantial investments in the water sector, Nigeria's water resources development and management continue to face numerous challenges. According to FGN (2004), Shiru et al. (2020) and Ngene et al. (2021), some of the enormous challenges bedevilling the Nigeria water sector include:

i. Rapid population growth, urbanization, rising standards of living, and dwindling financial resources.
ii. Watersheds and waterways degradation.
iii. Unregulated and indiscriminate water resources development.
iv. Inadequate management and insufficient data on water resources.
v. Poor catchment management and severe pollution threats from hazardous waste disposal into water bodies.
vi. Limited groundwater availability in areas underlain by crystalline rocks.
vii. Uncontrolled and excessive groundwater exploitation in areas underlain by sedimentary rock.
viii. Increasing drought and desertification.
ix. Financial mismanagement, poor accountability, and lack of performance transparency.
x. Insufficient stakeholder and expert engagement in water resources planning, management, and development.
xi. Inadequate, outdated, and unreliable water resources projects that fail to meet consumer demands.

Moreover, the challenges identified have several inherent consequences. According to Ngene et al. (2021) and Adeoti et al. (2024) such impact includes but not limited to the following:

i. Water scarcity and increasing water supply costs.
ii. Prolonged drought, flood, erosion, destruction of farmlands, leading to food insecurity and scarcity, destruction of farmlands and livelihoods, increased risk of water-borne diseases.
iii. Reduced life expectancy due to malaria, diarrhoea, guinea worm, gastroenteritis and other water-related illnesses.

Furthermore, Hassan et al. (2016) identified significant water challenges in Taraba state, Nigeria as a case study. Out of 14 water supply schemes in the state, only 2 were operational at the time of the investigation, while the remaining 12 were non-functional due to lack of operational inputs, consumables and malfunctioning equipment. The state's water supply was also compromised by poor reticulation networks that traversed polluted environments, inadequate quality control measures and the absence of a water testing laboratory in the state water supply agency. The Nigerian populace faces several limitations in accessing adequate and wholesome water supply, including inadequate infrastructure, poor funding, inability of the Nigerian Government to forecast and address inherent problems associated with poor water supply (Hassan et al., 2016).

9.8.3 Nigerian Government Past Efforts to Combat Water Insecurity

Water resources managers and institutions play a crucial role in ensuring adequate water availability to the populace. They operate at the federal, state, and local government levels. Specifically, the Federal Ministry of Water Resources (FMWR) is responsible for managing, developing, documenting, and monitoring water resources, as well as formulating national policies (Ngene et al., 2021). Twelve River Basin Development Authorities (RBDAs) were established to develop, operate, and manage water sources in their respective regions under the FMWR's supervision (Adeoti, 2020; USAID, 2010). Thirty-seven State Water Agencies (SWAs) exist in states, responsible for water quality control and maintaining water supply schemes in urban and rural areas (Olaniyi et al., 2023; USAID, 2010). However, the SWAs face significant challenges and some of these include inadequate financial capacity, lack of autonomy in operations, moribund and obsolete infrastructure, and operational inefficiencies (Ajibade et al., 2015; Ngene et al., 2021; USAID, 2010). Similarly, the 774 local governments in Nigeria struggle with financial constraints, hindering their ability to manage rural water supply schemes and ensure sanitation and hygiene (Olise et al., 2023; USAID, 2010).

Furthermore, the Nigerian government initiated various programmes to improve water supply schemes. Some of these schemes are National Borehole Programme (NBP), the Directorate of Food, Roads and Rural Infrastructure (DFRRI) and Partnerships with private organizations and international bodies such as World Bank, African Development Bank (AfDB), Department of International Development of the UK (Ajibade et al., 2015; Odife, 2022). Despite these efforts, the water sector remains underdeveloped, unsatisfactory and plagued with inadequate operational and management data, ineffective fund utilization, unprofessional water facility management, poor maintenance culture, unsustainable designs for water supply schemes, ineffective policies, lack of political will and technical limitations (Adeniran et al.,

2021; Ajibade et al., 2015; FGN, 2004; Ngene et al., 2021). Due to these inefficiencies, many households in Nigeria's urban centres have taken on the role of water and energy corporations, harnessing groundwater and providing energy for water access.

9.9 The Effects of Climate Change on Nigeria's Water Resources: Enhancing Adaptation and Resilience

9.9.1 Climate Change Impacts on Water Resources

Climate change has significantly impacted the quantity and quality of available water resources, with far-reaching consequences for human health, socio-economic development, and ecosystems (Mir et al., 2021). The effects of climate change on water resources are evidenced as disruption of the hydrologic cycle, altering patterns of precipitation, temperature, and drought, escalated conflicts between communities over limited water resources, and increased sediment transport which impairs water quality (Hunt et al., 2020; Qiu et al., 2023; Yang et al., 2021). These lead to challenges in water management and planning due to uncertainties surrounding water availability and quality (Herman et al., 2020). Increased temperature and evaporation resulting from climate change also leads to extreme hydrological events such as droughts and floods and increased agricultural, urban, and industrial water demand amidst water scarcity (Srivastava & Chinnasamy, 2021).

Many parts of the world, including Nigeria, are projected to experience shifts in precipitation patterns due to climate change (Shiru et al., 2020). Southern regions of Nigeria, historically known for high rainfall, are expected to face more intense rainfall events, whereas arid areas are forecasted to become even drier (Awolala et al., 2023; Ogunrinde et al., 2021a). Rainy seasons are anticipated to receive increased precipitation, while dry seasons will likely become hotter and more severe (Tabari, 2020).

Climate change has been reported to significantly impact the availability of surface water in rivers, lakes, natural and artificial reservoirs, as well as groundwater, which constitutes approximately 30% of the world's fresh water supply (Abdelhalim et al., 2020). Fluctuations in precipitation patterns affect the recharge of underground aquifers, which rely on percolating rainfall-runoff (Shyam et al., 2021). Moreover, more frequent and intense storms can lead to higher surface runoff, potentially carrying pollutants such as pesticides, fertilizers, herbicides, and industrial chemicals into water bodies (Wato et al., 2020). The combination of these pollutants with thermal pollution can contribute to eutrophication and the proliferation of harmful algae on water surfaces (Vasseghian et al., 2024). Increased surface runoff also poses risks to wastewater treatment facilities, potentially causing untreated or inadequately treated water to flow into water bodies, thereby contaminating them and rendering them unsuitable for domestic, agricultural, and industrial purposes (Jan et al., 2022; Tariq & Mushtaq, 2023). These environmental stressors can lead to a decrease in the total dissolved oxygen in water bodies, resulting in adverse effects on aquatic life, contaminate drinking water supplies to further deterioration of water quality and then facilitate the spread of waterborne diseases.

Climate change also affects aquatic ecosystems, biodiversity, and wetlands (Wang & Gu, 2021). Wetlands provide crucial ecosystem services, including water purification through natural filtration, flood regulation, and habitat biodiversity (Agaton & Guila, 2023). However, these vital services are increasingly threatened by changing precipitation patterns, rising temperatures, and sea level rise, all of which place wetlands at risk.

Sea level rise and saltwater intrusion into freshwater sources, particularly in coastal areas, have also been reported as consequences of climate change (Abd-Elaty et al., 2022). This intrusion can render freshwater unsuitable for human consumption and agricultural irrigation, with potentially severe impacts on human health and food security. The rising sea levels are driven by increasing global temperatures, which cause ice to melt and subsequently raise the volume of seawater (Allison et al., 2021; Rignot, 2022).

9.9.2 Sustainable Water Management and Climate Change Mitigation

The significant impacts of climate change on water resources can be mitigated through sustainable water management strategies, including the Integrated Water Resources Management (IWRM) approach, infrastructure improvements, and ecosystem-based adaptation (Basuki et al., 2022; Vollmer et al., 2022). IWRM takes a holistic approach to managing water resources by considering the interconnections between surface water, groundwater, ecosystems, and social, economic, and environmental factors to ensure sustainable water availability and quality (Armas Vargas et al., 2023). The IWRM approach aims to improve water use efficiency, reduce waste, and promote water reuse, recycling, and treatment (Koop et al., 2022; Kumar & Thakur, 2024), which can simultaneously reduce water pollution and health risks associated with untreated water for domestic, agricultural, and industrial purposes (Saravanan et al., 2021).

The benefits of IWRM are multifaceted, contributing to poverty reduction, food security, and human well-being. By ensuring equitable access to improved and clean water supplies, IWRM can enhance livelihoods and promote social equity (Ngene et al., 2021; Seigerman et al., 2023). The IWRM approach allows for optimised water use in agriculture, strengthened food security, stabilised food prices, reduced incidences of water-related conflicts and diseases, and enhanced ecosystem services (Bozorgzadeh & Mousavi, 2023). In addition, IWRM enhances human well-being, supports environmental sustainability, social equity as well economic development and generally improves water security by bolstering the resilience of water resources against climate-related disasters (Julio et al., 2021). These outcomes can stimulate economic growth, foster social cohesion, and promote environmental sustainability (Santos et al., 2023).

Furthermore, the effects of climate change on Nigeria's water resources can be mitigated through significant investment in the improvement of water infrastructure and ecosystem-based adaptation to enhance resilience (Leal Filho et al., 2021). Such infrastructure developments would improve water storage, conservation, and flood management, including the construction of dams, reservoirs, and levees to effectively manage water resources (Hovis et al., 2021; Sen, 2020). Additionally, the protection of existing ecosystems and the restoration of degraded ones, such as forests, vegetated buffer zones along riverbanks, and wetlands, would offer numerous benefits (Nair et al., 2024). These ecosystems provide natural water storage, filtration, and security, while also conserving biodiversity, sequestering carbon, and bolstering resilience against the impacts of climate change (Norman et al., 2022). They contribute to improved water quality in natural channels, provide habitats for wildlife, and reduce soil erosion and nutrient runoff (Verdonschot & Verdonschot, 2023).

9.9.3 Adaptation Strategies for Water Resources Management

Although the global water crisis, driven by climate change, population growth, and urbanisation, continues to escalate, it cannot be left unaddressed. In Nigeria, adopting strategies to ensure the long-term sustainability of water resources has become imperative. These strategies should include water-saving and conservation practices, long-term water harvesting and storage, water recycling and reuse, and the promotion and implementation of best land-use practices (Almulhim & Abubakar, 2023; Kumar & Thakur, 2024). Additionally, designing and constructing climate-resilient water infrastructure, introducing water education and literacy programmes, implementing schemes for water pollution reduction and control, and strengthening water governance and policy frameworks are essential (Leal Filho et al., 2022).

Given that agriculture is the largest consumer of freshwater globally, water-saving and conservation strategies must also include the education of farmers and agricultural practitioners (Amali et al., 2021). This orientation should promote the adoption of drip irrigation systems, which are more efficient in conserving water compared to surface or sprinkler irrigation methods (Kumar & Thakur, 2024). Additionally, the use of water-efficient crops and cropping systems should be encouraged to further conserve water (Liu & Song, 2020). Domestic and industrial water users must also be educated on the importance of adopting water-saving practices (Kumar & Thakur, 2024). It is crucial to repair all pipe leakages from municipal water supply networks and other water distribution systems to prevent unnecessary water loss.

The over-reliance on groundwater in Nigeria is a significant concern. To address this challenge, the use of alternative water sources should be encouraged, including desalinated, recycled, and harvested water. Rainwater harvesting and runoff collection should be prioritised by directing these waters into storage tanks or underground reservoirs (Doost et al., 2024). This approach ensures that rainwater and runoff are utilised economically during periods of water scarcity before they return to the natural waterways and continue in the hydrologic cycle. Additionally, Nigeria urgently needs robust and functional early warning and risk management systems to provide timely alerts on floods, droughts, and water quality issues (Awolala et al., 2022; Ufuoma et al., 2021). Implementing such systems would significantly reduce the number of people vulnerable to water-related disasters (Ufuoma et al., 2021).

Nigeria must also invest in the development of resilient water systems capable of withstanding unstable climate conditions, population growth, and the test of time (Jellason et al., 2021; Ndimele et al., 2024). These resilient systems should include modern, well-maintained water infrastructure that can absorb disturbances, handle extreme events, and adapt to changing conditions, all while ensuring sustained water availability and quality (Amin et al., 2023).

Furthermore, the protection, conservation, and restoration of natural ecosystems and biodiversity should be prioritised to secure the sustainability of water resources (Norman et al., 2022; Verdonschot & Verdonschot, 2023). Incorporating green infrastructure, such as wetlands, green roofs, and permeable pavements, can also be effective in improving stormwater management and enhancing water quality. Furthermore, the adoption of modern technologies, such as sensors, Internet of Things (IoT), data analytics, and automated water control systems, can facilitate real-time data collection, detect leaks, and improve the resilience of water infrastructure in Nigeria (Olatunde et al., 2024). In addition to these technological and infrastructural advancements, it is crucial for Nigeria to formulate and enact policies that enhance water rights, allocation, and distribution. These policies would help prevent conflicts arising from unequal water distribution, ensure fair access to water resources, especially during times of scarcity, and regulate the pollution of water resources, thereby protecting public health (Babuna et al., 2023).

9.10 Water Conservation Techniques to Combat Climate Change

9.10.1 Efficient Irrigation Systems

The primary goal of any efficient irrigation system is to optimise water usage, minimise water wastage, and promote sustainable agricultural practices while maximising crop yield (Ray & Majumder, 2024). Nigeria should adopt such technology for its water resource conservation efforts. This technology integrates various components that work together to deliver the precise amount of water to crops at the right time, reducing excess water use and minimising the environmental footprint (Liang et al., 2020).

An efficient irrigation system operates optimally due to the communication between various sensors, including soil moisture sensors, flow and pressure sensors, weather monitoring sensors, and microcontrollers (Jamroen et al., 2020; Liang et al., 2020). These systems can be managed through mobile applications for remote monitoring (Jamroen et al., 2020). The system also incorporates precision sprinkler nozzles, drip emitters, micro sprinklers, water harvesting and storage facilities, as well as water treatment and filtration systems (Fayed, 2020). The working principle of such an efficient irrigation system can be based on drip irrigation, sprinkler systems, centre pivot systems, or micro-irrigation techniques. Among these, the drip system is particularly effective, with well-designed and maintained systems achieving up to 90% efficiency (Fayed, 2020).

Utilising an efficient irrigation system for crop production can significantly increase crop yields while saving energy and reducing chemical and weed management costs (Lakhiar et al., 2024). Although the initial investment can be substantial and the design and construction require a high level of technical expertise, the benefits of water conservation far outweigh these limitations. Unlike traditional or surface irrigation methods, efficient irrigation systems are specifically designed to minimise water wastage and runoff, prevent soil erosion, and optimise both waters use efficiency and productivity (Kilemo, 2022). These systems can greatly enhance sustainable agriculture and efficient water utilisation, particularly in the northern arid and semi-arid regions of Nigeria, where water resources are scarce and dry spells are prolonged (Ogunrinde et al., 2021b). Moreover, by optimising water use and improving crop yields, efficient irrigation systems contribute to food security and environmental protection.

9.10.2 Mulching and Soil Conservation Techniques

Mulching is a soil and water conservation technique that involves covering the soil surface with organic or inorganic materials to improve soil health and enhance agricultural productivity (El-Beltagi et al., 2022; Mgolozeli et al., 2020). This method helps to reduce or control soil erosion, conserve soil moisture, regulate soil temperature, and suppress weed growth (El-Beltagi et al., 2022). Organic materials such as straw, wood shavings, crop residues, and grass clippings, as well as inorganic materials like plastic or polythene, can be used for mulching (Iqbal et al., 2020).

Mulching effectively reduces surface runoff, limits sediment transport, prevents the washing of agricultural chemicals into waterways and consequently prevents soil moisture loss as well as water quality degradation (Niziolomski et al., 2020). Additional soil conservation techniques that can mitigate soil erosion and sediment transport into waterways include contour farming, terrace construction, cover crop cultivation, crop rotation, agroforestry practices, conservation tillage, soil reclamation, gully stabilisation, and bioengineering (Craswell, 2020; Kumawat et al., 2020).

9.10.3 Water-Saving Technologies and Innovations

To tackle the escalating water crisis in Nigeria, the adoption of water-saving technologies and innovations is crucial. These innovations, water-efficient appliances and fixtures play a leading role in regulating and reducing excessive water consumption across domestic, industrial, and agricultural sectors. These technologies not only promote the sustainable use of limited water resources but also minimise environmental impacts by reducing water consumption and the energy required for treatment (Gupta et al., 2020; Yusuf et al., 2020).

Examples of innovative water-saving appliances and fixtures include low-flow showerheads, dual-flush toilets, highly efficient washing machines, water-saving dishwashers, smart irrigation systems, tankless water heaters, low-flow faucets and aerators, sensor-activated faucets, pressure-reducing valves, and water-efficient urinals (Wells & Vermeer, 2020). These technologies contribute to effective water management and conservation, making them essential tools in addressing the global water crisis (Tzanakakis et al., 2020; Weerasooriya et al., 2021).

Additionally, innovative technologies that can significantly enhance water conservation, saving, reuse, and recycling include greywater recycling systems, rainwater harvesting systems, smart water meters, and desalination technologies (Rodrigues et al., 2023; Varma, 2022). Greywater recycling systems enable the reuse of wastewater from sinks, showers, and washing machines for non-potable purposes, while rainwater harvesting systems collect and store rainwater for later use (Gautam & Dahiya, 2022; Saeedi & Goodarzi, 2020). Smart water meters offer the advantage of tracking consumer water usage and detecting leaks early, which helps in efficient water management and prompt issue resolution (Gupta et al., 2020). Desalination technologies, which convert saltwater into freshwater, are particularly beneficial for coastal areas of Nigeria where freshwater supplies are limited (Eke et al., 2020). Although desalination traditionally incurs high energy costs, advancements in modern technology are leading to more efficient and cost-effective solutions (Shokri & Fard, 2022).

9.10.4 Community-Based Water Management Projects

Water conservation and savings cannot be achieved solely through policies without active community involvement (Shahangian et al., 2022). To meet water conservation goals and objectives, it is essential for individuals to adopt water-saving behaviours and habits. Engaging the Nigerian populace in the planning, implementation, and monitoring of water resource conservation is crucial. This can be accomplished through community workshops, training sessions, school-based programmes, offering incentives for water conservation, and encouraging households to participate in public challenges and also make pledges (Moreno et al., 2021).

Community participation enhances water security, fosters social cohesion, supports environmental protection, reduces water pollution, and lowers water bills (Davids et al., 2021; Julio et al., 2021). Although such engagement may involve significant economic investment and require technical expertise and legal regulation, the benefits far outweigh these challenges.

9.10.5 *Future Directions for Water Conservation in Nigeria*

The southern part of Nigeria experiences substantial annual rainfall, in contrast to the northern regions, which receive significantly less precipitation. Despite this disparity, both regions face significant water scarcity issues. These challenges are exacerbated by population growth, urbanisation, industrialisation, and, most critically, climate change. Nigeria must focus on innovative and sustainable long-term solutions to address the growing problem of water scarcity, particularly during the annual dry season. The future direction of water conservation in Nigeria should be guided by the following strategies as highlighted by FGN (2004), Ajibade et al. (2015), Ehret et al. (2021), Ngene et al. (2021), Vushe (2021), Cui et al. (2021), Shahangian et al. (2022), Ariom et al. (2022), Ajijola et al. (2023) and Burt et al. (2023).

i. Formulation and Enforcement of Water Policies: The development and enforcement of comprehensive water policies and governance frameworks that address climate change, population growth, and increased agricultural demand is of utmost importance. These policies should foster synergy between the public and private sectors. To achieve these the water must be recognized as an economic resource which should be sustainably managed.
ii. Adoption of Innovative Technologies: Advanced technologies, appliances, and fixtures for efficient water management, distribution, and supply should be utilized.
iii. Community Engagement and Behavioural Change: Communities must be engaged and involved in water conservation efforts, build their capacity, and encourage positive behavioural changes towards water conservation goals. Public awareness on water resources management and conservation must be raised. Human resources in the water sector should also be trained and developed to enhance capacity and expertise.
iv. Promotion of Alternative Water Sources: Absolute dependence on groundwater should be discouraged by encouraging rainwater harvesting, water reuse and recycling, and groundwater recharge. Furthermore, existing gigantic water supply scheme should be decentralised, and local government authorities should be empowered with more responsibilities for water supply and sanitation, backed by adequate and monitored financial support.
v. Advancement of Efficient Irrigation Practices: Campaign against flood irrigation and promotion of drip irrigation systems usage, drought-resistant crops

cultivation, and efficient livestock watering systems are the ways to go in agricultural practices.

vi. Investment in Resilient Infrastructure: Heavy investment with transparency in climate-resilient water infrastructure and focus on the conservation and restoration of ecosystems.

vii. Funding Research and Development: Research and development on water conservation technologies, climate change impacts, and foster international research collaboration should be supported.

viii. Integration into Urban Planning: The incorporation of water conservation techniques into urban planning and development, such as integrating green buffers, green spaces, permeable pavements, and rainwater harvesting systems are important to achieve water conservation. Sustainable and effective operational and maintenance schemes for water supply and sanitation should be designed and implemented.

ix. Financing and Incentives: Financial investment in water conservation projects, and provision of incentives to households and businesses that adopt water-saving technologies is also a way to go. Accountability should also be ensured in water resources management and the water sector should also enjoy some level of financial autonomy.

x. Providing improved and stable power supply to support water treatment and distribution.

9.11 Conclusions, Recommendations and Limitations

This review explored the critical role of climate-smart water conservation strategies in Nigeria, revealing a complex network of challenges within the water sector, inadequate infrastructure, poor governance, and the impacts of climate change. The findings underscore the urgency of addressing escalating water scarcity and conservation issues exacerbated by climate change.

Although Nigeria possesses abundant water resources, many regions, particularly rural communities, continue to face challenges such as uneven water distribution, water scarcity, poor water quality, and inequitable access to safe drinking water and sanitation. Despite numerous government initiatives aimed at improving the water sector, these efforts have been hampered by operational and management shortcomings, inefficient use of funds, and a lack of political will. Climate change further exacerbates these challenges by disrupting hydrological cycles, intensifying conflicts among communities, farmers, and herders.

Addressing Nigeria's complex water scarcity issue requires a comprehensive, integrated, and sustainable water resources management strategy. The sustainable water management approach is therefore recommended to combine innovative technologies with robust policy frameworks, thus incorporating sustainable long- and short-term investments in climate-resilient water infrastructure. Support for research and development, promotion of water conservation practices and encouragement of alternative water sources are as well vital. Additionally, reducing groundwater over-reliance, implementing efficient irrigation systems, and adopting innovative water fixtures are essential measures to alleviate scarcity, enhance water security, and mitigate climate change's impact on dwindling resources.

Substantial financial investments in climate-resilient infrastructure are critical to mitigating climate change's effects on water availability. Moreover, promoting water conservation practices and ensuring prudent financial management within the water sector are crucial for securing the country's future water supply, ensuring sustainability and resilience for generations to come.

However, some limitations in this area of study primarily include the lack of adequate data to quantify Nigeria's water demand deficit, water demand and water supply, particularly in rural communities. Additionally, a significant portion of the Nigerian population lacks awareness about effective water conservation strategies thus, hindering the optimal utilization of the country's limited water resources.

Acknowledgements Not applicable

Funding No funds, grants, or other support was received for conducting this study.

Data Availability Statement No dataset is associated with this current study. The materials used for the current study are available on reasonable request from the author.

Ethical Clearance Not applicable.

Competing Interests The authors declare there are no competing interests or personal relations that could have influenced the study reported in this paper.

References

Abd-Elaty, I., Kushwaha, N. L., Grismer, M. E., Elbeltagi, A., & Kuriqi, A. (2022). Cost-effective management measures for coastal aquifers affected by saltwater intrusion and climate change. *Science of the Total Environment, 836*, 155656. https://doi.org/10.1016/j.scitotenv.2022.155656

Abdelhalim, A., Sefelnasr, A., & Ismail, E. (2020). Response of the interaction between surface water and groundwater to climate change and proposed megastructure. *Journal of African Earth Sciences, 162*, 103723. https://doi.org/10.1016/j.jafrearsci.2019.103723

Abdulla, F., Abdulla, C., & Eslamian, S. (2021). Concept and technology of rainwater harvesting. In S. Eslamian (Ed.), *Handbook of water harvesting and conservation: Basic concepts and fundamentals* (pp. 1–16). Wiley. https://doi.org/10.1002/9781119478911.ch1

Abegunrin, T. P., Odeniyi, J., & Onofua, O. E. (2014). Roof age effect on the quality of harvested rainwater and its health implication in a selected location, Southwest Nigeria. *International Journal of Water Resources and Environmental Engineering, 6*(10), 261–266. https://doi.org/10.5897/IJWREE2013.0525

Adekitan, A. A., Ogunsua, A. O., Martins, O., Awomeso, J. A., & Ojekunle, Z. O. (2017). Effects of untreated household water on the inhabitants of Oluyole Local Government Area, Oyo State Nigeria. *Scholarly Journal of Science Research and Essay, 6*(1), 18–26.

Adeniran, A., Daniell, K. A., & Pittock, J. (2021). Water infrastructure development in Nigeria: Trend, size, and purpose. *Water, 13*(17), 2416. https://doi.org/10.3390/w13172416

Adeoti, O. (2020). Constraints on data collection implementation at the river basin level in Nigeria. *Journal of Hydrology: Regional Studies, 32*, 100738. https://doi.org/10.1016/j.ejrh.2020.100738

Adeoti, O. S., Kandasamy, J., & Vigneswaran, S. (2024). Water infrastructure sustainability challenge in Nigeria: A detailed examination of infrastructure failures and potential solutions. *Water Supply, 24*(6), 2066–2076. https://doi.org/10.2166/ws.2024.127

Adesogan, S. O. (2014). Strategies and techniques of providing adequate and affordable potable water in rural areas of Nigeria. *International Journal of Water Resources and Environmental Engineering, 6*(1), 32–39. https://doi.org/10.5897/IJWREE2013.0418

Adesogan, S. O., & Sasanya, B. F. (2021). Water availability investigation and assessment in South Western Nigeria. *International Journal of Energy and Water Resources, 6*, 457–473. https://doi.org/10.1007/s4208-021-00136.2

Agaton, C. B., & Guila, P. M. C. (2023). Ecosystem services valuation of constructed wetland as a nature-based solution to wastewater treatment. *Earth, 4*(1), 78–92. https://doi.org/10.3390/earth4010006

Ahmed, S. F., Kumar, P. S., Kabir, M., Zuhara, F. T., Mehjabin, A., Tasannum, N., et al. (2022). Threats, challenges and sustainable conservation strategies for freshwater biodiversity. *Environmental Research, 214*, 113808. https://doi.org/10.1016/j.envres.2022.113808

Aikowe, J. O., & Mazancová, J. (2021). Barriers to water access in rural communities: Examining the factors influencing water source choice. *Water, 13*(19), 2755. https://doi.org/10.3390/w13192755

Ajibade, F. O., Adewumi, J. R., Ojo, O. M., Babatola, J. O., & Oguntuase, A. M. (2015, October 28–30). Issues, challenges and management of water supply and sanitation in Nigeria: an overview. In *Conference Proceedings on National Development Strategies Towards Sustainable Civil Infrastructure, International Civil Engineering Conference Held at Kwara Hotel*, Ilorin, Nigeria (pp. 19–34). Nigerian Institution of Civil Engineers.

Ajibade, F. O., Olajire, O. O., Ajibade, T. F., Fadugba, O. G., Idowu, T. E., Adelodun, B., Opafola, O. T., Lasisi, K. H., Adewumi, J. R., & Pham, Q. B. (2021). Groundwater potential assessment as a preliminary step to solving water scarcity challenges in Ekpoma, Edo State, Nigeria. *Acta Geophysica, 69*(4), 1367–1381. https://doi.org/10.1007/s11600-021-00611-8

Ajijola, S., Ibem, F. A., & Adetoye, E. M. (2023). Potentials of building information modeling (BIM) tools for water usage simulation and efficiency optimization in residential buildings in Lagos. *Architecture, 5*(1), 126–137.

Algur, K. D., Patel, S. K., & Chauhan, S. (2021). The impact of drought on the health and livelihoods of women and children in India: A systematic review. *Children and Youth Services Review, 122*, 105909. https://doi.org/10.1016/j.childyouth.2020.105909

Allison, I., Paul, F., Colgan, W., & King, M. (2021). Ice sheets, glaciers, and sea level. In W. Haeberli & C. Whiteman (Eds.), *Snow and ice-related hazards, risks, and disasters* (pp. 707–740). Elsevier. https://doi.org/10.1016/B978-0-12-817129-5.00013-5

Almulhim, A. I., & Abubakar, I. R. (2023). Developing a sustainable water conservation strategy for Saudi Arabian cities. *Groundwater for Sustainable Development, 23*(2023), 101040. https://doi.org/10.1016/j.gsd.2023.101040

Amali, A. A., Mersha, A. N., Nofal, E. R., Murray, K., Norouzi, S., Saboory, S., Salo, H., Chevuru, S. R., Sarai Tabrizi, M., Reddy, P. K., & Abdullahi, A. O. (2021). Non-conventional sources of agricultural water management: Insights from young professionals in the irrigation and drainage sector. *Irrigation and Drainage, 70*(3), 524–540. https://doi.org/10.1002/ird.2495

Amanambu, A. C., Obarein, O. A., Mossa, J., Li, L., Ayeni, S. S., Balogun, O., et al. (2020). Groundwater system and climate change: Present status and future considerations. *Journal of Hydrology, 589*, 125163. https://doi.org/10.1016/j.jhydrol.2020.125163

Amin, F., Dar, M. A., & Gupta, A. K. (2023). Sustainability through integrated resilience and risk management: Rivers and disasters in changing climate. In M. Pandey, A. K. Gupta, & G. Oliveto (Eds.), *River, sediment and hydrological extremes: Causes, impacts and management* (pp. 417–434). Springer Nature. https://doi.org/10.1007/978-981-99-4811-6_25

Andrés, L., Joseph, G., & Rana, S. (2021). The economic and health impacts of inadequate sanitation. In *Oxford research encyclopedia of environmental science*. Retrieved August 12, 2024, from https://oxfordre.com/environmentalscience/view/10.1093/acrefore/9780199389414.001.0001/acrefore-9780199389414-e-561

Antolini, F., Tate, E., Dalzell, B., Young, N., Johnson, K., & Hawthorne, P. L. (2020). Flood risk reduction from agricultural best management practices. *JAWRA Journal of the American Water Resources Association, 56*(1), 161–179. https://doi.org/10.1111/1752-1688.12812

Ariom, T. O., Dimon, E., Nambeye, E., Diouf, N. S., Adelusi, O. O., & Boudalia, S. (2022). Climate-smart agriculture in African countries: A review of strategies and impacts on smallholder farmers. *Sustainability, 14*(18), 11370. https://doi.org/10.3390/su141811370

Armas Vargas, F., Nava, L. F., Gómez Reyes, E., Olea-Olea, S., Rojas Serna, C., Sandoval Solís, S., & Meza-Rodríguez, D. (2023). Water and environmental resources: A multi-criteria assessment of management approaches. *Water, 15*(16), 2991.

Awolala, D. O., Mutemi, J., Adefisan, E., Antwi-Agyei, P., & Taylor, A. (2022). Profiling user needs for weather and climate information in fostering drought risk preparedness in Central-Southern Nigeria. *Frontiers in Climate, 4*, 787605. https://doi.org/10.3389/fclim.2022.787605

Awolala, D., Mutemi, J., Adefisan, E., Antwi-Agyei, P., Taylor, A., Muita, R., et al. (2023). Economic value and latent demand for agricultural drought forecast: Emerging market for weather and climate information in Central-Southern Nigeria. *Climate Risk Management, 39*, 100478. https://doi.org/10.1016/j.crm.2023.100478

Ayejoto, D. A., Egbueri, J. C., Agbasi, J. C., Omeka, M. E., Unigwe, C. O., Nwazelibe, V. E., et al. (2023). Influence of seasonal changes on the quality of water resources in southwestern Nigeria: A review. In J. C. Egbueri, J. O. Ighalo, & C. B. Pande (Eds.), *Climate change impacts on Nigeria: Environment and sustainable development* (pp. 423–447). Springer International Publishing. https://doi.org/10.1007/978-3-031-21007-5_22

Babuna, P., Yang, X., Tulcan, R. X. S., Dehui, B., Takase, M., Guba, B. Y., et al. (2023). Modeling water inequality and water security: The role of water governance. *Journal of Environmental Management, 326*, 116815. https://doi.org/10.1016/j.jenvman.2022.116815

Baggio, G., Qadir, M., & Smakhtin, V. (2021). Freshwater availability status across countries for human and ecosystem needs. *Science of the Total Environment, 792*, 148230. https://doi.org/10.1016/j.scitotenv.2021.148230

Basuki, T. M., Nugroho, H. Y. S. H., Indrajaya, Y., Pramono, I. B., Nugroho, N. P., Supangat, A. B., et al. (2022). Improvement of integrated watershed management in Indonesia for mitigation and adaptation to climate change: A review. *Sustainability, 14*(16), 9997. https://doi.org/10.3390/su14169997

Belhassan, K. (2021). Water scarcity management. In A. Vaseashta & C. Maftei (Eds.), *Water safety, security and sustainability: Threat detection and mitigation* (pp. 443–462). Springer International Publisher. https://doi.org/10.1007/978-3-030-76008-3_19

Besada, H., & Werner, K. (2015). An assessment of the effects of Africa's water crisis on food security and management. *International Journal of Water Resources Development, 31*(1), 120–133. https://doi.org/10.1080/07900627.2014.905124

Bozorgzadeh, E., & Mousavi, S. J. (2023). Water-constrained green development framework based on economically-allocable water resources. *Scientific Reports, 13*(1), 5306. https://doi.org/10.1038/s41598-023-31550-7

Burt, Z., Leal, S., Workman, J., McElroy, M., & Bouhia, H. (2023). The design of climate-adaptive water subsidies: Financial incentives for urban water conservation in Morocco. *Journal of Water, Sanitation and Hygiene for Development, 13*(6), 424–432. https://doi.org/10.2166/washdev.2023.236

Bwapwa, J. K. (2018). Review on main issues causing deterioration of water quality and water scarcity: A case study of South Africa. *Environmental Management and Sustainable Development, 7*(3), 14–34.

Cantonati, M., Poikane, S., Pringle, C. M., Stevens, L. E., Turak, E., Heino, J., et al. (2020). Characteristics, main impacts, and stewardship of natural and artificial freshwater environments: Consequences for biodiversity conservation. *Water, 12*(1), 260.

Chen, H. E., Chiu, Y. Y., Tsai, T. L., & Yang, J. C. (2020). Effect of rainfall, runoff and infiltration processes on the stability of foot slopes. *Water, 12*(5), 1229. https://doi.org/10.3390/w12010260

Craswell, E. T. (2020). Erosion control: Soil conservation. In B. D. Fath & S. E. Jorgensen (Eds.), *Managing soils and terrestrial systems* (pp. 407–414). CRC Press, Taylor and Francis Group.

Cui, M., Ferreira, F., Fung, T. K., & Matos, J. S. (2021). Tale of two cities: How nature-based solutions help create adaptive and resilient urban water management practices in Singapore and Lisbon. *Sustainability, 13*(18), 10427. https://doi.org/10.3390/su131810427

Danert, K., & Healy, A. (2021). Monitoring groundwater use as a domestic water source by urban households: Analysis of data from Lagos State, Nigeria and Sub-Saharan Africa with implications for policy and practice. *Water, 13*(4), 568. https://doi.org/10.3390/w13040568

Datta, K., Chakraborty, S., & Roychoudhury, A. (2025). Management of soil, waste and water in the context of global climate change. In H. S. Jatav, T. Minkina, S. K. Singh, B. Singh, & V. D. Rajput (Eds.), *Environmental nexus for resource management* (pp. 1–26). CRC Press.

Datta, S., Sinha, D., Chaudhary, V., Kar, S., & Singh, A. (2022). Water pollution of wetlands: A global threat to inland, wetland, and aquatic phytodiversity. In A. K. Rathoure (Ed.), *Handbook of research on monitoring and evaluating the ecological health of wetlands* (pp. 27–50). IGI Global. https://doi.org/10.4018/978-1-7998-9498-8

Davids, R., Rouget, M., Burger, M., Mahood, K., Ditlhale, N., & Slotow, R. (2021). Civic ecology uplifts low-income communities, improves ecosystem services and well-being, and strengthens social cohesion. *Sustainability, 13*(3), 1300. https://doi.org/10.3390/su13031300

Doost, Z. H., Alsuwaiyan, M., & Yaseen, Z. M. (2024). Runoff management based water harvesting for better water resources sustainability: A comprehensive review. *Knowledge-Based Engineering and Sciences, 5*(1), 1–45. https://doi.org/10.51526/kbes.2024.5.1.1-45

du Plessis, A. (2023). Water resources from a global perspective. In *South Africa's water predicament: Freshwater's unceasing decline* (pp. 1–25). Springer International Publishing. https://doi.org/10.1007/978-3-031-24019-5_1

Dunkerley, D. (2021). The importance of incorporating rain intensity profiles in rainfall simulation studies of infiltration, runoff production, soil erosion, and related landsurface processes. *Journal of Hydrology, 603*, 126834. https://doi.org/10.1016/j.jhydrol.2021.126834

Ehret, P. J., Hodges, H. E., Kuehl, C., Brick, C., Mueller, S., & Anderson, S. E. (2021). Systematic review of household water conservation interventions using the information–motivation–behavioral skills model. *Environment and Behavior, 53*(5), 485–519. https://doi.org/10.1177/0013916519896868

Eke, J., Yusuf, A., Giwa, A., & Sodiq, A. (2020). The global status of desalination: An assessment of current desalination technologies, plants and capacity. *Desalination, 495*, 114633. https://doi.org/10.1016/j.desal.2020.114633

El-Beltagi, H. S., Basit, A., Mohamed, H. I., Ali, I., Ullah, S., Kamel, E. A., et al. (2022). Mulching as a sustainable water and soil saving practice in agriculture: A review. *Agronomy, 12*(8), 188. https://doi.org/10.3390/agronomy12081881

Elhassnaoui, I., Shehata Wahba, M. A., Wolde, S. G., Mohamed, B. D., & Moumen, A. (2023). The past, current, and future of the Africa Green Revolution: The case study of Kenya, Morocco, and Nigeria. *World Water Policy, 9*(4), 869–892. https://doi.org/10.1002/wwp2.12135

Fahad, S., Adnan, M., Noor, M., Arif, M., Alam, M., Khan, I. A., et al. (2019). Major constraints for global rice production. In *Advances in rice research for abiotic stress tolerance* (pp. 1–22). Woodhead Publishing. https://doi.org/10.1016/B978-0-12-814332-2.00001-0

Fayed, M. (2020). Drip irrigation technology: Principles, design, and evaluation. In E. S. Omran & A. Negm (Eds.), *Technological and modern irrigation environment in Egypt: Best management practices & evaluation* (pp. 275–303). Springer International Publishing. https://doi.org/10.1007/978-3-030-30375-4_13

Federal Government of Nigeria (FGN). (2004, July). *National water policy*. Retrieved August 24, 2024 from https://faolex.fao.org/docs/pdf/nig158231.pdf

Gautam, D., & Dahiya, P. (2022). Strategic re-use and recycling of grey water through treatment systems for resource recovery. In V. Kunmar & J. Singh (Eds.), *Microbial technologies for wastewater recycling and management* (pp. 281–295). CRC Press.

Gelfan, A., Panin, A., Kalugin, A., Morozova, P., Semenov, V., Sidorchuk, A., et al. (2023). Hydroclimatic processes as the primary drivers of the Early Khvalynian transgression of the Caspian Sea: New developments. *EGUsphere, 2023*, 1–28. https://doi.org/10.5194/hess-28-241-2024

Gupta, A. D., Pandey, P., Feijóo, A., Yaseen, Z. M., & Bokde, N. D. (2020). Smart water technology for efficient water resource management: A review. *Energies, 13*(23), 6268. https://doi.org/10.3390/en13236268

Gürsu, H. (2024). Assessment of theoretical and test performance considerations of concentrated solar water purification system "parabosol" in underserved regions. *Sustainability, 16*(15), 6611. https://doi.org/10.3390/su16156611

Hassan, A. S., Hayatu, J. M., & Mohammed, I. U. (2016). An overview of water supply infrastructural challenges in Nigeria: A case study of Taraba State. *IOSR Journal of Mechanical and Civil Engineering, 13*(1), 46–51. https://doi.org/10.9790/1684-13154651

He, C., Liu, Z., Wu, J., Pan, X., Fang, Z., Li, J., & Bryan, B. A. (2021). Future global urban water scarcity and potential solutions. *Nature Communications, 12*(1), 4667. https://doi.org/10.1038/s41467-021-25026-3

He, Z., Jia, G., Liu, Z., Zhang, Z., Yu, X., & Xiao, P. (2020). Field studies on the influence of rainfall intensity, vegetation cover and slope length on soil moisture infiltration on typical watersheds of the Loess Plateau, China. *Hydrological Processes, 34*(25), 4904–4919. https://doi.org/10.1002/hyp.13892

Herman, J. D., Quinn, J. D., Steinschneider, S., Giuliani, M., & Fletcher, S. (2020). Climate adaptation as a control problem: Review and perspectives on dynamic water resources planning under uncertainty. *Water Resources Research, 56*(2), e24389. https://doi.org/10.1029/2019WR025502

Hovis, M., Hollinger, J. C., Cubbage, F., Shear, T., Doll, B., Kurki-Fox, J. J., Line, D., Fox, A., Baldwin, M., Klondike, T., & Lovejoy, M. (2021). Natural infrastructure practices as potential flood storage and reduction for farms and rural communities in the North Carolina coastal plain. *Sustainability, 13*(16), 9309. https://doi.org/10.3390/su13169309

Huang, Z., Nya, E. L., Rahman, M. A., Mwamila, T. B., Cao, V., Gwenzi, W., & Noubactep, C. (2021). Integrated water resource management: Rethinking the contribution of rainwater harvesting. *Sustainability, 13*(15), 8338. https://doi.org/10.3390/su13158338

Hunt, K. M., Turner, A. G., & Shaffrey, L. C. (2020). The impacts of climate change on the winter water cycle of the western Himalaya. *Climate Dynamics, 55*(7), 2287–2307. https://doi.org/10.1007/s00382-020-05383-3

Huo, J., Liu, C., Yu, X., Jia, G., & Chen, L. (2021). Effects of watershed char and climate variables on annual runoff in different climatic zones in China. *Science of the Total Environment, 754*, 142157. https://doi.org/10.1016/j.scitotenv.2020.142157

Ibebuchi, C. C., & Abu, I. O. (2023). Rainfall variability patterns in Nigeria during the rainy season. *Scientific Reports, 13*(1), 7888. https://doi.org/10.1038/s41598-023-34970-7

Iqbal, R., Raza, M. A. S., Valipour, M., Saleem, M. F., Zaheer, M. S., Ahmad, S., et al. (2020). Potential agricultural and environmental benefits of mulches—A review. *Bulletin of the National Research Centre, 44*, 1–16. https://doi.org/10.1186/s42269-020-00290-3

Islam, M. S. (2023). Groundwater: Sources, functions, and quality. In *Hydrogeochemical evaluation and groundwater quality* (pp. 17–36). Springer Nature Switzerland. https://doi.org/10.1007/978-3-031-44304-6_2

Isukuru, E. J., Opha, J. O., Isaiah, O. W., Orovwighose, B., & Emmanuel, S. S. (2024). Nigeria's water crisis: Abundant water, polluted reality. *Cleaner Water, 2*(2024), 100026. https://doi.org/10.1016/j.clwat.2024.100026

Jamroen, C., Komkum, P., Fongkerd, C., & Krongpha, W. (2020). An intelligent irrigation scheduling system using low-cost wireless sensor network toward sustainable and precision agriculture. *IEEE Access, 8*, 172756–172769. https://doi.org/10.1109/ACCESS.2020.3025590

Jan, I., Ahmad, T., Wani, M. S., Dar, S. A., Wani, N. A., Malik, N. A., & Tantary, Y. R. (2022). Threats and consequences of untreated wastewater on freshwater environments. In G. H. Dar, R. A. Bhat, H. Qadri, & K. R. Hakeem (Eds.), *Microbial consortium and biotransformation for pollution decontamination* (pp. 1–26). Elsevier. https://doi.org/10.1016/B978-0-323-91893-0.00009-2

Jellason, N. P., Conway, J. S., Baines, R. N., & Ogbaga, C. C. (2021). A review of farming challenges and resilience management in the Sudano-Sahelian drylands of Nigeria in an era of climate change. *Journal of Arid Environments, 186*, 104398. https://doi.org/10.1016/j.jaridenv.2020.104398

Jiang, S., Zhang, Q., Werner, A. D., Wellen, C., Hu, P., Sun, J., et al. (2020). Modelling the impact of runoff generation on agricultural and urban phosphorus loading of the subtropical Poyang Lake (China). *Journal of Hydrology, 590*, 125490. https://doi.org/10.1016/j.jhydrol.2020.125490

Jiang, W., & Yu, J. (2022). Impact of rainstorm patterns on the urban flood process superimposed by flash floods and urban waterlogging based on a coupled hydrologic–hydraulic model: A case study in a coastal mountainous river basin within southeastern China. *Natural Hazards, 112*(1), 301–326. https://doi.org/10.1007/s11069-021-05182-x

Julio, N., Figueroa, R., & Ponce Oliva, R. D. (2021). Water resources and governance approaches: Insights for achieving water security. *Water, 13*(21), 3063. https://doi.org/10.3390/w13213063

Kaletova, T., & Nemetova, Z. (2017). Determination of surface runoff from the modelled area. *Environment, Earth and Ecology, 1*(1), 61–66. https://doi.org/10.24051/eee/68610

Kilemo, D. B. (2022). The review of water use efficiency and water productivity metrics and their role in sustainable water resources management. *Open Access Library Journal, 9*(1), 1–21. https://doi.org/10.4236/oalib.1107075

Koop, S. H., Grison, C., Eisenreich, S. J., Hofman, J., & van Leeuwen, K. (2022). Integrated water resources management in cities in the world: Global solutions. *Sustainable Cities and Society, 86*, 104137. https://doi.org/10.1016/j.scs.2022.104137

Kumar, A., & Thakur, A. (2024). Industrial water conservation by water footprint and sustainable development goals. In S. A. Bandh & F. A. Malla (Eds.), *Current directions in water scarcity research* (Vol. 8, pp. 87–117). Elsevier. https://doi.org/10.1016/B978-0-443-23631-0.00007-8

Kumawat, A., Yadav, D., Samadharmam, K., & Rashmi, I. (2020). Soil and water conservation measures for agricultural sustainability. In R. S. Meena & R. Datta (Eds.), *Soil moisture importance* (pp. 23–44). IntechOpen Limited. https://doi.org/10.5772/intechopen.92895

Kummu, M., Guillaume, J. H., de Moel, H., Eisner, S., Flörke, M., Porkka, M., et al. (2016). The world's road to water scarcity: Shortage and stress in the 20th century and pathways towards sustainability. *Scientific Reports, 6*(1), 1–16. https://doi.org/10.1038/srep38495

Lachassagne, P., Dewandel, B., & Wyns, R. (2021). Hydrogeology of weathered crystalline/hard-rock aquifers—Guidelines for the operational survey and management of their groundwater resources. *Hydrogeology Journal, 29*(8), 2561–2594. https://doi.org/10.1007/s10040-021-02339-7

Lakhiar, I. A., Yan, H., Zhang, C., Wang, G., He, B., Hao, B., Han, Y., Wang, B., Bao, R., Syed, T. N., Chauhdary, J. N., & Rakibuzzaman, M. (2024). A review of precision irrigation water-saving technology under changing climate for enhancing water use efficiency, crop yield, and environmental footprints. *Agriculture, 14*(7), 1141. https://doi.org/10.3390/agriculture14071141

Leal Filho, W., Azeiteiro, U. M., Balogun, A. L., Setti, A. F. F., Mucova, S. A., Ayal, D., Totin, E., Lydia, A. M., Kalaba, F. K., & Oguge, N. O. (2021). The influence of ecosystems services depletion to climate change adaptation efforts in Africa. *Science of the Total Environment, 779*, 146414. https://doi.org/10.1016/j.scitotenv.2021.146414

Leal Filho, W., Totin, E., Franke, J. A., Andrew, S. M., Abubakar, I. R., Azadi, H., Nunn, P. D., Ouweneel, B., Williams, P. A., & Simpson, N. P. (2022). Understanding responses to climate-related water scarcity in Africa. *Science of the Total Environment, 806*, 150420. https://doi.org/10.1016/j.scitotenv.2021.150420

Lee, J., Perera, D., Glickman, T., & Taing, L. (2020). Water-related disasters and their health impacts: A global review. *Progress in Disaster Science, 8*, 100123. https://doi.org/10.1016/j.pdisas.2020.100123

Liang, Z., Liu, X., Xiong, J., & Xiao, J. (2020). Water allocation and integrative management of precision irrigation: A systematic review. *Water, 12*(11), 3135. https://doi.org/10.3390/w12113135

Liu, Y., & Song, W. (2020). Modelling crop yield, water consumption, and water use efficiency for sustainable agroecosystem management. *Journal of Cleaner Production, 253*, 119940. https://doi.org/10.1016/j.jclepro.2019.119940

Lukenga, W. (2015). Water resource management. In *Book boon, London* (pp. 18–19). Retrieved August 12, 2024, from https://d1wqtxts1xzle7.cloudfront.net/45638278/Water_resource_management-libre.pdf?1463303674

Marques, L. (2020). Water and soil. In *Capitalism and environmental collapse* (pp. 65–96). Springer International Publishing. https://doi.org/10.1007/978-3-030-47527-7

Marszałek, A., Dudziak, M., & Kudlek, E. (2023). Analysis of organic and inorganic compounds in rainwater from the highway and its treatment in membrane processes. *Desalination and Water Treatment, 305*, 217–226. https://doi.org/10.5004/dwt.2023.29836

Martínez-Santos, P., Martín-Loeches, M., Díaz-Alcaide, S., & Danert, K. (2020). Manual borehole drilling as a cost-effective solution for drinking water access in low-income contexts. *Water, 12*(7), 1981. https://doi.org/10.3390/w12071981

Mazurkiewicz, K., Jeż-Walkowiak, J., & Michałkiewicz, M. (2022). Physicochemical and microbiological quality of rainwater harvested in underground retention tanks. *Science of the Total Environment, 814*, 152701. https://doi.org/10.1016/j.scitotenv.2021.152701

Md Anawar, H., & Chowdhury, R. (2020). Remediation of polluted river water by biological, chemical, ecological and engineering processes. *Sustainability, 12*(17), 7017. https://doi.org/10.3390/su12177017

Mekonnen, M. M., & Hoekstra, A. Y. (2016). Four billion people facing severe water scarcity. *Science Advances, 2*(2), e1500323. https://doi.org/10.1126/sciadv.1500323

Mgolozeli, S., Nciizah, A. D., Wakindiki, I. I., & Mudau, F. N. (2020). Innovative pro-smallholder farmers' permanent mulch for better soil quality and food security under conservation agriculture. *Agronomy, 10*(4), 605. https://doi.org/10.3390/agronomy10040605

Mir, B. H., Kumar, R., Lone, M. A., & Tantray, F. A. (2021). Climate change and water resources of Himalayan region—Review of impacts and implication. *Arabian Journal of Geosciences, 14*, 1–14. https://doi.org/10.1007/s12517-021-07438-z

Mishra, R. K. (2023). Fresh water availability and its global challenge. *British Journal of Multidisciplinary and Advanced Studies, 4*(3), 1–78. https://doi.org/10.37745/bjmas.2022.0208

Molden, D. (2020). Scarcity of water or scarcity of management? *International Journal of Water Resources Development, 36*(2–3), 258–268. https://doi.org/10.1080/07900627.2019.1676204

Moreno, G. D., Schmidt, L. A., Ritchie, L. D., McCulloch, C. E., Cabana, M. D., Brindis, C. D., Green, L. W., Altman, E. A., & Patel, A. I. (2021). A cluster-randomized controlled trial of an elementary school drinking water access and promotion intervention: Rationale, study design, and protocol. *Contemporary Clinical Trials, 101*, 106255. https://doi.org/10.1016/j.cct.2020.106255

Moslen, M., Miebaka, C. A., & Izah, S. C. (2024). Occurrence and management of toxicity associated with parasitic pathogens in drinking water sources in rural areas in the Niger Delta, Nigeria. In S. Madhav, A. L. Srivastav, S. C. Izah, & E. V. Hullebusch (Eds.), *Water resources management for rural development* (pp. 275–286). Elsevier. https://doi.org/10.1016/B978-0-443-18778-0.00009-X

Mushtaq, B., Bandh, S. A., Shafi, S., Mushtaq, B., Bandh, S. A., & Shafi, S. (2020). Management of water resources. In *Environmental management: Environmental issues, awareness and abatement* (pp. 1–46). Springer. https://doi.org/10.1016/B978-0-443-18778-0.00009-X

Naik, P. K. (2017). Water crisis in Africa: Myth or reality? *International Journal of Water Resources Development, 33*(2), 326–339. https://doi.org/10.1080/07900627.2016.1188266

Nair, J. V., Verma, J., & Shinde, V. R. (2024). Enriching and maintaining the riparian buffer zone. In *Managing urban rivers* (pp. 161–180). Elsevier. https://doi.org/10.1016/B978-0-323-85703-1.00018-3

Nanekely, M. A. A. (2020). A regulatory directive technical framework for sustainable water management in the semi-arid climates. *University of Salford (United Kingdom).* Retrieved August 03, 2024, from https://www.proquest.com/pq1academic/dissertations-theses/regulatory-directive-technical-framework/docview/2636115357/sem-2?accountid=145057

Naves, A., Samper, J., Pisani, B., Mon, A., Dafonte, J., Montenegro, L., & García-Tomillo, A. (2021). Hydrogeology and groundwater management in a coastal granitic area with steep slopes in Galicia (Spain). *Hydrogeology Journal, 29*(8), 2655–2669. https://doi.org/10.1007/s10040-021-02349-5

Ndimele, P. E., Ojewole, A. E., Mekuleyi, G. O., Badmos, L. A., Agosu, C. M., Olatunbosun, E. S., Lawal, O. O., Shittu, J. A., Joseph, O. O., Ositimehin, K. M., Ndimele, F. C., & Ayodele, O. T. (2024). Vulnerability, resilience and adaptation of Lagos coastal communities to flooding. *Earth Science, Systems and Society, 4*, 10087. https://doi.org/10.3389/esss.2024.10087

Neto, S., & Camkin, J. (2022). A water bridge over a troubled world: The need for an intersectional approach. *World Water Policy, 8*(1), 6–8. https://doi.org/10.1002/wwp2.12075

Ngene, B. U., Nwafor, C. O., Bamigboye, G. O., Ogbiye, A. S., Ogundare, J. O., & Akpan, V. E. (2021). Assessment of water resources development and exploitation in Nigeria: A review of integrated water resources management approach. *Heliyon, 7*(1), 12–30. https://doi.org/10.1016/j.heliyon.2021.e05955

Niziolomski, J. C., Simmons, R. W., Rickson, R. J., & Hann, M. J. (2020). Efficacy of mulch and tillage options to reduce runoff and soil loss from asparagus interrows. *CATENA, 191*, 104557. https://doi.org/10.1016/j.catena.2020.104557

Norman, L. M., Lal, R., Wohl, E., Fairfax, E., Gellis, A. C., & Pollock, M. M. (2022). Natural infrastructure in dryland streams (NIDS) can establish regenerative wetland sinks that reverse desertification and strengthen climate resilience. *Science of the Total Environment, 849*, 157738. https://doi.org/10.1016/j.scitotenv.2022.157738

Odife, I. (2022). Was there light at the end of the tunnel? Interrogating rural healthcare delivery in Anambra state, Nigeria, 1976–1991. *Ohazurume-Unizik Journal of Culture and Civilization, 1*(1), 1–10.

Ogunrinde, A. T., Enaboifo, M. A., Olotu, Y., Pham, Q. B., & Tayo, A. B. (2021a). Characterization of drought using four drought indices under climate change in the Sahel region of Nigeria: 1981–2015. *Theoretical and Applied Climatology, 143*, 843–860. https://doi.org/10.1007/s00704-020-03453-4

Ogunrinde, A. T., Oguntunde, P. G., Akinwumiju, A. S., & Fasinmirin, J. T. (2021b). Evaluation of the impact of climate change on the characteristics of drought in Sahel Region of Nigeria: 1971–2060. *African Geographical Review, 40*(2), 192–210. https://doi.org/10.1080/19376812.2020.1814826

Olabode, O. F., & Comte, J. C. (2024). Water scarcity in the fast-growing megacity of Lagos, Nigeria and opportunities for managed aquifer recharge. *Wiley Interdisciplinary Reviews: Water, 11*(5), e1733. https://doi.org/10.1002/wat2.1733

Olalekan, A. S., Adewoye, S. O., Henry, S. O., Olaniyi, O. A., & Raimi, M. O. (2023). Comprehensive understanding of hydrogeochemical evaluation of seasonal variability in groundwater quality dynamics in the gold mining areas of Osun state, Nigeria. *International Journal of Hydrology, 7*(5), 206–220. https://doi.org/10.15406/ijh.2023.07.00359

Olaniyi, T. K., Nwankwo, N., & Idahose, U. E. (2023). Assessment of health, safety and environmental implications for water quality management in the Global South: A case study of Lagos,

Federal Republic of Nigeria. *International Journal of e-Healthcare Information Systems, 9*(1), 239–249. https://doi.org/10.20533/ijehis.2046.3332.2023.0032

Olatunde, T. M., Adelani, F. A., & Sikhakhane, Z. Q. (2024). A review of smart water management systems from Africa and the United States. *Engineering Science & Technology Journal, 5*(4), 1231–1242. https://doi.org/10.51594/estj.v5i4.1014

Olise, C. N., Emeh, I. E., & Amujiri, B. A. (2023). Water Sanitation and Hygiene (WASH) programme and the hygiene situation in Anambra State: A focus on Aguata and Anambra East local government areas. *African Renaissance, 20*(1), 215. https://hdl.handle.net/10520/ejc-aa_afren_v20_n1_a12

Omotosho, O. A., & Sangodoyin, A. Y. (2013). Production and utilization of cassava peel activated carbon in treatment of effluent from cassava processing industry. *Water Practice and Technology, 8*(2), 215–224. https://doi.org/10.2166/wpt.2013.023

Palermo, S. A., Maiolo, M., Brusco, A. C., Turco, M., Pirouz, B., Greco, E., et al. (2022). Smart technologies for water resource management: An overview. *Sensors, 22*(16), 6225. https://doi.org/10.3390/s22166225

Parums, D. V. (2021). Review articles, systematic reviews, meta-analysis, and the updated preferred reporting items for systematic reviews and meta-analyses (PRISMA) 2020 guidelines. *Medical Science Monitor: International Medical Journal of Experimental and Clinical Research, 27*, e934475-1. https://doi.org/10.12659/MSM.934475

Peña-Ramos, J. A., José López-Bedmar, R., Sastre, F. J., & Martínez-Martínez, A. (2022). Water conflicts in sub-Saharan Africa. *Frontiers in Environmental Science, 10*, 863903. https://doi.org/10.3389/fenvs.2022.863903

Poudel, D. D. (2020). Asta-Ja framework: A peaceful approach to food, water, climate, and environmental security coupled with sustainable economic development and social inclusion in Nepal. *Strategic Planning for Energy and the Environment, 39*(1–4), 243–318. https://doi.org/10.13052/spee1048-4236.391412

Qiu, J., Shen, Z., & Xie, H. (2023). Drought impacts on hydrology and water quality under climate change. *Science of the Total Environment, 858*, 159854. https://doi.org/10.1016/j.scitotenv.2022.159854

Raghunath, H. M. (2006). *Hydrology principles, analysis and design*. New Age International Limited, Publishers.

Rahman, A. (2021). Rainwater harvesting for sustainable developments: Non-potable use, household irrigation and stormwater management. *Water, 13*(23), 3460. https://doi.org/10.3390/w13233460

Rakib, M. A., Sasaki, J., Matsuda, H., Quraishi, S. B., Mahmud, M. J., Bodrud-Doza, M., Ullah, A. A., Fatema, K. J., Newaz, M. A., & Bhuiyan, M. A. (2020). Groundwater salinization and associated co-contamination risk increase severe drinking water vulnerabilities in the southwestern coast of Bangladesh. *Chemosphere, 246*, 125646. https://doi.org/10.1016/j.chemosphere.2019.125646

Rammal, M., & Berthier, E. (2020). Runoff losses on urban surfaces during frequent rainfall events: A review of observations and modeling attempts. *Water, 12*(10), 2777. https://doi.org/10.3390/w12102777

Ray, S., & Majumder, S. (2024). Water management in agriculture: Innovations for efficient irrigation. In P. Sil, P. Chhetri, S. Majumder, & D. T. Santos (Eds.), *Modern agronomy* (pp. 169–185). Santish Serial Publishing House.

Rey-Romero, D. C., Domínguez, I., & Oviedo-Ocaña, E. R. (2022). Effect of agricultural activities on surface water quality from Páramo ecosystems. *Environmental Science and Pollution Research, 29*(55), 83169–83190. https://doi.org/10.1007/s11356-022-21709-6

Richards, S., Rao, L., Connelly, S., Raj, A., Raveendran, L., Shirin, S., et al. (2021). Sustainable water resources through harvesting rainwater and the effectiveness of a low-cost water treatment. *Journal of Environmental Management, 286*, 112223. https://doi.org/10.1016/j.jenvman.2021.112223

Rignot, E. (2022). Sea level rise from melting glaciers and ice sheets caused by climate warming above pre-industrial levels. *Uspekhi Fizicheskikh Nauk, 192*(11), 1203–1213. https://doi.org/10.3367/UFNe.2021.11.039106

Ritchie, H., Eisma, J. A., & Parker, A. (2021). Sand dams as a potential solution to rural water security in drylands: Existing research and future opportunities. *Frontiers in Water, 3*, 651954. https://doi.org/10.3389/frwa.2021.651954

Rodrigues, A. M., Formiga, K. T. M., & Milograna, J. (2023). Integrated systems for rainwater harvesting and greywater reuse: A systematic review of urban water management strategies. *Water Supply, 23*(10), 4112–4125. https://doi.org/10.2166/ws.2023.240

Saeedi, I., & Goodarzi, M. (2020). Rainwater harvesting system: A sustainable method for landscape development in semiarid regions, the case of Malayer University campus in Iran. *Environment, Development and Sustainability, 22*(2), 1579–1598. https://doi.org/10.1007/s10668-018-0218-8

Saha, R., Mitran, T., Mukherjee, S., Das, I. C., & Kumar, K. V. (2021). Groundwater management for irrigated agriculture through geospatial techniques. In T. Mitran, R. S. Meena, & A. Chakraborty (Eds.), *Geospatial technologies for crops and soils* (pp. 455–488). Springer International Publishing. https://doi.org/10.1007/978-981-15-6864-0_13

Santos, E., Carvalho, M., & Martins, S. (2023). Sustainable water management: Understanding the socioeconomic and cultural dimensions. *Sustainability, 15*(17), 13074. https://doi.org/10.3390/su151713074

Saravanan, A., Kumar, P. S., Jeevanantham, S., Karishma, S., Tajsabreen, B., Yaashikaa, P. R., & Reshma, B. (2021). Effective water/wastewater treatment methodologies for toxic pollutants removal: Processes and applications towards sustainable development. *Chemosphere, 280*, 130595. https://doi.org/10.1016/j.chemosphere.2021.130595

Sarkar, A. (2020). Everyday practices of poor urban women to access water: Lived realities from a Nairobi slum. *African Studies, 79*(2), 212–231. https://doi.org/10.1080/00020184.2020.1781594

Sasanya, B. F., Ademola, A. A., & Adesogan, S. O. (2024). Spatiotemporal trend analysis of precipitation, temperature, and soil moisture in selected agrarian locations within Nigeria—A comparative study. *Arabian Journal of Geosciences, 17*(8), 228–240. https://doi.org/10.1007/s12517-024-12026-y

Sasanya, B. F., & Adesogan, S. O. (2021). Water scarcity—The driving force of under development in rural areas of developing countries. *African Scientific, Research and Innovation Council Journal of Natural Science, 1*(2021), 81–89.

Sasanya, B. F., & Adesogan, S. O. (2023). A simple technique to enhance runoff quality for water supply in a Nigerian rural community: A pilot scale study. *Sustainable Water Management, 9*(190), 1–10. https://doi.org/10.1007/s40899-023-00978-0

Seigerman, C. K., McKay, S. K., Basilio, R., Biesel, S. A., Hallemeier, J., Mansur, A. V., Piercy, C., Rowan, S., Ubiali, B., Yeates, E., & Nelson, D. R. (2023). Operationalizing equity for integrated water resources management. *JAWRA Journal of the American Water Resources Association, 59*(2), 281–298. https://doi.org/10.1111/1752-1688.13086

Sen, Z. (2020). Water structures and climate change impact: A review. *Water Resources Management, 34*(13), 4197–4216. https://doi.org/10.1007/s11269-020-02665-7

Shahangian, S. A., Tabesh, M., Yazdanpanah, M., Zobeidi, T., & Raoof, M. A. (2022). Promoting the adoption of residential water conservation behaviors as a preventive policy to sustainable urban water management. *Journal of Environmental Management, 313*, 115005. https://doi.org/10.1016/j.jenvman.2022.115005

Shan, V., Singh, S. K., & Haritash, A. K. (2020). Water crisis in the Asian countries: Status and future trends. In M. Kumar, F. Munoz-Arriola, H. Furumai, & T. Chaminda (Eds.), *Resilience, response, and risk in water systems: Shifting management and natural forcings paradigms* (pp. 173–194). Springer International Publisher. https://doi.org/10.1007/978-981-15-4668-6_10

Sharma, R., Vymazal, J., & Malaviya, P. (2021). Application of floating treatment wetlands for stormwater runoff: A critical review of the recent developments with emphasis on heavy metals and nutrient removal. *Science of the Total Environment, 777*, 146044. https://doi.org/10.1016/j.scitotenv.2021.146044

Shi, L., Ahmad, S., Shukla, P., & Yupho, S. (2021). Shared injustice, splintered solidarity: Water governance across urban-rural divides. *Global Environmental Change, 70*, 102354. https://doi.org/10.1016/j.gloenvcha.2021.102354

Shiru, M. S., Shahid, S., Shiru, S., Chung, E. S., Alias, N., Ahmed, K., Dioha, E. C., Sa'adi, Z., Salman, S., Noor, M., Nashwan, M. S., et al. (2020). Challenges in water resources of Lagos mega city of Nigeria in the context of climate change. *Journal of Water and Climate Change, 11*(4), 1067–1083. https://doi.org/10.2166/wcc.2019.047

Shokri, A., & Fard, M. S. (2022). A sustainable approach in water desalination with the integration of renewable energy sources: Environmental engineering challenges and perspectives. *Environmental Advances, 9*, 100281. https://doi.org/10.1016/j.envadv.2022.100281

Shukla, A. K., Ojha, C. S. P., Shukla, S., & Garg, R. D. (2021). Water quality challenges in Ganga River Basin, India. In M. S. Chauhan & C. S. P. Ojha (Eds.), *The Ganga River Basin: A hydrometeorological approach* (pp.1–19). Springer. https://doi.org/10.1007/978-3-030-60869-9_1

Shyam, G. M., Taloor, A. K., Singh, S. K., & Kanga, S. (2021). Sustainable water management using rainfall-runoff modeling: A geospatial approach. *Groundwater for Sustainable Development, 15*, 100676. https://doi.org/10.1016/j.gsd.2021.100676

Simpson, I. M., Winston, R. J., & Dorsey, J. D. (2023). Monitoring the effects of urban and forested land uses on runoff quality: Implications for improved stormwater management. *Science of the Total Environment, 862*, 160827. https://doi.org/10.1016/j.scitotenv.2022.160827

Sitterson, J., Knightes, C., Parmar, R., Wolfe, K., Muche, M., Avant, B., & Muche, M. (2018). An overview of rainfall-runoff model types. In *International Congress on Environmental Modelling and Software*. Retrieved August 3, 2024, from https://scholarsarchive.byu.edu/iemssconference/2018/Stream-C/41

Srivastava, A., & Chinnasamy, P. (2021). Developing village-level water management plans against extreme climatic events in Maharashtra (India)—A case study approach. In A. Vaseashta & C. Maftei (Eds.), *Water safety, security and sustainability: Threat detection and mitigation* (pp. 615–635). Springer International Publisher. https://doi.org/10.1007/978-3-030-76008-3_27

Suganthi, S., & Elango, L. (2020). Estimation of groundwater abstraction induced land subsidence by SBAS technique. *Journal of Earth System Science, 129*(1), 46. https://doi.org/10.1007/s12040-019-1298-z

Tabari, H. (2020). Climate change impact on flood and extreme precipitation increases with water availability. *Scientific Reports, 10*(1), 13768. https://doi.org/10.1038/s41598-020-70816-2

Tariq, A., & Mushtaq, A. (2023). Untreated wastewater reasons and causes: A review of most affected areas and cities. *International Journal of Chemical and Biochemical Sciences, 23*(1), 121–143.

Tay, C. K. (2017). Hydrogeochemical framework and factor analysis of fluoride contamination in groundwater within the Savelugu-Nanton district, northern Ghana. *West African Journal of Applied Ecology, 25*(1), 33–55.

Tengan, B. M., & Akoto, O. (2022). Comprehensive evaluation of the possible impact of roofing materials on the quality of harvested rainwater for human consumption. *Science of the Total Environment, 819*, 152966. https://doi.org/10.1016/j.scitotenv.2022.152966

Teston, A., Piccinini Scolaro, T., Kuntz Maykot, J., & Ghisi, E. (2022). Comprehensive environmental assessment of rainwater harvesting systems: A literature review. *Water, 14*(17), 2716. https://doi.org/10.3390/w14172716

Thomas, J., & Harden, A. (2008). Methods for the thematic synthesis of qualitative research in systematic reviews. *BMC Medical Research Methodology, 8*(45), 1–10. https://doi.org/10.1186/1471-2288-8-45

Tzanakakis, V. A., Paranychianakis, N. V., & Angelakis, A. N. (2020). Water supply and water scarcity. *Water, 12*(9), 2347. https://doi.org/10.3390/w12092347

Ufuoma, G., Sasanya, B. F., Abaje, P., & Awodutire, P. (2021). Efficiency of camera sensors for flood monitoring and warnings. *Scientific African, 13*(2021), 1–8. https://doi.org/10.1016/j.sciaf.2021.e00887

United Nations (UN). (2016). *The sustainable development goals (SDG) report*. Retrieved August 12, 2017, from https://unstats.un.org/sdgs/report/2016/The%20Sustainable%20Development%20Goals%20Report%202016.pdf

United State Agency for International Development (USAID). (2010, March). *Nigeria water and sanitation profile*. Retrieved August 12, 2017, from www.washplus.org/sites/default/files/nigeria.pdf

Varma, V. G. (2022). Water-efficient technologies for sustainable development. In A. Srivastav, S. Madhav, A. K. Bhardwaj, & E. Valsami-Jones (Eds.), *Current directions in water scarcity research* (Vol. 6, pp. 101–128). Elsevier. https://doi.org/10.1016/B978-0-323-91838-1.00009-9

Vasseghian, Y., Nadagouda, M. M., & Aminabhavi, T. M. (2024). Biochar-enhanced bioremediation of eutrophic waters impacted by algal blooms. *Journal of Environmental Management, 367*, 122044. https://doi.org/10.1016/j.jenvman.2024.122044

Verdonschot, P. F. M., & Verdonschot, R. C. M. (2023). The role of stream restoration in enhancing ecosystem services. *Hydrobiologia, 850*(12), 2537–2562. https://doi.org/10.1007/s10750-022-04918-5

Verlicchi, P., & Ghirardini, A. (2022). Contribution of land runoff to the release of pesticides into water bodies in arable areas. In M. S. Rodríguez-Cruz & M. J. Sánchez-Martín (Eds.), *Pesticides in soils: Occurrence, fate, control and remediation* (pp. 225–249). Springer International Publishing. https://doi.org/10.1007/698_2021_802

Vinci, G., Maddaloni, L., Mancini, L., Prencipe, S. A., Ruggeri, M., & Tiradritti, M. (2021). The health of the water planet: Challenges and opportunities in the Mediterranean area. An overview. *Earth, 2*(4), 894–919. https://doi.org/10.3390/earth2040052

Vollmer, D., Burkhard, K., Adem Esmail, B., Guerrero, P., & Nagabhatla, N. (2022). Incorporating ecosystem services into water resources management—Tools, policies, promising pathways. *Environmental Management, 69*(4), 627–635. https://doi.org/10.1007/s00267-022-01640-9

Vushe, A. (2021). Proposed research, science, technology, and innovation to address current and future challenges of climate change and water resource management in Africa. In S. Diop, P. Scheren, & A. Niang (Eds.), *Climate change and water resources in Africa: Perspectives and solutions towards an imminent water crisis* (pp. 489–518). Springer International Publishing. https://doi.org/10.1007/978-3-030-61225-2_21

Walker, S. J. (2022). Sustainable safe water and sanitation interventions in remote parts of Ghana. In J. Rajapakse (Eds.), *Safe water and sanitation for a healthier world: A global view of progress towards SDG 6* (pp. 75–95). Springer International Publishing. https://doi.org/10.1007/978-3-030-94020-1_5

Wang, W., Zhao, X., Cao, J., Li, H., & Zhang, Q. (2020). Barriers and requirements to climate change adaptation of mountainous rural communities in developing countries: The case of the eastern Qinghai-Tibetan Plateau of China. *Land Use Policy, 95*, 104354. https://doi.org/10.1016/j.landusepol.2019.104354

Wang, Y. S., & Gu, J. D. (2021). Ecological responses, adaptation and mechanisms of mangrove wetland ecosystem to global climate change and anthropogenic activities. *International Biodeterioration & Biodegradation, 162*, 105248. https://doi.org/10.1016/j.ibiod.2021.105248

Wato, T., Amare, M., Bonga, E., Demand, B. B. O., & Coalition, B. B. R. (2020). The agricultural water pollution and its minimization strategies—A review. *Journal of Resources Development and Management, 64*, 10–22. https://doi.org/10.7176/JRDM/64-02

Weatherl, R. K., Henao Salgado, M. J., Ramgraber, M., Moeck, C., & Schirmer, M. (2021). Estimating surface runoff and groundwater recharge in an urban catchment using a water balance approach. *Hydrogeology Journal, 29*(7), 2411–2428. https://doi.org/10.1007/s10040-021-02385-1

Weerasooriya, R. R., Liyanage, L. P. K., Rathnappriya, R. H. K., Bandara, W. B. M. A. C., Perera, T. A. N. T., Gunarathna, M. H. J. P., & Jayasinghe, G. Y. (2021). Industrial water conservation by water footprint and sustainable development goals: A review. *Environment, Development and Sustainability, 23*(9), 12661–12709. https://doi.org/10.1007/s10668-020-01184-0

Wells, W., & Vermeer, K. (2020). *Blueprint for greening affordable housing*. Island Press.

Yang, D., Yang, Y., & Xia, J. (2021). Hydrological cycle and water resources in a changing world: A review. *Geography and Sustainability, 2*(2), 115–122. https://doi.org/10.1016/j.geosus.2021.05.003

Yenehun, A., Nigate, F., Belay, A. S., Desta, M. T., Van Camp, M., & Walraevens, K. (2020). Groundwater recharge and water table response to changing conditions for aquifers at different physiography: The case of a semi-humid river catchment, northwestern highlands of Ethiopia. *Science of the Total Environment, 748*, 142243. https://doi.org/10.1016/j.scitotenv.2020.142243

Yusuf, A., Sodiq, A., Giwa, A., Eke, J., Pikuda, O., De Luca, G., Di Salvo, J. L., & Chakraborty, S. (2020). A review of emerging trends in membrane science and technology for sustainable water treatment. *Journal of Cleaner Production, 266*, 121867. https://doi.org/10.1016/j.jclepro.2020.121867

Zdeb, M., Zamorska, J., Papciak, D., & Słyś, D. (2020). The quality of rainwater collected from roofs and the possibility of its economic use. *Resources, 9*(2), 12. https://doi.org/10.3390/resources9020012

Zhao, J., Zhuang, M., Liu, Z., Zhao, C., Pullens, J. W., Liu, K., Harrison, M. T., & Yang, X. (2024). Climate-adaptive crop distribution can feed food demand, improve water scarcity, and reduce greenhouse gas emissions. *Science of the Total Environment, 944*, 173819. https://doi.org/10.1016/j.scitotenv.2024.173819

Chapter 10
Carbon Farming to Carbon Trading—Can Translating Science into Commerce Reverse Climate Change?

Siddhartha Shankar Bhattacharyya, Robert Ugochukwu Onyeneke, Fernanda Figueiredo Granja Dorilêo Leite, Emeka Emmanuel Osuji, Sikiru Ibrahim-Olesin, and Ahmed Khairul Hasan

Abstract Curbing atmospheric carbon dioxide (CO_2) concentrations requires in-depth understanding of the existing synergies between carbon farming and carbon trading practices. Carbon emission reduction, agricultural, economic, and environmental sustainability can be simultaneously achieved by transforming scientific knowledge into profitable ventures that fully benefit farmers and practitioners. In this chapter, we explore the possibilities of translating scientific knowledge of carbon farming into carbon trading, emphasizing the driving forces and factors that lead to sustainable carbon markets. More precisely, this chapter aims to delineate how carbon farming and ecosystem-based carbon market initiatives might be utilized to incentivize farmers and investors to embrace carbon-sequestering strategies for reversing global climate change. Carbon farming practices can sequester a considerable amount of carbon, however, improvements are needed in carbon trading practices to establish

---

S. S. Bhattacharyya (✉)
Soil and Crop Sciences, Texas A&M University, College Station, TX, USA
e-mail: ss_bhattacharyya@tamu.edu

Texas A&M AgriLife Research, College Station, TX, USA

R. U. Onyeneke · E. E. Osuji
Department of Agricultural Economics, Alex Ekwueme Federal University Ndufu-Alike, Ikwo, Nigeria
e-mail: robert.onyeneke@funai.edu.ng

F. F. G. D. Leite
Geochemistry Department (Graduate Program of Environmental Geochemistry), Chemistry Institute, Federal Fluminense University, Niterói, Rio de Janeiro, Brazil

S. Ibrahim-Olesin
Department of Agricultural Extension, Alex Ekwueme Federal University Ndufu-Alike, Ikwo, Nigeria

A. K. Hasan
Department of Agronomy, Bangladesh Agricultural University, Mymensingh, Bangladesh

R. U. Onyeneke et al. (eds.), *Energy Transition, Climate Action and Sustainable Agriculture*, https://doi.org/10.1007/978-3-031-83165-2_10

fair carbon pricing, address prevailing disparities among buyers, sellers, and extension workers, and formulate persuasive action plans, permits and trading conditions. In essence, increased regulatory support and the development of protocols at local, regional, and national level are required to monetize the carbon farming practices, sustain their benefits, and increase their adoption rate.

Keywords Climate finance · Carbon farming · Carbon markets · Carbon emissions · Climate change · Cap-and-trade · Emission trading

10.1 Introduction

Empirical findings provide a strong correlation between rising atmospheric carbon dioxide (CO_2) emissions and climate change (Gür, 2022; Kellogg & Schware, 2019; Peters et al., 2020; Rehman et al., 2021; Terrenoire et al., 2019; Wang et al., 2020a) and suggest adopting carbon farming practices (Nath et al., 2015) either or pricing CO_2 emissions (Verbruggen, 2021) to reduce the magnitude and volume of carbon emissions. Carbon farming refers to the diverse array of agricultural practices and land use systems that can sequester atmospheric CO_2 in soil, vegetations and/biomass of plants (Antle et al., 2018; Jansson et al., 2021; Nath et al., 2015). Successful carbon farming occurs when carbon losses are outweighed by carbon gains due to improved land management and/or conservation techniques (Kondvilkar & Thakare, 2021; Smith et al., 2014). Likewise, carbon trading, an integral part of carbon farming refers to the pricing of carbon dioxide (CO_2) emissions through different carbon trading systems including emissions trading system (ETS), global uniform carbon tax or other schemes (Smith et al., 2014; Verbruggen, 2021). Translating the potential linkages between carbon farming and trade is necessary if we want to limit global warming to 1.5 °C by 2050 and mitigate the catastrophic effects of climate change (Kondvilkar & Thakare, 2021). In essence, climate change mitigation requires significant scientific advances, socio-political changes and unified broad policy strategies that can address synergies and trade-offs between climate policy tools and environmental protection (Essl et al., 2017; Oliveira & Thorseth, 2016). Clearly, carbon farming and trading can both offer economic, environmental protection while mitigating climate change.

Studies have shown that different carbon farming strategies e.g., conservation agriculture, agroforestry, woody biomass crop stubble management, and conversion from annual to perennial crops or pastures can increase carbon stock in agricultural soils and decrease the amount of carbon emitted back into the atmosphere (Gonzalez-Sanchez et al., 2019; Khataza et al., 2017; Nath et al., 2015). Concrete examples from literatures show that conservation agriculture in African soils can mitigate 524 Tg CO_2 yr^{-1} (Gonzalez-Sanchez et al., 2019), three common agroforestry systems (hedgerows, shelterbelts and silvopasture) have decreased around 2566 MT CO_2 over 20 years in Central Alberta, Canada (An et al., 2022), and sustainable pasture management (intensification and diversification) can mitigate

1.47–3.48 Mg CO_2 ha^{-1} yr^{-1} in Brazil (Damian et al., 2021). These examples indicate that carbon farming may create an opportunity for farmers and investors to earn tradable carbon offsets through carbon trading (Nath et al., 2015). For instance, in Central Alberta, Canada, An et al. (2022) calculated that the price of soil stored carbon through agroforestry is around $102.7 billion over 20 years. In addition, Becker et al. (2013) demonstrated that carbon farming (large-scale plantation) reduced around 18T CO_2 yr^{-1} ha^{-1} over a 20-year period, and the trade value was 42–63 euro for per ton CO_2 in a hot, arid coastal location. These instances show that carbon farming is a viable source of income for farmers and is not dependent on specific soil-climatic conditions.

Under such context, market-based emission trading systems (ETS) are realistic choices for reducing carbon emissions since they are cost effective; provide flexible options and serves as the central nexus of international collaboration and coordination (Colmer et al., 2022; Qi et al., 2016). Growing number of countries are adopting and implementing these systems (Fig. 10.1) for their capabilities to achieve carbon neutrality. For instance, European Union ETS can offset 8–12% CO_2 emission in firms that abide by the protocols compared to unregulated firms (Colmer et al., 2022). Additionally, multi-regional computable general equilibrium (CGE) model under Intended Nationally Determined Contribution (INDC) emission reduction target for 2030 revealed that global ETS can significantly lower the cost of emission reduction in Annex I countries in United Nations Framework Convention on Climate Change and carbon prices in the global carbon market would be $29.83 ton^{-1} CO_2 under national ETS (Qi et al., 2016). However, "cap-and-trade systems" and "baseline-and-credit systems" are widely used as carbon trading systems (see Sects. 10.5.1 and 10.5.3) (OECD). Carbon tax is also used to cut down carbon emissions in some countries (see Sect. 10.5.2).

Cap-and-trade is considered one of the ideal strategies to decrease carbon emissions, and several countries have implemented GHG cap-and-trade systems in succession since 2005 (Moyes, 2008; Wu et al., 2016; Zhao et al., 2019). This system can help mitigate the negative economic consequences of climate change and is a crucial driver of carbon trading price and volume. For instance, Wu et al. (2016) estimated that this method would prevent 2.3–2.2% GDP loss in China in 2030. Furthermore, the carbon trading price and volume would be $69 ton^{-1} CO_2 and 6.2 MT of CO_2. Contrarily, carbon tax may reduce the competitiveness of sectors (e.g., agriculture), although emission reduction is evident (Mardones & Lipski, 2020). Studies have shown that carbon tax can also reduce farm net production and income (Mardones & Lipski, 2020; Olale et al., 2019). However, under cap-and-trade framework, both trading and sharing of carbon emission reduction expertise (knowledge) are beneficial to small and medium size enterprises for decreasing emissions. Moreover, higher incentives make the manufacturing enterprises more inclined to share their expertise with their strategic partners who have complementary information. Subsidies or prizes for information sharing can assist to cut carbon emissions and exhibit significant improvement of public policies addressing the problem (Yao et al., 2019). Additionally, confirm increment in financial incentives through legislation may incite more

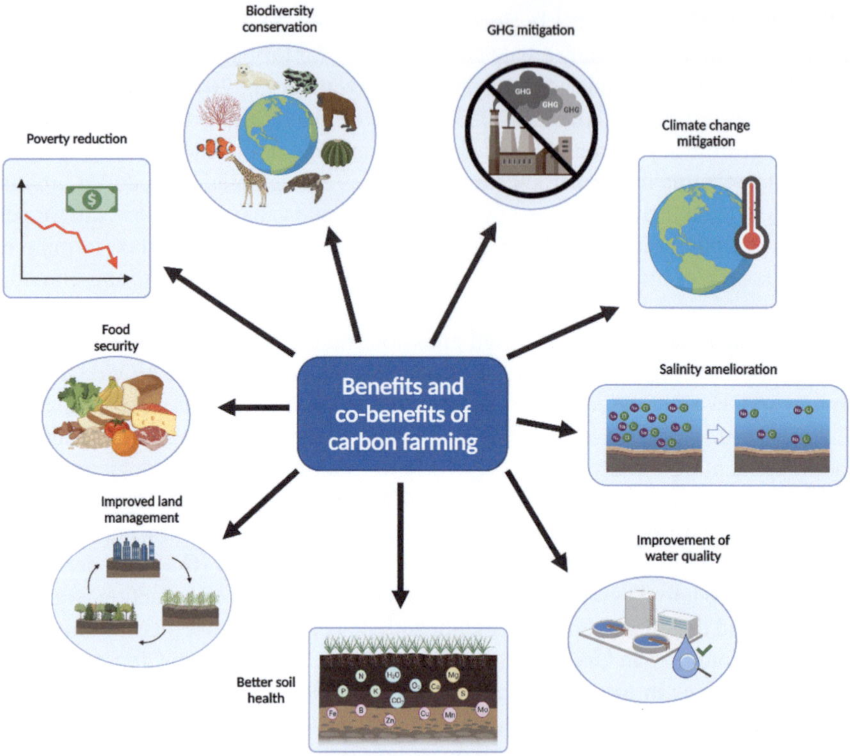

Fig. 10.1 Conceptual schematic representation of the benefits and co-benefits exhibited by carbon farming

adoption of carbon farming practices and substantial abatement of carbon emissions (Tang et al., 2018).

Among the fastest growing market mechanisms, carbon markets as nature-based solutions promise to spur investment in climate-friendly practices while curbing carbon emissions. Carbon markets have evolved from carbon taxes in the Nordic, Baltic and European countries, to the EU Emission Trading System (EU ETS), and later to the current scenario termed as "climate regime complex", which is diverse and comprises sub-national, national and different regions of the world (Brewer et al., 2016). Simply put, the carbon market developed as a tax which is levied on emitters of carbon through emitting facilities (Carattini et al., 2018). Carbon market, as a crucial tool to further enhance the economic reduction of CO_2 emission, however, is a tractability mechanism towards solving the problem of climate change, and seemingly the only required policy instrument of climate that would ensure the promotion of innovation-diffusion of identified technologies that is essential for decarbonization (Lilliestam et al., 2021). Sequel to the launching of the Carbon Market Platform under the German's G7 presidency in the year 2015, development of carbon market strategies, through international cooperation, and sustainable and worthy approaches to carbon pricing has enhanced political, institutional, and technology-based climate action. Overall, putting a price on carbon is an instrumental move in the delivery of soaring ambition of climate and provides support for the recovery of the green economy (ICAP, 2020).

However, these benefits can only be reaped with the proper regulation and input from those who live and work in the areas being affected by carbon offsetting. In this chapter, we bring together the body of knowledge on carbon farming and monetizing their benefits through carbon marketing initiatives. This study is unique in that it compares the existing linkages between carbon farming and carbon trading (note that carbon trading and carbon markets both are synonymous). More specifically, this chapter shows how carbon farming and ecosystem-based carbon market initiatives might be utilized to incentivize farmers and investors to embrace carbon-sequestering strategies for reversing global climate change. In addition, these strategies can help landowners generate revenue while encouraging them to conserve their natural resources by simultaneously reducing their CO_2 footprint.

10.2 Carbon Farming: Scientific Background, Socio-economic Aspects and Environmental Benefits

Carbon farming provides diverse benefits and co-benefits (Rosende, 2022; Fig. 10.1). The primary goals of carbon farming are to curb the carbon emission and increase the ecosystems services beneficial for farmers and farm productivity (Rosende, 2022). Some of the benefits include reduction in erosion and nutrient runoff, purification of surface and groundwater, increased biodiversity, restoration of heavily degraded soils, increased nutrient availability, and improved soil health (McDonald et al., 2021; Melillo & Gribkoff, 2021). A variety of carbon farming strategies that adhere to conservation agriculture principles such as crop residue retention, crop rotation, integrated nutrient management, and decreased soil disturbance have the potential to increase soil organic carbon (SOC) stocks (Chenu et al., 2019; Lin et al., 2013; Nath et al., 2015). The process of storing atmospheric carbon into soils by managing arable lands is termed as soil carbon sequestration (Sharma et al., 2021b). Studies have shown that 0.4% annual SOC stock increase in the 0–40 cm layer of the soil may counteract anthropogenic emissions and moderate climate change (Rumpel et al., 2020; Soussana et al., 2019). The increasing of soil carbon stocks in carbon farming practices occurs through the mechanisms of stabilization of soil organic matter (SOM), which increase the residence time of organic carbon in the soil (Dynarski et al., 2020; Lorenz & Lal, 2012). One of those mechanisms is the selective preservation of more recalcitrant organic structures (lignin, phenols, aliphatic components) that are more difficult to be consumed by soil microorganisms (Senesi et al., 2018). The stabilization and preservation of better-quality organic matter in the soil can be achieved with well managed pastures, reducing the emission of CO_2 into the atmosphere through the biological oxidation

(Assunção et al., 2019). Soil physical aggregation favoured the carbon preservation in Integrated Crop-Livestock system, as it presented aggregates with the largest weighted average diameter and greater C content in macro and micro aggregates (Sato et al., 2019). This results in an increase in the number of organic compounds and an increase in the amount of SOM and total organic carbon (Calil et al., 2016; Coser et al., 2018; Oliveira et al., 2018).

Properly managed pastures have great potential to keep the recalcitrant carbon located within the mineral fraction, due to the lack of soil disturbance, extensive root system and the maintenance of the aggregate structure (Assmann et al., 2014; Gmach et al., 2018). The rotation between crop and livestock (pastures) replacing crop monoculture caused an increase in the soil C stock from 66.4 to 73.0 Mg ha^{-1} after nine years of rotation (Carvalho et al., 2014). Another study found that after four years of conversion of degraded pasture to an agroforestry system, there was an increase in soil organic carbon stocks from 52.6 to 66.5 Mg ha^{-1}, being a sustainable alternative to recovery degraded soils (Coser et al., 2018). No-tillage practices favour the SOM accumulation, reduce GHG emission, and enhance physical, chemical and biological soil properties, due to the reduced soil disturbance and maintenance of crop residues (Campanha et al., 2019; de Figueiredo et al., 2018).

Carbon farming strategies can protect biodiversity, increase nutrient cycling, reduce nitrous oxide (N_2O) emissions and nitrate leaching, and serve as a key weapon in the fight against climate change (Almaraz et al., 2021). Carbon farming and soil management approaches have the ability to preserve biodiversity, prevent nutrient depletion, and improve ecosystem services (Polidoro et al., 2021; Sharma et al., 2021a). More diversified systems, such as agroforestry, promote biological nutrient cycling amongst plants, supporting greater production efficiency, increased organic matter input, and improved soil fertility over time (Nair, 2011; Thevathasan et al., 2012). Higher plant variability through species interactions improves soil aggregation and conservation, as does enhanced microbial population and carbon of microbial biomass (Levinski-Huf & Klein, 2018; Lisboa et al., 2014).

Carbon farming practices and sustainable agriculture can increase crop productivity and preserve native forests (Corbeels et al., 2016; Sharma et al., 2021b). Forests are strategically important to counteract climate change because they possess high potential for carbon sequestration in their above biomass (trunks and leaves) (Hui et al., 2015; Torres et al., 2017). Maintaining native forests sustainably without harvesting timber, by producing and preserving fruits, seeds, oil, and medicinal plants for local farmers, can boost environmental protection, carbon sequestration, and economic benefits (Ros-Tonen et al., 2008). Without necessarily affecting coffee productivity, the coffee crop, in conjunction with other tree species in the forest, can boost the supply of ecosystem services and give significant advantages, such as improved soil conservation, water quality, and biodiversity (Meylan et al., 2017). Carbon farming practices and sustainable agriculture are able to increase crop productivity and preserve native forests (Corbeels et al., 2016; Sharma et al., 2021a). Inadequate soil management in conventional production systems exacerbates negative impacts on agricultural productivity and in social aspects, thus, carbon farming provides an opportunity to enhance biodiversity, economic and social benefits (Altieri et al., 2017; Evans et al., 2015). Strategies to foster the adoption of sustainable agricultural practices lead to the improved and diversified farm income, and open perspectives for new markets and new consumers for the farmer's products (Lehner & Rosenberg, 2018; Li & Kallas, 2021). For example, Australian Government introduced carbon farming initiative (CFI) to improve farmers' livelihood and encouraging adoption of carbon farming practices such as, reforestation in cleared lands, improved animal feeding strategies to reduce livestock emissions, and improved fire management to reduce fires in savanna regions (Dumbrell et al., 2016).

Carbon farming and forest restoration approaches need thorough scientific understanding and technical competence in order to be implemented, providing job possibilities, and boost landowner participation (Liu & Zheng, 2021; Rezende et al., 2018). This can bring socioeconomic benefits too, such as improved infrastructure, cost and input optimization, increased profit, and additional revenue to sell carbon credits in green market schemes (Baumber et al., 2019; dos Reis et al., 2021; Kragt et al., 2017).

The development of sustainable farming practices and forest restoration enhances the public image of the producers for their customers and promotes positive effects for the farmers and across the society (Baumber et al., 2020; Rovere et al., 2018). They are also important for socio-cultural benefits, such as Indigenous community conservation and valorisation, and better livelihoods and community cohesion, improved physical and mental health, and aesthetics purposes (Baumber et al., 2019).

Combining farmers' training, technical assistance and changes in input use, those practices can improve farmers' lives, with increased food productivity while reducing greenhouse gases due to the "land sparing" effect, avoiding deforestation of native forests for food production (Bragança et al., 2022). Land sparing is a promising strategy to reduce GHG emissions from the Agriculture, Forestry and Other Land Use (AFOLU) sector around the world (Lamb et al., 2016). As a result, the emission of GHG due to fires and deforestation would decrease substantially, helping to combat climate change and preserve the environment. Avoiding deforestation has global mitigation potential of 3719 Tg CO_2-eq yr^{-1} (Griscom et al., 2017). In addition, the use of sustainable agriculture practices to improve soil organic matter, crop rotation, and increasing soil carbon stock can avoid the emission of 4817 Tg CO_2-eq yr^{-1} (Griscom et al., 2017). In addition, avoiding deforestation can also attract green investments, including existing mechanisms for Reducing Emissions from Deforestation and Forest Degradation (REDD+) under the United Nations United Nations Framework Convention on Climate Change (UNFCCC) (Stabile et al., 2020). This initiative seeks to provide financial incentives to developing countries to reduce GHG emissions associated with deforestation, and to promote sustainable forest management, conservation and enhancement of forest carbon stocks (Boissière et al., 2017; Strassburg et al., 2014). But carbon farming practices and restoration of degraded areas depend on institutions, information and incentives to promote all the environmental, social and economic benefits (Evans, 2018). Thus, strengthening the forest economy and empowering local people can create long lasting opportunities for restoration and improved rural and urban livelihoods (Garrett et al., 2021).

10.3 Means of Improving Carbon Farming Adoption

10.3.1 Inadequate Extension and Outreach of Carbon Farming's Benefits Must Be Addressed

Technical information dissemination on carbon farming practices is one of the main barriers and challenges that must be overcome, as many farmers don't have the knowledge about these practices and about the pros and cons of the adoption of carbon farming (Carrer et al., 2017; Sharma et al., 2021a). Morgan et al. (2015) identified that the major psychological factors that motivates adoption of carbon farming practices by farmers are knowledge, self-efficacy, skills to develop such practices, and financial incentives for the implementation.

Extension agents can provide rural extension services and information transfer to farmers, which can then spread to other farmers by oral communication leading to increased awareness of better yielding and more sustainable practices (Zu Ermgassen et al., 2018). Through the sharing of information and farmers' experiences, extension and outreach initiatives help to spread those practices. For example, early adopters of carbon farming may be able to disseminate information by sharing their success stories with their neighbours in order to boost adoption, but failure of one farmer can have a detrimental impact on the others, hence, demonstrative units and field trials must be handled properly (Dumbrell et al., 2016; Kragt et al., 2017). However, a lack of pioneer and innovative growers in a specific region might make carbon farming acceptance difficult (Mattila et al., 2022).

In this context, technical assistance for carbon farming begins with a visit from an extension specialist who gathers baseline data and creates management plans tailored to that farmer (Zu Ermgassen et al., 2018). Providing technical help to farmers who have already been taught, increases the adoption of sustainable agricultural practices (e.g., carbon farming) compared to those who just gets training (Bragança et al., 2022).

Other impediments to carbon farming implementation include high expenses of the methods, and farmers' concerns regarding yield and farm productivity (Sharma et al., 2021b). Many farmers are reluctant to new agricultural practices since they have been employing the same ones for several generations, therefore, they are wary of new technologies, innovations and ideas (Costa Jr. et al., 2019; Garrett et al., 2017). To implement a different and more sophisticated system, it is often essential to enhance financial management, infrastructure, logistics, inputs, and early investments (Carrer et al., 2020; Reis et al., 2020). As a result, increasing agricultural productivity while also benefiting the environment necessitates knowledge and technical assistance for farms of all sizes and socioeconomic levels (Alves et al., 2017). It is critical to establish initiatives, policies, and activities that strengthen farmers' and rural extension agents' complementary knowledge, benefitting both parties through demonstration units, field days, practical courses, and material published on booklets in an accessible language (Newton et al., 2016). Large and small landowners should be considered, with professional help utilizing the co-creation of knowledge with the rural producers, so that techniques may be tailored to each local situation (Garrett et al., 2017).

Carbon Action, a research effort aimed at increasing the adoption of carbon farming in Finland, provided training for farmers who were able to create outstanding carbon farming plans (Mattila et al., 2022). Best management and agricultural practices are supported by programs like ABC Cerrado in Brazil, which provide on-site customized technical assistance to farmers, encouraging them to adopt new technologies that are critical for promoting sustainable intensification (Bragança et al., 2022; Costa Jr. et al., 2019). The Carbon Farming Futures initiative, established in Australia, sponsors research trials and landholder outreach as part of the policy carbon farming initiative (CFI), which was critical to the country's widespread adoption of carbon farming practices (Dumbrell et al., 2016). However, many more programs and policies need be developed to improve farmers' access to technical help and extension, allowing them to apply their technical expertise to carbon farming techniques (Bragança et al., 2022; Knook et al., 2020; Newton et al., 2016).

10.3.2 Financial Subsidies from Government and Other Organizations Must Be Increased

Supporting the establishment of carbon farming and forest restoration requires accessibility to rural credit services, government funding and subsidies (Evans, 2018; Liu & Zheng, 2021; Zu Ermgassen et al., 2018) as well as efficient extension programs that can build rapport with farmers and extension workers. That support must be based on producer profiles and geographic contexts (Porgo et al., 2018). The initial disbursement of inputs (fertilizers, seeds, machineries, etc.), assurance of funding, favourable negotiation of terms and crop insurance play a critical role in the establishment of new agricultural practices and reforestation (Carrer et al., 2020). Inadequate financial assistance from government acts as one of the major obstacles to increasing sustainable climate-friendly agricultural practices adoption. For instance, farmers in the Brazilian Amazon are often unable to adopt agroforestry practices due to limited access to rural credit, particularly those lacking capital and technological innovation (Maia et al., 2021). However, farmers without credit constraints are more likely to adopt intensive agricultural practices (Ali et al., 2014). One method for increasing access to rural credit is to join agricultural associations, which can offer collateral and hire aged farmers (Carrer et al., 2013). Under such context, Souza Filho et al. (2019) revealed that the producer's credit and marketing policies affect crop productivity. As a result, the producer's economic performance may increase, as well as the adoption of new carbon farming technologies. Land use, credit, and marketing policies can also be used to increase the producer's benefits. These studies indicate that government subsidies and other economic tools are crucial to transform society away from fossil fuels towards a low-carbon economy (La Rovere, 2020; Sharma et al., 2021b).

Companies and government entities who seek to encourage climate-friendly practices might help propel a sustainable/green financing sector. As a result, businesses may benefit from producer finance while also improving the environmental, social, and economic quality of natural resources (Havemann et al., 2020). Carbon farming with sustainable agricultural intensification and forest restoration might be aided by public policy and financial measures such as Payment for Environmental Services (PES) and certification schemes (Alves-Pinto et al., 2017). Many other governance initiatives, such as proper institutional frameworks, policy stability, complementary limits on deforestation, land tenure security, systemic pricing signal, and so on, are needed to support reforestation and carbon farming activities (Evans, 2018).

10.3.3 Addressing Political and Market Uncertainties

Farmers' adoption of carbon farming practices is linked to agri-environmental legislations and incentives that encourage farmers to embrace such methods (Sharma et al., 2021b). One of the major issues for the upcoming years is to design regulations that will allow forest restoration with ecological aims while limiting competition with agricultural production and guaranteeing food security in order to meet the sustainable development goals (Crouzeilles et al., 2020; Soussana et al., 2019).

Sudden changes in policies and Government can cause political uncertainty, which, in turn, can impact farmers and stakeholders' trust, especially in long-term commitments (e.g., carbon farming adoption) that demand more political stability (Kragt et al., 2017). The relevant institutional frameworks, suitable political debate and governance, and enough information dissemination are necessary for the efficient execution of public policies connected to carbon farming (Sharma et al., 2021b).

Besides, a set of public policies is required to execute carbon farming economic development approaches that are consistent with the utilization of natural resources and forest restoration, laying the groundwork for a long-term economic transition to a low-carbon society (Alves et al., 2017; Arima et al., 2014). Studies have claimed that the loss of biodiversity and ecosystem services, because of deforestation in the Amazon rainforest, has an impact on the reduction of international funds for actions to prevent and monitor deforestation and can lead to the boycott of Brazilian products on the international market (Crouzeilles et al., 2020; Levis et al., 2020; Rajão et al., 2020). Thus, deforestation reduction requires tailored approaches and stricter laws that include better monitoring systems to detect even small-scale deforestation (Montibeller et al., 2020; Wang et al., 2020a, 2020b). Development of strategies and political dialogue should be addressed to promote a new model of agricultural growth, in which punitive measures are complemented by positive and financial incentives for producers, to achieve productive and sustainable rural development (Godar et al., 2014; Nepstad et al., 2014). Sustainable agriculture practices can improve productivity without destroying native forests to advance in agricultural production and could spare land for habitat conservation or restoration (Alves et al., 2017; Nepstad et al., 2019).

Science and society are increasingly recognizing that natural and human systems are inextricably linked, and that solutions must be integrated (Rezende et al., 2018). Low-interest loans, payment for ecosystem services, and low-carbon labelling are all examples of public–private partnerships that may add value to cattle ranching and encourage more sustainable production techniques (Maia et al., 2021). However, the carbon market for climate change solutions is sometimes referred to as an abstract amount of traded emission rights, which causes consumer uncertainty (Lohmann, 2010). Thus, to boost market confidence, additional measures are needed, such as the dissemination of clear, accessible, and trustworthy information regarding agricultural practices, production methods, and product certification (Evans, 2018). Consumers who perceive the foods they are buying are produced through carbon (organic) farming methods are willing to pay roughly 29.5% more for those foods (Li & Kallas, 2021). Thus, a robust Measuring, Reporting and Verification (MRV) system is essential for producers to be able to place their low carbon products in domestic and international markets that trust and value this attribute, currently or in the future (Boissière et al., 2017; Smith et al., 2020). But there is a lack of clear processes for measure and verify carbon sequestration and effective market mechanisms (Baumber et al., 2019). Consumers are looking for transparency and traceability of their products in all the supply chains to monitor indirect and direct suppliers, to check if they came from a deforested area. That should be independently audited and publicly reported (Zu Ermgassen et al., 2018). The MRV system to certify climate change mitigation strategies based on natural and farming solutions is not easy to implement, and has several scientific, technical and operational challenges that must be addressed in the following years (Smith et al., 2020).

10.4 Role of Strategic Alliances in the Adoption of Carbon Farming and Carbon Markets

Due to the increase in carbon emissions, various regions and countries around the world have begun to form mutually beneficial partnerships known as strategic alliances in order to increase atmospheric carbon storage in soil and reduce excess carbon emissions, and a number of such coalitions have been formed around the world (Chen & Ulya, 2019; Ji et al., 2017; Ontl & Schulte, 2012). These partnerships aim to absorb and store carbon while also reducing greenhouse gas emissions at the agricultural level (He et al., 2019; Melillo & Gribkoff, 2021). Strategic alliance impact the carbon farming and market by integrating cutting edge innovative technologies in controlling carbon emissions and accomplishing a safer environment free of carbon pollutions (McDonald et al., 2021; Zhang & Zhang, 2020). It seeks to control the influx of carbon into the atmosphere and the trading of saved carbon emissions among carbon producing firms. Around the world, different nations are progressively creating agro-innovative technologies and practices as well as carbon valuing instruments, in accomplishing carbon emission decrease objectives (Shen et al., 2021; Streck, 2021; Zhang & Zhang, 2020; Zhang et al., 2019).

For example, the Indonesian government developed the Low Carbon Development Initiative (LCDI) to integrate climate action into its development strategy. LCDI seeks to discover policies and investments that promote carbon farming and the reduction of carbon and GHG emissions. Under the initiative, the Ministry of National Development Planning undertook an examination of the economic and environmental impact of low-carbon development policies that might be incorporated into the country's National Medium-Term Development Plan 2020–2024. This is to limit human activities in relation to the carrying capacity of natural systems, to assure zero carbon emissions, and to promote the use of contemporary agro-technologies that do not exacerbate soil/land degradation (Silalahi, 2018). In addition, there are other collaboratives too that focus on promoting environmental benefits of carbon farming.

The Forest Carbon Partnership Facility (FCPF) is a global collaboration of governments, corporations, civil society, and indigenous peoples from 47 poor nations in Africa, Asia, Latin America, and the Caribbean (FCPF, 2022). It focuses on lowering emissions from deforestation and forest degradation, protecting forest carbon stocks, managing forests sustainably, and increasing forest carbon reserves in developing nations. Its operations are generally known as REDD+ (FCPF, 2022). The World Bank is the Trustee of the FCPF's readiness fund and carbon fund, and a Facility Management Team provides secretariat services. The team collaborates with other World Bank departments to help REDD+ nations formulate and implement recommendations. The initiative has focused on increasing readiness by assisting with governance, increasing transparency, and developing ties with indigenous groups and women, both of which play critical roles in resource preservation. As of December 2019, four Emission Reduction Payment Agreements had been signed, indicating that the initiative is progressing toward payment for ecosystem services implementation. The World Bank developed the Forest Investment Program (FIP) in 2009 to address the causes of deforestation and forest degradation by assisting developing nations in their efforts and encouraging sustainable forest management.

Examples of some global strategic alliances related to carbon farming and trading are presented in Table 10.1.

10.4.1 Impact of Carbon Trading Initiatives on Supply Chains

Significant carbon emissions come from supply chain practices. Sustainability in supply chain activities require integration of the environment and its surroundings in a way that can mitigate negative effects (Acquaye et al., 2017). Carbon tax collection is one of the sustainable supply chain management procedures that sets a decent cost for carbon emissions as evidenced in biomass-to-bio-oil supply chain (Sadeghi &

Table 10.1 Strategic alliances of different countries (carbon farming and carbon market)

Country	Strategic alliances	Goal/objectives	References
Indonesia	Low Carbon Development Initiative (LCDI)	LCDI aims to identify policies and investments that support carbon farming and reduction in carbon and GHG emissions. This is to reduce human activities relative to the carrying capacity of natural systems, ensure zero carbon emissions and adoption of modern agro-technologies devoid of soil/land exacerbations.	Silalahi (2018)
Africa, Asia, Latin America, and the Caribbean's	The Forest Carbon Partnership Facility (FCPF)	Focused on reducing emissions from deforestation and forest degradation; conserving forest carbon stocks; sustainably managing forests; and enhancing forest carbon stocks in developing countries.	FCPF (2022)
Netherlands	Strategic partnerships in green environment and carbon farming support	Optimizing carbon capture on working landscapes by implementing practices that are known to improve the rate at which CO_2 is removed from the atmosphere and stored in plant material and/or soil organic matter.	MEACP (2019)
In EU countries	New strategic business and alliance model for the farmers	Incentives for farmers to adopt farming practices that deliver a climate benefit at farm level. To support more widespread and effective carbon farming practices using EU and national funds.	McDonald et al. (2021) and WEF (2022)

(continued)

Table 10.1 (continued)

Country	Strategic alliances	Goal/objectives	References
Republic of Korea	Green New Deal and the Digital New Deal	To serve as a steppingstone to reach carbon neutrality by 2050. To create synergies between the Green New Deal and the Digital New Deal.	Government of the Republic of Korea (2020)
USA	(i) U.S. Climate Action Partnership (USEAP) framed by 10 organizations in U.S. and four non-public organizations (ii) BP America, DuPont, Alcoa, and various firms have lined up with the World Resource Institute (WRI), Washington, DC-based NGO (iii) Others incorporates, U.S. Climate Action Partnership, the Green Alliances, and the Greenhouse Gas Protocol Initiative. General Electric, Johnson and Johnson, and 11 unique firms have agreed with WRI to begin the Environment Northeast Partnership	This coalition tries to spread out a mandatory U.S. cap-and-trade program for lessening and control of carbon dioxide emissions. Establish carbon free climate and agreeable emissions exchanging advancements, drive greener environmental system/conditions, and ensure decrease of carbon dioxide emissions. Encourage frameworks for businesses to prosper in a carbon-obliged economy. Extend boundless energy advancements and increase corporate interest and markets for unlimited energy.	UNFCCC (2022), Li et al. (2022) and Choi et al. (2019)
Uganda	Uganda firms/companies formed alliance with Clean Development Mechanism (CDM)	Generation of carbon emission credits (income) and reduction in carbon emissions.	World Bank and Ecofys (2017)

(continued)

Table 10.1 (continued)

Country	Strategic alliances	Goal/objectives	References
China	(i) A few organizations have coordinated corporate obligation and maintainability into their tasks. For instance, Coca-Cola HBC, a main bottler of the Coca-Cola Company, went into concurrence with Chinese government to intentionally set a science-based discharge decrease target (ii) Public authority (Government) formed alliance with 7 carbon trading companies/firms in different pilot districts (iii) Formed alliance with firms and companies for establishment of national carbon trading market for power generation industry (iv) Alliance for Chinese Certified Emission Reduction (CCER) projects	Reduce carbon emissions across the total value chain by 25% by 2030 Generation of carbon emission credits (income) and reduction in carbon emissions. Renewable source of power generation and supply. Drive low carbon emissions.	Wang et al. (2017), Wu and Kung (2020) and Zhang and Zhang (2020)
Asia	Carbon market tasks in Asia have increased with the arrangement of the Asia–Pacific Carbon Market in partnership with the Forum for Carbon Pricing Mechanism with China, Japan and Korea, in cooperation drives	Reduction of public carbon emissions, balancing carbon exchanging frameworks. Environmental sustenance and harmonization of carbon trading across Asia.	ADB (2015)
Australia	Carbon emitting firms and industries had entered serious alliance with Australian Carbon Emission Reduction Trust Fund (ACERTF)	To achieve Australia's 2030 carbon emission reduction target (26–28%) by 2030 and to meet its nationally determined contributions.	Cybercom (2019)

(continued)

Table 10.1 (continued)

Country	Strategic alliances	Goal/objectives	References
United Kingdom	Carbon pricing and emission trading alliance has been formed with World Resource Institute (WRI)	To project and motivate for low-carbon technology innovations, development and general adoption and/or reception. To establish a low carbon and green economy.	Choi et al. (2019)
South Africa	Formation of strategic alliance with carbon emission industries and firms	To diminish its domestic/homegrown carbon emissions, to lessen carbon emissions, expand utilization of environmentally friendly power sources and reception of clean energy innovations, and uptake of domestic renewable energy projects.	ICAP (2020)

Haapala, 2019). Efficient supply chain requires government incentives and companies that implement low-carbon manufacturing such as minimum quantity lubrication (MQL) can generate significant carbon offsets and participate in cap-and-trade or voluntary carbon markets (Sarkar et al., 2018; Shen et al., 2021). Incorporation of such carbon trading policies ensures a viable pathway for economic growth by maintaining environmental sustainability (Shen et al., 2021). However, cap-and-trade mechanism is only sustainable when consumers prefer low-carbon products since retailers are more likely to adopt low-carbon practices even without manufacterer incentives (Ji et al., 2017). Studies suggest that a joint emission reduction strategy—where both manufacturers and retailers contribute—yields greater environmental and economic benefits than single-entity efforts (Liu et al., 2020). Wang et al. (2017) revealed that in a decentralized supply chain, retail prices and government imposed carbon-taxes are higher than in a centralized system and retail price rise in both supply chain structures are associated with the increase in carbon taxes. In addition to these, demand uncertainty is another important parameter that influences supply chains and associated carbon emission reduction. For example, government subsidies act as a natural coordination mechanism, ensuring that manufacturers and consumers respond effectively to green technology adoption (Cohen et al., 2016). Overall, carbon offset programs are designed to help greenhouse gas emitters meet emission caps by investing in external projects that reduce emissions. However, ensuring the credibility of these reductions remains a significant challenge. The experience of early offset systems, such as the United Nations' Clean Development Mechanism (CDM), highlighted risks such as non-additionality, inflated baseline estimates, and perverse incentives that may unintentionally increase emissions (Haya et al., 2020).

10.5 Summary of Carbon Trading Strategies

10.5.1 Cap-and-Trade

Cap-and-trade is a market based regulation system designed to curb emissions. It works by setting a cap on total emission, and firms trade emission allowances within this limit with a goal to internalize the true social cost of pollution by aligning firms' price-setting with the optimal level of emissions (Balbi, 2009; Frank, 2014). To enhance a time-to-time reduction of emission by a system, the set cap is reduced over time (Kachi et al., 2015). Within the set cap however, buying or receiving allowances of emissions by installations is done, and these allowances can be traded with each other as they are needed (European Commission, 2021). The limit is therefore placed on the total number of available allowances, and this ensures their value. Based on the trading system, an installation is expected to submit enough allowances to fully take care of its emissions each year, the absence of which makes it liable to a heavy fine (Partnership for Market Readiness, & International Carbon Action Partnership, 2016). For installations that reduce their emissions, the spare allowances can be kept covering for the needs of the future or be sold to other installations that need allowances (Guo et al., 2017). With trading however, emissions are cut at a minimal cost. The cap-and-trade has gained popularity and is in use by most carbon markets in the world including, Brazil California, European Union, India, Japan, Kazakhstan, New Zealand, Norway, Ontario, Quebec, RGGI states (United States), Republic of Korea, Tokyo, UK, and China, among others.

10.5.2 Carbon Tax

Carbon tax refers to fundamental market-based policy that aims to impose a tax on each unit of carbon released as pollution (Center for Climate & Energy Solutions, 2015). Wen et al. (2018) suggested that an increase in the price of carbon may serve as a better pricing strategy in a competitive market. This strategy is expected to be a push for firms to invest in carbon emission reduction activities in a way that will not lead to a high product price and reduced profit. However, carbon tax is levied on the emissions of carbon in the process of producing goods and services. The objective of this tax is to account for the costs of emissions of carbon which remain hidden during production processes (Parry, 2019). However, carbon tax has been identified to be a kind of Pigouvian tax that is levied on the activity of the market and may exert negative unintended consequences. Carbon tax has been implemented globally including the US, South Africa, Switzerland, Europe and Latin America (World Bank, 2019). The application of this concept to carbon farming is that farmers are required to pay tax for emitting beyond the limit of the allocated. Overall, well-designed carbon tax is an essential tool for mitigating climate change, encouraging green innovation, and fostering a transition toward a low-carbon economy. However, it should be complemented with policies to address economic and social concerns, ensuring a just and equitable transition for all.

10.5.3 Baseline-and-Credit Program

A baseline-and-credit program is a flexible, market-driven mechanism designed to reduce emissions by rewarding organizations that lower their emissions below a predetermined baseline. Unlike cap-and-trade systems, which impose a strict limit on emissions, these programs set individual baselines based on historical emissions or industry benchmarks. Companies that emit less than their baseline earn credits, which can be sold to entities exceeding their allowed emissions, thereby creating a financial incentive for sustainable practices. The effectiveness of baseline-and-credit systems depends on accurate baseline setting, robust monitoring, and strict verification to ensure genuine reductions. These programs are particularly beneficial in sectors where emissions reductions are difficult to regulate directly, as they encourage innovation and efficiency without enforcing rigid caps. However, challenges such as credit price volatility, potential over-allocation of baselines, and the risk of non-additional reductions must be carefully managed.

10.5.4 Performance Achieve and Trade (PAT)

The Performance Achieve and Trade (PAT) program is another market-based strategy focused on enhancing energy efficiency in energy-intensive industries while reducing carbon emissions. It operates by setting sector-specific energy efficiency targets, encouraging industries to improve their performance through technological advancements and process optimizations. Organizations that exceed their assigned targets earn tradable efficiency credits, while those failing to meet their goals must purchase these credits to comply with regulations.

10.5.5 Other Strategies

Designing emissions caps across states, countries and regions requires the consideration of factors, the likes of environmental profiles, economic profiles, and or the developmental level of the participating entity (Diniz Oliveira et al., 2021). However, taking into account climate partners with comparable ambitions and trends in emissions over the medium to long term becomes crucial to enhancing the predicted global decrease in emissions (Green et al., 2014; Kachi et al., 2015). This is because climate ambitions at different levels can hinder necessary link or prevent the reduction of emissions globally (Zetterberg, 2012). In addition, larger geographic and sectoral markets enhance heterogeneous options of abatement across entities that are already regulated (Diniz Oliveira et al., 2021), leading to increased efficiency gains through traded allowances, and addressing concerns related to competitiveness and carbon emissions (Görlach et al., 2015). With the advantage of linking one ETS to other systems to create a common market (Diniz Oliveira et al., 2021), implementing a worldwide ETS is feasible. This approach is known to be beneficial for preserving the climate and providing access to less expensive abatement solutions over time (Kossoy et al., 2015). This will become especially apparent if the number of connected ETS increases in what is known as "full bilateral linking" (Kachi et al., 2015).

T4 here has been bilateral and multilateral cooperation, and linkages of ETS across countries, regions and sub-regions as witnessed by China and the EU (Council of the European Union, 2015), EU and Switzerland (Council of the European Union, 2017), and the RGGI of the United States (RGGI, 2015). International carbon action partnership (ICAP) is a forum that only promotes discussions that are detailed and technical, on the issues of compatibility, design and linking of the cap-and-trade systems, comprising different nations under the ETS (Kachi et al., 2015). By linking ETS at the global level, converging the prices of carbon between different systems reduces competitiveness among entities operating under same market. At the same time, it enhances the advantage of meeting caps in systems with higher abatement costs by recognizing abatement undertaken in systems where costs are lower (Dellink et al., 2014).

10.6 Role of Competitiveness in Carbon Markets and Decarbonization

Competitiveness, as advanced by Berger (2007), connotes the ability of the economic sectors of different countries or emitting entities to maintain their shares of market, and ensure profitability. As identified by Hausotter et al. (2011), it is the capability of an emitting entity (either a sector or a subsector) to engage in competition with other entities outside the fold of ETS. The importance of competitiveness among carbon-emitting sectors for the achievement of the decarbonization needed as enshrined in the Paris Agreement cannot be over emphasized. OECD (2011) stressed that with the competitiveness of permitted markets, especially when large and unconstrained; there is the actualization of a total cap of emission at a possible lowest cost. According to Ellis et al. (2019), there is a need for a change over a given time in the relative competitiveness of emitting sectors, with the sectors that have high emission losing their shares of market or becoming non-existent, and the sectors with lower emissions gaining the shares of market. However, some ETS are majorly designed in a way that would avoid the leakage of carbon; a situation that arises when more polluting entities or technologies are gaining market shares (Nachtigall, 2019). The positivity of relative competitiveness reflects in gain of market share by lower emission sectors (Ellis et al., 2019), making it to be easily established that the competition is between the lower emission sectors and the higher emission sectors; with the former aiding deep decarbonization in line with the goals of temperature reduction of the Paris Agreement, and the latter working against it (Krishnan et al., 2022; Rissman et al., 2020) has identified that those economic sectors that have products or operations with high emissions would encounter employment, demand, and production costs effects, especially if the relative competitiveness is positive—with low emissions sectors gaining bigger market share, unless appropriate abatement technologies are deployed.

10.6.1 Steps to Achieve Competitiveness in a Decarbonization Era

As a part of the decarbonization process, Lash and Wellington (2007) found that enterprises that are aware of climate change threats while still looking for profit possibilities would have a competitive advantage over other competing organizations or entities. Firms' competitiveness is assumed to be influenced by carbon pricing legislation, international trade, market structure, and other policies and measures that impact carbon pricing (Hausotter et al., 2011). Available work (Tamiotti et al., 2009) has noted that, for most emitting firms, the policies of climate change have minor effects on their competitiveness. Furthermore, intrinsic factors such as costs of labor, costs of different input, and rates of currency exchange pose greater impacts on a firm's competitiveness than extrinsic factors such as price of carbon (Hausotter et al., 2011). As a result, competitive advantage can be achieved by firms through the development of their own climate strategies for competitiveness based on their exposure and contribution to climate change. On this, Lash and Wellington (2007) identified four steps which serve as framework for firms' competitive advantage to include:

10.6.1.1 Quantifying Firm's Carbon Footprint

This is achieved through; creating correct inventory of firm's GHG emissions, differentiating between indirect emission (emissions that comes from the consumption of energy by the firm) and direct emissions (firm's smokestack emission), and establishing and adjusting baselines of emissions, and evaluating best practices employed in reporting this information. One identified method for carrying out this task is with the use of GHG Protocol (WRI & WBCSD, 2004).

10.6.1.2 Assessing Firm's Risks and Opportunities that Are Related to Carbon

This is done by determining the possibility of a rise in the costs of transport in the delivery of firm's product in a carbon-constrained economy, determining the possibility of effects of climate change on firm's raw materials, identifying the direct and indirect financial impact of climate change through determining the percentage of profit derived from firm's products with either high or low carbon emission, and assessing the effects of climate change on firm's revenue and costs. Generally, determining the effects of climate risks on the cash flow and the costs of capital of the firm is dependent on the understanding of a firm's competitive strength in a decarbonized future.

10.6.1.3 Adapting a Firm's Business to Respond to Identified Risks and Opportunities

Responding to risks and opportunities that are identified by firms in a carbon constrained economy requires a proper adjustment of firm's activities to respond and overcome expected risks and take advantage of identified opportunities. With effects of climate risks on cash flow and firm's capital costs, response can be through an evident cut down in firm's energy consumption and emission of carbon, together with a total reinvention of parts of business to ways that would aid abatements of carbon and committing R&D funds to abatement technologies in the firm's production activities.

10.6.1.4 Doing Things Better Than the Competitors

In a competitive environment where having an upper hand against competitors, particularly high emission firms, is crucial to remain profitable in the trading system, outperforming these competitors through lowering firm's exposure to climate-related hazards and identifying commercial opportunities within those hazards becomes pertinent. Firms need to do better than their competitors by deploying technologies and envisioning the productions and services that address high emissions to gain the advantage of competitiveness.

10.6.2 Interlinkages Between Carbon Trading and Emerging Technologies for Energy Production

The need to address the problems stemming from the change in climate requires the adoption of technology, and a continual and supportive use of technological innovation. According to Saikku et al. (2017), the core of 'low-carbon' economy, a model of economy that is based on low consumption of energy, low emissions of GHG, and low pollution of the environment, is the adoption of low carbon technological innovation (Fig. 10.2). Counting the impacts of ETS on the innovation of low-carbon technologies, Lyu et al. (2020) has revealed that yearly increase in the innovations of low-carbon technologies has been possible due to the operation of the ETS, and much progress in this technology has been enhanced (Zheng et al., 2021; Zhou et al., 2020; Zhu et al., 2019). The progress is however due to the need for the emitting entities to change their production technologies from high emissions technologies to low emission technologies for a significant impact in emission reduction. For example, Northrup et al. (2021) has identified that the combination of electrification, digital agriculture, and microbial and crop genetics can enhance the reduction in the emission of greenhouse gases for about 71% which is equivalent to 1744 kg of CO_2 from agriculture in 15 years' time. Furthermore, Wang et al. (2021) also identified different technologies of renewable energy that ensure clean environmentally friendly energy and carbon neutrality, against the much dependence on energy from sources that are not renewable with numerous environmental impacts. They include the integrated framework for electricity and chemicals which is carbon dioxide neutral (Dutta et al., 2017), development of photovoltaic technology as a powerful method for harnessing solar energy (Kılkış et al., 2020), and nuclear energy which accounts for about 40% world electricity generation, and with the avoidance of about 1.7Gt of Carbon dioxide (Serp et al., 2014) among other technologies. However, with these outstanding results, proposals to strengthen the ETS to further spur technical innovation and adoption of emission abatement technology by enterprises on the ETS platform need to be bolstered.

10.7 Conclusions and Perspectives

In this chapter, we show that carbon markets/trading systems can play a role in promoting carbon farming practices by creating incentives for landowners to connect with buyers for the carbon stored in their soils and vegetation. Our major findings may be summarized as follows:

1. Carbon farming drives up demand for carbon offset credits generated by these practices. However, to accomplish this crucial goal, policy formation must take potential mitigation strategies into account.

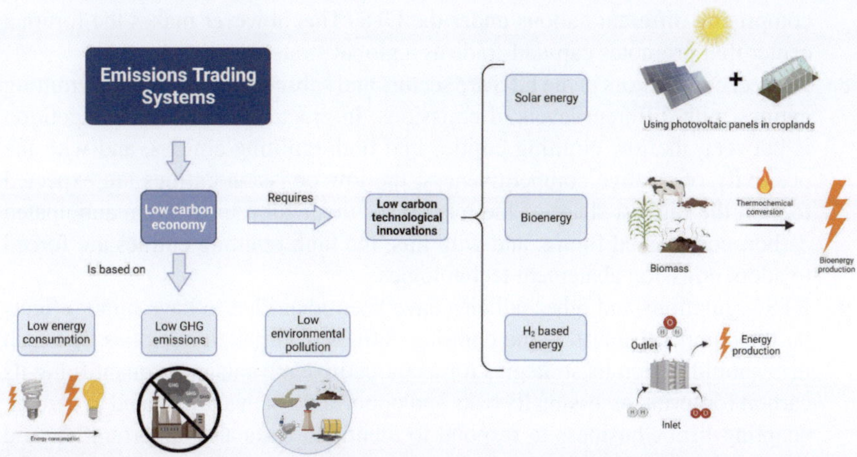

Fig. 10.2 Interlinkages among emissions trading systems, low carbon economy and innovations on emerging technologies for energy production

2. Lack of adequate extension services to convey the benefits and co-benefits of carbon farming, as well as political and market uncertainties remain as major barriers to the sustainability of carbon trading.
3. Various regions and countries throughout the world have begun to develop *mutually beneficial agreements* to prevent or reduce carbon emissions. Several crucial coalitions have been created around the world, according to published literature. These alliances are aimed at regulating carbon emissions and the human entities involved in carbon trading.
4. Several countries and regions have undertaken carbon-exchange schemes to create an empowered environment free of carbon emissions, restoring natural equilibrium.
5. Carbon trading reduces carbon emissions and transforms the whole environment into a more hospitable and impactful habitat for sustainable living; it improves overall air, water, and land quality, resulting in a carbon-free environment.
6. Carbon supply chain incorporates the environment and its surroundings into supply chain processes to ensure that deliverables are seamless and without negative consequences. In all carbon supply chain operations, organizations, enterprises, and individuals engaging in carbon emission trading are required to incorporate the best global supply chain practices and processes. The development of a workable carbon trade exchange scheme has necessitated persuasive action plans, permits, and trading conditions that can make a bridge among all stakeholders at the national, regional, and local levels.
7. The widely used competitive strategies in the carbon market are the cap-and-trade strategy and the carbon tax. The International Carbon Action Partnership (ICAP) is a forum that only promotes discussions that are detailed and technical, on the issues of compatibility, design and linking of the cap-and-trade systems,

comprising different nations under the ETS. This however makes the forum a center that promotes cap-and-trade as a global strategy.

8. Indirect competitors in the ETS are sectors and subsectors representing emitting entities with different levels of emissions. In practical sense, the competition is between the low emitting entities and high emitting entities, and with the positivity of relative competitiveness, the low emission entities are expected to gain the highest share of the market for the achievement of an anticipated carbon-constrained future, and with this, the high emitting entities are forced to adopt emission abatement technologies.
9. ETS regulations and other policies have been identified to have minor effects on the competitiveness of the emitting entities or firms, this implies that each firm should design its strategies for a competitive advantage by quantifying its carbon footprint, assessing its risks and opportunities that are related to carbon, adapting firm's business to respond to identified risks and opportunities, and doing things better than the competitors.
10. The carbon market has enhanced yearly progress in emission abatement technologies, and technological change has greatly reduced firms' carbon and other GHG emissions, this shows the effectiveness of the carbon market as a tool for deep decarbonization.

Author Contributions **Siddhartha Shankar Bhattacharyya**: Conceptualization, Literature search, Data Synthesis, Curation and Visualization, Investigation, Figures, Tables, Writing-Original Draft, Review and Editing, Supervision. **Robert Ugochukwu Onyeneke**: Literature Search, Data Synthesis, Curation and Visualization, Tables, Writing-Original Draft. **Fernanda Figueiredo Granja Dorilêo Leite**: Literature Search, Data Synthesis, Curation and Visualization, Writing-Original Draft. **Emeka Emmanuel Osuji**: Data Synthesis, Curation and Visualization, Writing-Original Draft. **Sikiru Ibrahim-Olesin**: Data Synthesis, Curation and Visualization, Writing-Original Draft. **Ahmed Khairul Hasan**: Tables, Writing-Original Draft, Review and Editing.

Funding No funds, grants, or other support was received.

Data Availability Statement This article does not contain any newly produced data. All relevant data are included within the article.

Conflict of Interest(s) The authors declare that they have no known competing financial interests or personal relationships that could have appeared to influence the work reported in this paper.

Competing Interests Robert Onyeneke is an editor of this handbook and was not involved in the editorial review, or the decision to publish this article. All authors declare that there are no competing interests.

Ethical Clearance Not applicable.

References

Acquaye, A., Feng, K., Oppon, E., Salhi, S., Ibn-Mohammed, T., Genovese, A., & Hubacek, K. (2017). Measuring the environmental sustainability performance of global supply chains: A multi-regional input-output analysis for carbon, sulphur oxide and water footprints. *Journal of Environmental Management, 187*, 571–585. https://doi.org/10.1016/j.jenvman.2016.10.059

ADB (Org.). (2015). *Emissions trading schemes and their linking: Challenges and opportunities in Asia and the Pacific*. Asian Development Bank.

Ali, D. A., Deininger, K., & Duponchel, M. (2014). Credit constraints and agricultural productivity: Evidence from rural Rwanda. *The Journal of Development Studies, 50*(5), 649–665. https://doi.org/10.1080/00220388.2014.887687

Almaraz, M., Wong, M. Y., Geoghegan, E. K., & Houlton, B. Z. (2021). A review of carbon farming impacts on nitrogen cycling, retention, and loss. *Annals of the New York Academy of Sciences, 1505*(1), 102–117. https://doi.org/10.1111/nyas.14690

Altieri, M. A., Nicholls, C. I., & Montalba, R. (2017). Technological approaches to sustainable agriculture at a crossroads: An agroecological perspective. *Sustainability, 9*(3), Art. 3. https://doi.org/10.3390/su9030349

Alves, B. J., Madari, B. E., & Boddey, R. M. (2017). Integrated crop–livestock–forestry systems: Prospects for a sustainable agricultural intensification. *Nutrient Cycling in Agroecosystems, 108*(1), 1–4. https://doi.org/10.1007/s10705-017-9851-0

Alves-Pinto, H. N., Latawiec, A. E., Strassburg, B. B. N., Barros, F. S. M., Sansevero, J. B. B., Iribarrem, A., Crouzeilles, R., Lemgruber, L., Rangel, M. C., & Silva, A. C. P. (2017). Reconciling rural development and ecological restoration: Strategies and policy recommendations for the Brazilian Atlantic Forest. *Land Use Policy, 60*, 419–426. https://doi.org/10.1016/j.landusepol.2016.08.004

An, Z., Bork, E. W., Duan, X., Gross, C. D., Carlyle, C. N., & Chang, S. X. (2022). Quantifying past, current, and future forest carbon stocks within agroforestry systems in central Alberta, Canada. *GCB Bioenergy, 14*(6), 669–680. https://doi.org/10.1111/gcbb.12934

Antle, J. M., Zhang, H., Mu, J. E., Abatzoglou, J., & Stöckle, C. (2018). Methods to assess between-system adaptations to climate change: Dryland wheat systems in the Pacific Northwest United States. *Agriculture, Ecosystems & Environment, 253*, 195–207. https://doi.org/10.1016/j.agee.2017.03.017

Arima, E. Y., Barreto, P., Araújo, E., & Soares-Filho, B. (2014). Public policies can reduce tropical deforestation: Lessons and challenges from Brazil. *Land Use Policy, 41*, 465–473. https://doi.org/10.1016/j.landusepol.2014.06.026

Assmann, J. M., Anghinoni, I., Martins, A. P., Costa, S. E. V. G. de A., Cecagno, D., Carlos, F. S., & Carvalho, P. C. de F. (2014). Soil carbon and nitrogen stocks and fractions in a long-term integrated crop–livestock system under no-tillage in southern Brazil. *Agriculture, Ecosystems & Environment, 190*, 52–59. https://doi.org/10.1016/j.agee.2013.12.003

Assunção, S. A., Pereira, M. G., Rosset, J. S., Berbara, R. L. L., & García, A. C. (2019). Carbon input and the structural quality of soil organic matter as a function of agricultural management in a tropical climate region of Brazil. *Science of the Total Environment, 658*, 901–911. https://doi.org/10.1016/j.scitotenv.2018.12.271

Balbi, S. (2009). Cap-and-trade systems versus carbon taxes: A reflection in view of Copenhagen 2009. *SSRN Electronic Journal*. https://doi.org/10.2139/SSRN.1498454

Baumber, A., Metternicht, G., Cross, R., Ruoso, L.-E., Cowie, A. L., & Waters, C. (2019). Promoting co-benefits of carbon farming in Oceania: Applying and adapting approaches and metrics from existing market-based schemes. *Ecosystem Services, 39*, 100982. https://doi.org/10.1016/j.ecoser.2019.100982

Baumber, A., Waters, C., Cross, R., Metternicht, G., Simpson, M., Baumber, A., Waters, C., Cross, R., Metternicht, G., & Simpson, M. (2020). Carbon farming for resilient rangelands: People, paddocks and policy. *The Rangeland Journal, 42*(5), 293–307. https://doi.org/10.1071/RJ20034

Becker, K., Wulfmeyer, V., Berger, T., Gebel, J., & Münch, W. (2013). Carbon farming in hot, dry coastal areas: An option for climate change mitigation. *Earth System Dynamics, 4*(2), 237–251. https://doi.org/10.5194/esd-4-237-2013

Berger, T. (2007). Concepts of national competitiveness. In *International assessment of research and development in brain-computer interfaces* (Chap. 1). Retrieved December 18, 2007, from https://citeseerx.ist.psu.edu/viewdoc/download?doi=10.1.1.1085.8829&rep=rep1&type=pdf

Blanc, S., Accastello, C., Bianchi, E., Lingua, F., Vacchiano, G., Mosso, A., & Brun, F. (2019). An integrated approach to assess carbon credit from improved forest management. *Journal of Sustainable Forestry, 38*(1), 31–45. https://doi.org/10.1080/10549811.2018.1494002

Boissière, M., Herold, M., Atmadja, S., & Sheil, D. (2017). The feasibility of local participation in measuring, reporting and verification (PMRV) for REDD+. *PLoS ONE, 12*(5), e0176897. https://doi.org/10.1371/journal.pone.0176897

Bragança, A., Newton, P., Cohn, A., Assunção, J., Camboim, C., de Faveri, D., Farinelli, B., Perego, V. M. E., Tavares, M., Resende, J., de Medeiros, S., & Searchinger, T. D. (2022). Extension services can promote pasture restoration: Evidence from Brazil's low carbon agriculture plan. *Proceedings of the National Academy of Sciences, 119*(12), e2114913119. https://doi.org/10.1073/pnas.2114913119

Brewer, T. L., Derwent, H., Błachowicz, A., & Grubb, M. (2016). *Carbon market clubs and the new Paris Regime*. World Bank. https://doi.org/10.1596/25768

Calil, F. N., Lima, N. L., Silva, R. T., Moraes, M. D. A. de, Barbosa, P. V. G., Lima, P. A. F., Brandao, D. C., Silva-Neto, C. de M. e, Carvalho, H. C. de S., & Nascimento, A. dos R. (2016). Biomass and nutrition stock of grassland and accumulated litter in a silvopastoral system with Cerrado species. *African Journal of Agricultural Research, 11*(38), 3701–3709. https://doi.org/10.5897/AJAR2016.11369

Campanha, M. M., de Oliveira, A. D., Marriel, I. E., Gontijo Neto, M. M., Malaquias, J. V., Landau, E. C., de Albuquerque Filho, M. R., Ribeiro, F. P., & de Carvalho, A. M. (2019). Effect of soil tillage and N fertilization on N_2O mitigation in maize in the Brazilian Cerrado. *Science of the Total Environment, 692*, 1165–1174. https://doi.org/10.1016/j.scitotenv.2019.07.315

Carattini, S., Carvalho, M., & Fankhauser, S. (2018). Overcoming public resistance to carbon taxes. *Wires Climate Change, 9*(5), e531. https://doi.org/10.1002/wcc.531

Carrer, M. J., de Souza Filho, H. M., & Batalha, M. O. (2017). Factors influencing the adoption of Farm Management Information Systems (FMIS) by Brazilian citrus farmers. *Computers and Electronics in Agriculture, 138*, 11–19. https://doi.org/10.1016/j.compag.2017.04.004

Carrer, M. J., Maia, A. G., de Mello Brandão Vinholis, M., & de Souza Filho, H. M. (2020). Assessing the effectiveness of rural credit policy on the adoption of integrated crop-livestock systems in Brazil. *Land Use Policy, 92*, 104468. https://doi.org/10.1016/j.landusepol.2020.104468

Carrer, M. J., Silveira, R. L. F. da, Filho, H. M. de S., & Vinholis, M. de M. B. (2013). Factors influencing beef cattle farmers use of risk management instruments in the State of Sao Paulo, Brazil. *Ciencia Rural, 43*(2), 370–377.

Carvalho, J. L. N., Raucci, G. S., Frazão, L. A., Cerri, C. E. P., Bernoux, M., & Cerri, C. C. (2014). Crop-pasture rotation: A strategy to reduce soil greenhouse gas emissions in the Brazilian Cerrado. *Agriculture, Ecosystems & Environment, 183*, 167–175. https://doi.org/10.1016/j.agee.2013.11.014

Center for Climate and Energy Solutions. (2015). *Market mechanisms: Understanding the options*. https://www.c2es.org/wp-content/uploads/2015/04/market-mechanisms-brief.pdf

Chen, C.-K., & Ulya, M. A. (2019). Analyses of the reward-penalty mechanism in green closed-loop supply chains with product remanufacturing. *International Journal of Production Economics, 210*, 211–223. https://doi.org/10.1016/j.ijpe.2019.01.006

Chenu, C., Angers, D. A., Barré, P., Derrien, D., Arrouays, D., & Balesdent, J. (2019). Increasing organic stocks in agricultural soils: Knowledge gaps and potential innovations. *Soil and Tillage Research, 188*, 41–52. https://doi.org/10.1016/j.still.2018.04.011

Choi, T.-M., Ma, C., Shen, B., & Sun, Q. (2019). Optimal pricing in mass customization supply chains with risk-averse agents and retail competition. *Omega, 88*, 150–161. https://doi.org/10.1016/j.omega.2018.08.004

Cohen, M. C., Lobel, R., & Perakis, G. (2016). The impact of demand uncertainty on consumer subsidies for green technology adoption. *Management Science, 62*(5), 1235–1258. https://doi.org/10.1287/mnsc.2015.2173

Colmer, J., Martin, R., Muûls, M., & Wagner, U. J. (2022). *Does pricing carbon mitigate climate change? Firm-level evidence from the European Union emissions trading scheme* (SSRN Scholarly Paper No 4026889). https://papers.ssrn.com/abstract=4026889

Corbeels, M., Marchão, R. L., Neto, M. S., Ferreira, E. G., Madari, B. E., Scopel, E., & Brito, O. R. (2016). Evidence of limited carbon sequestration in soils under no-tillage systems in the Cerrado of Brazil. *Scientific Reports, 6*(1), Art. 1. https://doi.org/10.1038/srep21450

Coser, T. R., de Figueiredo, C. C., Jovanovic, B., Moreira, T. N., Leite, G. G., Cabral Filho, S. L. S., Kato, E., Malaquias, J. V., & Marchão, R. L. (2018). Short-term buildup of carbon from a low-productivity pastureland to an agrisilviculture system in the Brazilian savannah. *Agricultural Systems, 166*, 184–195. https://doi.org/10.1016/j.agsy.2018.01.030

Costa Jr., N. B. da, Baldissera, T. C., Pinto, C. E., Garagorry, F. C., Moraes, A. de, & Carvalho, P. C. de F. (2019). Public policies for low carbon emission agriculture foster beef cattle production in southern Brazil. *Land Use Policy, 80*, 269–273. https://doi.org/10.1016/j.landusepol.2018.10.014

Council of the European Union. (2015). *EU-China joint statement on climate change.* https://www.consilium.europa.eu/en/press/press-releases/2015/06/29/eu-china-climate-statement/

Council of the European Union. (2017). *Non-legislative acts: International agreements.* Official Journal of the European Union. https://eur-lex.europa.eu/legal-content/EN/TXT/PDF/?uri=CELEX:32017D2240&from=EN

Crouzeilles, R., Beyer, H. L., Monteiro, L. M., Feltran-Barbieri, R., Pessôa, A. C. M., Barros, F. S. M., Lindenmayer, D. B., Lino, E. D. S. M., Grelle, C. E. V., Chazdon, R. L., Matsumoto, M., Rosa, M., Latawiec, A. E., & Strassburg, B. B. N. (2020). Achieving cost-effective landscape-scale forest restoration through targeted natural regeneration. *Conservation Letters, 13*(3), e12709. https://doi.org/10.1111/conl.12709

Cybercom. (2019). *World Bank ProGreen.* Openaid. https://openaid.se/en/activities/SE-0-SE-6-14242A0101-GGG-41010

Damian, J. M., da Silva Matos, E., e Pedreira, B. C., de Faccio Carvalho, P. C., Premazzi, L. M., Williams, S., Paustian, K., & Cerri, C. E. P. (2021). Predicting soil C changes after pasture intensification and diversification in Brazil. *CATENA, 202*, 105238. https://doi.org/10.1016/j.catena.2021.105238

de Figueiredo, C. C., de Oliveira, A. D., dos Santos, I. L., Ferreira, E. A. B., Malaquias, J. V., de Sá, M. A. C., de Carvalho, A. M., & dos Santos, J. de D. G. (2018). Relationships between soil organic matter pools and nitrous oxide emissions of agroecosystems in the Brazilian Cerrado. *Science of the Total Environment, 618*, 1572–1582. https://doi.org/10.1016/j.scitotenv.2017.09.333

Dellink, R., Jamet, S., Chateau, J., & Duval, R. (2014). Towards global carbon pricing: Direct and indirect linking of carbon markets. *OECD Journal: Economic Studies, 2013*(1), 209–234. https://doi.org/10.1787/eco_studies-2013-5k421kk9j3vb

Diniz Oliveira, T., Costa Gurgel, A., & Tonry, S. (2021). Potential trading partners of a Brazilian emissions trading scheme: The effects of linking with a developed region (Europe) and two developing regions (Latin America and China). *Technological Forecasting and Social Change, 171*, 120947. https://doi.org/10.1016/j.techfore.2021.120947

dos Reis, J. C., Rodrigues, G. S., de Barros, I., Ribeiro Rodrigues, R. de A., Garrett, R. D., Valentim, J. F., Kamoi, M. Y. T., Michetti, M., Wruck, F. J., Rodrigues-Filho, S., Pimentel, P. E. O., & Smukler, S. (2021). Integrated crop-livestock systems: A sustainable land-use alternative for food production in the Brazilian Cerrado and Amazon. *Journal of Cleaner Production, 283*, 124580. https://doi.org/10.1016/j.jclepro.2020.124580

Dumbrell, N. P., Kragt, M. E., & Gibson, F. L. (2016). What carbon farming activities are farmers likely to adopt? A best–worst scaling survey. *Land Use Policy, 54*, 29–37. https://doi.org/10.1016/j.landusepol.2016.02.002

Dutta, A., Farooq, S., Karimi, I. A., & Khan, S. A. (2017). Assessing the potential of CO_2 utilization with an integrated framework for producing power and chemicals. *Journal of CO_2 Utilization, 19*, 49–57. https://doi.org/10.1016/j.jcou.2017.03.005

Dynarski, K. A., Bossio, D. A., & Scow, K. M. (2020). Dynamic stability of soil carbon: Reassessing the "permanence" of soil carbon sequestration. *Frontiers in Environmental Science, 8*. https://doi.org/10.3389/fenvs.2020.514701

Ellis, J., Nachtigall, D., & Venmans, F. (2019). *Carbon pricing and competitiveness: Are they at odds?* OECD. https://doi.org/10.1787/f79a75ab-en

Essl, F., Hulme, P. E., Jeschke, J. M., Keller, R., Pyšek, P., Richardson, D. M., Saul, W.-C., Bacher, S., Dullinger, S., Estévez, R. A., Kueffer, C., Roy, H. E., Seebens, H., & Rabitsch, W. (2017). Scientific and normative foundations for the valuation of alien-species impacts: Thirteen core principles. *BioScience, 67*(2), 166–178. https://doi.org/10.1093/biosci/biw160

European Commission. (2021). *Carbon farming (climate action).* European Commission. https://climate.ec.europa.eu/eu-action/forests-and-agriculture/sustainable-carbon-cycles/carbon-farming_en

Evans, M. C. (2018). Effective incentives for reforestation: Lessons from Australia's carbon farming policies. *Current Opinion in Environmental Sustainability, 32*, 38–45. https://doi.org/10.1016/j.cosust.2018.04.002

Evans, M. C., Carwardine, J., Fensham, R. J., Butler, D. W., Wilson, K. A., Possingham, H. P., & Martin, T. G. (2015). Carbon farming via assisted natural regeneration as a cost-effective mechanism for restoring biodiversity in agricultural landscapes. *Environmental Science & Policy, 50*, 114–129. https://doi.org/10.1016/j.envsci.2015.02.003

FCPF. (2022). *About the FCPF | Forest Carbon Partnership Facility*. https://www.forestcarbonpartnership.org/about

Frank, C. (2014). Pricing carbon: A carbon tax or cap-and-trade? *Brookings*. https://www.brookings.edu/blog/planetpolicy/2014/08/12/pricing-carbon-a-carbon-tax-or-cap-and-trade/

Garrett, R. D., Cammelli, F., Ferreira, J., Levy, S. A., Valentim, J., & Vieira, I. (2021). Forests and sustainable development in the Brazilian Amazon: History, trends, and future prospects. *Annual Review of Environment and Resources, 46*(1), 625–652. https://doi.org/10.1146/annurev-environ-012220-010228

Garrett, R. D., Niles, M. T., Gil, J. D. B., Gaudin, A., Chaplin-Kramer, R., Assmann, A., Assmann, T. S., Brewer, K., de Faccio Carvalho, P. C., Cortner, O., Dynes, R., Garbach, K., Kebreab, E., Mueller, N., Peterson, C., Reis, J. C., Snow, V., & Valentim, J. (2017). Social and ecological analysis of commercial integrated crop livestock systems: Current knowledge and remaining uncertainty. *Agricultural Systems, 155*, 136–146. https://doi.org/10.1016/j.agsy.2017.05.003

Gmach, M.-R., Dias, B. O., Silva, C. A., Nóbrega, J. C. A., Lustosa-Filho, J. F., & Siqueira-Neto, M. (2018). Soil organic matter dynamics and land-use change on Oxisols in the Cerrado, Brazil. *Geoderma Regional, 14*, e00178. https://doi.org/10.1016/j.geodrs.2018.e00178

Godar, J., Gardner, T. A., Tizado, E. J., & Pacheco, P. (2014). Actor-specific contributions to the deforestation slowdown in the Brazilian Amazon. *Proceedings of the National Academy of Sciences, 111*(43), 15591–15596. https://doi.org/10.1073/pnas.1322825111

Gonzalez-Sanchez, E. J., Veroz-Gonzalez, O., Conway, G., Moreno-Garcia, M., Kassam, A., Mkomwa, S., Ordoñez-Fernandez, R., Triviño-Tarradas, P., & Carbonell-Bojollo, R. (2019). Meta-analysis on carbon sequestration through conservation agriculture in Africa. *Soil and Tillage Research, 190*, 22–30. https://doi.org/10.1016/j.still.2019.02.020

Görlach, B., Mehling, M., & Llm, E. R. (2015). *Designing institutions; structures and mechanisms to facilitate the linking of emissions trading schemes.* Ecologic Institute. http://119.78.100.173/C666//handle/2XK7JSWQ/2710

Government of the Republic of Korea. (2020). *2050 carbon neutral strategy of the Republic of Korea towards a sustainable and green society*. https://unfccc.int/sites/default/files/resource/LTS1_RKorea.pdf

Green, J. F., Sterner, T., & Wagner, G. (2014). A balance of bottom-up and top-down in linking climate policies. *Nature Climate Change, 4*(12), Art. 12. https://doi.org/10.1038/nclimate2429

Griscom, B. W., Adams, J., Ellis, P. W., Houghton, R. A., Lomax, G., Miteva, D. A., Schlesinger, W. H., Shoch, D., Siikamäki, J. V., Smith, P., Woodbury, P., Zganjar, C., Blackman, A., Campari, J., Conant, R. T., Delgado, C., Elias, P., Gopalakrishna, T., Hamsik, M. R., et al. (2017). Natural climate solutions. *Proceedings of the National Academy of Sciences, 114*(44), 11645–11650. https://doi.org/10.1073/pnas.1710465114

Guo, P., Zhu, D., Lam, J., & Li, V. (2017). The future of wind energy development in China: Drivers and barriers. In *Wind energy engineering: A handbook for onshore and offshore wind turbines*.

Gür, T. M. (2022). Carbon dioxide emissions, capture, storage and utilization: Review of materials, processes and technologies. *Progress in Energy and Combustion Science, 89*, 100965. https://doi.org/10.1016/j.pecs.2021.100965

Hausotter, T., Steuwer, S., & Taenzler, D. (2011). *Competitiveness and linking of emission trading systems*. https://www.osti.gov/etdeweb/biblio/21403154

Havemann, T., Negra, C., & Werneck, F. (2020). Blended finance for agriculture: Exploring the constraints and possibilities of combining financial instruments for sustainable transitions. *Agriculture and Human Values*. https://doi.org/10.1007/s10460-020-10131-8

Haya, B., Cullenward, D., Strong, A. L., Grubert, E., Heilmayr, R., Sivas, D. A., & Wara, M. (2020). Managing uncertainty in carbon offsets: Insights from California's standardized approach. *Climate Policy, 20*(9), 1112–1126. https://doi.org/10.1080/14693062.2020.1781035

He, P., He, Y., & Xu, H. (2019). Channel structure and pricing in a dual-channel closed-loop supply chain with government subsidy. *International Journal of Production Economics, 213*, 108–123. https://doi.org/10.1016/j.ijpe.2019.03.013

Hui, D., Deng, Q., & Tian, H. (2015). *Climate change and carbon sequestration in forest ecosystems* (pp. 1–40). https://doi.org/10.1007/978-1-4614-6431-0_13-2

ICAP. (2020). *Emissions trading worldwide: ICAP status report 2020*. International Carbon Action Partnerships. https://icapcarbonaction.com/en/publications/emissions-trading-worldwide-icap-status-report-2020

Jansson, C., Faiola, C., Wingler, A., Zhu, X.-G., Kravchenko, A., de Graaff, M.-A., Ogden, A. J., Handakumbura, P. P., Werner, C., & Beckles, D. M. (2021). Crops for carbon farming. *Frontiers in Plant Science, 12*. https://doi.org/10.3389/fpls.2021.636709

Ji, J., Zhang, Z., & Yang, L. (2017). Carbon emission reduction decisions in the retail-/dual-channel supply chain with consumers' preference. *Journal of Cleaner Production, 141*, 852–867. https://doi.org/10.1016/j.jclepro.2016.09.135

Kachi, A., Unger, C., Böhm, N., Stelmakh, K., Haug, C., & Frerk, M. (2015). *Linking emissions trading systems: A summary of current research* (pp. 1–20). International Carbon Action Partnership.

Kellogg, W. W., & Schware, R. (2019). *Climate change and society: Consequences of increasing atmospheric carbon dioxide*. Routledge. https://doi.org/10.4324/9780429048739

Khataza, R. R. B., Hailu, A., Kragt, M. E., & Doole, G. J. (2017). Estimating shadow price for symbiotic nitrogen and technical efficiency for legume-based conservation agriculture in Malawi. *Australian Journal of Agricultural and Resource Economics, 61*(3), 462–480. https://doi.org/10.1111/1467-8489.12212

Kılkış, Ş, Krajačić, G., Duić, N., Rosen, M. A., & Al-Nimr, M. A. (2020). Advances in integration of energy, water and environment systems towards climate neutrality for sustainable development. *Energy Conversion and Management, 225*, 113410. https://doi.org/10.1016/j.enconman.2020.113410

Knook, J., Eory, V., Brander, M., & Moran, D. (2020). The evaluation of a participatory extension programme focused on climate friendly farming. *Journal of Rural Studies, 76*, 40–48. https://doi.org/10.1016/j.jrurstud.2020.03.010

Kondvilkar, N., & Thakare, R. (2021). Carbon farming: For climate-smart agriculture and environmental security. In R. Bhatt, R. S. Meena, & A. Hossain (Eds.), *Input use efficiency for food and environmental security* (pp. 241–268). Springer Nature. https://doi.org/10.1007/978-981-16-5199-1_8

Kossoy, A., Peszko, G., Oppermann, K., Prytz, N., Gilbert, A., Klein, N., Lam, L., & Wong, L. (2015). *Carbon pricing watch 2015* [Brief]. World Bank. https://openknowledge.worldbank.org/handle/10986/21986

Kragt, M. E., Dumbrell, N. P., & Blackmore, L. (2017). Motivations and barriers for Western Australian broad-acre farmers to adopt carbon farming. *Environmental Science & Policy, 73*, 115–123. https://doi.org/10.1016/j.envsci.2017.04.009

Krishnan, M., Samandari, H., Woetzel, J., Smit, S., Pacthod, D., Pinner, D., Naucler, T., Tai, H., Farr, A., Wu, W., & Imperato, D. (2022). *Economic sectors with high-emissions products or operations would face substantial effects on demand, production costs, and employment*. https://www.mckinsey.com/capabilities/sustainability/our-insights/sectors-are-unevenly-exposed-in-the-net-zero-transition

La Rovere, E. L. (2020). The potential contribution of emerging economies to stop dangerous climate change. The case of Brazil. *WIREs Climate Change, 11*(1), e614. https://doi.org/10.1002/wcc.614

Lamb, A., Green, R., Bateman, I., Broadmeadow, M., Bruce, T., Burney, J., Carey, P., Chadwick, D., Crane, E., Field, R., Goulding, K., Griffiths, H., Hastings, A., Kasoar, T., Kindred, D., Phalan, B., Pickett, J., Smith, P., Wall, E., et al. (2016). The potential for land sparing to offset greenhouse gas emissions from agriculture. *Nature Climate Change, 6*(5), Art. 5. https://doi.org/10.1038/nclimate2910

Lash, J., & Wellington, F. (2007). Competitive advantage on a warming planet. *Harvard Business Review*. https://hbr.org/2007/03/competitive-advantage-on-a-warming-planet

Lehner, P., & Rosenberg, N. (2018). *Promoting climate-friendly agriculture for the benefit of farmers, rural communities, and the environment* (SSRN Scholarly Paper ID 3264459). Social Science Research Network. https://papers.ssrn.com/abstract=3264459

Levinski-Huf, F., & Klein, V. A. (2018). Organic matter and physical properties of a Red Latosol under an integrated crop-livestock-forestry system. *Pesquisa Agropecuária Tropical, 48*, 316–322. https://doi.org/10.1590/1983-40632018v4852737

Levis, C., Flores, B. M., Mazzochini, G. G., Manhães, A. P., Campos-Silva, J. V., Borges de Amorim, P., Peroni, N., Hirota, M., & Clement, C. R. (2020). Help restore Brazil's governance of globally important ecosystem services. *Nature Ecology & Evolution, 4*(2), Art. 2. https://doi.org/10.1038/s41559-019-1093-x

Li, H., Li, R., Shang, M., Liu, Y., & Su, D. (2022). Cooperative decisions of competitive supply chains considering carbon trading mechanism. *International Journal of Low-Carbon Technologies, 17*, 102–117. https://doi.org/10.1093/ijlct/ctab085

Li, S., & Kallas, Z. (2021). Meta-analysis of consumers' willingness to pay for sustainable food products. *Appetite, 163*, 105239. https://doi.org/10.1016/j.appet.2021.105239

Lilliestam, J., Patt, A., & Bersalli, G. (2021). The effect of carbon pricing on technological change for full energy decarbonization: A review of empirical ex-post evidence. *Wires Climate Change, 12*(1), e681. https://doi.org/10.1002/wcc.681

Lin, B. B., Macfadyen, S., Renwick, A. R., Cunningham, S. A., & Schellhorn, N. A. (2013). Maximizing the environmental benefits of carbon farming through ecosystem service delivery. *BioScience, 63*(10), 793–803. https://doi.org/10.1525/bio.2013.63.10.6

Lisboa, F. J. G., Chaer, G. M., Fernandes, M. F., Berbara, R. L. L., & Madari, B. E. (2014). The match between microbial community structure and soil properties is modulated by land use types and sample origin within an integrated agroecosystem. *Soil Biology and Biochemistry, 78*, 97–108. https://doi.org/10.1016/j.soilbio.2014.07.017

Liu, C., & Zheng, H. (2021). How social capital affects willingness of farmers to accept low-carbon agricultural technology (LAT)? A case study of Jiangsu, China. *International Journal of Climate Change Strategies and Management, 13*(3), 286–301. https://doi.org/10.1108/IJCCSM-09-2020-0100

Liu, X., Du, W., & Sun, Y. (2020). Green supply chain decisions under different power structures: Wholesale price vs. revenue sharing contract. *International Journal of Environmental Research and Public Health, 17*(21), Art. 21. https://doi.org/10.3390/ijerph17217737

Lohmann, L. (2010). Uncertainty markets and carbon markets: Variations on Polanyian themes. *New Political Economy, 15*(2), 225–254. https://doi.org/10.1080/13563460903290946

Lorenz, K., & Lal, R. (2012). Cropland soil carbon dynamics. In R. Lal, K. Lorenz, R. F. Hüttl, B. U. Schneider, & J. von Braun (Eds.), *Recarbonization of the biosphere: Ecosystems and the global carbon cycle* (pp. 303–346). Springer Netherlands. https://doi.org/10.1007/978-94-007-4159-1_14

Lyu, X., Shi, A., & Wang, X. (2020). Research on the impact of carbon emission trading system on low-carbon technology innovation. *Carbon Management, 11*(2), 183–193. https://doi.org/10.1080/17583004.2020.1721977

Maia, A. G., Eusebio, G. dos S., Fasiaben, M. do C. R., Moraes, A. S., Assad, E. D., & Pugliero, V. S. (2021). The economic impacts of the diffusion of agroforestry in Brazil. *Land Use Policy, 108*, 105489. https://doi.org/10.1016/j.landusepol.2021.105489

Mardones, C., & Lipski, M. (2020). A carbon tax on agriculture? A CGE analysis for Chile. *Economic Systems Research, 32*(2), 262–277. https://doi.org/10.1080/09535314.2019.1676701

Mattila, T. J., Hagelberg, E., Söderlund, S., & Joona, J. (2022). How farmers approach soil carbon sequestration? Lessons learned from 105 carbon-farming plans. *Soil and Tillage Research, 215*, 105204. https://doi.org/10.1016/j.still.2021.105204

McDonald, H., Frelih-Larsen, A., Keenleyside, C., Lóránt, A., Duin, L., Andersen, S. P., Costa, G., Aubert, G., & Hiller, N. (2021). *Carbon farming, making agriculture fit for 2030.*

MEACP. (2019). *Long term strategy on climate mitigation.* Ministry of Economic Affairs and Climate Policy. https://ec.europa.eu/clima/sites/lts/lts_nl_en.pdf

Melillo, J., & Gribkoff, E. (2021). *Soil-based carbon sequestration.* MIT Climate Portal. https://climate.mit.edu/explainers/soil-based-carbon-sequestration

Meylan, L., Gary, C., Allinne, C., Ortiz, J., Jackson, L., & Rapidel, B. (2017). Evaluating the effect of shade trees on provision of ecosystem services in intensively managed coffee plantations. *Agriculture, Ecosystems & Environment, 245*, 32–42. https://doi.org/10.1016/j.agee.2017.05.005

Montibeller, B., Kmoch, A., Virro, H., Mander, Ü., & Uuemaa, E. (2020). Increasing fragmentation of forest cover in Brazil's Legal Amazon from 2001 to 2017. *Scientific Reports, 10*(1), Art. 1. https://doi.org/10.1038/s41598-020-62591-x

Morgan, M. I., Hine, D. W., Bhullar, N., & Loi, N. M. (2015). Landholder adoption of low emission agricultural practices: A profiling approach. *Journal of Environmental Psychology, 41*, 35–44. https://doi.org/10.1016/j.jenvp.2014.11.004

Moyes, T. E. (2008). Greenhouse gas emissions trading in New Zealand: Trailblazing comprehensive cap and trade. *Ecology Law Quarterly, 35*, 911.

Nachtigall, D. (2019). Dynamic climate policy under firm relocation: The implications of phasing out free allowances. *Environmental and Resource Economics, 74*(1), 473–503. https://doi.org/10.1007/s10640-019-00326-x

Nair, P. K. R. (2011). Agroforestry systems and environmental quality: Introduction. *Journal of Environmental Quality, 40*(3), 784–790. https://doi.org/10.2134/jeq2011.0076

Nath, A. J., Lal, R., & Das, A. K. (2015). Managing woody bamboos for carbon farming and carbon trading. *Global Ecology and Conservation, 3*, 654–663. https://doi.org/10.1016/j.gecco.2015.03.002

Nepstad, D., McGrath, D., Stickler, C., Alencar, A., Azevedo, A., Swette, B., Bezerra, T., DiGiano, M., Shimada, J., Seroa da Motta, R., Armijo, E., Castello, L., Brando, P., Hansen, M. C., McGrath-Horn, M., Carvalho, O., & Hess, L. (2014). Slowing Amazon deforestation through public policy and interventions in beef and soy supply chains. *Science, 344*(6188), 1118–1123. https://doi.org/10.1126/science.1248525

Nepstad, L. S., Gerber, J. S., Hill, J. D., Dias, L. C. P., Costa, M. H., & West, P. C. (2019). Pathways for recent Cerrado soybean expansion: Extending the soy moratorium and implementing integrated crop livestock systems with soybeans. *Environmental Research Letters, 14*(4), 044029. https://doi.org/10.1088/1748-9326/aafb85

Newton, P., Gomez, A. E. A., Jung, S., Kelly, T., Mendes, T. de A., Rasmussen, L. V., Reis, J. C. dos, Rodrigues, R. de A. R., Tipper, R., van der Horst, D., & Watkins, C. (2016). Overcoming barriers to low carbon agriculture and forest restoration in Brazil: The Rural Sustentável project. *World Development Perspectives, 4*, 5–7. https://doi.org/10.1016/j.wdp.2016.11.011

Northrup, D. L., Basso, B., Wang, M. Q., Morgan, C. L. S., & Benfey, P. N. (2021). Novel technologies for emission reduction complement conservation agriculture to achieve negative emissions from row-crop production. *Proceedings of the National Academy of Sciences, 118*(28), e2022666118. https://doi.org/10.1073/pnas.2022666118

OECD. (2011). *Emission permits and competition.* Organization for Economic Cooperation and Development Policy Roundtables. https://one.oecd.org/document/DAF/COMP(2010)35/en/pdf

Olale, E., Yiridoe, E. K., Ochuodho, T. O., & Lantz, V. (2019). The effect of carbon tax on farm income: Evidence from a Canadian province. *Environmental and Resource Economics, 74*(2), 605–623. https://doi.org/10.1007/s10640-019-00337-8

Oliveira, J. de M., Madari, B. E., Carvalho, M. T. de M., Assis, P. C. R., Silveira, A. L. R., de Leles Lima, M., Wruck, F. J., Medeiros, J. C., & Machado, P. L. O. de A. (2018). Integrated farming systems for improving soil carbon balance in the southern Amazon of Brazil. *Regional Environmental Change, 18*(1), 105–116. https://doi.org/10.1007/s10113-017-1146-0

Oliveira, R. V., & Thorseth, M. (2016). Ethical implications of co-benefits rationale within climate change mitigation strategy. *Etikk i Praksis—Nordic Journal of Applied Ethics, 2*, Art. 2. https://doi.org/10.5324/eip.v10i2.1942

Ontl, T., & Schulte, L. (2012). Soil carbon storage. *Nature Education Knowledge, 3*(10), 35.

Parry, I. W. (2019). Putting a price on pollution: Carbon-pricing strategies could hold the key to meeting the world's climate stabilization goals. *Finance & Development, 56*(004).

Partnership for Market Readiness, & International Carbon Action Partnership. (2016). *Emissions trading in practice: A handbook on design and implementation* [Technical Paper]. World Bank. https://doi.org/10.1596/23874

Peters, G. P., Andrew, R. M., Canadell, J. G., Friedlingstein, P., Jackson, R. B., Korsbakken, J. I., Le Quéré, C., & Peregon, A. (2020). Carbon dioxide emissions continue to grow amidst slowly emerging climate policies. *Nature Climate Change, 10*(1), Art. 1. https://doi.org/10.1038/s41558-019-0659-6

Polidoro, J. C., Freitas, P. L. de, Hernani, L. C., Anjos, L. H. C. dos, Rodrigues, R. de A. R., Cesário, F. V., Andrade, A. G. de, & Ribeiro, J. L. (2021). Potential impact of plans and policies based on the principles of conservation agriculture on the control of soil erosion in Brazil. *Land Degradation & Development*. https://doi.org/10.1002/ldr.3876

Porgo, M., Kuwornu, J. K. M., Zahonogo, P., Jatoe, J. B. D., & Egyir, I. S. (2018). Credit constraints and cropland allocation decisions in rural Burkina Faso. *Land Use Policy, 70*, 666–674. https://doi.org/10.1016/j.landusepol.2017.10.053

Qi, T., Weng, Y., Zhang, X., & He, J. (2016). An analysis of the driving factors of energy-related CO_2 emission reduction in China from 2005 to 2013. *Energy Economics, 60*, 15–22. https://doi.org/10.1016/j.eneco.2016.09.014

Rajão, R., Soares-Filho, B., Nunes, F., Börner, J., Machado, L., Assis, D., Oliveira, A., Pinto, L., Ribeiro, V., Rausch, L., Gibbs, H., & Figueira, D. (2020). The rotten apples of Brazil's agribusiness. *Science, 369*(6501), 246–248. https://doi.org/10.1126/science.aba6646

Rehman, A., Ma, H., Ahmad, M., Irfan, M., Traore, O., & Chandio, A. A. (2021). Towards environmental sustainability: Devolving the influence of carbon dioxide emission to population growth, climate change, forestry, livestock and crops production in Pakistan. *Ecological Indicators, 125*, 107460. https://doi.org/10.1016/j.ecolind.2021.107460

Reis, J. C. dos, Kamoi, M. Y. T., Latorraca, D., Chen, R. F. F., Michetti, M., Wruck, F. J., Garrett, R. D., Valentim, J. F., Rodrigues, R. de A. R., & Rodrigues-Filho, S. (2020). Assessing the economic viability of integrated crop–livestock systems in Mato Grosso, Brazil. *Renewable Agriculture and Food Systems, 35*(6), 631–642. https://doi.org/10.1017/S1742170519000280

Rezende, C. L., Scarano, F. R., Assad, E. D., Joly, C. A., Metzger, J. P., Strassburg, B. B. N., Tabarelli, M., Fonseca, G. A., & Mittermeier, R. A. (2018). From hotspot to hopespot: An opportunity for the Brazilian Atlantic Forest. *Perspectives in Ecology and Conservation, 16*(4), 208–214. https://doi.org/10.1016/j.pecon.2018.10.002

RGGI. (2015). *Regional Greenhouse Gas Initiative (RGGI): An emission trading case study*. https://www.edf.org/sites/default/files/rggi-ets-case-study-may2015.pdf

Rissman, J., Bataille, C., Masanet, E., Aden, N., Morrow, W. R., Zhou, N., Elliott, N., Dell, R., Heeren, N., Huckestein, B., Cresko, J., Miller, S. A., Roy, J., Fennell, P., Cremmins, B., Koch Blank, T., Hone, D., Williams, E. D., de la Rue du Can, S., et al. (2020). Technologies and policies to decarbonize global industry: Review and assessment of mitigation drivers through 2070. *Applied Energy, 266*, 114848. https://doi.org/10.1016/j.apenergy.2020.114848

Rosende, J. (2022). *Value beyond carbon: What are carbon farming co-benefits?* EAgronom. https://eagronom.com/en/blog/what-are-carbon-farming-co-benefits/

Ros-Tonen, M. A. F., van Andel, T., Morsello, C., Otsuki, K., Rosendo, S., & Scholz, I. (2008). Forest-related partnerships in Brazilian Amazonia: There is more to sustainable forest management than reduced impact logging. *Forest Ecology and Management, 256*(7), 1482–1497. https://doi.org/10.1016/j.foreco.2008.02.044

Rovere, E. L. L., Wills, W., Grottera, C., Dubeux, C. B. S., & Gesteira, C. (2018). Economic and social implications of low-emission development pathways in Brazil. *Carbon Management, 9*(5), 563–574. https://doi.org/10.1080/17583004.2018.1507413

Rumpel, C., Amiraslani, F., Chenu, C., Garcia Cardenas, M., Kaonga, M., Koutika, L.-S., Ladha, J., Madari, B., Shirato, Y., Smith, P., Soudi, B., Soussana, J.-F., Whitehead, D., & Wollenberg, E. (2020). The 4p1000 initiative: Opportunities, limitations and challenges for implementing soil organic carbon sequestration as a sustainable development strategy. *Ambio, 49*(1), 350–360. https://doi.org/10.1007/s13280-019-01165-2

Sadeghi, J., & Haapala, K. R. (2019). Optimizing a sustainable logistics problem in a renewable energy network using a genetic algorithm. *OPSEARCH, 56*(1), 73–90. https://doi.org/10.1007/s12597-019-00356-5

Saikku, L., Tainio, P., Hildén, M., Antikainen, R., Leskinen, P., & Koskela, S. (2017). Diffusion of solar electricity in the network of private actors as a strategic experiment to mitigate climate change. *Journal of Cleaner Production, 142*, 2730–2740. https://doi.org/10.1016/j.jclepro.2016.11.003

Sarkar, B., Omair, M., & Choi, S.-B. (2018). A multi-objective optimization of energy, economic, and carbon emission in a production model under sustainable supply chain management. *Applied Sciences, 8*(10), Art. 10. https://doi.org/10.3390/app8101744

Sato, J. H., de Figueiredo, C. C., Marchão, R. L., de Oliveira, A. D., Vilela, L., Delvico, F. M., Alves, B. J. R., & de Carvalho, A. M. (2019). Understanding the relations between soil organic matter fractions and N_2O emissions in a long-term integrated crop–livestock system. *European Journal of Soil Science*, 12819. https://doi.org/10.1111/ejss.12819

Senesi, G. S., Martin-Neto, L., Villas-Boas, P. R., Nicolodelli, G., & Milori, D. M. B. P. (2018). Laser-based spectroscopic methods to evaluate the humification degree of soil organic matter in whole soils: A review. *Journal of Soils and Sediments, 18*(4), 1292–1302. https://doi.org/10.1007/s11368-016-1539-6

Serp, J., Allibert, M., Beneš, O., Delpech, S., Feynberg, O., Ghetta, V., Heuer, D., Holcomb, D., Ignatiev, V., Kloosterman, J. L., Luzzi, L., Merle-Lucotte, E., Uhlíř, J., Yoshioka, R., & Zhimin, D. (2014). The molten salt reactor (MSR) in generation IV: Overview and perspectives. *Progress in Nuclear Energy, 77*, 308–319. https://doi.org/10.1016/j.pnucene.2014.02.014

Sharma, M., Kaushal, R., Kaushik, P., & Ramakrishna, S. (2021a). Carbon farming: Prospects and challenges. *Sustainability, 13*(19), Art. 19. https://doi.org/10.3390/su131911122

Sharma, S., Rana, V. S., Prasad, H., Lakra, J., & Sharma, U. (2021b). Appraisal of carbon capture, storage, and utilization through fruit crops. *Frontiers in Environmental Science, 9*, 785531007.

Shen, L., Wang, X., Liu, Q., Wang, Y., Lv, L., & Tang, R. (2021). Carbon trading mechanism, low-carbon E-commerce supply chain and sustainable development. *Mathematics, 9*(15), Art. 15. https://doi.org/10.3390/math9151717

Silalahi, P. (2018). *Press release: Partnership towards Indonesia's science-based low carbon development | New climate economy.* Commission on the Economy and Climate. https://newclimateeconomy.net/content/press-release-partnership-towards-indonesias-science-based-low-carbon-development

Smith, P., Bustamante, M., Ahammad, H., Clark, H., Dong, H., Elsiddig, E., Haberl, H., Harper, R., House, J., Jafari, M., et al. (2014). Agriculture, forestry and other land use (AFOLU). In *Climate change 2014: Mitigation of climate change. Contribution of working group III to the fifth assessment report of the Intergovernmental Panel on Climate Change* (Chap. 11, pp. 811–922).

Smith, P., Soussana, J.-F., Angers, D., Schipper, L., Chenu, C., Rasse, D. P., Batjes, N. H., van Egmond, F., McNeill, S., Kuhnert, M., Arias-Navarro, C., Olesen, J. E., Chirinda, N., Fornara, D., Wollenberg, E., Álvaro-Fuentes, J., Sanz-Cobena, A., & Klumpp, K. (2020). How to measure, report and verify soil carbon change to realize the potential of soil carbon sequestration for atmospheric greenhouse gas removal. *Global Change Biology, 26*(1), 219–241. https://doi.org/10.1111/gcb.14815

Soussana, J.-F., Lutfalla, S., Ehrhardt, F., Rosenstock, T., Lamanna, C., Havlík, P., Richards, M., Wollenberg, E. (Lini), Chotte, J.-L., Torquebiau, E., Ciais, P., Smith, P., & Lal, R. (2019). Matching policy and science: Rationale for the '4 per 1000—Soils for food security and climate' initiative. *Soil and Tillage Research, 188*, 3–15. https://doi.org/10.1016/j.still.2017.12.002

Souza Filho, H. M., Carrer, M. J., Saes, M. S. M., Gomes, L. A. de V., & Nicolella, A. C. (2019). Performance heterogeneity and strategic orientation: An analysis of small farmers of an agrarian reform project in Brazil. *Land Use Policy, 86*, 23–30. https://doi.org/10.1016/j.landusepol.2019.04.018

Stabile, M. C. C., Guimarães, A. L., Silva, D. S., Ribeiro, V., Macedo, M. N., Coe, M. T., Pinto, E., Moutinho, P., & Alencar, A. (2020). Solving Brazil's land use puzzle: Increasing production and slowing Amazon deforestation. *Land Use Policy, 91*, 104362. https://doi.org/10.1016/j.landusepol.2019.104362

Strassburg, B. B. N., Latawiec, A. E., Barioni, L. G., Nobre, C. A., da Silva, V. P., Valentim, J. F., Vianna, M., & Assad, E. D. (2014). When enough should be enough: Improving the use of current agricultural lands could meet production demands and spare natural habitats in Brazil. *Global Environmental Change, 28*, 84–97. https://doi.org/10.1016/j.gloenvcha.2014.06.001

Streck, C. (2021). How voluntary carbon markets can drive climate ambition. *Journal of Energy & Natural Resources Law, 39*(3), 367–374. https://doi.org/10.1080/02646811.2021.1881275

Tamiotti, L., Teh, R., Kulaçoğlu, V., Olhoff, A., Simmons, B., & Abaza, H. (2009). *Trade and climate change: A report by the United Nations Environment Programme and the World Trade Organization.* UNEP/World Trade Organization.

Tang, K., Hailu, A., Kragt, M. E., & Ma, C. (2018). The response of broadacre mixed crop-livestock farmers to agricultural greenhouse gas abatement incentives. *Agricultural Systems, 160*, 11–20. https://doi.org/10.1016/j.agsy.2017.11.001

Terrenoire, E., Hauglustaine, D. A., Gasser, T., & Penanhoat, O. (2019). The contribution of carbon dioxide emissions from the aviation sector to future climate change. *Environmental Research Letters, 14*(8), 084019. https://doi.org/10.1088/1748-9326/ab3086

Thevathasan, N. V., Gordon, A. M., Bradley, R., Cogliastro, A., Folkard, P., Grant, R., Kort, J., Liggins, L., Njenga, F., Olivier, A., Pharo, C., Powell, G., Rivest, D., Schiks, T., Trotter, D., Van Rees, K., Whalen, J., & Zabek, L. (2012). Agroforestry research and development in Canada: The way forward. In P. K. R. Nair & D. Garrity (Eds.), *Agroforestry—The future of global land use* (pp. 247–283). Springer Netherlands. https://doi.org/10.1007/978-94-007-4676-3_15

Tong, W., Mu, D., Zhao, F., Mendis, G. P., & Sutherland, J. W. (2019). The impact of cap-and-trade mechanism and consumers' environmental preferences on a retailer-led supply Chain. *Resources, Conservation and Recycling, 142*, 88–100. https://doi.org/10.1016/j.resconrec.2018.11.005

Torres, C. M. M. E., Jacovine, L. A. G., de Olivera, N., Neto, S., Fraisse, C. W., Soares, C. P. B., de Castro Neto, F., Ferreira, L. R., Zanuncio, J. C., & Lemes, P. G. (2017). Greenhouse gas emissions and carbon sequestration by agroforestry systems in southeastern Brazil. *Scientific Reports, 7*(1), 16738. https://doi.org/10.1038/s41598-017-16821-4

UNFCCC. (2022). *UNFCCC partners with the air carbon exchange to promote carbon offsetting*. United Nations—Climate Change. https://unfccc.int/news/unfccc-partners-with-the-aircarbon-exchange-to-promote-carbon-offsetting

Verbruggen, A. (2021). *Pricing carbon emissions: Economic reality and utopia*. Routledge. https://doi.org/10.4324/9781003173816

Wang, C., Wang, W., & Huang, R. (2017). Supply chain enterprise operations and government carbon tax decisions considering carbon emissions. *Journal of Cleaner Production, 152*, 271–280. https://doi.org/10.1016/j.jclepro.2017.03.051

Wang, F., Harindintwali, J. D., Yuan, Z., Wang, M., Wang, F., Li, S., Yin, Z., Huang, L., Fu, Y., Li, L., Chang, S. X., Zhang, L., Rinklebe, J., Yuan, Z., Zhu, Q., Xiang, L., Tsang, D. C. W., Xu, L., Jiang, X., et al. (2021). Technologies and perspectives for achieving carbon neutrality. *The Innovation, 2*(4), 100180. https://doi.org/10.1016/j.xinn.2021.100180

Wang, L., Vo, X. V., Shahbaz, M., & Ak, A. (2020a). Globalization and carbon emissions: Is there any role of agriculture value-added, financial development, and natural resource rent in the aftermath of COP21? *Journal of Environmental Management, 268*, 110712. https://doi.org/10.1016/j.jenvman.2020.110712

Wang, Y., Ziv, G., Adami, M., de Almeida, C. A., Antunes, J. F. G., Coutinho, A. C., Esquerdo, J. C. D. M., Gomes, A. R., & Galbraith, D. (2020b). Upturn in secondary forest clearing buffers primary forest loss in the Brazilian Amazon. *Nature Sustainability, 3*(4), 290–295. https://doi.org/10.1038/s41893-019-0470-4

WEF. (2022). *EU organizations join forces with farmers to restore nature*. World Economic Forum. https://www.weforum.org/agenda/2022/01/european-organizations-join-forces-with-farmers-to-fight-climate-change-and-restore-nature/

Wen, W., Zhou, P., & Zhang, F. (2018). Carbon emissions abatement: Emissions trading vs consumer awareness. *Energy Economics, 76*, 34–47. https://doi.org/10.1016/j.eneco.2018.09.019

World Bank. (2019). *State and trends of carbon pricing 2019* [Serial]. World Bank Group. https://doi.org/10.1596/978-1-4648-1435-8

World Bank, & Ecofys. (2017). *Carbon pricing watch 2017* [Serial]. World Bank. https://doi.org/10.1596/978-1-4648-1129-6

WRI, & WBCSD. (2004). *The greenhouse gas protocol. A corporate accounting and reporting standard*. World Resources Institute and World Business Council for Sustainable Development. https://www.wbcsd.org/contentwbc/download/2670/33469/1

Wu, R., Dai, H., Geng, Y., Xie, Y., Masui, T., & Tian, X. (2016). Achieving China's INDC through carbon cap-and-trade: Insights from Shanghai. *Applied Energy, 184*, 1114–1122. https://doi.org/10.1016/j.apenergy.2016.06.011

Wu, T., & Kung, C.-C. (2020). Carbon emissions, technology upgradation and financing risk of the green supply chain competition. *Technological Forecasting and Social Change, 152*, 119884. https://doi.org/10.1016/j.techfore.2019.119884

Yao, X., Huang, R., & Song, M. (2019). How to reduce carbon emissions of small and medium enterprises (SMEs) by knowledge sharing in China. *Production Planning & Control, 30*(10–12), 881–892. https://doi.org/10.1080/09537287.2019.1582096

Zetterberg, L. (2012). Linking the emissions trading systems in EU and California. *The Mistra Indigo Program.*

Zhang, F., & Zhang, Z. (2020). The tail dependence of the carbon markets: The implication of portfolio management. *PLoS ONE, 15*(8), e0238033. https://doi.org/10.1371/journal.pone.0238033

Zhang, J., Cao, Q., & He, X. (2019). Contract and product quality in platform selling. *European Journal of Operational Research, 272*(3), 928–944. https://doi.org/10.1016/j.ejor.2018.07.023

Zhao, D., Wang, W., & Luo, Z. (2019). Global ETS operation and their merits and demerits. In D. Zhao, W. Wang, & Z. Luo (Eds.), *A brief overview of China's ETS pilots: Deconstruction and assessment of Guangdong's greenhouse gas emission trading mechanism* (pp. 11–36). Springer. https://doi.org/10.1007/978-981-13-1888-7_2

Zheng, Y., Sun, X., Zhang, C., Wang, D., & Mao, J. (2021). Can emission trading scheme improve carbon emission performance? Evidence from China. *Frontiers in Energy Research, 9*. https://doi.org/10.3389/fenrg.2021.759572

Zhou, D., Liang, X., Zhou, Y., & Tang, K. (2020). Does emission trading boost carbon productivity? Evidence from China's pilot emission trading scheme. *International Journal of Environmental Research and Public Health, 17*(15), Art. 15. https://doi.org/10.3390/ijerph17155522

Zhu, J., Fan, Y., Deng, X., & Xue, L. (2019). Low-carbon innovation induced by emissions trading in China. *Nature Communications, 10*(1), Art. 1. https://doi.org/10.1038/s41467-019-12213-6

Zu Ermgassen, E. K. H. J., Alcântara, M. P. de, Balmford, A., Barioni, L., Neto, F. B., Bettarello, M. M. F., Brito, G. D., Carrero, G. C., Florence, E. D. A. S., Garcia, E., Gonçalves, E. T., Da Luz, C. T., Mallman, G. M., Strassburg, B. B. N., Valentim, J. F., & Latawiec, A. (2018). Results from on-the-ground efforts to promote sustainable cattle ranching in the Brazilian Amazon. *Sustainability, 10*(4), Art. 4. https://doi.org/10.3390/su10041301

Chapter 11
Adoption Intensity of Digital Climate-Smart Agricultural Techniques (D-CSA) Among Dry-Season Female Vegetable Farmers in Nigeria: A Pathway to Climate-Resilient Farming

Igwe Ikenna Ukoha, Christopher Chiedozie Eze, Maryann Nnenna Osuji, Christopher Ogbuji Echereobia, and Emeka Emmanuel Osuji

Abstract Changes in climate have had an impact on humanity, most significantly farmers, by lowering the yield levels of their crops. The slow adoption of digital climate-smart agricultural technologies needed to mitigate the effect of climate change among farmers frequently frustrates technological growth and promotional efforts by agricultural stakeholders. This study investigated the adoption intensity of Digital Climate-Smart Agricultural Techniques (D-CSA) among dry-season female vegetable farmers in Imo State, investigated their level of awareness of the D-CSA technology applications available to them, and assessed the various barriers hindering their adoption of the D-CSA technologies. The Ordinary Least Squares (OLS) regression analysis was applied to 120 dry-season female vegetable farmers. We find new evidence from the results that age, relationship with extension officers, and the level of education attained by the dry-season female vegetable farmers were the major drivers of their adoption intensity of D-CSA technologies. In addition, we observed that there was a low level of awareness of the various applications of D-CSA techniques among the farmers. Furthermore, we observed that the main obstacle faced by the farmers in adopting D-CSA technologies was a lack of access to mobile phones and reliable internet. These issues were compounded by the inability to use D-CSA

I. I. Ukoha (✉) · C. C. Eze · M. N. Osuji
Department of Agribusiness, Federal University of Technology Owerri, Owerri, Nigeria
e-mail: igwe.ukoha@futo.edu.ng

C. O. Echereobia
Department of Crop Science Technology, Federal University of Technology Owerri, Owerri, Nigeria

E. E. Osuji
Department of Agricultural Economics, Alex Ekwueme Federal University Ndufu-Alike, Ikwo, Ebonyi State, Nigeria

R. U. Onyeneke et al. (eds.), *Energy Transition, Climate Action and Sustainable Agriculture*, https://doi.org/10.1007/978-3-031-83165-2_11

technologies and a general lack of trust in digital information among others. The findings demonstrate the necessity for improved extension services, user-friendly technologies tailored to meet the peculiarities of dry-season female vegetable farmers in order to overcome these challenges and promote a better roadmap towards a climate-resilient farming system in Nigeria.

Keywords Climate change · Adoption intensity · Digital climate-smart agricultural technologies · Female farmers · Dry-season

11.1 Introduction

Climate is one of the key factors in agricultural productivity. According to the World Bank (2024), the impact of climate change and insufficient food and nutrition supplies are two of our most pressing development issues. Based on the projections of the statistics presented in Abu et al. (2022), the global population is expected to exceed eight billion people by 2023 and roughly ten billion individuals by the year 2050. Hence, to satisfy the demands for adequate nutrition in emerging countries, the digital era of farming is inevitable. Historically, rising food production has been associated with agricultural growth and the unsustainable exploitation of resources such as land. This generates a vicious spiral, resulting in higher carbon emissions (World Bank, 2024). The Nigerian agricultural sector remains a pillar of its economy, considerably enhancing its economic output (Korgbeelo, 2022). Farming in Nigeria is vital to providing food security and livelihoods for farmers. A significant portion of the gross domestic product, or GDP, of emerging economies emanates from the farming industry. Dry-season vegetable farming is very important in the agricultural industry since it contributes to family sustenance and revenue production in seasons of minimal rainfall. According to Robert et al. (2021), dry-season vegetable farming in Nigeria is a significant component of the country's economy, providing food security, income, and employment opportunities, especially during the off-season of rainfall agriculture. However, according to Adebisi et al. (2022), vegetable farming in Nigeria is highly susceptible to variations in the climate. The precariousness of farming activities in Nigeria to climate change poses a significant threat to food security as it is extremely dependent on rain-fed agriculture, making it especially vulnerable to the adverse effects of climate change. This variability in climate is worsened by inconsistencies in weather patterns, fluctuations in rainfall, and fluctuations in temperature, amongst other factors.

As stated by the World Bank (2024), climate-smart agriculture (CSA) is a collection of farming techniques and technology that increase production, improve adaptability, and minimize greenhouse gas emissions. Climate-smart agriculture (CSA) has evolved as a comprehensive strategy for achieving food safety and environmentally friendly development while mitigating climate change concerns. The use of digital technology is revolutionizing several industries, notably agribusiness. Utilizing contemporary technology to promote sustainable growth in agricultural domains including the cultivation of crops, soil observation, animal farming and handling, and the fishing industry is known as "digital agriculture" (Degila et al., 2023). However, Digital Climate-Smart Agriculture (D-CSA) addresses climate change concerns by combining modern digital technology with sustainable farming methods. This method seeks to increase production and adaptability and minimize emissions of greenhouse gases while maintaining an adequate supply of food (Mercy Corps AgriFin, 2020). D-CSA employs digital technologies such as mobile applications in farm management activities, mobile apps for pest and disease notifications, and weather information apps, among others, to assist farmers in boosting their production levels and adaptability while confronting the effects of climate change, as well as aiding in the reduction of the negative effects associated with climate change. Several measures should be proposed to mitigate the effects of the changing climate on crop yields. D-CSA technologies that incorporate the upsides of an environmentally friendly rise in crop yield, advancement of robust farming as well as food-secure frameworks, and minimizing greenhouse gas pollution from farming operations demonstrate great promise. Experts in the field of climate change are responsible for enlightening authorities on successes and challenges faced by farmers in adopting D-CSA technologies. According to Chidiebere-Mark and Nwaebo (2023), women perform seventy percent of agricultural labour in Nigeria and are important players in the nutritional crop value chain. According to FAO and ECOWAS (2018) and FAO (2019), in Africa (Nigeria inclusive), women are more engaged in crop cultivation than their male counterparts, as seventy-three percent of the farmers engaged in the cultivation of vegetables are women, sixteen percent of the farmers involved in postharvest operations are women, and fifteen percent of the farmers engaged in agroforestry activities are females. According to Faniyi et al. (2018), FAO (2019), and Deji (2021), in the nation of Nigeria, females contribute much more to the national agricultural sector than their male counterparts, making them important players in the agricultural development of Nigeria.

Even though women vegetable farmers are involved in playing an integral part in Nigerian food production, they frequently encounter a variety of obstacles that restrict their yield, economic viability, and accessibility to advances, notably the advances in D-CSA that's currently revolutionizing the field of modern-day farming. Female vegetable farmers face several challenges to using digital technology, mostly due to individual, technological, societal, and institutional restrictions. According to Shivani and Kashyap (2024) and Ugwuja and Ayodeji (2021), many women lack the eagerness to use digital technology tools, which is exacerbated by inadequate instruction by service providers about these digital advances.

D-CSA also makes loans and banking services more accessible. Dry-season female vegetable farmers may obtain advances, make financial transactions, and manage funds with the use of electronic payment methods like mobile banking services. According to Ugwuja and Adesope (2021), the interaction of female vegetable farmers with digital technology is further complicated by problems like cyber threats and failed transactions with digital financial services.

Nigeria's advancement in achieving the 2030 Agenda for SDGs, adopted at the United Nations (UN) in September 2015 namely: ending poverty (SDG1), zero hunger (SDG2), goal 5 (gender equality) and sustainable consumption and production (SDG12); Nigeria's advancement in the achievement of the National Agricultural Technology and Innovation Plan (NATIP), namely Promoting Innovation and Technological Integration (Objective 1) and Enhancing Agricultural Productivity through Technological Advancements (Objective 3); in addition Nigeria's advancement in the achievement of African Union (AU) Agenda 2063 of the African Union- (Aspiration 1; Goal 5) Modernization of agriculture for increased productivity and (Aspiration 1; Goal 7) production environmentally sustainable and climate-resilient economies and communities, all align with the development of dry-season agriculture through the adoption of D-CSA techniques (Smidt & Jokonya, 2021).

Despite the potential benefits of D-CSA, the level of awareness and adoption of digital technology among dry-season female vegetable farmers remains limited. According to Gumbi et al. (2023), despite gradual increases in access and uptake of various digital solutions such as farm management software, data analytics tools, smart irrigation systems, and mobile phones, various factors, such as a lack of digital literacy, inadequate infrastructure, and high costs, pose significant barriers. Muriithi et al. (2021b) were of the opinion that challenges hindering the adoption of digital technologies among smallholder farmers include financial constraints, a lack of information, and the existence of a well-organized middleman structure in the agri-food supply chain. Additionally, Mohammed et al. (2019) found that factors like fluctuations in prices and low farm output due to poor adoption of modern production practices contribute to the limited uptake of digital technologies among farmers in Nigeria.

Many research studies carried out in the utilization of D-CSA techniques by farmers concentrated on expanded outcomes across genders rather than considering the particular difficulties and obstacles faced by women farmers (Gumbi and Twinomurinzi, 2023; Sissoko, 2023; Serote et al., 2021; Smidt & Jokonya, 2021; Ouinsou & Nonvide, 2023). Studies on the specific difficulties that women encounter in the adoption of D-CSA, especially while growing vegetables during the dry seasons, specifically in Nigeria, have received less attention. Previous research has largely ignored the peculiar difficulties encountered by female vegetable farmers who have limited financial means, poor levels of digital literacy, confining gender stereotypes, and insufficient assistance with extension, which significantly impede women's ability to use digital tools. In addition, the highly sensitive nature of dry-season vegetable production in Nigeria to real-time information on the climate, pests, and marketplaces was substantially disregarded by previous research. To address this research gap, this study aims to answer the following questions: (1) What are the sociodemographic profiles of dry-season female vegetable farmers that may influence the intensity of the widespread adoption of D-CSA technologies? (2) What are the major factors influencing the adoption intensities of D-SA technologies by the dry-season female vegetable farmers? (3) What is the level of awareness of the dry-season female vegetable farmers in the study area on the applications of D-CSA technologies? (4) What are the main factors militating against the use of D-CSA techniques and technologies by the farmers? This article aims to evaluate to determine the socioeconomic characteristics of the respondents, ascertain the major factors influencing the adoption intensities of D-SA technologies by the dry-season female vegetable farmers, examine the level of awareness of the respondents in the study area on the applications of D-CSA technologies, and ascertain the main factors militating against the use of D-CSA techniques and technologies by the farmers. We employed the use of descriptive statistics to present the socioeconomic profile of the respondents, including the main factors militating against the use of D-CSA techniques and technologies by the farmers. The OLS regression analysis was employed to ascertain the major factors influencing the adoption intensities of D-SA technologies by the dry-season female vegetable farmers. A Likert scale was used to measure the level of awareness of the dry-season female vegetable farmers in the study area on the applications of D-CSA technologies. The rest of this paper is grouped as follows: The second section describes the methodology, which includes empirical processes, data gathering, variables evaluated, and analytic procedures. The third section discusses the empirical findings of the study, while the fourth section gives conclusions and policy recommendations, limitations, and proposes areas for further research.

11.2 Methodology

11.2.1 Description of the Study Area

This study on the evaluation of the adoption intensity of D-CSA techniques among Nigerian dry-season female vegetable farmers was specifically carried out in Owerri, Orlu and Okigwe agricultural zones of Imo State, south-east Nigeria. According to Adikaibe (2024), the state is located between latitude 5045′ and 6035′ north of the equator and longitude 6035′ and 7028′ east of the Greenwich Meridian. With a population of 3,934,899 and an area of around 5067.20 km^2, Imo State has an average population density of approximately 725 km^2 and a large number of subsistence farmers (Nigeria Population Commission (NPC), 2006; and the National Bureau of Statistics (NBS), 2007). Imo State has three agricultural zones, namely Okigwe, Owerri, and Orlu (Ukoha, 2017). The primary activities of families in the region are education, agriculture, trade, and government jobs.

11.2.2 Sampling Procedures and Sample Size Determination

A multi-stage sampling procedure was adopted in the selection of the respondents. Firstly, the three agricultural zones were purposively selected for the study to ensure equitable distribution of all zones in the state and due to the farming activities of female vegetable farmers in Imo state (Emodi et al., 2020; Enwerem et al., 2022). Secondly, 1 Local Government Area (LGA) from each of the agricultural zones was purposively selected, namely, Owerri West, Orlu and Okigwe LGAs. Thirdly, 135 dry-season female vegetable farmers were selected randomly from a sampling frame of 559 registered female vegetable farmers from the ADP office in the state with the help of extension officers and trained enumerators. To ensure great precision in our estimations, we used Yamane's formula in selecting a sample size of 135 respondents with a 7.5 percent margin of error from a sample frame of 559 dry-season female vegetable farmers in the study area. However, exactly 120 dry-season female vegetable farmers of the selected sample size derived from Yamane's formula returned a valid questionnaire that was usable for the advancement of the research. The questionnaires that were chosen ended up being returned, yielding an 88.9% percentage compliance rate.

Based on the specification of Yamane's formula, the study derived a representative sample size from a population with a margin of error of 7.5%, as represented below:

$$n = \frac{\mathrm{N}}{1+\mathrm{N}(\mathrm{e}^2)} \quad (11.1)$$

where:

$n =$ Sample size.

$N =$ Population size.

$e =$ Margin of error (7.5%).

Considering the time limitations and budgetary constraints of the study, we made use of a margin of error of (7.5%). However, in order to prevent the underrepresentation of any zone, the number of sample sizes from each representative agricultural zones (Okigwe, Orlu and Owerri agricultural zones) was calculated using probability proportional to size (PPS). This is presented in Eq. (11.2) below:

$$n_i = \frac{nN_I}{\Sigma N} \quad (11.2)$$

where, n_i = number of sample size of female dry season vegetable farmers, n = sample size determined (135), N_I = number of farmers in each zone, N = target population

The number of sample size per selected LGA calculated by PPS is presented in Table 11.1.

Table 11.1 Dry season vegetable farmers available per selected LGAs and number of sample size determined by PPS

S/N	Selected LGAs	Population size	Sample size
1	Owerri	189	46
2	Okigwe	208	50
3	Orlu	162	39
4	Total	559	135

Source Author's computation

11.2.3 Method of Data Collection

The study was carried out using primary data. The primary data were gathered through field investigations and surveys with organized questionnaires and interview schedules. This Research was conducted, and necessary data gathered between the months of January and April 2024 with the help of well-trained enumerators.

11.2.4 Method of Data Analysis

11.2.4.1 Definition of the Variables Specified in the OLS Regression Model

Table 2 is a summary of the variables inputted in the OLS regression model for the analysis of the determinants of adoption intensity of the D-CSA technologies by the sampled dry-season female vegetable farmers in the study area.

The dependent variable is the adoption intensity for the *ith* dry-season female vegetable farmers (AI_i) is measured as a continuous variable as expressed in Eq. (11.3) in the manuscript. Some sociodemographic factors, such as membership in social organizations, household size, educational level of farm operator, and age of the female vegetable farmers, as well as other factors that could influence the adoption intensity, such as extension contacts, farm size, and farming experience, were the selected independent variables for this study. Previous research (Addai et al., 2023; Djibo & Maman, 2019; Kiconco et al., 2022; Subedi et al., 2019) utilized a comparable data set as well.

It is hypothesized that farmers who are members of social organizations are more likely to adopt more D-CSA technologies. This is due to the fact that membership in social organizations encourages networking and discussions within their platforms based on knowledge gained from the use of technologies, thereby encouraging other farmers to adopt new technologies. According to Subedi et al. (2019), participation in social groups and the adoption of technological advances in agriculture are positively correlated.

For the purposes of this research, the size of the household is defined as the number of persons living collectively in a residence at the period of the research and eating from the same catering arrangement. It is hypothesized that larger household sizes of farmers have a big impact on how well agricultural technologies are adopted. According to Kiconco et al. (2022), with the consideration of the heavy reliance on family labour in agricultural practices, agricultural innovations are more likely to be adopted by bigger families since they are frequently labour-intensive.

The number of years a farmer has been in school is used to determine the education variable. Farmers who have received basic education are better equipped to arrive at intelligent decisions. It is predicted that farmers with higher levels of education would be more likely to employ D-CSA technologies in their farming activities.

Sari et al. (2023) contended that more educated farmers will be more inclined to adopt agricultural technologies. This finding aligns with an expanding corpus of literature that emphasizes the value of education and a flexible mindset in embracing new technologies. Farmers could be more receptive to advances as a result of their exposure to modern education and forward-looking mindset. According to Ngema et al. (2018), a literate farmer ought to consider farming with ideas gained from exposure to expertise.

With regard to age, Abegunde et al. (2020) were of the view that, while older farmers are anticipated to have more agricultural expertise and an increased likelihood of accumulating more social and physical capital, the direction of the impact of age on the amount of D-CSA adoption cannot be predicted. However, aging is also linked to a reduction in physical prowess, a rise in risk aversion, and a decrease in strategic thinking.

According to Manda et al. (2020), empirical evidence suggests that cooperative membership is highly linked with the adoption of agricultural technologies in developing countries. A study conducted by Khan et al. (2022) suggests that agricultural cooperative service can play an essential role in accelerating the adoption of agricultural technologies by smallholders. Similarly, Djibo and Maman (2019); Subedi et al. (2019) claimed that social network effects from cooperative membership can influence farmers decisions and that farmers engage, exchange, and learn technological advances in agriculture by being members of cooperative associations.

The adoption of a new technology refers to the "process by which an innovation is communicated through certain channels over time among the members of the social system" (Xu et al., 2022). It starts with acquiring information (awareness) about the new technology, whether through media advertisements, extension agents, or social networks. According to Gao and Shi (2021), extension contacts can greatly improve the task-technology fit; that is, agricultural technology will accurately fit farmers requirements and demands.

The dataset also includes farm size as an independent variable. It is calculated as the total area of land under cultivation in a production year, expressed in acres. D-CSA methods are anticipated to be increasingly adopted by farmers with bigger farms as they are more likely to have the size economies required to justify the cost of subscribing to new agricultural technologies. According to Brown et al. (2020), farmers will choose the technology that minimizes their costs, given their size. Most farmers with large farm sizes, excluding small farm owners, will invest in new technology in response to a lower output price.

A farmer's farming experience is the duration of time they have been in the farming sector since they began to make their own farming decisions, and it is usually expressed in years of operation. The possibility of accumulating the social and physical capital required for the adoption of D-CSA technologies increases with farming experience. According to Onyeneke et al. (2018), the development of social and physical capital can provide farmers with increased knowledge and the ability to implement suggested or novel farming methods.

11.2.4.2 Model Specification

Previous research has employed several estimation methods such as the endogenous switching regression framework (Agbenyo et al., 2022) probit econometric model, (Belay et al., 2023), Multinomial logistic (MNL) regression (Muriithi et al., 2021a) the multivariate probit model and the Heckpoisson regression model (Miine et al., 2023) to analyze the adoption of one particular or several agricultural technologies such as climate-smart agricultural practices technologies, digital agricultural practices including the adoption intensities of these practices, respectively. For the purpose of this research, the Ordinary Least Square Regression (OLS) will be the perfect method to model the factors influencing the adoption intensity of D-CSA technologies among dry-season female farmers. Hence, the adoption intensity of D-CSA was investigated using the ordinary least square estimation technique due to the fact that the dependent variable [adoption intensity for the ith dry-season female vegetable farmers (AI_i) is measured as a continuous variable].

For this reason, we detailed the estimation method based on the relationship is expressed in Eq. 11.3

$$AI_i = \beta_0 + \beta_i X_i + \varepsilon_i \tag{11.3}$$

where:

AI_i is the adoption intensity (Dependent variable).

The adoption intensity was determined by the formula:

$$AI_i = \frac{NTA}{TNTA} \times \frac{100}{1} \tag{11.4}$$

where; AI_i = Adoption Intensity intensity for the ith dry-season female vegetable farmer, *NTA* is the number of D-CSA techniques/technologies adopted i^{th} dry-season female vegetable farmers, and $TNTAisthet$ otal D-CSA techniques/technologies available to the ith dry-season female vegetable farmers. $\beta_i X_i$ Is the vector of controlled variables consisting of socio-demographic and economic characteristics of female dry-season vegetable farmers. These independent variables comprise extension contacts (EXT), membership of social organizations (MSO), farm size (FSIZE), farming experience (FEXP), household size (HHS), educational level of female farmer (EDU), age of the female vegetable farmers (AGE). The coefficients β_i provide an estimate of the marginal impact of controlled variables on adoption intensity of D-CSA. $\mathcal{E}_i$ is an independently and identically distributed error term. Equation 11.3 is transformed into the double log as:

$$\begin{aligned} \ln AI_i = {} & \beta_0 + \beta_1 \ln \text{EXT}_i + \beta_2 \text{MSO}_i + \beta_3 \ln \text{FSIZE}_i \\ & + \beta_4 \ln \text{FEXP}_i + \beta_5 \ln \text{HHS}_i + \beta_6 \ln \text{EDU}_i + \beta_7 \ln \text{AGE}_i + \varepsilon_i \end{aligned} \tag{11.5}$$

Upon estimating the double log model in Eq. (11.5), the coefficients β_i ($\beta_1 \beta_2 \beta_3 \ldots\ldots \beta_8$) are used to ascertain how the independent variables X_i (EXT, MSO,

FSIZE, FEXP, HHS, EDU, and AGE) affect the dependent variable (AI_i). The elasticity of dependent variable (AI_i) with respect to your X_i variable is represented by the coefficients β_i in a double-log model. The coefficient β_i to put it another way, is the expected percentage variation of the dependent variable (AI_i) for every percentage change in independent variable X_i. This agrees with the research carried out by Baidoo et al. (2017). However, the variable membership of social organizations (MSO_i) couldn't be transformed in the double-log function due to their categorical nature.

11.2.5 Measurement of Female Vegetable Farmers Awareness and Factors Militating Against the Use of D-CSA Techniques

The extent of respondents' awareness of D-CSA techniques and technologies was successfully gauged using a 5-point Likert scale. This was achieved by asking female vegetable farmers to rate their familiarity with various D-CSA techniques and technologies, and the female vegetable farmers responded by indicating how much they knew about them. Descriptive statistics were also used in the assessment of factors militating against the use of D-CSA techniques and technologies.

11.3 Results and Discussions

The descriptive statistics and estimates from the OLS regression that were utilized for the analysis are shown in this section.

11.3.1 Descriptive Statistics

The descriptive statistics of dry-season female farmers Socio economic and characteristics and other factors that influence the adoption intensities of D-CSA technologies are presented in Table 11.3.

The result in Table 11.3 indicates that the farmers had a mean number of 1.32 extension visits annually, indicating that female farmers in the study area are usually carried along by extension agents as regards to the use of technologies and modern farming practices. According to Gao and Shi (2021), extension contacts can greatly improve the task-technology fit; that is, agricultural technology will accurately fit farmers requirements and demands. It was further observed that most of the sampled respondents (73.3%) were members of social organizations and groups. This is an indication that the farmers will gain from the series of networking activities and

Table 11.2 Variables in the OLS regression model for the analysis of the determinants of the adoption intensity of the D-CSA technologies by the sampled dry-season female vegetable farmers and their expected outcomes

Variable	Explanation and measurement type	Anticipated result (+/−)	Type of variable
Adoption intensity (AI) (dependent variable)	The degree to the D-CSA technique is used by the ith dry-season female vegetable farmer once it has been adopted. (%)	+ /_	Continuous
Extension contacts (EXT)	Number of times visited by an extension agent yearly	+	Continuous
Membership of social organizations (MSO)	Whether the farmer is a member of a social organization (yes = 1; otherwise = 0)	+	Dummy
Farm size (FSIZE)	Measures as the size of the farm in hectares	+	Continuous
Farming experience (FEXP)	Number of years spent in farming	+	Continuous
Household size of the farmer (HHS)	Number of people eating from same catering arrangement in farmers household	+	Continuous
Educational level of female farmer (EDU)	Number of years in school by farmer	+	Continuous
Age (AGE)	Age of farmer in years	+ /_	Continuous

Source Computed by the author from field survey data 2024

discussions on modern new practices and use of technologies, which is a major identity of social organizations. The mean farm size and farming experience were 2.1 ha and 8.2 years, respectively. This implies that most of the sampled female farmers in the study area had a considerable level of farming experience and limited farmlands for their farming activities. According to Akpan et al. (2019), this could be attributed to traditional land restrictions, urbanization, and land fragmentation reasons, which allow limited ownership rights of land to women. According to Ukoha et al. (2020), female small-scale farmers are renowned for growing food mainly for their household's consumption. In addition, the results show that the mean household size was 7 persons. Larger household sizes of farmers have a big impact on how well agricultural technologies are adopted, and at the same time, it comes with challenges on resource allocation. Given the significant reliance on family labour in agricultural techniques, Kiconco et al. (2022) contend that larger families are more likely to adopt agricultural innovations since they are often labour-intensive. The result further shows that 25% of the farmers had no formal education, while 75% had different levels of education (primary, secondary, and tertiary). It is common knowledge that

Table 11.3 The descriptive statistics of dry-season female farmers Socio economic and characteristics and other factors that influence the adoption intensities of D-CSA technologies

Variables	Frequency	Percentage	Mean
Extension contacts (EXT)			1.32
Membership of social organizations			
Yes	88	73.3	
No	32	26.7	
Farm size (FSIZE)			2.1
Mean			
Farming experience (FEXP)			8.2
Household size (HHS)			7
Educational level of female farmer (EDU)			
Non- formal education	30	25.0	
Primary	44	36.7	
Secondary education	33	27.5	
Tertiary education	13	10.8	
Age of the female vegetable farmers (AGE)			48.04

Source Computed by the author from field survey data 2024

farmers could be more receptive to technological advances as a result of their exposure to modern education and forward-looking mindset. The results show that the mean age of the sampled female vegetable farmers was 48.04 years. The average age of around 48 years suggests that the farming population is economically engaged. Abegunde et al. (2020) were of the view that older farmers are anticipated to have more agricultural expertise and an increased likelihood of accumulating more social and physical capital.

11.3.2 Factors Influencing the Adoption Intensity of D-CSA Techniques and Technologies Among Dry-Season Female Farmers

The estimates of regression results on factors influencing the adoption intensity of D-CSA techniques and technologies among female dry-season vegetable farmers are presented in Table 11.4.

From the Table 11.4, he double-log function was chosen as the lead equation based on the values of the coefficient of multiple determination (R^2), F-statistics, the number of significant variables and the signs on the variables as the conform to a priori expectations. The result in Table 11.4 showed that the coefficient of multiple determinations (R^2) is 0.703 which implies that about 70.3% of the variability in adoption intensity of D-CSA technologies for the female dry-season farmers in the

Table 11.4 Estimates of regression results on factors influencing the adoption intensity of D-CSA techniques and technologies among female dry-season vegetable farmers in the study area

Variables	Linear function	Exponential function	Double-log function (+)	Semi-log function
Intercept	15.185 (1.359)	3.151 (5.462)	−0.241 (−0.114)	−88.778 (−1.860)
Extension contacts (X_1)	1.783 (2.791)*	0.091 (2.766)*	1.210 (3.525)*	3.163 (2.471)*
Membership of social organization (X_2)	−0.095 (−0.742)	−0.010 (−1.581)***	−0.185 (−0.563)	1.295 (0.174)
Farm size (X_3)	−0.027 (−0.012)	0.075 (0.649)	−0.034 (−0.284)	−2.966 (−1.092)
Farming experience (X_4)	0.102 (1.100)	0.006 (1.213)	0.081 (1.272)	1.153 (0.804)
Household size (X_5)	0.291 (−0.716)	−0.034 (−1.605)	−0.137 (−0.627)	1.056 (0.215)
Educational level (X_6)	0.001 (2.135)***	−3.524E-005 (−1.236)	1.191 (2.257)**	−4.586 (−2.401)**
Age of the female vegetable farmers (X_7)	0.000 (5–862) *	−1.214E-005 (−4.386)*	−0.524 (−4.339)*	−12.830 (−4.712)*
R^2	0.700	0.653	0.703	0.651
F-value	12.344*	9.966	12.508	9.871
Durbin-Watson	2.036	2.121	2.098	2.037
Standard error	5.15611	0.26627	0.24651	5.56083

* = significant at 0.01 level (1%) ** = significant at 0.05 level (5%) *** = significant at 0.10 level (10%)
+ = lead equation
Note The figures in parentheses represent the *t*-test
Source Field survey data, 2024

study area is caused by the specific explanatory variables. The results show that membership of social organizations, educational level and age of the respondents are the significant determinants to the adoption intensity of D-CSA technologies.

The age of the female vegetable farmers was significant (1%), which suggests that elderly farmers are less inclined to accept D-CSA techniques and technologies, whereas younger farmers are inclined to be open to new ideas. Initiatives to encourage D-CSA techniques and technologies ought to consider elderly farmers' unique issues to improve acceptance. Aryal et al. (2018) and Serote et al. (2021) explain the negative relationship between age and adoption as follows: as farmers age, they become increasingly reluctant to take risks and do not have the stamina to effectively carry out agricultural activities with short-term planning horizons. Conversely, younger farmers tend to be more adaptive and eager to accept new technology (Patle et al., 2019).

The extension contact was significant (1%) and positive, which emphasizes the role of extension services for agriculture purposes in boosting the use of technology. Female vegetable farmers who benefit from more regular contact with extension personnel are more inclined to use D-CSA techniques and technologies, highlighting the importance of strengthening and expanding these programs. Prior studies by Sissoko (2023) and Ambali and Georgantzís (2021) found that extension contact had a favorable and substantial impact on female farmers' use of improved agricultural innovations. Extension personnel are responsible for informing agricultural producers and processors about new and improved digital innovations in farming. Thus, strengthening extension contact and providing targeted agricultural extension systems can boost the utilization of D-CSA techniques and technologies, ultimately improving agricultural production and environmental sustainability.

The education level of the dry-season female vegetable farmers was significant (5%) and positive, suggesting that better-educated female vegetable farmers will be more inclined to use D-CSA techniques and technologies. Boosting educational possibilities and giving specific training to female vegetable farmers will help them accept technology more effectively, increasing agricultural output and environmental sustainability. According to Ouinsou and Nonvideo (2023), elementary, secondary, and university education all have an important role in promoting rural communities to use digital technology. Similarly, Jena and Tigga (2023) found that farmers' extent of education had a positive association with their mindset toward using digital technologies in agriculture. Encouraging education across all stages, as well as improving the availability of technological services, is critical for encouraging producers to use D-CSA techniques and technologies.

11.3.3 Level of Awareness of the Applications of D-CSA Techniques and Technologies

The distribution of respondents according to their level of awareness of the applications of D-CSA techniques and technologies is presented in Table 11.5.

As can be seen from Table 11.5, female vegetable farmers are typically not very knowledgeable about different applications of digital technologies. For instance, almost half of the female vegetable farmers (44.2%) do not know how to use online tools to obtain meteorological data, and a sizable portion (98.3%) do not know how to use digital platforms to obtain agricultural inputs through electronic vouchers and receive subsidies. Furthermore, 95% of female vegetable farmers do not know how to use digital tools to identify pests and diseases, and 86.7% do not know about mobile applications for farm management chores. Female farmers' knowledge is similarly low, as seen by their use of digital platforms for communication with agricultural extension agents and their use of online platforms to access crop market prices (of which 70 and 55.8%, respectively) lack awareness. 80.8% of female vegetable farmers are not familiar with online training and capacity-building initiatives.

Table 11.5 Distribution of respondents according to their level of awareness of the applications of D-CSA techniques and technologies

D-CSA techniques and technologies applications	Not aware (%)	Slightly aware (%)	Aware (%)	Very aware (%)	Total (%)
Accessing weather information online (e.g., websites, apps)	53 (44.2%)	34 (28.3%)	19 (15.8%)	14 (11.7%)	120 (100.0%)
Digital platforms for receiving subsidies and accessing agricultural inputs through electronic vouchers	118 (98.3%)	–	1 (0.8%)	1 (0.8%)	120 (100.0%)
Utilizing digital tools for pest and disease identification	114 (95.0%)	–	1 (0.8%)	5 (4.2%)	120 (100.0%)
Employing mobile applications for farm management tasks	104 (86.7%)	–	7 (5.8%)	9 (7.5%)	120 (100.0%)
Communication with agricultural extension agents through digital platforms (e.g., Facebook, WhatsApp)	84 (70.0%)	5 (4.2%)	11 (9.2%)	20 (16.7%)	120 (100.0%)
Using online platforms to access market prices for crops	67 (55.8%)	17 (14.2%)	18 (15.0%)	18 (15.0%)	120 (100.0%)
Online training and capacity building (participating in online agricultural training programs or webinars)	97 (80.8%)	3 (2.5%)	6 (5.0%)	14 (11.7%)	120 (100.0%)

Source Field Survey Data, 2024

These results have important ramifications for extension services and policies pertaining to agriculture. The relatively low degree of awareness points to the necessity for focused educational initiatives and training sessions to raise female farmers' knowledge and proficiency with D-CSA techniques and technologies. Their general farming efficacy, production, and availability of essential information may all be improved by this. To improve innovation distribution, the awareness gap may be closed by using digital communication channels and bolstering extension services. Additionally, raising female farmers' knowledge of digital resources for market pricing data and subsidies will help them make better choices, which will eventually increase yields from agriculture and generate income.

11.3.4 Factors Militating Against the Use of D-CSA Techniques and Technologies

The distribution of respondents according to the factors militating against their use of D-CSA techniques and technologies in the study area is presented in Table 11.6.

Table 11.6 Factors militating against the use of D-CSA techniques and technologies

Challenges	Frequency	Percentage	Ranking
Lack of access to mobile phones or reliable internet	120	18.43	1st
Difficulty using or understanding digital information/tool	115	17.67	2nd
Lack of trust or confidence in digital information	104	15.98	4th
Cost of mobile data or internet access	114	17.51	3rd
Language barrier (information not available local language)	83	12.75	5th
Cultural norms or beliefs discouraging technology use	59	9.06	6th
Lower levels of education or literacy skills	56	8.60	7th
TOTAL	651	100	

Source Field Survey Data, 2024, Note: multiple responses were recorded

Table 11.6 illustrates some of the barriers that dry-season female vegetable farmers face while attempting to adopt D-CSA techniques and technologies. The biggest obstacles are lack of access to mobile phones or reliable internet (18.43%). Most digital technologies in agriculture are powered with smart phones and high-speed internet services. When these are not available to farmers it will be very difficult for extension agents to teach the farmers how to apply such technologies. According to Kyle et al. (2021) subscribers with low-speed internet and lack of functional smart phones will face challenges utilizing digital technologies. Problems associated with the adoption of D-CSA technologies are also compounded by the inability to use or comprehend digital technologies (17.67%) and a general lack of trust in digital information (15.98%). D-CSA technologies should be user friendly, and awareness of this technologies should be created to boost farmers confidence. It is common knowledge that farmers will be discouraged from subscribing to technologies that are not eased to apply. Additionally, language barriers, with 12.75% of female vegetable farmers facing issues due to information not being available in local languages, and while (9.06%) of the farmers face adoption challenged due to the fact that their cultural norms discourage the use of technologies which further further hinder their level of adoption. Lower levels of education or literacy skills (8.60%) also play a role in limiting the effective use of D-CSA techniques and technologies. Education plays a major role in the adoption of D-CSA technologies. Knowledge gained from ICT trainings and classroom interactions can enable the farmer better to understand and adopt the working of D-CSA techniques. This agrees with the works of Huang et al. (2022) who opined that educated farmers have a higher advantage over non-educated farmers in the adoption of ICT skills.

Overcoming these obstacles will have a big impact on dry-season female farmers' adoption of D-CSA techniques and technologies. The provision of focused training, data cost subsidies, and increased accessibility to mobile technology and internet services can all aid in overcoming financial and technological challenges. User trust may be increased by making sure digital platforms are reliable and accessible in their native tongues. Bettering educational opportunities and addressing cultural attitudes are also critical to creating a more welcoming atmosphere for the use of digital technologies. By addressing these issues, governments and development organizations may encourage a wider usage of D-CSA techniques and technologies, which will enhance agricultural output.

11.4 Conclusion Policy Recommendations and Limitations

Recent literature has shown that the studies on the adoption intensities of D-CSA technologies have not been painstakingly investigated in most developing nations like Nigeria. In addition, research on the factors influencing adoption intensities of D-CSA technologies, its awareness, and the factors militating against their adoption does not exist in Nigeria. The present study made use of descriptive statistics to ascertain the socioeconomic characteristics of the sampled respondents. The adoption intensity model and the OLS regression model were employed to ascertain the factors influencing D-CSA technologies by dry-season female vegetable farmers. The Likert scale was deployed to ascertain the level of awareness of D-CSA technologies by the respondents.

From the descriptive statistics that were used to ascertain the socioeconomic characteristics, we observed that the respondents had an average of 1.32 extension visits annually; most of the sampled respondents (73.3%) were members of social organizations and groups, while the mean farm size and farming experience were 2.1 ha and 8.2 years, respectively. In addition, the mean household size was 7 persons. Furthermore, it was observed that 25% of the farmers had no formal education, while 75% had different levels of education [primary (36.7%), secondary (36.7%), and tertiary (10.8%)]. Finally, we observed that the respondents had a mean age of 48.04 years. The findings from the OLS regression indicate that the age of the female farmers, their relationship with agricultural extension agents, and the level of education attained by the dry-season female vegetable farmers were the major drivers of their adoption intensity of D-CSA. The study further highlights the insufficient level of awareness of D-CSA technologies by the respondents. Furthermore, we observed that limited access to smart phones and internet services, non-user-friendliness of D-CSA techniques, lack of trust in digital information, language barriers, and cultural norms that discourage technology use played major roles in limiting the effective use of D-CSA technologies by the dry-season female vegetable farmers in the study area.

The paper offers a few policy recommendations. For instance, the outcome of this research can be used to inform the government and policymakers, as well as other relevant stakeholders, about the negative effects of climate change, which is critical in promoting the level of adoption of D-CSA technologies, including the best practices in integrating the D-CSA technologies into the Nigerian farming system. Priority needs to be devoted to improving extension, as frequent visits by extension agents may be highly helpful in integrating the adoption of D-CSA techniques among dry-season farmers in Nigeria. Secondly, government subsidies will go a long way in making available necessary gadgets and phones to farmers, as their affordability will encourage the adoption of more climate-smart digital technologies. We also propose that relevant agencies encourage vendors to provide user-friendly technologies; in addition, extension agents should sensitize the farmers on the importance of climate technologies and organize tailored trainings for farmers on the use of D-CSA technologies in languages that will be understood by the farmers irrespective of their age.

Even though these research outcomes may prove essential in understanding the challenges to shifting to a climate-resilient farming system in the study region, this study has certain limitations, which subsequent investigations should resolve. The present research did not classify/categorize the farming population's use of the D-CSA as low, medium, or high. Hence, it is expected that feature research should categorize the farming community as high, medium, or low users or adopters of D-CSA technology. In addition, due to funding restrictions and instability, this study concentrated solely on three agricultural zones in Imo State, Nigeria. Future research might include more agricultural zones in and even the entirety of the nation, in consideration of the fact that a variety of vital factors not covered in our study may impact farmers' decisions to embrace agricultural D-CSA technology. Furthermore, the current research did not examine the influence of the farmers adoption intensity on important areas of agriculture such as the welfare of farmers, production efficiency, and farmer poverty reduction. Finally, the present research addresses female dry-season vegetable producers; however, subsequent studies may include male farmers.

Acknowledgements We are grateful to the Federal University of Technology, Owerri, Nigeria, for providing a conducive environment to carry out this research.

Funding The authors state that there was no funding.

Data Availability Statement The primary data utilized in this study is available and can be shared upon request.

Ethical Clearance Not applicable.

Conflict of Interest The authors state that they have no conflicts of interest.

References

Abu, N. S., Bukhari, W. M., Ong, C. H., Kassim, A. M., Izzuddin, T. A., Sukhaimie, M. N., Norasikin, M. A., & Rasid, A. F. A. (2022). Internet of things applications in precision agriculture: a review. *Journal of Robot Control (JRC), 3*, 338–347.

Adebisi, L. O., Adebisi, O. A., Jonathan, A., Oludare, O. T., & Odum, E. E. B. (2022). Effect of climate smart agricultural practices on food security among farming households in kwara state, north-central Nigeria. *Pesquisa Agropecuária Tropical, 52*. https://doi.org/10.1590/1983-40632022v5270538

Abegunde, V. O., Sibanda, M., & Obi, A. (2020). Determinants of the adoption of climate-smart agricultural practices by small-scale farming households in king Cetshwayo District Municipality, South Africa. *Sustainability, 12*(1), 195. https://doi.org/10.3390/su12010195

Agbenyo, W., Jiang, Y., Jia, X., Wang, J., Ntim-Amo, G., Dunya, R., Siaw, A., Asare, I., & Twumasi, M. A. (2022). Does the adoption of climate-smart agricultural practices impact farmers' income? Evidence from Ghana. *International Journal of Environmental Research and Public Health. 19*(7), 3804. https://doi.org/10.3390/ijerph19073804

Addai, K., Temoso, O., & Ng'ombe, J. (2023). Heterogeneous effects of agricultural technology adoption on smallholder household welfare in Ghana. *Journal of Agricultural and Applied Economics, 55*(2), 283–303. https://doi.org/10.1017/aae.2023.16

Akpan, S. B., Udo, U. J., & Akpan, P. J. (2019). Analysis of the gross margins and commercialization of manure and fertilizer-based waterleaf (talinum triangulare) farmers in Nigeria. *Agricultural and Resource Economics: International Scientific E-Journal, 5*(4), 5–31. https://doi.org/10.51599/are.2019.05.04.01

Ambali, O., & Georgantzís, N. (2021). Improved rice technology adoption: the role of spatially-dependent risk preference. *Agriculture, 11*(8), 691. https://doi.org/10.3390/agriculture11080691

Aryal, J. P., Rahut, D. B., Maharjan, S., & Erenstein, O. (2018). Factors affecting the adoption of multiple climate-smart agricultural practices in the Indo-Gangetic Plains of India. *Natural Resources Forum, 42*(3), 141–158. https://doi.org/10.1111/1477-8947.12152

Adikaibe, P. (2024). Sustainable economic analysis of oil palm value chain in Imo state Nigeria: Opportunities and challenges. *Iop Conference Series Earth and Environmental Science, 1364*(1), 012018. https://doi.org/10.1088/1755-1315/1364/1/012018

Baidoo, S., Yusif, H., & Anwar, U. (2017). The effect of smallholder livestock production on income of farm households in Northern Ghana. *Journal of Science and Technology (Ghana), 36*(3), 8–19. https://doi.org/10.4314/just.v36i3.2

Belay, A. D., Kebede, W. M., & Golla, S. Y. (2023). Determinants of climate-smart agricultural practices in smallholder plots: Evidence from Wadla district, northeast Ethiopia. *International Journal of Climate Change Strategies and Management, 15*(5), 619–637. https://doi.org/10.1108/IJCCSM-06-2022-0071

Brown, W. M., Ferguson, S., & Viju-Miljusevic, C. (2020). Farm size, technology adoption, and agricultural trade reform: Evidence from Canada. *Journal of Agricultural Economics, 71*(3), 676–697. https://doi.org/10.1111/1477-9552.12372

Chidiebere-Mark, N. M., & Nwaebo, C. P. (2023). Gender gap in land rights and access to agricultural production in Okigwe agricultural zone Imo State, Nigeria. *Journal of Agriculture and Food Sciences, 21*(1), 163–174. https://doi.org/10.4314/jafs.v21i1.11

Degila, J., Tognisse, I. S., Honfoga, A. C., Houetohossou, S. C. A., Sodedji, F. A. K., Avakoudjo, H. G. G., Tahi, S. P. G., & Assogbadjo, A. E. (2023). *A survey on digital agriculture in five West African FAOs achieving zero hunger: The critical role of investments in social protection and agriculture.* https://oacps-ri.eu/wp-content/uploads/Paper1_Survey-in-5-countries_13-01067.pdf

Deji, O. F. (2021). *Gender implications of farmers' indigenous climate change adaptation strategies along the agriculture value chain in Nigeria African Handbook of Climate Change Adaptation*, pp. 1811–1834. https://doi.org/10.1007/978-3-030-45106-6_13

Djibo, O., & Maman, N. (2019). Determinants of agricultural technology adoption: Farm households evidence from Niger. *Journal of Development and Agricultural Economics, 11*(1), 15–23. https://doi.org/10.5897/jdae2018.0998

Emodi, A. I., Nwokolo, C. I., & Obiorah, J. A. (2020). Training needs of melon (<i>citrillus colocynthis</i> (l.) schrad) farmers in Okigwe agricultural zones of Imo state, Nigeria. *Journal of Agricultural Extension, 24*(4), 62–71. https://doi.org/10.4314/jae.v24i4.7

Enwerem, V. A., Osuagwu, O. C., Okereke-Ejiogu, E., Ajaero, J. O., Osugiri, I. I., & Ohajianya, D. (2022). Effects of agricultural extension services on leaf vegetables farmers use of agrochemicals in Imo state, Nigeria. *Asian Journal of Agriculture and Food Sciences, 10*(4). https://doi.org/10.24203/ajafs.v10i4.7007

Faniyi, E. O., Deji, O. F., Alabi, D. L., & Ijigbade, J. O. (2018). Soil and water conservation capabilities of male and female vegetable farmers on micro-veg project sites in southwestern Nigeria. *Journal of Agriculture Extern, 22*(3). https://doi.org/10.4314/jae.v22i3.12

FAO and ECOWAS Commission. (2018). *The national gender profile of agriculture and rural livelihoods in Nigeria.* Country Gender Assessment Series, Abuja, pp. 92.

FAO. (2019) *FAO and the 2030 agenda for sustainable development.* https://doi.org/10.18356/ea058ce2-en

Gao, J. J., & Shi, Q. H. (2021). Research on the reduction and control strategy of agricultural chemical inputs from the perspective of supply. *Xinjiang State Farms Econ. 9*, 1–13.

Gumbi, N., Gumbi, L., & Twinomurinzi, H. (2023). Towards sustainable digital agriculture for smallholder farmers: A systematic literature review. *Sustainability, 15*(16), 12530. https://doi.org/10.3390/su151612530

Jena, P., & Tigga, A. S. (2023). Digital competency in the agriculture sector: An Outlook in the Indian Context. *International Journal of Environment and Climate Change, 13*(5), 224–230. https://doi.org/10.9734/ijecc/2023/v13i51763

Huang, J., Su, L., & Liu, X. (2022). Facilitating inclusive use of ICTs in rural china. *Emerging-Economy State and International Policy Studies,* 197–211. https://doi.org/10.1007/978-981-19-5542-6_15

Khan, N., Ray, R. L., Kassem, H. S., Ihtisham, M., Siddiqui, B. N., & Zhang, S. (2022). Can cooperative supports and adoption of improved technologies help increase agricultural income? Evidence from a recent study. *Land, 11*(3), 361. https://doi.org/10.3390/land11030361

Kiconco, S., Babu, S., & Akankwasa, K. (2022). Adoption patterns and intensity for multiple banana technologies in Uganda. *Sustainability, 14*(23), 15986. https://doi.org/10.3390/su142315986

Korgbeelo, D. (2022). Contribution of agriculture to the development of the nigerian economy. *South Asian Research Journal of Humanities and Social Sciences.* https://doi.org/10.36346/sarjhss.2022.v04i03.004

Kyle, M. A., Blendon, R. J., Findling, M. G., & Benson, J. M. (2021). Telehealth use and satisfaction among u.s. households: results of a national survey. *Journal of Patient Experience, 8.* https://doi.org/10.1177/23743735211052737

Manda, J., Khonje, M. G., Alene, A. D., Tufa, A. H., Abdoulaye, T., Mutenje, M., Setimela, P., Manyong, V. J. T. F., & Change, S. (2020). Does cooperative membership increase and accelerate agricultural technology adoption? Empirical evidence from Zambia. *Technology of Forecasting Social Change, 158*, 120160.

Miine, L. K., Akorsu, Angela, D., Boampong, O., & Bukari, S. (2023). Drivers and intensity of adoption of digital agricultural services by smallholder farmers in Ghana. *Heliyon, 9(*12), 1–14, e23023. https://doi.org/10.1016/j.heliyon.2023.e23023

Mohammed, S. B.. T., N. P. B., Gracen, V., Dzidzienyo, K. D., M., N. I. F., Umar, M. L., & Abdulmalik, R. O. (2019). Factors determining the adoption of improved cowpea varieties among smallholder farmers in the dry savannas of Nigeria. *Journal of Agripreneurship and Sustainable Development, 2*(1), 132–143. https://doi.org/10.59331/jasd.v2i1.58

Muriithi, L., Onyari, C., Mogaka, H., Gichimu, B., Gatumo, G., & Kwena, K. (2021a). Adoption determinants of adapted climate smart agriculture technologies among smallholder farmers in machakos, makueni, and kitui counties of Kenya. *Journal of Agricultural Extension, 25*(2), 75–85. https://doi.org/10.4314/jae.v25i2.7

Muriithi, D. I., Wambua, B.N., & Omoke, K. (2021b). Constraints and opportunities for greenhouse farming technology as an adaptation strategy to climate variability by smallholder farmers of Nyandarua County, Kenya. *East African Journal of Science, Technology, and Innovation, 2.* https://doi.org/10.37425/eajsti.v2i.333

Ngema, P., Sibanda, M., & Musemwa, L. (2018) Household food security status and its determinants in Maphumulo Local Municipality, South Africa. *Sustainability, 10*, 3307.

Onyeneke, R. U., Igberi, C. O., Uwadoka, C. O., & Aligbe, J. O. (2018) Status of climate-smart agriculture in southeast Nigeria. *GeoJournal, 83*, 333–346.

Mercy Corps AgriFin. (2020). *Digital Climate Smart Agriculture—Mercy Corps AgriFin.* https://www.mercycorpsagrifin.org/digital-climate-smart-agriculture/

National Bureau of Statistics (NBS). (2007). *National Bureau of Statistics Official Gazette (FGP 71/52007/2,500(OL24): Legal Notice on Publication of the Details of the Breakdown of the*

National and State Provisional Totals, 2006 Census. www.nigerianstat.gov.ng. Accessed 28 February, 2016.

Nigeria Population Commission (NPC). (2006). Nigeria Population Commission, Nigeria Federal Government Initiative of individual head count by gender. Spread, State by State. In: MOFINEWS; (accessed 28 February, 2016). Jan-Feb,2007, 6(3): Nigeria.

Ouinsou, C. A., & Nonvide, A. M. (2023). *Determinants of Digital Technology Adoption by Agricultural Households in Benin*. Available at: https://doi.org/10.2139/ssrn.4583101

Patle, G. T., Kumar, M., & Khanna, M. (2019). Climate-smart water technologies for sustainable agriculture: A review. *Journal of Water and Climate Change, 11*(4), 1455–1466. https://doi.org/10.2166/wcc.2019.257

Smidt, H. J., & Jokonya, O. (2021). Factors affecting digital technology adoption by small-scale farmers in agriculture value chains (AVCs) in South Africa. *Information Technology for Development, 28*(3), 558–584. https://doi.org/10.1080/02681102.2021.1975256

Robert, K., Kara, H. A., Adamu, M. M., & Muhammad, B. (2021). Analysis of dry season vegetable production among kiri dam users in adamawa state, nigeria. *Journal of Agripreneurship and Sustainable Development, 4*(4), 173–179. https://doi.org/10.59331/jasd.v4i4.268

Sarie, F., Mohammad, W., Jamin, N. S., & Ramlan, W. (2023). The influence of demographic factors, farmer knowledge, and motivational factors on the adoption of agricultural technology innovation: a case study on dairy farmers in south Bangka. *West Science Agro, 1*(01), 28–35. https://doi.org/10.58812/wsa.v1i01.374

Serote, B., Mokgehle, S., Plooy, C., Mpandeli, S., Nhamo, L., and Senyolo, G. (2021). Factors influencing the adoption of climate-smart irrigation technologies for sustainable crop productivity by smallholder farmers in arid areas of South Africa. *Agriculture, 11*(12), 1222. https://doi.org/10.3390/agriculture11121222

Sissoko, P. (2023). Effect of adoption of climate-smart agriculture technologies on cereal production, food security, and food diversity in central Mali. *Agriculture, 13*(6), 1196. https://doi.org/10.3390/agriculture13061196

Shivani, J., & Kashyap, S. K. (2024). Measure constraints faced by farm women using ICT tools by the garett ranking method. *Advances in Research*. https://doi.org/10.9734/air/2024/v25i41101

Subedi, S., Ghimire, Y. N., Adhikari, S., Devkota, D., Shrestha, J., Poudel, H., & Sapkota, B. (2019). Adoption of certain improved varieties of wheat (Triticum aestivum l.) in seven different provinces of Nepal. *Archives of Agriculture and Environmental Science, 4*(4), 404–409. https://doi.org/10.26832/24566632.2019.040406

Ugwuja, V. C., & Adesope, O. M. (2021). *Cyber risks in microfinance digitization: exposures and preventions among female headed farm households in Southern Nigeria*. https://doi.org/10.24018/EJFOOD.2021.3.3.291

Ukoha, I. I., Henri-Ukoha, A., Ibeagwa, O., Essien, U. A., & Osuji, M. N. (2017). Determinants of Rural Farmers Preference for Cash-Less Transactions in Imo State. *Review of Agricultural and Applied Economics (RAAE)*, Faculty of Economics and Management, Slovak Agricultural University in Nitra, vol. 20(1), April.

Ukoha, I. I., Mejeha, R. O., Nwaru, J., Ibeagwa, O. B., & Maduike, I. A. (2020). Explicating the prevalence of women farmers' deterrence from microfinance banks loans in south-east nigeria: A censored tobit regression model of analysis. *International Journal of Applied Economics, Finance and Accounting, 7*(2), 82–89. https://doi.org/10.33094/8.2017.2020.72.82.89

World Bank. (2024). *Climate-Smart Agriculture. World Bank*. https://www.worldbank.org/en/topic/climate-smart-agriculture

Xu, Z., Li, J., & Ma, J. (2022). Impacts of extension contact on the adoption of formulated fertilizers and farm performance among large-scale farms in rural China. *Land, 11*(11), 1974. https://doi.org/10.3390/land11111974

Chapter 12
Application of Blockchain Technology in Carbon Footprint Traceability and Carbon Credit in the Animal Industry

Jalil Ghassemi Nejad, Majid Lotfi Ghahroud, Robert Ugochukwu Onyeneke, and Iman Janghorban Esfahani

Abstract Reducing carbon emissions by tracing the carbon footprint in animal and agri-food products is critical in the face of climate change. To do so, we must first be able to calculate the carbon footprint in agri-food, animals and their products. The carbon footprint should then be recorded in an immutable, reliable, transparent, and non-changeable manner. There are several methods for calculating the carbon footprint. Among these, the use of blockchain technology is a novel approach that is not well understood. In this letter, we provided an overview of blockchain technology and assessed the feasibility of using it to track carbon footprints in order to plan strategies to mitigate the effects of rising carbon emissions. We also clarified the distinction between a carbon footprint and a carbon credit and highlighted the applicability of blockchain technology may be used in both cases. Finally, future research will investigate the extent to which blockchain contributes to carbon footprint and the extent to which blockchain technology might aid in carbon footprint traceability.

J. G. Nejad
Genebiotech Co., Ltd. Godeung, Sinwonsa-Ro, Gyeryong-MyeonChungcheongnam-Do, Gongju-Si 32619, Republic of Korea
e-mail: jalilgh@genebiotech.co.kr

M. L. Ghahroud
The State University of New York SUNY, Korea (SUNY Korea), Incheon, South Korea

R. U. Onyeneke
Department of Agricultural Economics, Alex Ekwueme Federal University Ndufu-Alike, Ikwo, Nigeria
e-mail: robert.onyeneke@funai.edu.ng

I. J. Esfahani (✉)
Glopex Co., Ltd., R&D Center, 1301-1302, 156 Gwanggyo-ro, Yeongtong-gu, Suwon-Si, Gyeonggi-Do, Republic of Korea
e-mail: john@glopex.kr; john@twoens.com

Twoens Co., Ltd., R&D Center, GeumGang Penterium IX Tower B2064, Dongtancheomdansaneop 1-Ro 27,, Hwaseong-Si 18469, Gyeonggi-Do, Republic of Korea

R. U. Onyeneke et al. (eds.), *Energy Transition, Climate Action and Sustainable Agriculture*, https://doi.org/10.1007/978-3-031-83165-2_12

12.1 Introduction

The overall amount of greenhouse gases (GHG) that a person, business, or activity produces is referred to as their "carbon footprint" (Ju et al., 2022a, 2022b; Flachowsky and Kamphues, 2012). By retaining heat in the Earth's atmosphere, these gases, which also include CO_2, CH_4, and other gases, cause climate change. Carbon dioxide equivalent (CO_2e), a common unit of measurement for the global warming potential of various greenhouse gases, is used to calculate a carbon footprint (Rosado da Cruz et al., 2020; Xuan et al., 2020). On the other hand, a carbon credit is a trading unit that indicates a decrease in greenhouse gas emissions (Woo et al., 2020). People, companies, or governments can earn carbon credits by engaging in actions that decrease greenhouse gas emissions or promote carbon sequestration. After then, carbon credits can be purchased and sold on a market to reduce greenhouse gas emissions (Ju et al., 2022a, 2022b; Woo et al., 2020).

Animal productions result in significant global emissions of GHG including CO_2. Hence, carbon footprints were developed to compare emissions from different diets derived from animals (Flachowsky and Kamphues, 2012). The carbon footprint for animal-based food is influenced by a number of variables including animal species, production method, animal nutrition, animal performance level, system boundaries, and output/endpoints of production (Flachowsky and Kamphues, 2012). Carbon footprints consider the GHG potential of climate-relevant gases and are expressed in terms of one gram or kilogram of CO_2 equivalent (CO_2eq) per product unit.

12.1.1 Blockchain Technology, Multi-faceted Applications and Its Role in Enhancing Carbon Market Transparency and Trust

Blockchain technology is not condensed in bitcoin, and bitcoin is only a small part of it. Blockchain technology employs a combination of technologies with a long history in computer science and commercial applications (Ge et al., 2017). Blockchain can track everything from green supply chains to emissions reductions, facilitate the exchange of green energy, and so on. Numerous programs and startups are utilizing the technology (Krithika, 2022; Ge et al., 2017). Blockchains are consist of public/private key cryptography, cryptographic hash functions, database technologies, particularly distributed databases, consensus algorithms, and decentralized processing are among the component technologies (Krithika, 2022). The primary goal is to achieve database consistency and integrity in a distributed decentralized database. However, it is difficult to trace the complete supply chain of "green" products. Blockchain promises a paradigm shift in how we approach carbon footprint tracking and carbon credit methods because it is a decentralized and unchangeable ledger. By design, blockchain uses cryptographic consensus techniques to preserve data integrity, making changes or tampering nearly difficult. This addresses issues with data transparency and accuracy, which are essential for reputable carbon credit sales and emissions reporting. Additionally, the decentralized structure of blockchain encourages stakeholder engagement while ensuring data confidentiality through encryption methods.

As nations, businesses, and individuals struggle to meet their emissions-reduction targets, carbon markets are rapidly expanding (Hu et al., 2020). However, transparency is one of the issues that participants in carbon markets face. In that sense, serious concerns have been raised about how carbon market trading is managed, including issues of trust, transparency, and double counting of GHG emission reductions (Rosado da Cruz et al., 2020; Wiedmann and Minx, 2007). To address these issues, an emerging digital technology known as blockchain has recently emerged and is being used in carbon trading markets to improve and revolutionize market structure by increasing transparency, lowering transaction costs, and fostering trust. The technology enables secure resource transfers and transaction documentation through a centralized database. Blockchain technology can help by monitoring and reporting emission reduction trading, eliminating double counting issues, improving finance flows, and fostering trust (Ju et al., 2022a, 2022b; Rosado da Cruz et al., 2020). Despite being relatively new, blockchain technology has the potential to help address the challenges of transparency in carbon trading in the near future (Menson and Jain, 2022).

12.1.2 Integration of Blockchain Technology with Carbon Footprint: Enhancing Transparency and Traceability

Building consumer trust and preventing greenwashing, eliminating double counting, effective tracking of carbon markets, streamlining and accelerating carbon trading, enhancing carbon emission trading, facilitating clean energy trading and ensuring appropriate use of funds, and increasing climate finance flows are just a few of the ways blockchain technology can improve carbon trading transparency (Hu et al., 2020; Ju et al., 2022a, 2022b; Xuan et al., 2020). There are various advantages of merging blockchain technology with carbon footprint. To accomplish this, these components should be clarified as tools that aid in comprehending the application of blockchain technology in measuring and tracing the carbon footprint of food and animal goods (Muzumdar et al., 2022; Zhao et al., 2022; Hua et al., 2020). These integrations must be clarified. We have provided an overview of how blockchain technology can be integrated with carbon footprints below (Fig. 12.1).

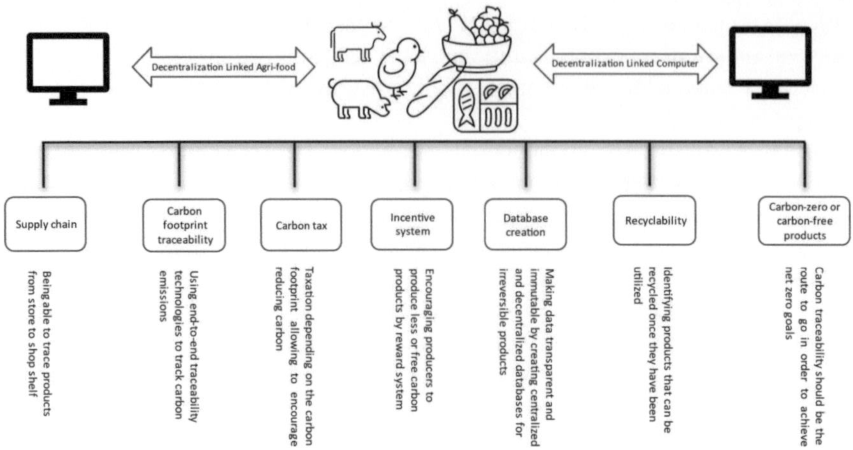

Fig. 12.1 An infographic summary of blockchain technology's integration with carbon footprint

12.1.3 Enhancing Supply Chain Sustainability Through Blockchain Technology for Carbon Footprint Tracing and Supply Chain Transpare

To accomplish sustainability targets, supply chains have the most space for improvement (Ju et al., 2022a, 2022b; Tian, 2016). More than 80% of greenhouse gas emissions are attributed to supply chain impacts, as are more than 90% of the impacts on air, land, water, biodiversity, and geological resources (Flachowsky and Kamphues, 2012; Flachowsky & Hachenberg, 2009). Inefficiency in the supply chain affects more than only emissions, environmental repercussions, and the planet. Because supply chains consume resources on a vast scale, they account for a disproportionately big amount of global carbon emissions (Rosado da Cruz et al., 2020). What makes a supply chain sustainable varies greatly depending on the company, product, and procedure (Saberi et al., 2019). Overall, it is critical for businesses to assess each stage of their supply chain and explore potential efficiencies (Tan et al., 2019; Tian, 2016). Unfortunately, it can be difficult for firms that outsource, especially if the companies with whom they deal are not completely transparent about their measurement and reporting. Retailers and food processors are already exploring for ways to cut emissions throughout their supply chains to save money, capitalize on market opportunities, and fulfill new trading rules (Krithika, 2022). Furthermore, unrelated agricultural operations and land-use changes now account for more than half of the CO_2 produced by agri-food systems in some regions, while it has more than doubled in emerging nations over the last three decades (Krithika, 2022). The food supply chain is on track to overtake farming and land use as the greatest contributor to GHGs emissions from the agri-food system in many countries (Tubiello et al., 2022). According to the latest estimate, of the 16.5 billion tonnes of GHG emissions from global total agri-food systems in 2019, 7.2 billion tonnes originated from within the farm gate, 3.5 billion from land use change, and 5.8 billion from supply-chain activities (Tubiello et al., 2022). Given these considerations, blockchain technology offers a new platform for addressing supply chain concerns that is transparent, immutable, irreversible, cost effective, and trustworthy (Menson and Jain, 2022).

Each year, roughly 13.7 billion metric tons of CO_2e are emitted globally through the food supply chain, representing about 26% of total anthropogenic greenhouse gas emissions (Gaillac & Marbach, 2021). A life cycle assessment is used to calculate a food's carbon footprint. It is a systematic and quantitative examination of the environmental impact of each stage of a product's life cycle. There are now some smartphone apps (e.g. GreenSwapp) that operate as a carbon–neutral grocery store, tracking the carbon footprint of the food items to buy, recommending lower-impact products that can reduce your cart's footprint, and offsetting the rest to make them carbon–neutral. Decentralization, transparency, open source, autonomy, anonymity, and immutability are all concepts embraced by blockchain technology (Menson & Jain, 2022). Tian et al. (2019) suggests a decentralized solution based on IoT and blockchain technology to aid with food supply chain tracking. They outlined a set of benefits and drawbacks of blockchain databases and propose BigChainDB, a novel concept that combines the benefits of distributed databases and blockchain. In this manner, the author addresses blockchain issues such as low throughput, excessive latency, and scalability (Tian, 2016). According to Galvez et al. (2018), the use of blockchain has many benefits in traceability, such as allowing all stakeholders to check the entire history and current location of a product; ensuring geographic biological origin by collecting information from sensors and storing it directly into the blockchain. Caro et al. (2018) offer a blockchain-based decentralized system for agri-food supply chain management and traceability. The method offered may integrate data produced and consumed by IoT devices. Overall, blockchain technology has the potential to provide a new method to carbon tracing in agri-food products. Blockchain technology is an excellent solution for ensuring the transparency of the entire supply chain and making data on carbon emissions and carbon footprints trustworthy (Ju et al., 2022a, 2022b).

Livestock production contributes about 7.1 gigatonnes CO_2eq per year of the world's GHG emissions, representing 14–16% of global total emissions (Barthelmie, 2022; Gerber et al., 2013). As more businesses offset their emissions, they are essentially enacting a voluntary tax that aids in the avoidance, reduction, or removal of carbon emissions elsewhere. Some consumers are willing to pay more for products that have reduced and offset their emissions, which is not dissimilar to the likely results of mandating carbon neutrality through carbon taxation. This trend is supported by the growing awareness of and demand for carbon–neutral products. Any taxation must be implemented in such a way that it does not disproportionately harm people in financial need, which has never been more crucial given the current economic downturn. The reality is that some high-carbon foods, such as meat and dairy, are not only high in calories and nutrients but also play an important social and cultural function in our civilizations. Demanding societal change overnight is extremely difficult (Liu et al., 2019; Tian, 2016). Considering the issues of rising agri-food taxes, every effort should be applied with caution to avoid societal reaction and create transparency, which can be done by employing blockchain technology to record the tax database.

We acknowledge the importance of a strong theoretical foundation and the opportunity to combine information flow theory and supply chain transition theory to better understand how blockchain technology might facilitate the switch to lower GHG emissions in the food system. By using these theories as a guide, we can clarify the ways in which the intrinsic characteristics of blockchain, such as transparency and traceability, might efficiently enable the tracking and management of carbon footprints throughout the food supply chain. This article could provide a more complete perspective on the possible role of blockchain technology in allowing sustainable transitions within the food industry, ultimately leading to lower carbon emissions, by combining these theoretical frameworks with practical observations.

12.1.4 Blockchain-Powered Incentive Programs for Carbon–Neutral Agri-Food Consumption

Developing an incentive program as a reward system for consumers who utilize carbon–neutral or recycled agri-food products will help to motivate individuals to participate in the lower carbon emission program. While charges on producers may operate as an incentive to improve production practices, such taxes may jeopardize livelihoods (Muzumdar et al., 2022, Ge et al., 2017). Livestock producers, without a question, play an important part in mitigating global warming, but they must be protected from imports of cheaper, high-emission animal products that may avoid domestic carbon levies. Subsidies for more sustainable foods, which either practice low-carbon, regenerative farming or aim to completely replace high-carbon foods, may also be a preferable policy (Ge et al., 2017). Rather than relying simply on punitive measures, revenue from carbon pricing should be utilized to positively incentivize farmers, support sustainable practices, and boost carbon removal on farms through agroforestry and soil carbon sequestration. In these ways, blockchain technology can aid in the development of an automated incentive system by tracking the use of agricultural products in various areas and offering rewards.

12.1.5 Empowering Transparency and Trust Through Blockchain Technology and Exploring Blockchain's Potential for Recyclability and Low-Emission Products

In the presence of blockchain technology, all participants have access to the data, everything is made open and available, and the records are unchangeable and cannot be altered or deleted. It adheres to the guiding concepts of leadership, responsibility, openness, adaptability, accessibility, usefulness, manageability, and sustainability

(Krithika, 2022). These features are facilitated by distributed ledgers and the data integration with transparency in the supply chain (Krithika, 2022; Menson & Jain, 2022; Liu et al., 2021). Although there are some issues that need to be resolved, such as the ease of real-time data access, the potential for hacking manipulation, the cost of data storage, etc., blockchain technology is still in its early stages of development and has the potential to find solutions even though the most pressing ones have already been addressed. Technically speaking, blockchain is a trustworthy database that is jointly maintained through a decentralized, trustless mechanism. By employing cryptography, this might generate blocks across any number of system nodes (Tian, 2016). Each block contains the information from all transactions that took place in the system over a specific period of time, and it may produce a digital fingerprint that can be used to confirm the accuracy of the data and establish a connection with the following block (Bruce, 2013). The blockchain may contain a sizable number of such blocks. Every block has a hash of the one before it, and they are connected to one another in a straight line and chronological sequence (Feng, 2016; Bruce, 2013). On the other hand, currently, a large portion of compliance data and information is audited by reliable third parties and kept either on paper or in a centralized database. It is well known that these approaches have numerous informational flaws, including the high cost and inefficiency of paper-based processes, as well as fraud, corruption, and error both on paper and in IT systems. Food safety, food quality, and sustainability are seriously threatened by these informational issues, which show that current systems for increasing transparency and trust have not been able to address or, in some cases, have made the problem of low transparency and trust in agri-food chains worse. These phenomena can be solved using a decentralized database provided by blockchain technology.

Recyclability and finding products that are carbon-free or less emitted carbon are also other major issues that need to be addressed by using blockchain technology (Rosado da Cruz et al., 2020; Saberi et al., 2019). Carbon footprint traceability can be used to locate recyclable products and companies that create fewer carbon emissions. Blockchain technology and other next-generation technologies and innovations could help find solutions for improved food and livestock production, food production decarbonization and carbon sequestration, and plant-based and lower-emission protein production for alternative food products (Ju et al., 2022a, 2022b). However, it is unclear to what extent blockchain technology can help with these issues.

12.1.6 Leveraging Blockchain for Transparent Carbon Footprint Management and Emission Reductions

Since finding and calculating the carbon footprint of food products is complex and requires the cooperation of all the stakeholders of the food supply chain, a record-keeping system for recording and tracking carbon footprint while preserving privacy for the related parties is needed (Shakhbulatov et al., 2019). Blockchain applications can help the animal industry to track and then reduce carbon emissions through 7 categories including supply chain management, energy, recycling, emission cuts, non-profits, carbon tax, create incentives. We breakdown these categories in brief in Table 12.1.

Based on the future thinker's organization, these 7 categories can help the environment and animal industry to reduce carbon emissions. Also, blockchain technology can apply to build energy performance measurement, reporting, and verification (MRV) and the carbon credit market (Woo et al., 2021). In the fields of agriculture and animal science, the term "carbon credits" refers to a system in which farmers, ranchers, and other agricultural and animal product manufacturers can acquire credits by putting policies in place that lessen greenhouse gas emissions or boost carbon sequestration on their properties. After that, these credits can be sold to businesses or other entities seeking to offset their own greenhouse gas emissions (Patel et al., 2021). In the fields of agriculture and animal science, there are numerous activities that can result in carbon credits. By using conservation tillage techniques, for instance, farmers can gain credits for lowering the quantity of CO_2 emitted into the atmosphere as a result of soil erosion and decomposition. By implementing strategies that lower CH_4 emissions from cattle, such as altering the sort of feed they receive or putting manure management systems in place, ranchers can earn credits. In general, the purpose of carbon credits in agriculture and animal science is to offer producers a financial incentive to pursue ecologically beneficial techniques that lower GHG emissions (Patel et al., 2021). This can support efforts to cut greenhouse gas emissions globally and lessen the effects of climate change. Since the Kyoto Protocol was ratified in 1997, many countries have tried to diminish their GHG emissions by implementing various reduction schemes. Carbon credit trading, one of these programs,

Table 12.1 Blockchain applications in the environment, agriculture and animal industry

Blockchain applications	Explanations
Supply chain management	- Track products from source to store shelf - Reduce carbon footprint and unsustainable practices
Energy	- Increase efficiency with P2P electrical grids - Improve access to power in poor areas or areas hit by natural disasters
Recycling	- Encourage recycling with a tokenized reward system - Track and evaluate e impact of recycling programs
Emission cuts	- Track countries' compliance with emissions goals - Reduce fraud and manipulation (for ex. Carbon credit)
Non-profits	- Track where donations are going - Decrease inefficiency and bureaucracy in charity
Carbon tax	- Calculate tax on products based on carbon footprint - Create a reputation system for companies based on emissions
Create incentives	- Reward sustainable practices - Make climate impact visible to stakeholders

has been used in many countries to comply with the protocol (Woo et al., 2020) but it needs a new solution such as blockchain technology to be more transparent and precise.

We seek to address several crucial issues related to blockchain implementation in the context of carbon footprint traceability and carbon credit within the animal industry, acknowledging the value of a thorough review. First of all, we acknowledge that a blockchain system needs ways to correct incorrect data inputs. Although blockchain's immutability is a benefit, processes for data modification and error correction must be handled to assure accuracy. Second, the importance of maintaining personal privacy in an open blockchain economy cannot be overstated. Further study should go deeper into the methods of encryption that can be used to protect private information while maintaining accountability and traceability. Third, there is a legitimate concern about the accuracy of the data entered onto the blockchain. We suggest investigating techniques that can increase the reliability of recorded data, including consensus protocols and data validation procedures. Furthermore, it is crucial to consider the issue of interoperability across various blockchain networks and the governance frameworks that oversee these networks.

By offering a secure and transparent mechanism to track, verify, and exchange carbon credits, blockchain technology has the potential to play a big role in the carbon credit agriculture and animal industry. A decentralized platform for storing and verifying the carbon credits obtained by farmers, ranchers, and other producers might be developed using blockchain technology in this context. With the use of this platform, it will be possible to keep track of the carbon credits produced by techniques, such as conservation tillage or manure management, and to confirm that they are appropriately recorded and exchanged. Additionally, a safe and open marketplace for the purchase and sale of carbon credits might be developed using blockchain technology. In addition to offering a safe and transparent means to track the trade of carbon credits and ensuring that they are being utilized to offset actual greenhouse gas emissions, this might make it easier for producers to connect with potential customers. Overall, the adoption of more sustainable practices in agriculture and animal production could be accelerated with the use of blockchain technology in the carbon credit agriculture and animal industry. This technology has the potential to increase transparency, decrease fraud, and improve the efficiency of the carbon credit market.

12.2 Conclusions

Global warming and climate change are the significant challenges of the new century and carbon footprint is one of the most important key performance indicators (KPIs) of environmental Protection. In this article, we showed a new tool, blockchain technology, and its applications for reducing carbon emissions. The size of the carbon management market is expected to grow to significantly over the coming years and many companies offer solutions to trade carbon. Thus, there are several innovative methods such as carbon credit. In conclusion, a carbon footprint is the entire amount of greenhouse gases that a person, business, or activity produces, whereas a carbon credit is a reduction in emissions that may be exchanged on the market. Given the increasing interest on the use of blockchain technology in different aspect of agriculture and animal industry, blockchain applications for tracing the carbon footprint of the animal industry can help us by increasing transparency and trust and firms can track and monitor their activities. By tokenization, particularly based on carbon credit, we not only can access a new type of asset, but we can help firms to have a great contribution toward the world's carbon footprint reduction. There are various theoretical benefits of integrating blockchain technology with carbon emissions, including but not limited to transparency, cost reduction, enhanced carbon emission trading, speed, improved climate funding flows, and improved report tracking. Future research would be interested in determining how and to what extent the carbon footprint can be quantified in a transparent manner, as well as preserving the information for future use. The other key components of blockchain technology including cryptographic signature, permissioned vs. un-permission blockchains or ledgers, and the use of smart contracts and their possible application in the agri-food industry and carbon footprint can be utilized in further studies.

Acknowledgements The authors would like to convey their sincere appreciation to our institutions/organizations. We would also like to thank our coworkers and collaborators for their suggestions and criticism throughout the course of this study. We warmly thank Glopex for their dedication to science and their conviction that this research is crucial.

Funding This research received no external funding.

Data Availability Statement Data is available upon reasonable request.

Ethical Clearance There is no ethical issue to be reported.

Conflict of Interest Robert Onyeneke is an editor of this handbook and was not involved in the editorial review, or the decision to publish this article. All authors declare that there are no competing interests.

References

Barthelmie, R. J. (2022). Impact of dietary meat and animal products on GHG footprints: The UK and the US. *Climate, 10*(3), 43. https://doi.org/10.3390/cli10030043

Bruce, J. D. (2013). Purely P2P Crypto-currency with finite mini-blockchain. http://diyhpl.us/~bryan/papers2/bitcoin/Purely%20P2P%20crypto-currency%20with%20finite%20mini-blockchain.pdf

Caro, M. P., Ali, M. S., Vecchio, M. & Giaffreda, R. (2018). Blockchain-based traceability in agri-food supply chain management: a practical implementation. In: *2018 IoT Vertical and Topical Summit on Agriculture - Tuscany (IOT Tuscany)* (pp. 1–4). https://doi.org/10.1109/IOT-TUSCANY.2018.8373021

Feng, T. (2016). An agri-food supply chain traceability system for China based on RFID & blockchain technology. In: *Proceedings of the 2016 13th International Conference on Service Systems and Service Management (ICSSSM)* (pp. 1–6). Kunming: China.

Flachowsky, G., & Kamphues, J. (2012). Carbon footprints for food of animal origin: what are the most preferable criteria to measure animal yields? *Animals (Basel).*, *2*(2), 108–126. https://doi.org/10.3390/ani2020108

Flachowsky, G., & Hachenberg, S. (2009). CO_2-footprints for food of animal origin—Present stage and open questions. *Journal of Construction Protection Food Safety, 4*, 190–198. https://doi.org/10.1007/s00003-009-0481-6

Gaillac, R., & Marbach, S. (2021). The carbon footprint of meat and dairy proteins: A practical perspective to guide low carbon footprint dietary choices. *Journal of Cleaner Production, 321*, 128766. https://doi.org/10.1016/j.jclepro.2021.128766

Galvez, J. F., Mejuto, J. C., & Simal-Gandara, J. (2018). Future challenges on the use of blockchain for food traceability analysis. *TrAC Trends in Analytical Chemistry, 107*, 222–232. https://doi.org/10.1016/j.trac.2018.08.011

Ge, L., Brewster, C., Spek, J., Smeenk, A., Top, J., van Diepen, F., Klaase, B., Graumans, C., & de Ruyter de Wildt, M. (2017). *Blockchain for agriculture and food: Findings from the pilot study. (Wageningen Economic Research report; No. 2017–112). Wageningen Economic Research.* https://doi.org/10.18174/426747

Gerber, P. J., Steinfeld, H., Henderson, B., Mottet, A., Opio, C., Dijkman, J., Falcucci, A., & Tempio, G. (2013). *Tackling climate change through livestock—A global assessment of emissions and mitigation opportunities.* Food and Agriculture Organization of the United Nations (FAO), Rome.

Hu, Z., Du, Y., Rao, C., & Goh, M. (2020). Delegated proof of reputation consensus mechanism for blockchain-enabled distributed carbon emission trading system. *IEEE Access, 8*, 214932–214944. https://doi.org/10.1109/ACCESS.2020.3041689

Ju, C., Shen, Z., Bao, F., Weng, P., Xu, Y., & Xu, C. (2022a). A novel credible carbon footprint traceability system for low carbon economy using blockchain technology. *International Journal of Environmental Research and Public Health, 19*, 10316. https://doi.org/10.3390/ijerph191610316

Ju, C., Li, G., Bao, F., Gao, T., & Zhu, Y. (2022b). Social relationship prediction integrating personality traits and asymmetric interactions. *Frontiers in Psychology, 13*, 778722. https://doi.org/10.3389/fpsyg.2022.778722

Krithika, L. B. (2022). Survey on the applications of blockchain in agriculture. *Agriculture (Basel)* 12, 9. https://doi.org/10.3390/agriculture12091333

Liu, W., Shao, X.-F., Wu, C.-H., & Qiao, P. (2021). A systematic literature review on applications of information and communication technologies and blockchain technologies for precision agriculture development. *Journal of Cleaner Production, 298*, 126763. https://doi.org/10.1016/j.jclepro.2021.126763

Liu, K. H., Chang, S. F., Huang, W. H., & Lu, I. C. (2019). The framework of the integration of carbon footprint and blockchain: using blockchain as a carbon emission management tool. In: A. Hu, M. Matsumoto, T. Kuo, & S. Smith (Eds.), *Technologies and eco-innovation towards sustainability I*. Springer, Singapore. https://doi.org/10.1007/978-981-13-1181-9_2

Menon, S., & Jain, K. (2022). Blockchain technology for transparency in agri-food supply chain: use cases, limitations, and future directions. *IEEE Transmission English Management, 2022*, 1–15. https://doi.org/10.1109/TEM.2021.3110903

Muzumdar, A., Modi, C., & Vyjayanthi, C. (2022). A permissioned blockchain enabled trustworthy and incentivized emission trading system. *Journal of Cleaner Production, 349*, 131274. https://doi.org/10.1016/j.jclepro.2022.131274

Patel, N., Shukla, A., Tanwar, S., & Singh, D. (2021). KRanTi: Blockchain-based farmer's credit scheme for agriculture-food supply chain. *Transmission of Emergency Telecommunication Technology*, e4286. https://doi.org/10.1002/ett.4286

Rosado da Cruz, A., Santos, F., Mendes, P., & Cruz, E. (2020). Blockchain-based traceability of carbon footprint: a solidity smart contract for ethereum. In *Proceedings of the 22nd International Conference on Enterprise Information Systems*, vol 2: ICEIS, ISBN 978–989–758–423–7; ISSN 2184–4992, pages 258–268. https://doi.org/10.5220/0009412602580268

Saberi, S., Kouhizadeh, M., Sarkis, J., & Shen, L. (2019). Blockchain technology and its relationships to sustainable supply chain management. *International Journal of Production Research, 57*(7), 2117–2135. https://doi.org/10.1080/00207543.2018.1533261

Shakhbulatov, D., Arora, A., Dong, Z., & Rojas-Cessa, R. (2019). Blockchain implementation for analysis of carbon footprint across food supply chain. In *2019 IEEE international conference on blockchain (blockchain)* (pp. 546–551). IEEE.

Tan, B., Yan, J., Chen, S., & Liu, X. (2019). The impact of blockchain on food supply chain: The case of walmart. *In International Conference on Smart Blockchain.* https://doi.org/10.1007/978-3-030-05764-0_18

Tian, F. (2016). An agri-food supply chain traceability system for China based on RFID & blockchain technology. In: *2016 13th international conference on service systems and service management (ICSSSM)* (pp. 1–6). https://doi.org/10.1109/ICSSSM.2016.7538424

Tubiello, F. N., Karl, K., Flammini, A., Gütschow, J., Obli-Laryea, G., Conchedda, G., Pan, X., Qi, S. Y., Heiðarsdóttir, H. H., Wanner, N., Quadrelli, R., Souza, L. R., Benoit, P., Hayek, M., Sandalow, D., Contreras, E. M., Rosenzweig, C., Moncayo, J. R., Conforti, P., & Torero, M. (2022). Pre- and post-production processes increasingly dominate greenhouse gas emissions from agri-food systems. *Earth System Science Data, 14*, 1795–1809. https://doi.org/10.5194/essd-14-1795-2022

Wiedmann, T., & Minx, J. (2007). A definition of 'carbon footprint'. In: *Ecological economics research trends* (p. 1–11).

Woo, J., Ashish T. A., Jiaxuan, L., Wolfgang, D. R., Charles, J. K., & Alireza, S. (2020). Blockchain: a theoretical framework for better application of carbon credit acquisition to the building sector. In: *Construction research congress 2020: infrastructure systems and sustainability* (pp. 885–894). American Society of Civil Engineers.

Woo, J., Fatima, R., Kibert, C. J., Newman, R. E., Tian, Y., & Srinivasan, R. S. (2021). Applying blockchain technology for building energy performance measurement, reporting, and verification (MRV) and the carbon credit market: A review of the literature. *Building and Environment, 205*, 108199.

Xuan, D., Ma, X., & Shang, Y. (2020). Can China's policy of carbon emission trading promote carbon emission reduction? *Journal of Cleaner Production, 270*, 122383. https://doi.org/10.1016/j.jclepro.2020.122383

Zhao, C., Sun, J., Gong, Y., Li, Z., & Zhou, P. (2022). Research on the blue carbon trading market system under blockchain technology. *Energies, 15*, 3134. https://doi.org/10.3390/en15093134

Chapter 13
Can We Communicate Climate Change Differently? Lessons from Bayelsa State, Nigeria

Lasisi Raimi, Blessing Onyinyechukwu Umunnah, and Merosobo Lawson-Ikuru

Abstract This study addresses the need for a fundamental change in the communication of climate change in Nigeria's Niger Delta, with empirical insights from Bayelsa State. Present formal methods, characterized by academic discourses, radio, and television discussions, inadvertently exclude non-literate persons, who are the most susceptible to the effects of climate change. Through in-depth interviews, the study seeks to uncover the challenges faced by the non-literate populations due to these formal, high-level communication methods and to explore the potential effectiveness of informal approaches. By engaging participants in in-depth interviews, the research assesses the effectiveness of informal strategies such as visual aids, interactive activities like townhall sessions, as well as storytelling, and town crying in local languages. These informal approaches demonstrate an understanding of the cultural context and individual experiences within the local communities. After a detailed thematic analysis of interview data, the study reveals that existing approaches to climate change knowledge pose challenges that hinder the ability of the non-literate populations to fully comprehend the associated nature, causes, and impacts of the problem. The research advocates for a more inclusive and symbolic communication approach that integrates both formal and informal methods. By focusing on the possibilities of mainstreaming informal climate change communication approaches into current modes of knowledge frameworks, the study contributes to the development of strategies that could inform a robust system of climate change awareness, thereby increasing the adaptive ability of non-literate people.

L. Raimi (✉)
Department of Sociology and Anthropology, Federal University Otuoke, Otueke, Bayelsa State, Nigeria
e-mail: lasisirr@fuotuoke.edu.ng

B. O. Umunnah
Department of Journalism and Media Studies, Federal University Otuoke, Otueke, Bayelsa State, Nigeria

M. Lawson-Ikuru
Department of Linguistics, Federal University Otuoke, Otueke, Bayelsa State, Nigeria

R. U. Onyeneke et al. (eds.), *Energy Transition, Climate Action and Sustainable Agriculture*, https://doi.org/10.1007/978-3-031-83165-2_13

Keywords Communication · Climate change · Formal method · Informal methods · Non-literate population · Bayelsa State

13.1 Introduction

Within the broad framework of environmental literacy, climate literacy, which houses the concept of climate change communication and, by extension, education, continues to maintain a huge gravity for multi-disciplinary concerns. Pan et al. (2023) share the view that climate literacy is significantly influenced by the level of education that people have about it and not their experience of it. Perhaps this is a pointer to the fact that most people could have been swallowing the bitter pills of climate change in ignorance.

Sadly, over the years, academic and policy communications on the nature, causes, and consequences of climate change have tended to be too high-level (Raimi & Bieh, 2020). This is because, most often, the discussions (and in most cases, with a limited sense of communication) surrounding climate change have been mostly at the technical level with little effort to communicate outcomes to a larger more inclusive population. In a world that stood at 7.3 billion people in 2015, one publication by the Yale Program on climate change presented a troubling statistic suggesting that nearly 2 billion adults lack the knowledge of climate change, with developing countries being disproportionately affected (Leiserowitz et al., 2015).

While there is no doubting the fact that there is a lot of climate change education going on around the world, existing methods that are mostly characterized by academic discourses often convened through conferences and media communications tend to exclude non-literate persons, especially in rural communities. Interestingly, these persons are considered to be more susceptible to the effects of climate change, basically because their climate reality is nested in a lack of concept for the problem that they face. According to Leiserowitz et al. (2015), when people fall short of the knowledge that they are at risk, they become mostly at greater risk in the face of the phenomenon that they are ignorant about. In line with this, the Climate Change Council of Australia's article published in 2016 (as cited in Raimi & Bieh, 2020) notes that being able to adapt to climate change is a function of the level of understanding that people have about the issue. Hence, with the current approach to communicating climate change, several non-literate people are edged out of critical knowledge that they need to stay ahead of the many challenges that this presents daily.

Although the field of climate change communication has continued to expand in recent years with multi-disciplinary appeal for how various sub-groups of the human population respond to the problem, the discussion on how non-literate people, especially in rural areas, view and handle climate change has been largely relegated to the background. For several reasons, the focus has been on how climate change communication affects young people, with studies by Corner et al. (2015) and Ojala and Lakew (2017) leading the way in this regard. Their justification for focusing on younger people is because this cohort consist of the leaders and the decision-makers of tomorrow, hence the need to understand how climate change is communicated to them. Apart from drawing legitimacy for their research focus from the clarion call to focus on young populations around the world if we are to realize the Sustainable Development Goals (SDGs), studies of this kind are further validated by their futuristic approach to problem solving. However, one major downside is the inability of these studies to, within their study population, address the issue of non-literate young people who are still very much around in most developing countries of the world. One recent study in Belgium by Anciaux et al. (2023) revealed that this critical population is now, more than ever before, maintaining some unenviable distance when it comes to climate change discussions. Anciaux et al.'s study is by far more relevant to us in Nigeria, where young people are more focused on other issues that put food on their tables.

In Nigeria, the research focus on climate change communication has been rather malnourished, even though the loss is quickly gained more abundantly in the discussion on climate change adaptation for mostly rural communities. Studies by Adamaagashi et al. (2023), as well as Abdulssalam and Mohammed (2023), among others, have collectively expanded our knowledge of climate change adaptation, with associated findings suggesting that their study populations are aware of climate change. In fact, we can conveniently argue that by combining Adamaagashi et al.'s content analysis and Abdulssalam and Mohammed's inferential analysis, these scholars taken together have significantly expanded our understanding of adaptation to climate change by rural communities in Nigeria. However, the problem is that none of these scholars dealt with the most vulnerable population, which is the non-literate group. Even though Olusola's (2007) publication fourteen years ago can be viewed as a clarion call to change how we communicate environmental problems around the world, and particularly Nigeria.

Although the United Nations Educational, Scientific, and Cultural Organization (UNESCO) 2023 rankings (as cited in Everybodywiki, 2024) place Bayelsa State at 14th position out of 36 states in Nigeria, with a literacy rate of 62%, which is above the national average of 50.6%, this seemingly positive data masks underlying concerns. The fact that 38% of the population remains non-literate and primarily resides in rural areas, is worrisome. These individuals, often directly experiencing the impacts of climate change, lack the literacy skills that could empower them with crucial knowledge and adaptation strategies. To these people, the language of packaging climate change communication can make the concept foreign and elite-oriented and, by extension, subject them to a lived reality of surviving outside the global conception of the problem. This strongly aligns with Leiserowitz et al.'s (2015) submission that people can live with a problem without necessarily having a clear understanding of it. It is on the strength of this, that our current study is focused on addressing the question, "Can we communicate climate change differently? with lessons from Bayelsa State, Nigeria. Essentially, the study attempts to answer the following questions: What is the knowledge level of climate change among the non-literate population in Bayelsa State? What is the experience of non-literate people in terms of how climate change knowledge be communicated currently? How would non-literate people prefer that climate change is communicated to them? Therefore, the objectives are to (1) examine the knowledge level of climate change among the non-literate population in Bayelsa State, (2) document how climate change is currently being communicated to non-literate people, and (3) determine how non-literate people would prefer that climate change be communicated to them.

13.2 Conceptual Framework

The conceptual framework presented as Fig. 13.1 below expresses two things, which are (1) the description or conceptualization of the variables associated with the study, and (2) the visual presentation of the link between the key variables in the study.

Drawing from Fig. 13.1, it is easy to see that all the related concepts have been operationally defined. However, it is necessary to convey the hidden meaning of the diagram, especially the interface between the concepts. Having established the role of climate change communication in the introductory section, the diagram shows that conventional models of communicating climate change to the public in the study area are either knowingly or unknowingly directed (represented by the tick downward arrow) towards literate populations, while edging out non-literate populations (indicated by the broken downward arrow). This is due to the methods adopted, such as conferences, workshops, journal articles, mainstream and social media, among others. Additionally, informal methods such as visual aids, interactive activities like arts, storytelling, town crying, and townhall sessions in local languages, have

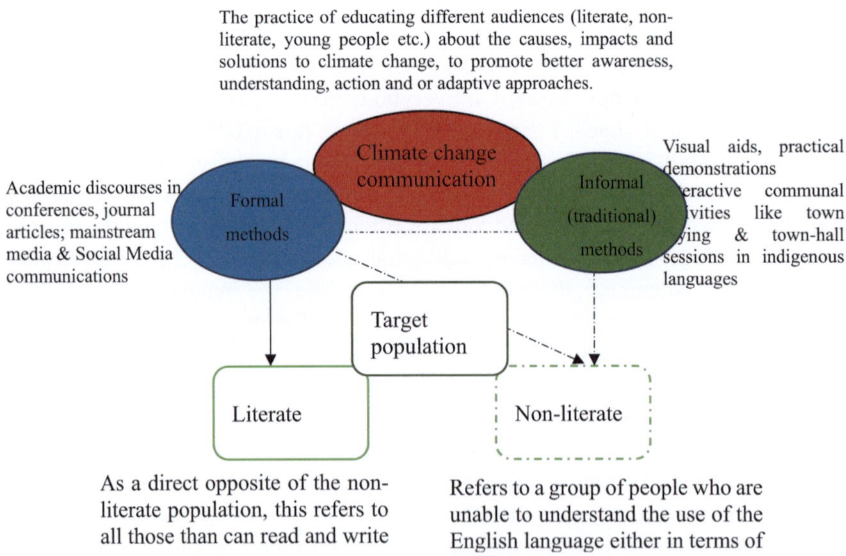

Fig. 13.1 Framework for current climate change communication in Bayelsa State

been mostly detached from the current climate change communication models. This clearly shows that the current climate change communication model is not robust enough to integrate the vast majority of non-literate population in the study area.

13.3 Theoretical Literature Framework

13.3.1 Symbolic Interactionism

This study is framed by the symbolic interactionist perspective, especially Herbert Blumer's variant. Blumer's views on the relevance of symbolic interaction as a focus of communication exchange are quite relevant when we take cognizance of the central role of meaning based on conceptual knowledge gained through interactions. According to him, people act based on the meanings that things have for them, and these meanings are derived from social interactions and modified consistently through interpretations. Hence, this clearly translates to some kind of communicative dynamism that suggests, in our context, that people's understanding of climate change is shaped by the meaning they derive from the symbols and interactions with their natural environment. The interactionist perspective is critical to understanding how people come about their knowledge of climate change and how they interact with others in the context of this knowledge. Berglund (2019) describes a situation where climate change problems are constructed based on the knowledge and interaction that people have of the issues. In this sense, people who engage in climate change discourse as front-liners, such as climate change activists, tend to have awakened knowledge, and this informs their subsequent presentation of the issue, while others maintain a false consciousness of the non-existence of the problem because of limited knowledge. Extending Berglund's (2019) views, Raimi and Bieh (2020) point to how climate change interactions isolate a section of people due to the non-symbolic approach to the communication of climate change, which contradicts the call to increase public knowledge of the existence, causes, and harms of climate change (Goldberg et al., 2021). Collectively, these authors draw attention to the need for climate change literacy to be all-encompassing for effective climate change knowledge dissemination.

The relevance of symbolic interactionism to our study is nested in its ability to address the importance of understanding how non-literate people interpret and make sense of climate change information, especially through the kind of interactions they have had and within their language context. In other words, the theory presents an opportunity to assess the efficacy of the existing climate change communication approaches in relation to non-literate people who do not understand the English language. Beyond this, it hopes to provide policy recommendations on how best to scale up climate change knowledge for non-literate people in Bayelsa State, the Niger Delta, and Nigeria in general.

13.4 Empirical Literature

Climate change, which is mostly considered the alteration in the natural balance of the climate or average weather conditions over a period of time, continues to pose a huge global worry. To say that climate change and the associated concerns present a posture of urgency for policymakers is to be mild about the challenge that it poses for the world (Alkhajar & Luthfia, 2021). In realization of this, we can argue here that the world is not short of activities that tend to communicate the causes and consequences of climate change, as well as possible interventions to people. This notwithstanding, the issue is ensuring that information about how to manage climate change effects gets to everyone in a manner that does not suggest that there is communication marginalization, especially as it relates to certain groups of people in society. The reason for this stems from the need to ensure that people have the right knowledge that can facilitate their ability to adapt to climate change and its associated challenges. Based on this, climate change literacy has gradually become an attraction for international, regional, and national policy framing, albeit with differential limitations.

A review of related empirical literature reveals that the approaches that involve face-to-face communication are often more persuasive and effective at influencing personal behaviour than mass-media communication (Moser, 2010; Moser & Dilling, 2007). Evidence in the literature shows that most of the attempts to discuss climate change have almost always tilted towards high-level approaches (Raimi & Bieh, 2020) that are mostly tied to technical conferences, workshops, and seminars with a mass media communication approach that easily excludes uneducated or non-literate populations. Olusola (2007) notes that a closer look at how climate change messages are communicated shows a clear alienation of a significant number of the population who are most affected by these issues. Although it can be argued that climate change itself maintains a high gravity for all forms of technicalities, especially because, by its nature, climate change emerged as a serious issue requiring high-level responses, the bigger problem is ensuring knowledge inclusion among diverse and, especially, vulnerable populations.

While there are efforts to trickle down climate change communication to non-literate populations across the world, research has shown that in most cases, these approaches have been bedeviled by prejudices that limit such attempts. One study by Cass (2020) that compared the use of Tok Pisin in Papua New Guinea (PNG) and Pidgin English in Nigeria to communicate climate change revealed huge limitations associated with how these languages are perceived and utilized. According to Cass, both Nigerian Pidgin and Tok Pisin in PNG are common lingua franca among the population, but they both face challenges as legitimate languages for communication, especially in the context of climate change. Yet, other researchers have proven the effectiveness of the use of indigenous languages in communicating climate change to non-literate populations when compared to the English language.

Although without categorizing their sample as non-literate population, Cherotich et al. (2012), in their study on how climate change information is accessed in Kenya, found that most of the elderly people in rural areas prefer to gain access to climate change knowledge through indigenous languages when compared to other mediums. On their part, Olatumile and Tunde-Awe (2019) found that the application of indigenous languages in disseminating knowledge has a significant impact on rural people's understanding of the concept and challenges of climate change. Interestingly, these researchers found that non-literate people who gained knowledge of climate change through indigenous languages were better placed to adapt than people of the same population who were communicated through the English language. Within this stream of studies on climate change communication to rural people, farmers, and fisherfolk seems to have more appeal.

Studies have shown that rural farmers and fisherfolk need to have timely information on climate change, as this is necessary to help them adapt to its impact on livelihoods (Leiserowitz et al., 2015; Pan et al., 2023). Sadly, in most parts of the developing world, highly formal approaches to communicating climate change still present some limitations with disparities in knowledge and perception on the issue (McGahey & Lumosi, 2018). In one study by Sen et al. (2021), language communication was found to be a strong barrier to accessing climate information among farmers in Vietnam, making it difficult for a significant number of these people to fully grasp the concept and challenges associated with climate change.

13.5 Methodology

This study employs the Interpretive Phenomenological Analysis (IPA) given its suitability for exploring how individuals make sense of their personal and interactive lifeworld. Its focus is more on the lived experiences of participants as well as how they interpret such experiences. By using the IPA model for their study, Kemkes and Akeman (2019) have shown how it can be very useful in driving in-depth methodological understanding of narrative accounts related to climate change. The study targeted non-literate people who were mostly farmers and fisherfolk. Data was collected from 41 participants that were purposively selected from four rural communities in two Local Government Areas (Ogbia & Nembe) in the state. The interviews were conducted by the researchers themselves within a space of two months, spanning from May to June 2024. To ensure effective communication as well as to elicit detailed narratives, the interviews were conducted in Pidgin English and local language, depending on the preference and language proficiency of the participants. The interview focused on eliciting detailed narratives about their understanding of climate change, their experiences with current communication methods, and their suggestions for more effective strategies to disseminate information on climate change in their various communities. All interviews were audio-recorded, transcribed, and translated into the English language for analysis. The data were analyzed using the IPA iterative process, which involves multiple transcript readings and identification of themes, sub-themes, patterns, and interpretations.

13.6 Results

13.6.1 *Theme 1: Perceptions and Understanding of Climate Change*

From the interviews, the views of the participants on their perception and understanding of climate change led to the emergence of three key sub-themes, which are awareness, misconceptions, and gaps in understanding and impact on livelihoods and adaptation strategies. Regarding awareness, it is evident that most of the participants became aware of climate change through experiential learning and community interactions rather than any formal communication channels. This clearly suggests the near absence of meaningful formal interaction on climate change that is targeted at non-literate people in the study area. While a good number of the participants shared similar views, one of them captured it better when she noted that:

> Na watin we see for our farm dey tell us say weather don change o! Rain no dey fall like before, sun too much, flood also too much. We dey hear am from our children and for village meeting sometimes, but nobody don come teach us wetin e mean

(We know that climate has changed from what we see on our farms. The rain does not fall like it used to, and the sun is too intense, flooding is also more frequent. We sometimes hear about climate change from our children and, in most cases, in village meetings, but no one has come to teach us what it means). *IDI Participant/Female/ 45 years old.*

While the above excerpt from the interviews suggests some level of awareness, it is safe to argue that this level of awareness is quite rudimentary in the context of the technicalities associated with climate change. This understanding led to further probing the causes and implications of climate change. As would be expected, despite being aware of changes in weather patterns, a huge sense of misconceptions about the causes and broader implications of climate change exists amongst participants. The view of one of the participants on this is presented below:

"Me I believe say na juju dey cause all these wahala for us. This too much flood mean say our juju them dey vex for us" (I believe it is witchcraft that is causing all these problems for us. The incessant flooding suggests that our gods or deities are angry at us). *IDI Participant/Female/52 years old.*

One other participant noted that:

"For me o, I believe say the world don dey end because dem talk am say when the world don dey end, different bad things go dey happen for the world. This weather wey dey change so, no dey like this before" (For me, I believe that the world is ending because they say that when the world is ending, many bad things will happen in the world. This change in weather was not like this before). *IDI Participant/Male/61 years old.*

Although the excerpts above reflect the views of two of the participants, it is safe to submit that these views resonated with a good number of the participants. This clearly suggests that the way people conceive and understand a thing would largely determine what meaning they give to it and in the context of this study, the finding here corroborates Leiserowitz et al.'s (2015) submission that awareness of climate change does not necessarily translate to deep understanding of its causes and consequences. Interestingly, even though by our nature, human beings are super adaptive, this level of misconception could play a strong role in slowing down the adaptive capacities of non-literate people when it comes to climate change.

13.6.2 Theme 2: Communication of Climate Change Knowledge

Just like the first theme discussed above, the second theme on communication of climate change knowledge to non-literate people generated three sub-themes that guided the analysis here. These sub-themes are channels of communication, accessibility and inclusivity, and effectiveness of communication. Thus, drawing from the interviews, participants primarily receive information about climate change through informal channels such as community meetings and discussions at home with their children who are literate. This clearly shows that formal channels like radio, television, or written materials like newspapers or even social media posts are less effective due to literacy barriers. One of the participants noted that:

"Even people wey dey manage dey tell me, say something dey wey dey cause the weather to change say dem dey hear am for radio and television. Some of my children dey tell me but I no dey hear am by myself because I no even get radio, and even if I get, I no dey understand am". (Even people that endeavour to explain to me, often tell me that there is something that causes the climate to change and that they heard it from the radio and television. Some of my children also told me, but I have never heard it myself because I don't have radio, and even if I have it, I do not understand the English language). *IDI Participant/Female/47 years old.*

Although evidence suggests that news broadcasting in Pidgin English is high in Bayelsa state (Egbilika & Zuokemefa, 2021), non-literate people who are mostly found in rural farm and fishing communities are still relatively marginalized. Hence, it is the argument here that the current methods of communicating climate change knowledge are not sufficiently inclusive of non-literate people. In addition to the significant reliance on written materials and technical language, inadequate facilities, such as power supply in rural areas, unite to exclude those who cannot read from understanding the science of climate change. This then leads to the issue of accessibility and inclusivity in the climate change communication process. One of the participants noted that "even when dem talk all their big grammar, we no dey understand wetin dem talk" (Even when they speak all their complex language, we do not understand what they are saying). *IDI Participant/Male/52 years old.*

Again, the discussion on accessibility and inclusivity necessarily leads to the attempt to worry about effective communication. Discussions with interviewees present some concerns that point to the ineffectiveness of current communication approaches, especially because of inadequateness in the comprehensive understanding of climate change amongst the participants that were interviewed. Generally speaking, a good number of the participants had a shortage of information concerning the technicalities associated with climate change, which is strongly linked to the misconceptions that they have. One of the participants captured this succinctly when she noted that:

"We dey see as weather dey change, and we dey even feel am, but we no really sabi wetin e mean. Dem no dey talk am for the wey we fit understand am. Na only few things we go hear from our people and our pikin them." (We see and feel that the climate is changing, but we don't really know what it means. They don't explain it in a way that we can understand it. We only learn a few things from our own people and children). *IDI Participant/Female/40 years old.*

Following the excerpts from the interviews presented above, our argument here is that the current mode of communicating climate change knowledge to non-literate people is characterized by significant barriers. The identified gaps in effective communication of climate change related knowledge call for inclusive approaches that take cognizance of the symbolic realities associated with the context of the language of the people. This leads to the question, "Can we communicate climate change differently?" Our lessons are presented briefly under theme three below.

13.6.3 Theme 3: Preferred Communication Methods for Climate Change

The interview outcomes that led to the clustering of conceptual categories under theme three present an opportunity to discuss the preferred methods of climate change communication under two sub-themes. These are communication by demonstration and community engagement through townhall and town-crier approaches. The participants in our study expressed a strong preference for visual aids and practical demonstrations using pidgin English or local language as effective methods for communicating climate change. According to most of the participants, these tools can go a long way to bridge the gap between complex (high-level) or technical scientific concepts and everyday experiences. According to one of the participants:

"If dem fit use picture and show us how the weather dey change, e go better. We go fit see with our eyes wetin dey happen and we go come know am well well." (If they can use images to show us how climate is changing, it would be better. We could see with our own eyes what is happening and understand it better). *IDI Participant/Female/40 years old.*

Another of the participants felt more disposed to larger gatherings where such visual presentations or demonstrations can be held, believing that they are better placed to learn in large groups than at individual level. The participant noted that:

"Me I go learn about this problem well well if dem fit call us for townhall meeting come use picture with our language explain am well. This one go good well well. And den we go fit ask questions wey go help us understand." (I think I can learn about the problem better if they can call a townhall meeting and explain to us in our indigenous language, it would be good. We could ask questions and understand everything). *IDI Participant/Male/46 years old.*

From the above discussions, the use of visual aids to demonstrate climate change-related issues in local languages can go a long way to bridge the current communication gap. However, one other method that resonated with a large number of the participants in our study is the use of mobile town-criers who use indigenous language and pidgin to communicate climate change to non-literate people. In this regard, one of the participants had this to say:

"For me o! town criers fit waka for village, talk for Pidgin, and our language make everybody hear. Dem fit explain wetin dey happen and wetin we go do to manage am. So many people go come know wetin this climate change be. No be just people wey go school suppose know am na." (In my own view, town-criers can go around the village using Pidgin English and or our indigenous language to ensure that everybody gets the information. They can explain what is causing it and what can be done to adapt to the situation. A lot of people would then know what climate change is. It is not only those that are educated that should know about climate change). *IDI Participant/Female/54 years old.*

13.7 Discussion

The findings of this study show significant barriers in the communication of climate change knowledge to the non-literate population in Bayelsa State. This is because these individuals, who are mostly located in the rural areas of the state, primarily receive information through informal channels such as community meetings and interactions with knowledgeable community members. However, these sources are often inadequate in providing comprehensive and accurate climate change information that can facilitate a progressive system of knowledge for this set of people. So, community meetings and interactions with knowledgeable community members represent sources of climate change knowledge dissemination to non-literate population, there is the understanding among participants in this study that these are largely limited. For instance, community meetings are not held frequently, and even when they are held, issues of climate change do not form part of the discussions. Our finding corroborates that of Raimi and Bieh (2020) and Cass (2020), who noted that several rural communities in the Niger Delta region are often isolated from having access to climate change information due to the formal ways that related information is disseminated. The reliance on high-level and technical communication for climate change issues associated with formal means of communication tends to exclude non-literate individuals, leading to fragmented and oftentimes incorrect understanding of climate change. This strongly supports the finding of Leiserowitz et al.'s (2015) that people can go on living with a problem without necessarily understanding the associated dangers due to a general lack of knowledge or half-truth about the problem. This goes to show that making climate change communication technical can go a long way to reinforce inequalities, which could further deepen the dangers that most vulnerable rural populations already face.

The study also shows that while non-literate populations are relatively aware of climate change through their daily experiences and communal interactions, their understanding is often incomplete and influenced by indigenous beliefs that tend to impose some misconceptions. These misconceptions further undermine their ability to align with scientific strategies that are aimed at limiting the impacts of climate change on human activities. Most of the participants in this study believe that climate change is caused by the anger of their gods showing a significant sense of insufficient knowledge or lack of understanding concerning the cause(s) of climate change. This finding contradicts that of Ojomo et al. (2015) who found in their study in Southern Nigeria that over 95% of their sample believe that climate change is caused by human activities. This contradiction may be as a result of the literacy level of their sample having used ministry workers and students who are largely open to formal means of communication that our sample are mostly isolated from or do not understand at all. However, the finding from a study conducted in the United States by De Bruin et al. (2024) shows that even within a literate population, the understanding of some terminologies associated with climate change presented huge communication gaps among their sample. This finding corroborates ours by showing that understanding terminologies related to climate change affects people's perception and responses. This has significant implication for our study, especially because the established weakness in effective communication of the problem to non-literate people means that these individuals do not receive accurate information about the popularly held concepts, scientific causes and consequences of climate change. As a result, their adaptation strategies are based on trial and error rather than informed decisions, and this presents a strong recipe for increased vulnerability to climate change impacts. This calls for a more inclusive approach to climate change knowledge dissemination (Raimi & Bieh, 2020).

Additionally, the study found that the preferred communication method for non-literate people is visual aids alongside practical demonstrations using Pidgin and indigenous languages. Participants believe that using visual aids and practical demonstrations in Pidgin and local languages can present them a symbolic and cultural context to better understand the meaning and consequences of climate change. Hence, complex climate change information can be conveyed with ease and in an accessible manner, allowing such individuals to see and understand the impacts firsthand. This aligns significantly with the finding of Cass (2020) that using local languages and culturally useful communication strategies or methods can be very important in disseminating knowledge about climate change to communities that do not understand mainstream English language. While Cass' (2020) study further drew attention to the challenge of relying on local languages, especially because it would mean that climate change information is cascaded in multiple indigenous languages, our finding goes beyond just the language of communication to an understanding of the mediums of communication. Our study has proven that beyond Cass' understanding of the complexities associated with deploying multiple languages, even when languages are used through formal mediums like radios, televisions, newspapers, non-literate populations still struggle to have access due to several other intervening concerns such as electricity supply in rural areas, especially in Nigeria. In this regard, town-hall sessions and the town crier approach are highly regarded as effective methods that can convey information to a broad audience that include non-literate people. This is particularly effective when Pidgin English is utilized, especially as it is widely understood by both literate and non-literate populations in Bayelsa state and Nigeria. Interestingly, participants believe that this approach can be very effective and could encourage a sense of community involvement and collective problem-solving which is typical of communal settings like the study area.

13.8 Conclusion, Recommendations, and Limitations

This study has dealt with the question "Can we communicate climate change differently?" as a fallout of our concern for how climate change is communicated to non-literate populations in Bayelsa State. Following the findings of this study, it is our submission here that there are significant barriers to how climate change is communicated to non-literate people who primarily rely on informal channels coupled with the exclusionary nature of formal communication approaches. These barriers result in a fragmented and often inadequate understanding of climate change, as influenced by indigenous beliefs and misconceptions.

Drawing from the findings and conclusion, the following suggestions have been designed to inform policy action and improve climate change communication for non-literate people.

a. There is the need to mainstream communication approaches that are culturally relevant and accessible to non-literate populations into existing climate change policy in Bayelsa and Nigeria in general. While Nigeria's climate change policy highlights the need to assist communities in becoming aware of the impacts of climate change, it is short on strategies on how to do this, especially for non-literate populations. Hence, the climate change policy should be reviewed to, but not limited to, the use of visual aids, practical demonstrations, and community-based approaches such as town-hall sessions and towncriers to scale up the knowledge levels of rural, non-literate people. These should be conducted in Pidgin English and or indigenous languages to increase the chances of conveying significant meanings to non-literate people.
b. Non-governmental organizations and civil society organizations operating within climate change communication spaces in rural communities in Bayelsa State and other parts of Nigeria should leverage existing community structures and knowledgeable individuals to serve as climate change information ambassadors. Training such community members in accurate climate change science and

communication techniques can help bridge the gap between technical (high-level) knowledge and indigenous understanding, thereby providing a relatable source of information for non-literate people. This can also be done in such a manner that the communication of scientific jargon associated with climate change is done using indigenous language and Pidgin English.

c. Lastly, Wilson's (2005) trado-modern method of communication can be explored and applied in this circumstance. This will allow for the integration of indigenous and scientific models of climate change communication for a more comprehensive and accurate understanding of causes, consequences or impacts, and adaptation strategies. Collaboration with indigenous Non-Governmental Organizations, community leaders, and educational institutions within such localities can catalyze a robust framework for communicating climate change to rural people, especially the non-literate ones.

Although well-rounded in its relative findings, this study encountered some limitations that should be acknowledged. Primarily, it was confined to two Local Government Areas (Ogbia and Nembe) in Bayelsa State, which imposes some limitations regarding the generalizability of its findings in relation to other regions of Nigeria or beyond. Additionally, the sample size of 41 participants, while sufficient for a study of this nature, may not have elicited a broader perspective of non-literate people in the area. Lastly, even though the use of IPA provided opportunity for a rich insight into the participants' lived experiences, it nevertheless obscured the relevance of statistical generalization. Future research could address these limitations, especially by expanding the spatial scope and considering a mixed-methods approach.

Acknowledgements The authors would like to express our sincere gratitude to all those who contributed in one way or another to the success of this study. We are grateful to the AfriExtract team at the University of Groningen, Netherlands, for providing the initial (non-funding) academic push to undertake this study. We are mostly indebted to Mr. Moses Ademe Ogiramkopu, who functioned as our tour guide to some of the communities where we held the interviews. Our sincere gratitude goes to all the study participants in Ogbia and Nembe Local Government Areas, who were more than willing to share their lived experiences and perspectives with us. To the community chiefs who shared valuable insights with us, we are grateful for your time and knowledge. Your cooperation was largely instrumental in facilitating as well as ensuring that our fieldwork was smooth.

Funding The authors did not receive any specific funding or grant from any agency for the purpose of conducting this study. This study was conducted as part of the academic responsibilities of the authors as demanded by the universities where we work, without any external financial incentives or influence.

Data Availability Statement We affirm that the data associated with the findings of this study are available from the corresponding author following reasonable request. Due to privacy and ethical restrictions, the data are held in confidence as they contain information that may compromise the privacy of the participants. However, if the need arises, the data may be shared for the sole purpose of collaborative research, and this will be subject to appropriate data sharing agreements that clearly comply with established ethical guidelines.

Ethical Clearance This work was done in accordance with the ethical guidelines associated with research that involves human participation. Consequently, before undertaking this research, we got ethical approval from the Research Ethics Committee housed under the Research and Quality Assurance unit of the Federal University Otuoke, Bayelsa State, Nigeria. Additionally, informed consent was secured for all participants prior to engaging them in the study. All interviewees were assured of confidentiality and anonymity, with a caveat that they have the right to withdraw from the study at any time if they so wish. All information collected were held in strict confidence and have been used only for the purpose of this study.

Conflict of Interest We declare that as authors, we have no known competing or conflict of interest either financially or personally that could have appeared to have influenced the views represented in this study.

References

Abdulssalam, B., & Mohammed, D. (2023). Determinants of adaptation strategies to climate change among farming communities along river Benue in Adamawa State, Nigeria. *Arid Zone Journal of Basic and Applied Research, 2*(1), 91–104. https://doi.org/10.55639/607.7475

Adamaagashi, I. P., Adedeji, O. B., & Jephthah, A. O. (2023). Investigating the adaptive practices implemented by rural communities in Nigeria to alleviate the impacts of climate change on agricultural sectors. *International Journal of Agriculture and Earth Science, 9*(4), 28–46.

Alkhajar, E. N. S., & Luthfia, A. R. (2021). Communicating climate change policy to people in rural areas. *IOP Conference Series: Earth and Environmental Science, 824*(1), 012114. https://doi.org/10.1088/1755-1315/824/1/012114

Anciaux, A., Cougnon, L.-A., Ducol, L., & Catellani, A. (2023). Youth, communication & climate: A pluridisciplinary analysis of distancing strategies in response to climate change among Belgian youth. *Youth, 3*(4), 1150–1173. https://doi.org/10.3390/youth3040073

Berglund, K. (2019). *There is no alternative: A symbolic interactionist account of Swedish climate activists (Master's thesis).* Lund University. Retrieved from https://lup.lub.lu.se/luur/download?func=downloadFile&recordOId=8990902&fileOId=8990903

Cass, P. (2020). Some aspects of climate change communication and effectiveness in PNG. *Pacific Journalism Review—Te Koakoa, 26*(1), 148–163. https://doi.org/10.24135/pjr.v26i1.1106

Cherotich, V. K., Saidu, O., & Bebe, B. O. (2012). Access to climate change information and support services by the vulnerable groups in semi-arid Kenya for adaptive capacity development. *African Crop Science Journal, 20*(2), 169–180. Retrieved from https://tspace.library.utoronto.ca/bitstream/1807/47584/1/cs12036.pdf

Corner, A., Roberts, O. R., Chiari, S., Völler, S., Mayrhuber, E. S., Mandl, S., & Monson, K. (2015). How do young people engage with climate change? The role of knowledge, values, message framing, and trusted communicators. *Wiley Interdisciplinary Reviews Climate Change, 6*(5), 523–534. https://doi.org/10.1002/wcc.353

De Bruin, W. B., Kruke, L., Sinatra, G. M., & Schwarz, N. (2024). Should we change the term we use for "climate change"? Evidence from a national U.S. terminology experiment. *Climatic Change, 177*(8). https://doi.org/10.1007/s10584-024-03786-3

Egbilika, J. C., & Zuokemefa, P. A. (2021). Indigenous language reporting: Language revitalization. *Multidisciplinary Journal of Academic Research and Development, 1*(2), 41–46.

EverybodyWiki. (2024, July 21). *List of Nigerian states by literacy rate.* Retrieved from https://en.everybodywiki.com/List_of_Nigerian_states_by_literacy_rate

Goldberg, M. H., Gustafson, A., Rosenthal, S. A., & Leiserowitz, A. (2021). Shifting Republican views on climate change through targeted advertising. *Nature Climate Change, 11*(7), 573–577. https://doi.org/10.1038/s41558-021-01070-1

Kemkes, R. J., & Akerman, S. (2019). Contending with the nature of climate change: Phenomenological interpretations from northern Wisconsin. *Emotion, Space and Society, 33*, 100614. https://doi.org/10.1016/j.emospa.2019.100614

Leiserowitz, A., Maibach, E., Roser-Renouf, C., Feinberg, G., & Rosenthal, S. (2015). *Climate change in the American mind:* October 2015–Yale University and George Mason University. Yale Program on Climate Change Communication.

McGahey, D. J., & Lumosi, C. K. (2018). Climate change communication for adaptation: Mapping communication pathways in semi-arid regions to identify research priorities. *Journal of Sustainable Development in Africa, 20*(1), 87–107. Retrieved from https://www.jsd-africa.com/#:~:text=The%20Journal%20of%20Sustainable%20Development

Ojomo, E., Elliott, M., Amjad, U., & Bartram, J. (2015). Climate change preparedness: A knowledge and attitudes study in Southern Nigeria. *Environments, 2*(4), 435–448. https://doi.org/10.3390/environments2040435

Moser, S. C., & Dilling, L. (Eds.). (2007). *Creating a climate for change: Communicating climate change and facilitating social change.* Cambridge University Press.

Moser, S. C. (2010). Communicating climate change: History, challenges, process and future directions. *Wiley Interdisciplinary Review Climate Change, 1*(1), 31–53. https://doi.org/10.1002/wcc.11

Olatumile, A., & Tunde-Awe, B. (2019). Effectiveness of the use of indigenous language in disseminating climate change information for sustainable environmental development in Nigeria. *American International Journal of Humanities, Arts and Social Sciences, 1*(2), 7–16. Retrieved from https://www.acseusa.org/journal/index.php/aijhass

Olusola, O. (2007). Dynamics of indigenous language in environmental communication. *Lagos Papers in English Studies, 1*(1), 228–235.

Ojala, M., & Lakew, Y. (2017). *Young people and climate change communication.* University Press.

Raimi, L., & Bieh, N. N. (2020). Town-crying for knowledge dissemination on climate change to non-literate rural people in Nigeria. *International Journal of Scientific Research in Humanities, Legal Studies & International Relations, 5*(1), 155–168.

Sen, L. T., Bond, J., Dung, N. T., Hung, H. G., Mai, N. T., & Phuong, H. T. (2021). Farmers' barriers to the access and use of climate information in the mountainous regions of Thừa Thiên Huế province, Vietnam. *Climate Services, 24.* https://doi.org/10.1016/j.cliser.2021.100267

Pan, W. L., Fan, R., Pan, W., Ma, X., Hu, C., Fu, P., & Su, J. (2023). The role of climate literacy in individual response to climate change: Evidence from China. *Journal of Cleaner Production, 405,* Elsevier. https://doi.org/10.1016/j.jclepro.2023.136874

Wilson, D. (2005). *New perspectives in applied communication.* Stirling Horden Publishers Ltd.

Chapter 14
Climate Change Risks and Adaptation Among Cassava Value Chain Actors in Southeast Nigeria

Chukwudi Loveday Njoku, Chibuzo Uzoma Izuogu, Igwe Oscar Obasi, Gilian Chiedozie Azuamairo, Sikiru Ibrahim-Olesin, Onyekachi Okwe Agwu, and Ngozi Ekunyi Umeh

Abstract The study assessed climate change awareness, perceptions, adaptation strategies, and constraints among cassava value chain actors in Southeast Nigeria. A multi-stage sampling procedure was used to select 360 respondents, with data collected through structured questionnaires and analyzed using frequency and mean counts. Results indicated a high awareness of climate change risks, such as extreme heat ($\bar{x} = 4.16$), weather fluctuation ($\bar{x} = 4.13$), and increased rainfall intensity ($\bar{x} = 4.10$). Key perceived effects included changes in rainfall ($\bar{x} = 3.31$), environmental degradation ($\bar{x} = 3.26$), and product scarcity ($\bar{x} = 3.18$). Adaptation strategies included seeking early warning information ($\bar{x} = 3.44$), using inorganic fertilizer ($\bar{x} = 3.40$), and adjusting planting and harvesting calendars ($\bar{x} = 3.37$). However, constraints such as high labour costs ($\bar{x} = 3.79$), poor access to market information ($\bar{x} = 3.71$), and lack of timely inputs ($\bar{x} = 3.64$) hindered effective adaptation. Strengthening climate services, improving access to finance and inputs, promoting sustainable practices, and enhancing infrastructure and policy support are recommended to improve resilience.

Keywords Climate change risks · Adaptation strategies · Cassava value chain · Southeast Nigeria

C. L. Njoku (✉) · C. U. Izuogu · S. Ibrahim-Olesin · O. O. Agwu · N. E. Umeh
Department of Agricultural Extension, Alex Ekwueme Federal University Ndufu-Alike, Ikwo, Ebonyi State, Nigeria
e-mail: chukzy4now1980@gmail.com

G. C. Azuamairo
Michael Okpara University of Agriculture Umudike, Umudike, Abia State, Nigeria

I. O. Obasi
Michael Okpara University of Agriculture Umudike, TETFUND Center of Excellence in Root and Tuber Crops Research, Umudike, Nigeria

R. U. Onyeneke et al. (eds.), *Energy Transition, Climate Action and Sustainable Agriculture*, https://doi.org/10.1007/978-3-031-83165-2_14

14.1 Introduction

Food security forms the bedrock of sustainable economic development, yet climate change poses a significant threat to the stability of food crop production globally, with cassava being no exception (Olowoyeye et al., 2023). Cassava (Manihot esculenta) is the third most vital source of calories in Africa's tropical and subtropical regions, following rice and maize (Scaria et al., 2024). It is extensively cultivated across many Sub-Saharan African countries, with Africa currently producing half of the world's cassava supply (FAO, 2020). As the leading global producer, Nigeria significantly contributes to this output, followed by Southeast Asia, Brazil, Indonesia, Thailand, and Vietnam (Ikuemonisan et al., 2020). Over 90% of rural households in Nigeria regularly consume cassava, underscoring its importance. However, cassava production in Nigeria faces increasing threats from climate change-related events (Osuji et al., 2023).

Cassava value chain actors, including producers, processors, marketers, and input dealers, play a crucial role in ensuring that cassava is efficiently and effectively delivered to consumers in good condition. Producers grow cassava, processors transform the raw cassava into products like flour or gari, marketers handle the sale of cassava products, and input dealers supply essential items such as seeds, fertilisers, and pesticides for cassava farming. All these actors contribute to the sustainability of the cassava supply chain. However, climate change poses substantial challenges to them. Producers face reduced tuber size and quality, water stress, and increased production risks (Nasir & Toth, 2022). Processors face higher operational costs due to inconsistent raw material supply and extreme weather damage (Ali et al., 2023). Marketers struggle with logistical issues and price instability, while input dealers are burdened by increased costs and supply chain disruptions (Bateh, 2024).

Climate change jeopardises agriculture, particularly rain-fed agriculture, which relies on favourable climatic conditions for productivity (Adamaagashi et al., 2023). Presently, seasonal variations and changes due to climate change threaten cassava yields globally, regionally, and locally (Osuji et al., 2023). The agricultural sector, especially cassava and maize crops, is vulnerable to these impacts, as evidenced by declining yields due to rising temperatures, changing rainfall patterns, and extreme weather events such as droughts, floods, and storms (Anyaegbu et al., 2023).

Globally, climate change threatens the environment, crop yields, and food security. Ensuring a sustainable climate, increasing crop yields, and securing food involves identifying significant long-term impacts and developing strategies to mitigate these effects (Annappa et al., 2023). There are key knowledge gaps in determining the specific climatic variables and greenhouse gases causing the most damage to cassava yields and in identifying solutions for climate action. Although various studies have explored the impacts of climate change on crops, few have attributed these effects to specific climatic variables, such as rainfall and temperature, and greenhouse gases, such as carbon dioxide (CO2), nitrous oxide (N2O), and methane (CH4), particularly concerning cassava yield in Nigeria. Understanding the individual impacts of these variables is essential for developing insights into mitigating their negative effects on yield and food security (Feng et al., 2021).

In Nigeria, climate change has been reported to disrupt cassava yields, outputs, and quality, leading to food shortages and reduced supply (Osuji et al., 2023). Extreme weather events, including rising temperatures, altered precipitation patterns, changes in relative humidity, and windstorms, are typically responsible for reduced cassava productivity (Kumar et al., 2022). The rapid deterioration of cassava tubers in the field and post-harvest affects farmers' income and profitability (Enesi et al., 2022). Reduced rainfall causes tubers to shrivel, decreasing size and market value, while increased rainfall promotes pests and diseases that reduce photosynthetic activity (Osuji et al., 2024). Projections indicate that Nigeria's cassava production will continue to decline as climate change persists (Osuji et al., 2023). Nigeria is the sixth least prepared and the 53rd most vulnerable nation to adapt to climate change (National Council on Climate Change [NCCC], 2021).

This situation necessitates a comprehensive strategy and awareness-raising among Nigerian rural farming households on climate change adaptation (Hassan & Knight, 2023). To improve its position in the global adaptation index, Nigeria urgently needs to adopt proven adaptation solutions to scale up cassava production and mitigate climate impacts (Osuji et al., 2023). Recently, cassava farmers in Southeast Nigeria have experienced a decline in cassava production, resulting in decreased yields, reduced quality, and loss of economic value, leading to hunger and economic hardship (Igberi et al., 2022; Ikuemonisan et al., 2020). Unprecedented changes in climate seriously threaten their land productivity.

The growing population in Southeast Nigeria is increasing the demand for food production, with cassava struggling to meet this demand as adverse climatic conditions reduce yields (Osuji et al., 2023). Addressing these issues highlights the research gap and motivates this study on climate risks and adaptations among cassava value chain actors in Southeast Nigeria. The broad objective of this study was to assess climate change risks and adaptation among cassava value chain actors in Southeast Nigeria. Specifically, the study assessed the awareness of climate-related risks within the cassava value chain, evaluate actors' perceptions of climate change effects on cassava, analyse the effectiveness of adaptation strategies utilised by respondents, and identify the constraints hindering the adoption of adaptation measures.

14.2 Methodology

The study was conducted in Southeast Nigeria, which comprises five states: Abia, Anambra, Ebonyi, Enugu, and Imo. Southeast Nigeria lies between Latitude 4° 30′ and 7° 00′ North and Longitude 5° 30′ and 9° 30′ East, covering a land area of approximately 29,525 km^2 with an estimated population of 24,067,008 (Ukwuaba & Ileka, 2024). This region experiences a tropical climate characterised by distinct wet and dry seasons. Recent climate studies indicate that Southeast Nigeria faces high variability in rainfall and temperature, with increasing occurrences of extreme weather events such as heavy rainfall and prolonged dry spells (Anyaegbu et al., 2023). The study area is endowed with fertile, well-drained soils, predominantly sandy loam, which, along with the climate, supports robust agricultural activities. The rainy season spans from April to October, while the dry season extends from November to March. Annual rainfall ranges from 2000 to 2500 mm, with relative humidity surpassing 75% throughout the year, peaking above 90% during the rainy season, and the mean annual temperature hovers around 28 °C (Ukwuaba & Ileka, 2024). This climatic variability significantly impacts agriculture, particularly the cassava value chain. Cassava (Manihot esculenta) is a vital staple crop in Southeast Nigeria, contributing significantly to food security and livelihoods. Its resilience to drought and poor soil conditions makes it crucial for subsistence farming (Skendžić et al., 2021; FAO, 2020). However, the changing climate has adverse effects on cassava yields and quality, making it imperative to study its response to climate variations and develop adaptive strategies (Singh, 2024).

This research utilised a cross-sectional survey design and a multi-stage sampling procedure to evaluate the climate change risks and adaptations among cassava value chain actors in Southeast Nigeria. In the initial stage, three states—Abia, Ebonyi, and Imo—were purposively selected based on their proximity, resources, convenience for the researcher, and extensive cassava cultivation. In the subsequent stage, four Local Government Areas (LGAs) were randomly selected from each state: Ikwuano, Isiala Ngwa, Bende, and Umuahia South in Abia; Okigwe, Ehime Mbano, Ikeduru, and Obowo in Imo; and Ivo, Afikpo, Ikwo, and Ohaozara in Ebonyi. In the final stage, 75% of cassava value chain actors (input dealers, producers, processors, and marketers) were sampled within each LGA, yielding a total of 360 respondents. Participant lists were sourced from the Value Chain Development Department of the Ministries of Agriculture, community representatives, extension agents, and trade associations.

Data collection was executed using a structured questionnaire administered by trained enumerators through field visits between March and June 2023. The study objectives were addressed through descriptive statistics, including frequencies and mean scores. Awareness of climate-related risks among cassava value chain actors was operationalised by asking respondents to indicate their awareness of 11 climate risk statements. Responses were scored as "Aware" = 0 and "Not Aware" = 1. For assessing climate-related risks, respondents rated the extent to which 10 climate-related risks affected them on a 5-point Likert scale (Very high = 5, High = 4, Moderate = 3, Low = 2, Very low = 1), with a decision rule of 3.00 calculated as $(5 + 4 + 3 + 2 + 1)/5 = 3.00$. Mean responses of 3.00 and above indicated significant risk factors, while values below 3.00 indicated lower risk.

Perception of climate change effects was evaluated by asking actors to indicate their agreement with 11 perception statements on a 4-point Likert scale (Strongly agree = 4, Agree = 3, Disagree = 2, Strongly disagree = 1). The cut-off was set at 2.50, calculated as $(4 + 3 + 2 + 1)/4 = 2.50$. Mean responses of 2.50 and above indicated high climate change perception, while those below 2.50 indicated otherwise. Adaptation strategies were assessed by asking respondents to indicate their level of agreement with 19 adaptation strategy statements using the same 4-point Likert scale, with a cut-off of 2.50. Mean responses of 2.50 and above indicated a strategy was being practiced, while scores below 2.50 signified the contrary.

Constraints faced by cassava value chain actors were measured by asking respondents to rate their agreement with 20 constraint statements on a 4-point Likert scale (Very serious = 4, Serious = 3, Slightly serious = 2, Not serious = 1). The cut-off was also set at 2.50, calculated as $(4 + 3 + 2 + 1)/4 = 2.50$. Mean scores of 2.50 and above indicated significant constraints, while scores below 2.50 indicated otherwise.

14.3 Results and Discussion

14.3.1 Actors' Awareness of Climate-Related Risks

The analysis of Table 14.1 demonstrates that cassava value chain actors exhibit a high level of awareness regarding climate-related risks. The top three risks identified are extreme heat ($\bar{x} = 4.16$), weather fluctuations ($\bar{x} = 4.13$), and increased intensity of rainfall ($\bar{x} = 4.10$). These results suggest that extreme heat is perceived as the most critical climate-related risk, followed closely by weather fluctuations and increased rainfall intensity.

The heightened awareness of extreme heat aligns with findings from other research conducted in Nigeria, where extreme weather conditions have been identified as major threats to agriculture (Assouto & Houngbeme, 2023; Emenyonu et al., 2020). Similar concerns regarding weather fluctuations and increased rainfall intensity have been observed in studies by the Food and Agriculture Organization (FAO, 2020), which cited these factors as key challenges in agricultural practices. These similarities underscore the growing recognition of climate variability's impact on agricultural systems.

The high awareness scores can be attributed to the direct and immediate effects of these climate risks on cassava production. For instance, extreme heat reduces cassava tuber size and quality, which has been corroborated by Assouto and Houngbeme (2023), who noted that extreme weather conditions, particularly heat, negatively affect tuber development and overall crop quality. Likewise, fluctuations in weather and increased rainfall intensity disrupt crop growth cycles, increase pest and disease outbreaks, and thereby directly reduce yields and productivity. According to the FAO (2020), irregular weather patterns and excessive rainfall can cause soil erosion, nutrient leaching, and heightened disease incidence, all of which threaten cassava cultivation.

These findings highlight the significance of addressing these risks through well-targeted adaptation strategies. The high awareness levels among cassava value chain actors suggest a strong need for interventions aimed at enhancing resilience and ensuring sustained productivity in the face of climate change (Olaosebikan et al., 2023; Suryani et al., 2023).

Table 14.1 Actors' awareness of climate-related risks

Climate-related risks	Producers ($\bar{x}$)	Processors ($\bar{x}$)	Marketers ($\bar{x}$)	Input dealers ($\bar{x}$)	Pooled ($\bar{x}$)
Increased intensity of rainfall	4.00	4.30	4.07	4.05	4.10
Prolonged dry season/ drought	3.92	4.10	4.06	3.79	3.97
Frequent floods	3.25	3.70	3.54	2.66	3.29
Increased heat/ temperature	3.91	4.10	4.10	3.79	3.98
Severe windstorm	3.53	3.72	3.60	3.11	3.49
Unpredictability of rainfall	3.48	3.65	3.58	3.23	3.48
Late onset of rain	3.99	4.10	3.83	3.57	3.87
Early cessation of rainfall	3.58	4.17	3.94	3.03	3.68
Weather fluctuation	4.23	4.17	4.10	4.00	4.13
Extreme heat	4.29	4.20	4.05	4.08	4.16
Total mean	38.18	40.21	38.87	35.31	38.15
Grand mean	3.82	4.02	3.89	3.53	3.82

Field Survey, 2023

14.3.2 Actors' Perception of Climate-Related Risks

The data from Table 14.2 reveal that cassava value chain actors perceive various climate-related risks with differing levels of concern. The highest pooled mean perceptions are associated with changes in rainfall ($\bar{x} = 3.31$), environmental degradation ($\bar{x} = 3.26$), and increased soil leaching due to flooding ($\bar{x} = 3.17$). These findings suggest that changes in rainfall patterns, environmental degradation, and the impact of flooding on soil quality are major concerns for the actors.

These perceptions align with previous studies that underscore the impact of rainfall variability on agricultural productivity (Assouto & Houngbeme, 2023; Ayinde et al., 2022; Osuji et al., 2023). Similarly, environmental degradation has been widely recognized as an increasing challenge in agriculture, contributing to the deterioration of soil health and reductions in crop yields (FAO, 2020; Omodara et al., 2023). The concern over increased soil leaching due to flooding is consistent with research that highlights the exacerbation of soil erosion and nutrient loss caused by extreme weather events (Emenyonu et al., 2020; Pedersen et al., 2024).

The actors' heightened perceptions of these risks can be attributed to their significant influence on agricultural productivity. Variations in rainfall can disrupt crop growth and yield, while environmental degradation leads to declining soil quality and agricultural output. Moreover, soil leaching, intensified by flooding, reduces soil fertility, further threatening production. Ayinde et al. (2022) similarly reported that

Table 14.2 Actors perception of climate-related risks

Perception of climate change	Producers (x̄)	Processors (x̄)	Marketers (x̄)	Input dealers (x̄)	Pooled (x̄)
Change in temperature	3.10	3.13	3.18	2.87	3.07
Seasonal fluctuations	3.03	3.03	3.08	3.13	3.07
Change in rainfall	3.28	3.27	3.27	3.40	3.31
Increased soil leaching due to flooding	3.30	3.10	3.05	3.22	3.17
Environmental degradation	3.13	3.23	3.25	3.43	3.26
Poor yield/output	3.13	3.17	3.15	3.28	3.18
Poor quality of output	3.12	3.10	3.07	3.22	3.13
Product price fluctuation	2.95	3.27	3.22	3.27	3.18
Scarcity of product	3.23	3.30	3.30	2.90	3.18
Reduced income	3.15	3.10	3.07	3.13	3.11
Increased risks and uncertainties	3.25	3.17	3.17	2.90	3.12
Total mean	34.67	34.87	34.81	34.75	34.78
Grand mean	3.15	3.17	3.16	3.16	3.16

Field Survey, 2023

farmers are increasingly concerned about rainfall inconsistency due to its adverse effect on cassava yields.

These findings underscore the necessity for adaptive measures to address these challenges and improve resilience within the cassava value chain. The actors' concerns highlight the urgency of developing targeted adaptation strategies to mitigate the impacts of climate change, ensuring food security and sustainable cassava production in Southeast Nigeria.

14.3.3 Actors' Climate Change Adaptation Strategies Practiced

Table 14.3 reveals that cassava value chain actors practised several climate change adaptation strategies, with the highest pooled mean scores observed for seeking early warning information about climate risks ($\bar{x} = 3.44$), planting improved cassava varieties ($\bar{x} = 3.35$), and using inorganic fertilizers ($\bar{x} = 3.40$). These findings suggest that actors proactively address climate challenges by gathering information, adopting resilient agricultural practices, and enhancing soil fertility.

Seeking early warning information corresponds with studies that stress the importance of timely climate data for effective adaptation (Osuji et al., 2023; Pedersen et al.,

Table 14.3 Actors' climate change adaptation strategies practiced

Adaptation strategies	Producers (x̄)	Processors (x̄)	Marketers (x̄)	Input dealers (x̄)	Pooled (x̄)
Seeking early warning information about climate risks	3.23	3.63	3.66	3.22	3.44
Planting trees on the farms to serve as shade	3.33	3.43	3.31	3.10	3.29
Planting improved cassava varieties	3.33	3.40	3.54	3.12	3.35
Buying insurance	3.07	3.43	3.45	2.77	3.18
Crop diversification (planting different crops)	3.16	3.50	3.27	3.02	3.24
Livelihood diversification	3.38	3.13	3.23	3.25	3.25
Soil and water conservation techniques	3.35	3.27	3.35	3.07	3.26
Adjusting planting and harvesting calendar	3.34	3.60	3.42	3.12	3.37
Irrigation	3.31	3.20	3.18	2.72	3.10
Using weather forecasting tools	3.32	3.17	3.24	2.40	3.03
Use of inorganic fertilizer	3.38	3.70	3.45	3.05	3.40
Use of organic fertilizer	3.42	3.40	3.21	3.02	3.26
Use of pesticides	3.41	3.27	3.18	3.07	3.23
Storing processed cassava	3.14	3.07	2.91	2.85	2.99
Selling processed cassava	3.38	3.27	3.22	3.42	3.32
Early harvesting of mature cassava tubers	3.42	3.33	3.33	3.30	3.35
Buying a small quantity of cassava	3.19	3.03	2.91	3.03	3.04
Livelihood diversification	3.31	3.00	3.08	2.90	3.07
Buying insurance	3.59	3.17	3.19	2.70	3.16
Total mean	63.06	63.0	62.13	57.13	61.33
Grand mean	3.32	3.32	3.27	3.01	3.23

Field Survey, 2023
Actors' Climate Change Adaptation Strategies

2024). The practise of improved cassava varieties is consistent with research demonstrating the role of high-yield, resilient crops as a key adaptation strategy (Assouto & Houngbeme, 2023). Similarly, the use of inorganic fertilizers underscores the importance of enhancing soil fertility to maintain productivity in the face of changing climate conditions (FAO, 2020; Omodara et al., 2023).

These practices reflect a practical response to the impacts of climate change on cassava production. Seeking early warning information aids in making timely, informed decisions, planting improved varieties enhances crop yield and resilience, and using inorganic fertilizers ensures better soil health. Olaosebikan et al. (2023) also highlight that adopting resilient crop varieties and efficient resource management is essential for building resilience in the cassava value chain. These strategies demonstrate the actors' proactive adaptation efforts and emphasize the need for ongoing support to effectively implement these practices, ensuring sustained resilience to climate risks.

Table 14.4 shows that cassava value chain actors face several constraints in adapting to climate change, with the highest pooled mean scores observed for the high cost of fertilizer ($\bar{x} = 3.53$), lack of timely access to inputs ($\bar{x} = 3.64$), and the incidence of pests and diseases ($\bar{x} = 3.39$). These findings highlight key barriers, including financial limitations, logistical challenges, and biological threats.

The high fertilizer cost is consistent with studies indicating that financial barriers are a major obstacle to adopting climate-smart agricultural practices (Osuji et al., 2023; Pedersen et al., 2024). Similarly, the lack of timely access to inputs corroborates research emphasizing the need for timely agricultural input availability to support effective adaptation (Assouto & Houngbeme, 2023; Canevari-Luzardo, 2019). The incidence of pests and diseases as a significant constraint supports findings that biological threats substantially affect agricultural productivity and adaptation efforts (FAO, 2020; Baffour-Ata et al., 2024).

These constraints illustrate the practical challenges faced by cassava value chain actors. The high cost of fertilizer restricts the ability to invest in essential inputs, while delays in access to inputs disrupt the agricultural calendar, impacting planting and harvesting schedules. Pests and diseases pose ongoing risks to crop health and yield. Research by Ifeanyi-Obi and Issa (2013) similarly highlights how such barriers can impede effective adaptation strategies. Addressing these constraints through enhanced financial support, improved input logistics, and better pest management strategies is crucial to improving the resilience of cassava production in response to climate change.

14.4 Conclusions, Recommendations and Limitations

The findings show that cassava value chain actors are highly aware of climate-related risks, with key concerns surrounding extreme heat, weather fluctuations, and intense rainfall. Commonly adopted adaptation strategies include seeking early warning information, planting improved cassava varieties, and using inorganic fertilizers. However, adaptation efforts are hindered by challenges such as high fertilizer costs, limited access to inputs, and pest and disease incidences.

To mitigate these climate-related risks, it is crucial to enhance support for adopting early warning systems, resilient cassava varieties, and sustainable soil fertility practices. Promoting climate-resilient agricultural techniques will help cassava value chain actors address extreme heat, weather variability, and rainfall changes. Additionally, improving access to timely inputs and information, as well as providing

Table 14.4 Actors' climate change adaptation constraints

Constraints	Producers ($\bar{x}$)	Processors ($\bar{x}$)	Marketers ($\bar{x}$)	Input dealers ($\bar{x}$)	Pooled ($\bar{x}$)
High cost of fertilizer	3.52	3.83	3.18	3.60	3.53
Lack of timely access to inputs	3.40	3.78	3.59	3.78	3.64
Inadequate land	3.33	3.48	3.27	3.40	3.37
Inadequate capital	3.21	3.60	3.05	3.40	3.32
Incidence of pests and diseases	3.56	3.47	3.25	3.27	3.39
Flood	2.58	3.30	2.56	3.17	2.90
Scorching sun/heat	3.74	3.10	3.07	3.28	3.30
High cost of labour	3.39	3.53	3.22	3.03	3.79
Inadequate climate information	3.15	3.67	3.30	3.48	3.40
Poor extension service	3.53	3.32	3.52	3.52	3.47
Inadequate storage facility	3.48	3.00	3.35	3.10	3.23
Poor access to market information on prices	3.43	3.65	4.48	3.27	3.71
Poor access to markets	3.32	3.85	3.28	3.30	3.44
High cost of transportation	3.38	3.13	3.43	3.10	3.26
Poor power supply	3.60	3.40	3.38	3.17	3.39
Lack of awareness of climate change effect	3.39	3.57	3.22	3.35	3.38
Poor access road	3.30	3.48	3.56	3.65	3.50
Price fluctuation	3.23	3.30	3.61	3.18	3.33
Barriers to entry in the cassava marketing industry	2.21	2.73	2.52	2.95	2.60

(continued)

Table 14.4 (continued)

Constraints	Producers (x̄)	Processors (x̄)	Marketers (x̄)	Input dealers (x̄)	Pooled (x̄)
Competition from foreign brands	2.31	2.47	2.02	2.58	2.35
Total mean	65.06	67.66	64.86	65.58	66.30
Grand mean	3.25	3.38	3.24	3.28	3.32

Field survey, 2023

technical assistance, will strengthen adaptation efforts and support productivity in cassava farming.

The study's limitations include potential biases due to self-reported data, which might have led to overestimating awareness and adaptation practices. The sample may not fully represent all cassava value chain actors in Nigeria, limiting the generalizability of the findings. Inadequate rural infrastructure and accessibility issues could have affected data completeness. The dynamic nature of climate change impacts and rapid adaptation adjustments may not have been fully captured. Limited financial and technical resources also restricted the implementation and sustainability of adaptation strategies.

Policy implications include prioritizing investments in agricultural extension services, improving access to inputs and climate information, and developing policies that support the affordability of fertilizers and pest management solutions. This will enable cassava value chain actors to better respond to climate change, enhancing food security and livelihoods.

Acknowledgements We appreciate the support and contributions of all participants and institutions involved in this study.

Funding This research received funding from the Michael Okpara University of Agriculture TETFund Center of Excellence in Root and Tuber Crops.

Data Availability Statement Data supporting the findings are available from the corresponding author upon reasonable request.

Ethical Clearance Respondents and other stakeholders were engaged, and their participation was assured of complete confidentiality, with responses used solely for research purposes.

Conflict of Interest The authors declare no conflict of interest.

References

Adamaagashi, I., Nzechie, O., Obiorah, J., & Idakwoji, A. A. (2023). Analyzing the critical impact of climate change on agriculture and food security in Nigeria. *International Journal of Agriculture and Environmental Sciences, 9*(4), 1–27.

Ali, I., Arslan, A., Tarba, S., & Mainela, T. (2023). Supply chain resilience to climate change inflicted extreme events in agri-food industry: The role of social capital and network complexity. *International Journal of Production Economics*, 264 https://doi.org/10.1016/j.ijpe.2023.108968

Annappa, N. N., Bhavya, N., Kasturappa, G., & Murthy, R. K. (2023). Climate change's threat to agriculture: Impacts, challenges, and strategies for a sustainable future. In S. Kushwaha (Ed.), *Climate change and agriculture* (vol. 9, pp. 115–136). AkiNik Publications. https://doi.org/10.22271/ed.book.2395

Anyaegbu, C. N., Okpara, K. E., Taweepreda, W., Akeju, D., Techato, K., Onyeneke, R. U., Poshyachinda, S., & Pongpiachan, S. (2023). Impact of climate change on cassava yield in Nigeria: An autoregressive distributed lag bound approach. *Agriculture, 13*(1), 80. https://doi.org/10.3390/agriculture13010080

Assouto, A., & Houngbeme, D.J.-L. (2023). Access to credit and agricultural productivity: Evidence from maize producers in Benin. *Cogent Economics & Finance, 11*(1), 2196856. https://doi.org/10.1080/23322039.2023.2196856

Ayinde, A. F. O., Johnston, P. A., Olujimi, O. O., Dasgupta, P., & Akerele, D. (2022). Perception of cassava-based farmers to climate variability in the rainforest and derived savannah biomes of Nigeria. *Sarhad Journal of Agriculture, 38*(5), 43–52. https://doi.org/10.17582/journal.sja/2022/38.5.43.52

Baffour-Ata, F., Boakye, L., Acquah, L. E., Brown, S. B., Kafui, J. D., Marfo, A. A., Acheampong, P., & Wheagar, S. (2024). Barriers confronting smallholder cassava farmers in the adoption and utilization of climate-smart agriculture in the Afigya Kwabre South District, Ghana. *Climate Resilience and Sustainability*. https://doi.org/10.1002/cli2.77

Bateh, D. (2024). Supply chain disruptions and their effect on suppliers and consumers in the marketplace. *Forum for Economic and Financial Studies*, *2*(1), 343. https://doi.org/10.59400/fefs.v2i1.343

Canevari-Luzardo, L. M. (2019). Value chain climate resilience and adaptive capacity in micro, small, and medium agribusiness in Jamaica: A network approach. *Regional Environmental Change, 19*, 2535–2550. https://doi.org/10.1007/s10113-019-01561-0

Enesi, R. O., Pypers, P., Kreye, C., Tariku, M., Six, J., & Hauser, S. (2022). Effects of expanding cassava planting and harvesting windows on root yield, starch content and revenue in southwestern Nigeria. *Field Crops Research, 286*, 108639. https://doi.org/10.1016/j.fcr.2022.108639

Emenyonu, C. A., Eze, C. C., & Ejike, O. U. (2020). Factors influencing cassava farmers' climate change risk perception in Anambra State Nigeria. *American Journal of Climate Change, 9*(3), 217–227. https://doi.org/10.4236/ajcc.2020.93014

Food and Agriculture Organization of the United Nations. (2020). *Cassava in tropical Africa: A reference manual*. FAO.

Feng, S., Hao, Z., Zhang, X., & Hao, F. (2021). Changes in climate-crop yield relationships affect risks of crop yield reduction. *Agricultural and Forest Meteorology, 304–305*, 108401. https://doi.org/10.1016/j.agrformet.2021.108401

Hassan, B. A., & Knight, J. (2023). Adaptation to climate change and variability by farming households in North-Central Nigeria. *Sustainability, 15*(23), 16309. https://doi.org/10.3390/su152316309

Ifeanyi-Obi, C. C., & Issa, F. O. (2013). Barriers faced by cassava farmers in adapting to climate change in Oron Agricultural Zone of Akwa Ibom State. *IOSR Journal of Agriculture and Veterinary Science (IOSR-JAVS, 4*(6), 19–26.

Igberi, C. O., Osuji, E. E., Anuli, R. O., Paul, C. O., & Ngozi, O. E. (2022). Climate smart adaptive measure of yellow cassava, linkages, and implications in Southeast Nigeria. *Agrociencia, 56*(3), 1–32.

Ikuemonisan, E. S., Mafimisebi, T. E., Ajibefun, I., & Adenegan, K. (2020). Cassava production in Nigeria: Trends, instability and decomposition analysis (1970–2018). *Heliyon, 6*(10), e05089. https://doi.org/10.1016/j.heliyon.2020.e05089

Kumar, L., Chhogyel, N., Gopalakrishnan, T., Hasan, M. K., Jayasinghe, S. L., Kariyawasam, C. S., Kogo, B. K., & Ratnayake, S. (2022). Chapter 4—Climate change and future of agri-food production. In *Future foods: Global trends, opportunities, and sustainability challenges* (pp. 49–79). https://doi.org/10.1016/B978-0-323-91001-9.00009-8

Nasir, M. W., & Toth, Z. (2022). Effect of drought stress on potato production: A review. *Agronomy, 12*(3), 635. https://doi.org/10.3390/agronomy12030635

National Council on Climate Change (NCCC). (2021). *About the National Council on Climate Change (NCCC): Climate Change Act of 2021*. Prepared by Halima Bawa-Bwari, Bala Najega Rufai, & Olajumoke Salako. National Council on Climate Change. https://natccc.gov.ng/publications/Nigeria%20Climate%20Adaptation%20Country%20Compacts.pdf

Olaosebikan, O., Bello, A., Utoblo, O., Okoye, B., Olutegbe, N., Garner, E., & Madu, T. (2023). Stressors and resilience within the cassava value chain in Nigeria: Preferred cassava variety traits and response strategies of men and women to inform breeding. *Sustainability, 15*(10), 1–18. https://doi.org/10.3390/su15107837

Olowoyeye, T., Abegunrin, G., & Alamuoye, T. (2023). Impact of climate change on sustainable food production in Nigeria. In *Scientific Applications Research eConference SciAppRes-eCon2023* (vol. 1), June 21–23, 2023.

Omodara, O. D., Ige, O. A., Oluwasola, O., Oyebanji, A. T., & Afape, O. O. (2023). Factors influencing cassava farmers' choice of climate change adaptation practices and its effect on cassava productivity in Osun State, Nigeria. *FARA Research Report, 7*(23), 274–288. https://doi.org/10.59101/frr072323

Osuji, E. E., Tim-Ashama, A. C., Nwose, R. N., Orji, J. E., Anyanwu, U. G., Nwachukwu, E. U., Anyiam, K. H., Umeh, N. E., Eze, E. U., Iwezor-Magnus, D. N., & Osang, E. A. (2024). Effect of climate change on sustainability of cassava production in Ebonyi State, Nigeria. *Sumerianz Journal of Agriculture and Veterinary*, 7(2), 5–14. https://doi.org/10.47752/sjav.72.5.14

Osuji, E. E., Igberi, C. O., & Ehirim, N. C. (2023). Climate change impacts and adaptation strategies of cassava farmers in Ebonyi State Nigeria. *Journal of Agricultural Extension, 27*(1), 35–48. https://doi.org/10.4314/jae.v27i1.4

Pedersen, S. M., Erekalo, K. T., Christensen, T., Denver, S., Gemtou, M., Fountas, S., Isakhanyan, G., Rosemarin, A., Ekane, N., Puggaard, L., Nertinger, M., Brinks, H., Puško, D., & Adrián, J. B. (2024). Drivers and barriers to climate-smart agricultural practices and technologies adoption: Insights from stakeholders of five European food supply chains. *Smart Agricultural Technology, 8*, 100478. https://doi.org/10.1016/j.atech.2024.100478

Scaria, S. S., Balasubramanian, B., Meyyazhagan, A., Gangwar, J., Jaison, J. P., Kurian, J. T., Pushparaj, K., Pappuswamy, M., Park, S., & Joseph, K. S. (2024). Cassava (Manihot esculenta Crantz)—A potential source of phytochemicals, food, and nutrition: An updated review. *eFood, 5*(1), e127. https://doi.org/10.1002/efd2.127

Singh, P. (2024). *Role of climate smart agriculture (CSA) in addressing challenges of food security and climate change* (Technical Report). https://doi.org/10.13140/RG.2.2.13965.35040/1

Skendžić, S., Zovko, M., Živković, I. P., Lešić, V., & Lemić, D. (2021). The impact of climate change on agricultural insect pests. *Insects, 12*(5), 440. https://doi.org/10.3390/insects12050440

Suryani, A., Masyhuri, M., Waluyati, L. R., & Utami, A. W. (2023). Risk analysis on the cassava value chain in Central Lampung Regency. *AGRARIS: Journal of Agribusiness and Rural Development Research, 9*(2), 150–173. https://doi.org/10.18196/agraris.v9i2.333

Ukwuaba, I. C., & Ileka, C. M. (2024). Determinants of the extent of use of climate-smart agricultural practices by farm households in Southeast Nigeria. *European Modern Studies Journal, 8*(2), 43–57. https://doi.org/10.59573/emsj.8(2).2024.5

Chapter 15
Socioeconomic Drivers of Climate Change Resilience Among Cassava Value Chain Actors in Southeast Nigeria

Chukwudi Loveday Njoku, Lovina Enyidiya Odoemelam, Chibuzo Uzoma Izuogu, Igwe Oscar Obasi, Gillian Chidozie Azuamairo, Ikechukwu Chiemena Chinaka, Ngozi Ekunyi Umeh, and Onyekachi Okwe Agwu

Abstract This study examined the socioeconomic drivers of climate change resilience among cassava value chain actors in Southeast Nigeria. It assessed their socioeconomic characteristics, climate change awareness, adaptation practices, and the factors influencing resilience. Using a cross-sectional survey and multi-stage sampling, data were collected through structured questionnaires and analysed using frequencies, percentages, mean statistics, and simple regression analysis. Results showed that actors were middle-aged (41–50 years), predominantly male, married, part of large households (6–10 members), with primary education, extensive farming experience (11–20 years), and income between N51,000 and N100,000. They frequently engaged in extension contacts (0.06) and cooperative memberships. High

C. L. Njoku (✉) · C. U. Izuogu · I. C. Chinaka · N. E. Umeh · O. O. Agwu
Department of Agricultural Extension, Alex Ekwueme Federal University Ndufu-Alike, Ikwo, Ebonyi State, Nigeria
e-mail: chukzy4now1980@gmail.com

L. E. Odoemelam
Department of Agricultural Extension and Rural Development, Michael Okpara University of Agriculture, Umudike, Nigeria

I. O. Obasi
TETFund Center of Excellence in Root and Tuber Crops Research and development, Michael Okpara University of Agriculture Umudike, Umudike, Abia State, Nigeria

G. C. Azuamairo
Department of Agribusiness, Michael Okpara University of Agriculture, Umudike, Abia State, Nigeria

I. C. Chinaka
Department of Crop Science, Alex Ekwueme Federal University Ndufu-Alike, Ikwo, Nigeria

O. O. Agwu
Department of Agricultural Economics, Alex Ekwueme Federal University Ndufu-Alike, Ikwo, Nigeria

R. U. Onyeneke et al. (eds.), *Energy Transition, Climate Action and Sustainable Agriculture*, https://doi.org/10.1007/978-3-031-83165-2_15

climate change awareness was noted, especially regarding increased rainfall intensity (91.9%), rainfall unpredictability (91.4%), and late onset of rains (91.7%). Key adaptation strategies included seeking early warning information ($\bar{x} = 3.44$), using inorganic fertilizer ($\bar{x} = 3.40$), and planting improved cassava varieties ($\bar{x} = 3.35$). Socioeconomic factors influencing resilience included age ($\beta = 0.02$), marital status ($\beta = -0.05$), education ($\beta = 0.10$), experience ($\beta = 0.04$), income ($\beta = 0.05$), and extension contacts ($\beta = 0.06$). The study concluded that promoting sustainable practices and improving access to climate information are essential for enhancing resilience. Strengthening extension services, promoting cooperative memberships, and enhancing education and training on climate adaptation are recommended to improve resilience.

Keywords Socioeconomic drivers · Climate change · Resilience · Cassava value chain

15.1 Introduction

This study examined the socioeconomic drivers of climate change resilience within Southeast Nigeria's cassava value chain (VC). Resilience, as defined by Bowen et al. (2020), refers to the capacity of households to prepare for, cope with, and adapt to shocks in ways that protect their well-being and prevent them from falling into or remaining in poverty. Cassava is crucial for both food security and climate change adaptation (Olaosebikan et al., 2023; Omodara et al., 2023), thanks to its ability to withstand various stressors.

The cassava value chain involves smallholder farmers, processors, traders, and other actors who rely on cassava as a key livelihood source. Smallholder farmers rely on cassava as a cash crop, selling fresh tubers and processed products such as gari, starch, and flour (Molnar, 2022). This crop is essential for meeting household needs and investing in other agricultural activities (Stringer et al., 2020). Processors add value by converting cassava into higher-value products like garri, fufu, and starch, which often command higher market prices, providing additional income (Kahinde & Aboaba, 2020). Traders facilitate income generation by purchasing cassava from farmers or processors and distributing it to wholesalers, retailers, and other market outlets, earning income through price differentials and margins (Ettah et al., 2020; Opondo et al., 2022).

The cassava value chain also creates employment opportunities, benefiting farmers, processors, traders, transporters, input suppliers, and service providers (Darko-Koomson et al., 2020). In response to climate variability, Nigerian farmers have increasingly adopted improved cassava varieties (Osuji et al., 2023). Cassava's resilience to extreme weather conditions and adaptability to various ecosystems and soil types make it a valuable alternative to more climate-sensitive crops like potatoes, rice, and maize (Vernooy, 2022; Yu et al., 2022).

Climate change poses a significant threat to agricultural productivity and food security in Southeast Nigeria. While cassava is resilient and able to grow in less fertile soils with minimal input, it still faces challenges such as resource scarcity, land-use conflicts, and climate variability. The cassava value chain plays an essential role in income generation, economic well-being, and employment opportunities, but its sustainability is at risk without adequate adaptation measures. Socioeconomic factors such as access to resources, market dynamics, and socio-economic conditions influence the ability of cassava value chain actors to cope with and adapt to these challenges. Understanding these socioeconomic drivers is essential for developing effective policies and interventions that support the resilience of cassava farmers in Southeast Nigeria.

The broad objective of this study was to examine the socioeconomic drivers of climate change resilience among cassava value chain actors in Southeast Nigeria. The specific objectives were to: describe the socioeconomic characteristics of cassava value chain actors; assess actors' climate change awareness; profile the choice of adaptation practices used to combat climate change; and identify socioeconomic factors influencing the choice of climate change adaptation.

15.2 Methodology

15.2.1 Study Area

The study was conducted in Southeast Nigeria, comprising five states: Abia, Anambra, Ebonyi, Enugu, and Imo. This region is located between Latitude 4° 30′ and 7° 00′ North and Longitude 5° 30′ and 9° 30′ East, covering a land area of approximately 29,525 km^2 with an estimated population of 24,067,008 people (Ukwuaba & Ileka, 2024). Southeast Nigeria experiences a tropical climate with distinct wet (April to October) and dry (November to March) seasons. The annual rainfall ranges from 2000 to 2500 mm, and relative humidity exceeds 75%, peaking above 90% during the rainy season. The mean temperature is around 28 °C (Ukwuaba & Ileka, 2024). The region has fertile, well-drained sandy loam soils, conducive to agriculture.

The study area faces high climate variability, with increasing extreme weather events such as heavy rainfall and prolonged dry spells (Anyanlade et al., 2022). These climatic changes significantly impact agriculture, particularly cassava production, which is a major staple and livelihood source for millions in the region. Cassava (Manihot esculenta), known for its resilience to drought and poor soil conditions, is crucial for both food security and income generation. However, climate variability poses challenges to cassava yield and quality, necessitating the development of adaptive strategies to build resilience within the cassava value chain. Understanding the socioeconomic drivers of climate resilience among cassava value chain actors in this region is vital for ensuring sustainable agricultural practices, food security, and economic stability (FAO, 2020; Singh, 2024; Skendžić et al., 2021).

15.2.2 Research Design

This study used a cross-sectional survey design and a multi-stage sampling procedure to assess climate change risks and adaptation among cassava value chain actors in Southeast Nigeria.

15.2.3 Sampling Procedure

15.2.3.1 Selection of States and Local Government Areas

In the first stage, three states—Abia, Ebonyi, and Imo—were purposively selected based on proximity, resource availability, convenience, and their significant involvement in cassava cultivation. In the second stage, four Local Government Areas (LGAs) were randomly selected from each state: Ikwuano, Isiala Ngwa, Bende, and Umuahia South in Abia; Okigwe, Ehime Mbano, Ikeduru, and Obowo in Imo; and Ivo, Afikpo, Ikwo, and Ohaozara in Ebonyi.

15.2.3.2 Sampling of Actors

In the final stage, 75% of actors in each category—input dealers, producers, processors, and marketers—were sampled within each LGA, leading to a total sample size of 360 respondents. Participant lists were obtained from the Value Chain Development Department in the Ministries of Agriculture, community representatives, extension agents, and trade associations.

15.2.4 Data Collection and Analysis

Data were collected using structured questionnaires administered by trained enumerators during field visits between March and June 2023. The socioeconomic characteristics of cassava value chain actors were analysed using frequencies and percentages. Actors' climate change awareness was assessed using percentages, while adaptation strategies were profiled using mean statistics. Socioeconomic factors influencing resilience were determined using simple regression analysis, based on an established adaptation of previous works that had successfully employed regression to examine climate resilience factors.

15.2.5 *Operationalisation of Variables*

15.2.5.1 Socioeconomic Variables

The socioeconomic variables were operationalised as follows:

- Age: Categorised into five groups: 20–30, 31–40, 41–50, 51–60, and 61–70 years.
- Sex: Recorded as Male or Female.
- Marital Status: Divided into Single, Married, Widow, and Divorced.
- Household Size: Classified into 2–5, 6–10, 11–15, and 16–20 members.
- Educational Level: Measured by the highest education attained, including no formal education, primary education, or secondary education.
- Farming Experience: Grouped into four ranges: 1–10, 11–20, 21–30, and 31–40 years.
- Income: Segmented into five bands: 10,000–50,000, 51,000–100,000, 101,000–150,000, 151,000–200,000, and 201,000–250,000 Naira.
- Extension Contacts: Measured by the number of interactions with extension services in the past year, categorised as 1–5, 6–10, 11–15, and 16–20 times.
- Cooperative Membership: Identified as Yes or No, indicating whether the respondent was a member of a cooperative society.

15.2.5.2 Awareness and Adaptation Strategies

Awareness of climate change among cassava value chain actors was measured by asking respondents to indicate their awareness of 11 specific climate risk statements using a multiple-response format, where "Aware" responses received a score of 0 and "Not Aware" responses received a score of 1. Adaptation strategies were evaluated by asking respondents to express their agreement with 19 adaptation strategy statements on a 4-point Likert scale: strongly agree (4), agree (3), disagree (2), and strongly disagree (1). A cut-off mean score of 2.50 was established, calculated as:

$$\text{Cut-off Mean Score} = (4 + 3 + 2 + 1)/4 = 2.50$$

Mean scores of 2.50 and above indicated that a strategy was practiced, while scores below 2.50 suggested it was not.

15.2.6 *Dependent and Independent Variables*

Resilience was the dependent variable and was estimated from adaptation strategies using a 4-point Likert scale, with an average score of approximately 3.25, reflecting the actors' capacity to adapt to and recover from climate-related impacts.

The independent variables influencing resilience were measured and operationalised as follows:

- Age was recorded in years and segmented into various age ranges.
- Sex was coded as Male (0) or Female (1).
- Marital status was classified into Single (0), Married (1), Widow (2), and Divorce (3).
- Household size indicated the number of people living in the household.
- Education level was measured by the total years of formal schooling completed.
- Experience was quantified by the number of years in the value chain, with categories of 1–10, 11–20, 21–30, and 31–40 years.
- Income was assessed based on estimated average monthly earnings over the past year, with ranges from 10,000–50,000 Naira to 201,000–250,000 Naira.
- Extension contacts were measured by the frequency of interactions with extension services, categorized into 1–5, 6–10, 11–15, and 16–20 times.
- Cooperative membership was recorded as Yes (1) or No (0).

15.2.7 Theoretical Framework

The study was grounded in Resilience Theory, which focuses on the capacity of systems to withstand and adapt to shocks (Bowen et al., 2020). This theoretical framework guided the investigation into how socioeconomic factors influence resilience within the cassava value chain. The selection of independent variables was based on their relevance to understanding and enhancing resilience among cassava value chain actors.

15.3 Results and Discussion

15.3.1 Socioeconomic Characteristics of Cassava Value Chain Actors

The largest group of cassava value chain actors are aged 41–50 (38.6%), indicating the predominance of middle-aged individuals, likely due to their experience, physical capability, and financial stability for farming investments. Omodara et al. (2023) found a similar trend, with a mean age of 45 years, highlighting this age group's role in agricultural productivity.

Males make up 60.6% of the actors, while females account for 39.4%, suggesting gender disparity due to cultural norms and better access to resources for men. Omodara et al. (2023) reported a male dominance of 78% in cassava farming, supported by Aboajah et al. (2018) on cultural constraints affecting women's participation.

The majority (72.2%) of actors are married, providing stability and resources necessary for farming. This aligns with Ibrahim (2022), who noted marital status as a significant factor in agricultural enterprise management. Household sizes of 6–10 members (43.9%) indicate the reliance on family labour for productivity. Omodara et al. (2023) observed similar household sizes, and Unaeze and Ihunwo (2021) emphasised their importance for adaptation strategies.

Primary education is the most common level of education among actors (39.7%). While higher education may improve adoption of advanced practices, Omodara et al. (2023) found that basic education facilitates climate-smart practices, consistent with Awotide et al. (2019).

A significant proportion (38.9%) of actors have 11–20 years of farming experience, which aids in effective management and adaptation. This is supported by Omodara et al. (2023) and Aboajah et al. (2018), who noted that experienced farmers are better at adopting adaptation practices. Most actors (39.4%) earn between 51,000 and 100,000 Naira, suggesting a low to middle-income bracket, which Omodara et al. (2023) found to influence adaptive capacity, as also reinforced by Ibrahim (2022).

Most actors (64.2%) have 1–5 extension contacts, highlighting the importance of regular extension services in disseminating best practices. Omodara et al. (2023) reported that frequent extension interactions improve cassava productivity. Only 40.8% of actors are cooperative members, which limits access to credit and information. Omodara et al. (2023) indicated that cooperative membership is positively linked with adopting climate-smart practices, as it enhances resource access and support (Table 15.1).

15.3.2 Actors' Awareness of Climate Change

The actors identified three primary climate risks: increased rainfall intensity (91.9%), unpredictability of rainfall (91.4%), and late onset of rain (91.7%). The high awareness of increased rainfall intensity highlighted its significant impact on agricultural productivity, mainly due to soil erosion and waterlogging. This finding aligns with Guido et al. (2020), who reported that increased rainfall intensity negatively affected agricultural stability and productivity, showing similar perceptions of this risk among different farming communities.

Unpredictability of rainfall was another major concern, affecting the actors' ability to plan effectively and leading to uncertainties in crop yields. This is consistent with Osei-Acheampong and Opoku (2024), who found that unpredictable rainfall significantly hindered farmers' adaptive capacity, especially in rain-fed agricultural systems. The high awareness among cassava value chain actors in this study reflects similar challenges noted by Osei-Acheampong and Opoku (2024), underlining the broad impact of rainfall unpredictability on farming activities.

Additionally, the late onset of rainfall was identified as a key risk, causing delays in planting and adversely affecting crop growth. This finding mirrors the observations of Tajudeen et al. (2022), who emphasized that delays in rainfall onset can significantly

Table 15.1 Socioeconomic characteristics of cassava value chain actors

Socio-economic characteristics	Frequency	Percentages
Age (years) 20–30	41	11.4
31–40	75	20.8
41–50	139	38.6
51–60	76	21.1
61–70	20	5.6
Sex		
Male	218	60.6
Female	142	39.4
Marital status		
Single	39	10.8
Married	260	72.2
Widow	46	12.8
Divorce	15	4.2
Household size		
2–5	93	25.8
6–10	158	43.9
11–15	85	23.6
16–20	24	6.7
Educational level		
No formal education	59	16.4
Primary education	143	39.7
Secondary education	115	31.9
Experience (Years)		
1–10	104	28.9
11–20	140	38.9
21–30	85	23.6
31–40	31	8.6
Income (Naira)		
10,000–50,000	85	23.6
51,000–100,000	142	39.4
101,000–150,000	82	22.8
151,000–200,000	23	6.4
201,000–250,000	30	8.3
Extension contacts (No. of times)		
1–5	231	64.2
6–10	97	26.9
11–15	23	6.4
16–20	09	2.5

(continued)

Table 15.1 (continued)

Socio-economic characteristics	Frequency	Percentages
Cooperative membership		
Yes	147	40.8
No	213	59.2

Table 15.2 Actors' awareness of climate change

Climate risks	Frequencies	Percentages
Increased intensity of rainfall	331	91.9
Prolonged dry season/drought	284	78.9
Frequent floods	267	74.2
Increased heat/temperature	310	86.1
Severe windstorm	281	78.1
Unpredictability of rainfall	329	91.4
Late onset of rain	330	91.7
Early cessation of rainfall	282	78.3
Weather fluctuation	327	90.8
Extreme heat	291	80.8

reduce crop yield, threatening overall productivity. The recognition of this issue among the actors aligns with similar concerns raised by Tajudeen et al. (2022), reinforcing the importance of timely rainfall for agricultural success.

Overall, the actors' awareness of increased rainfall intensity, unpredictability, and late onset of rain as climate risks indicates a comprehensive understanding of these challenges. These findings are supported by existing literature (Guido et al., 2020; Osei-Acheampong & Opoku, 2024; Tajudeen et al., 2022), which consistently links these climate risks to significant threats to agricultural productivity and resilience. This alignment with previous studies highlights both the local and widespread relevance of these climatic challenges to agriculture (Table 15.2).

15.3.3 Actors' Adaptation Strategies Practiced

The top three adaptation strategies among cassava value chain actors were seeking early warning information about climate risks ($\bar{x} = 3.44$), using inorganic fertilizers ($\bar{x} = 3.40$), and planting improved cassava varieties ($\bar{x} = 3.35$). Seeking early warning information was the most common strategy, underscoring the importance of timely data in mitigating climate impacts. This finding aligns with Osuji et al. (2023), who noted that access to climate information services equips farmers to implement effective strategies, thereby reducing crop failure risks.

The use of inorganic fertilizers was also prevalent, valued for their quick action in boosting crop yields. However, Krasilnikov et al. (2022) and Urra et al. (2019) cautioned against the environmental risks associated with over-reliance on fertilizers, emphasizing the necessity for balanced use alongside sustainable practices to maintain soil health. Planting improved cassava varieties ranked third, reflecting a proactive approach to adapting to climate conditions through high-yield and disease-resistant crops. Grigorieva et al. (2023) similarly highlighted the importance of improved varieties for enhancing resilience to climate challenges, as they offer better resistance to pests, diseases, and adverse conditions.

In summary, these adaptation strategies—seeking early warning information, using inorganic fertilizers, and planting improved varieties—are crucial for building resilience among cassava value chain actors. The findings are consistent with previous research (Grigorieva et al., 2023; Krasilnikov et al., 2022; Osuji et al., 2023; Urra et al., 2019), emphasizing the roles of information access, sustainable fertilizer use, and improved crop varieties in fostering agricultural resilience to climate change (Table 15.3).

Table 15.3 Actors' adaptation strategies practiced

Adaptation strategies	$\sum f(x)$	$\bar{x}$ (n = 360)
Early warning info	311	3.44
Shade trees planting	297	3.29
Improved varieties	304	3.35
Crop diversification	295	3.18
Livelihood diversification	294	3.21
Soil/water conservation	296	3.25
Adjusting calendar	304	3.26
Irrigation	284	3.37
Weather forecasting	280	3.10
Inorganic fertilizer	306	3.03
Organic fertilizer	295	3.40
Pesticides	293	3.26
Storing cassava	270	3.23
Selling cassava	298	2.99
Early harvesting	302	3.32
Buying in small quantities	274	3.35
Buying insurance	291	3.04
Total mean		61.33
Grand mean		3.23

15.3.4 Socioeconomic Factors Influencing Actors' Resilience

The socioeconomic factors influencing resilience among cassava value chain actors included age, marital status, education, experience, income, and extension contacts. Age ($\beta = 0.42, p < 0.01$) positively influenced resilience, indicating that older actors, with more accumulated expertise and knowledge, were better equipped to cope with climate challenges. This finding is consistent with Majnarić et al. (2021), Van Orden et al. (2021), and Eriksen et al. (2021), who similarly noted that older individuals often possess better adaptive capabilities due to their experience and social connections, which help mitigate vulnerabilities.

Marital status ($\beta = 0.29, p < 0.05$) also positively influenced resilience, suggesting that family support enhances adaptive capacity. This observation aligns with Yang et al. (2022), Qiu et al. (2021), and Yulita et al. (2020), who found that household support mechanisms help build resilience among individuals facing socioeconomic challenges.

Education level ($\beta = 0.35$, $p < 0.01$) was another significant factor. Educated actors tend to access more information and technology, improving their adaptive capacity, a finding supported by Gashure (2024), Fadina and Barjolle (2018), and Ojo et al. (2022), who observed that educated farmers adopted more effective climate adaptation strategies.

Experience ($\beta = 0.47$, $p < 0.01$) was positively associated with resilience, as experienced actors had likely developed practical coping strategies through prior encounters with climate challenges. This outcome is supported by Kyire et al. (2023) and Mustapha (2023), who observed that experience contributed significantly to adopting successful risk management strategies in Nigeria.

Income ($\beta = 0.51$, $p < 0.01$) significantly enhanced resilience, indicating that actors with higher income levels could invest in adaptation strategies, a finding consistent with Pan et al. (2023) and Lazurko and Pinter (2022), who similarly reported that financial stability supported resilience investments in industrial parks.

Finally, extension contacts ($\beta = 0.39, p < 0.01$) positively influenced resilience, underscoring the importance of advisory services in disseminating knowledge and best practices. Antwi-Agyei and Stringer (2021) and Kamruzzaman et al. (2021) similarly emphasized the role of extension services in enhancing farmers' adaptive strategies by providing crucial information for climate adaptation (Table 15.4).

The study finds that cassava value chain actors are primarily middle-aged males, mostly with primary education and moderate-income levels. Their marital status and farming experience enhance their stability and adaptive capacity, equipping them to address climate-related challenges. Actors demonstrate a strong awareness of climate risks, including increased rainfall intensity, unpredictability, and late onset of rains, recognizing the threats these pose to agricultural productivity.

To adapt, these actors primarily seek early warning information, use inorganic fertilizers, and plant improved cassava varieties, highlighting the significance of timely data, effective fertilization, and resilient crops. Age, marital status, education, experience, income, and extension contacts significantly influence resilience,

Table 15.4 Socioeconomic factors influencing actors' resilience

Variables	Coefficient (β)	Standard error	t-value	*p*-value
Constant	2.0	0.5	4.0	0.0001
Age	0.02	0.01	2.0	0.045
Sex	– 0.05	0.03	– 1.67	0.095
Marital status	0.10	0.05	2.0	0.050
Household size	0.03	0.02	1.5	0.140
Education level	0.08	0.03	2.67	0.010
Experience	0.04	0.02	2.0	0.045
Income	0.05	0.02	2.5	0.015
Extension contacts	0.06	0.03	2.0	0.045
Cooperative membership	0.07	0.04	1.75	0.085

with older, educated, and financially stable individuals exhibiting greater adaptive capabilities.

The study recommends enhancing education and training for cassava value chain actors to improve their understanding of climate change and sustainable practices. Strengthening access to timely climate information services, promoting cooperative membership for resource access, and facilitating frequent extension services are essential for ongoing support. Additionally, investing in research for climate-resilient cassava varieties and sustainable fertilizer use is crucial. Financial support mechanisms, such as subsidies or microcredit schemes, should be implemented to aid adaptive strategies. Community engagement in decision-making processes for climate adaptation and establishing monitoring frameworks to evaluate the effectiveness of these strategies over time is also advised.

These findings underscore the need for policymakers to prioritise educational initiatives and extension services in agricultural development programmes. Integrating timely climate information access into agricultural policies will enhance farmers' resilience. Supporting cooperative structures can improve resource sharing and access to credit, fostering financial stability within the farming community. Furthermore, research funding should focus on developing climate-resilient agricultural technologies to secure food production amidst climate change.

The study's limitations include a small sample size that may affect generalisability and reliance on self-reported data, which could introduce biases in assessing awareness and adaptation strategies. The cross-sectional design limits the ability to establish causal relationships, and the focus on a specific geographic area may restrict the applicability of results to other regions. Additionally, individual variations in experiences and practices among cassava value chain actors may not be fully captured.

Acknowledgements The authors extend their gratitude to all individuals, organizations, and institutions that contributed to the success of this study.

Funding This research received funding from the MOUAU TETFund on Root Crops Research and Development.

Data Availability Statement The data supporting the findings of this study can be made available upon reasonable request to the corresponding author.

Ethical Clearance Ethical approval for this research was obtained from the relevant ethics review board.

Conflict of Interest The authors declare that there are no conflicts of interest associated with this study.

References

Aboajah, F. N., Onjewu, S. S., Chia, J. I., & Okeme, S. (2018). Socio-economic determinants of cassava production in Benue State, Nigeria. *International Journal of Environment, Agriculture and Biotechnology, 3*(2), 550. https://doi.org/10.22161/ijeab/3.2.32

Antwi-Agyei, P., & Stringer, L. C. (2021). Improving the effectiveness of agricultural extension services in supporting farmers to adapt to climate change: Insights from northeastern Ghana. *Climate Risk Management, 32*, 100304. https://doi.org/10.1016/j.crm.2021.100304

Awotide, B. A., Abdoulaye, T., Alene, A., & Manyong, V. (2019). Socio-economic factors and smallholder cassava farmers' access to credit in South-Western Nigeria. *Tropicultura, 37*(1), 1–10. https://doi.org/10.25518/2295-8010.262

Ayanlade, A., Oluwaranti, A., Ayanlade, O. S., Borderon, M., Sterly, H., Sakdapolrak, P., Jegede, M. O., Weldemariam, L. F., & Ayinde, A. F. O. (2022). Extreme climate events in sub-Saharan Africa: A call for improving agricultural technology transfer to enhance adaptive capacity. *Climate Services, 27*, 100311. https://doi.org/10.1016/j.cliser.2022.100311

Bowen, T. V., Del Ninno, C., Andrews, C., Coll-Black, S., Gentilini, U., Johnson, K., Kawasoe, Y., Kryeziu, A., Maher, T. M., & Williams, A.M. (2020). *Adaptive Social Protection: Building Resilience to Shocks (Vol. 2): Overview (English).* International Development in Focus. World Bank Group. http://documents.worldbank.org/curated/en/503571598004338945/Overview

Darko-Koomson, S., Aidoo, R., & Abdoulaye, T. (2020). Analysis of cassava value chain in Ghana: Implications for upgrading smallholder supply systems. *Journal of Agribusiness in Developing and Emerging Economies, 10*(2), 217–235. https://doi.org/10.1108/JADEE-05-2019-0066

Eriksen, S., Schipper, E. L. F., Scoville-Simonds, M., Vincent, K., Adam, H. N., Brooks, N., Harding, B., Khatri, D., Lenaerts, L., Liverman, D., Mills-Novoa, M., Mosberg, M., Movik, S., Muok, B., Nightingale, A., Ojha, H., Sygna, L., Taylor, M., Vogel, C., & West, J. J. (2021). Adaptation interventions and their effect on vulnerability in developing countries: Help, hindrance or irrelevance? *World Development, 141*, 105383. https://doi.org/10.1016/j.worlddev.2020.105383

Ettah, I. O., Emmanuel, E. A., Ikutal, A., & Ubi, G. M. (2020). Analysis of marketing margins and efficiency of cassava-based product in Cross River Central Agricultural Zone, Nigeria. *Annual Research & Review in Biology, 34*(5), 1–7. https://doi.org/10.9734/arrb/2019/v34i530165

Fadina, A. M. R., & Barjolle, D. (2018). Farmers' adaptation strategies to climate change and their implications in the Zou Department of South Benin. *Environments, 5*(1), 15. https://doi.org/10.3390/environments5010015

Food and Agriculture Organization of the United Nations. (2020). *Cassava in tropical Africa: A reference manual*. FAO.

Gashure, S. (2024). Adaptation strategies of smallholder farmers to climate variability and change in Konso. *Ethiopia. Scientific Reports, 14*, 19203. https://doi.org/10.1038/s41598-024-70047-9

Grigorieva, E., Livenets, A., & Stelmakh, E. (2023). Adaptation of agriculture to climate change: A scoping review. *Climate, 11*(10), 202. https://doi.org/10.3390/cli11100202

Guido, Z., Zimmer, A., Lopus, S., Hannah, C., Gower, D., Waldman, K., Krell, N., Sheffield, J., Caylor, K., & Evans, T. (2020). Farmer forecasts: Impacts of seasonal rainfall expectations on agricultural decision-making in Sub-Saharan Africa. *Climate Risk Management, 30*, 100247. https://doi.org/10.1016/j.crm.2020.100247

Ibrahim, A. T. (2022). Socioeconomic analysis of cassava value chain in Ekiti State, Nigeria. *International Journal of Advanced Economics, 4*(1). https://doi.org/10.51594/ijae.v4i1.291

Kamruzzaman, M., Daniell, K. A., Chowdhury, A., & Crimp, S. (2021). The role of extension and advisory services in strengthening farmers' innovation networks to adapt to climate extremes. *Sustainability, 13*(4), 1941. https://doi.org/10.3390/su13041941

Kehinde, A. L., & Aboaba, K. O. (2020, January). Analysis of value addition in the processing of cassava tubers to "garri" among cottage level processors in southwestern Nigeria. In *Paper presented at the 5th International Conference of the African Association of Agricultural Economists*. Addis Ababa, Ethiopia, September 23–26, 2016

Krasilnikov, P., Taboada, M. A., & Amanullah. (2022). Fertilizer use, soil health and agricultural sustainability. *Agriculture, 12*(4), 462. https://doi.org/10.3390/agriculture12040462

Kyire, S. K. C., Kuwornu, J. K. M., Bannor, R. K., Apiors, E. K., & Martey, E. (2023). Perceived risk and risk management strategies under irrigated rice farming: Evidence from Tono and Vea irrigation schemes—Northern Ghana. *Journal of Agriculture and Food Research, 12*, 100593. https://doi.org/10.1016/j.jafr.2023.100593

Lazurko, A., & Pinter, L. (2022). Infrastructure investments for resilience: Opportunities, barriers, and a future research agenda from the Orange-Senqu River Basin. *Climate Risk Management, 35*, 100393. https://doi.org/10.1016/j.crm.2021.100393

Majnarić, L. T., Bosnić, Z., Guljaš, S., Vučić, D., Kurevija, T., Volarić, M., Martinović, I., & Wittlinger, T. (2021). Low psychological resilience in older individuals: An association with increased inflammation, oxidative stress, and the presence of chronic medical conditions. *International Journal of Molecular Sciences, 22*(16), 8970. https://doi.org/10.3390/ijms22168970

Mustapha, B. (2023). Risk management practice and organizational performance: The mediating role of business model innovation. *Journal of Law and Sustainable Development, 11*(4), e892. https://doi.org/10.55908/sdgs.v11i4.892

Ojo, T. O., Ogundeji, A. A., & Emenike, C. U. (2022). Does adoption of climate change adaptation strategy improve food security? A case of rice farmers in Ogun State Nigeria. *Land, 11*(11), 1875. https://doi.org/10.3390/land11111875

Olaomo, O. K. C. D., & Molnar, J. J. (2022). Building an inclusive value chain: Gender participation in cassava marketing and processing in Nigeria. *Environmental Science & Ecology: Current Research, 3*(1078). https://doi.org/10.54026/ESECR/1078

Olaosebikan, O., Bello, A., Utoblo, O., Okoye, B., Olutegbe, N., Garner, E., Teeken, B., Bryan, E., Forsythe, L., Cole, S., Kulakow, P., Egesi, C., Tufan, H., & Madu, T. (2023). Stressors and resilience within the cassava value chain in Nigeria: Preferred cassava variety traits and response strategies of men and women to inform breeding. *Sustainability, 15*(10), 7837. https://doi.org/10.3390/su15107837

Omodara, O. D., Ige, O. A., Oluwasola, O., Oyebanji, A. T., & Afape, O. O. (2023). Factors influencing cassava farmers' choice of climate change adaptation practices and its effect on cassava productivity in Nigeria. *Heliyon, e14563*. https://doi.org/10.1016/j.heliyon.2023.e14563

Opondo, F., Mshenga, P., & Louw, A. (2022). Analysis of marketing margins for cassava farmers and traders in Siaya County, Kenya. *Journal of Social Sciences, Education and Humanities, 1*(2), 16–32. Retrieved from http://journals.laikipia.ac.ke/index.php/jsseh/article/view/1/2

Osei-Acheampong, B., & Opoku, M. S. (2024). Impact of the rural enterprise programme interventions on smallholder farmers' climate adaptation strategies: Qualitative evidence from Ghana's transitional agro-ecological zone. *Community Development*, 1–26. https://doi.org/10.1080/15575330.2024.2403009

Osuji, E. E., Igberi, C. O., & Ehirim, N. C. (2023). Climate Change Impacts and Adaptation Strategies of Cassava Farmers in Ebonyi State Nigeria. *Journal of Agricultural Extension, 27*(1), 35–48. https://doi.org/10.4314/jae.v27i1.4

Pan, S.-C., Hu, T.-S., You, J.-X., & Chang, S.-L. (2023). Characteristics and influencing factors of economic resilience in industrial parks. *Heliyon, 9*(4), e14812. https://doi.org/10.1016/j.heliyon.2023.e14812

Qiu, Y., Huang, Y., Wang, Y., Ren, L., Jiang, H., Zhang, L., & Dong, C. (2021). The role of socioeconomic status, family resilience, and social support in predicting psychological resilience among Chinese maintenance hemodialysis patients. *Frontiers in Psychiatry, 12*, 723344. https://doi.org/10.3389/fpsyt.2021.723344

Singh, P. (2024). *Role of climate smart agriculture (CSA) in addressing challenges of food security and climate change* (Technical Report). https://doi.org/10.13140/RG.2.2.13965.35040/1

Skendžić, S., Zovko, M., Živković, I. P., Lešić, V., & Lemić, D. (2021). The impact of climate change on agricultural insect pests. *Insects, 12*(5), 440. https://doi.org/10.3390/insects12050440

Stringer, L. C., Fraser, E. D. G., Harris, D., Lyon, C., Pereira, L., Ward, C. F. M., & Simelton, E. (2020). Adaptation and development pathways for different types of farmers. *Environmental Science & Policy, 104*, 174–189. https://doi.org/10.1016/j.envsci.2019.10.007

Tajudeen, T. T., Omotayo, A., Ogundele, F. O., & Rathbun, L. C. (2022). The effect of climate change on food crop production in Lagos State. *Foods, 11*(24), 3987. https://doi.org/10.3390/foods11243987

Ukwuaba, I. C., & Ileka, C. M. (2024). Determinants of the extent of use of climate-smart agricultural practices by farm households in Southeast Nigeria. *European Modern Studies Journal, 8*(2), 43–57. https://doi.org/10.59573/emsj.8(2).2024.5

Unaeze, H. C., & Ihunwo, O. N. (2021). Determinants of climate change adaptation strategies used by compound farmers in Emohua Local Government Area of Rivers State. *Greener Journal of Agricultural Sciences, 11*(1), 26–31. https://gjournals.org/GJAS

Urra, J., Alkorta, I., & Garbisu, C. (2019). Potential benefits and risks for soil health derived from the use of organic amendments in agriculture. *Agronomy, 9*(9), 542. https://doi.org/10.3390/agronomy9090542

Van Orden, K. A., Bower, E., Lutz, J., Silva, C., Gallegos, A. M., Podgorski, C. A., Santos, E. J., & Conwell, Y. (2021). Strategies to promote social connections among older adults during "social distancing" restrictions. *American Journal of Geriatric Psychiatry, 29*(8), 816–827. https://doi.org/10.1016/j.jagp.2020.05.004

Vernooy, R. (2022). Does crop diversification lead to climate-related resilience? Improving the theory through insights on practice. *Agroecology and Sustainable Food Systems, 46*(6), 877–901. https://doi.org/10.1080/21683565.2022.2076184

Yang, C., Gao, H., Li, Y., Wang, E., Wang, N., & Wang, Q. (2022). Analyzing the role of family support, coping strategies and social support in improving the mental health of students: Evidence from post COVID-19. *Frontiers in Psychology, 13*, 1064898. https://doi.org/10.3389/fpsyg.2022.1064898

Yu, T., Mahe, L., Li, Y., Wei, X., Deng, X., & Zhang, D. (2022). Benefits of crop rotation on climate resilience and its prospects in China. *Agronomy, 12*(2), 436. https://doi.org/10.3390/agronomy12020436

Yulita, Muhamad, S., Kulub Abdul Rashid, N., Hussain, N. E., Mohamad Akhir, N. H., & Ahmat, N. (2020). Resilience as a moderator of government and family support in explaining entrepreneurial interest and readiness among single mothers. *Journal of Business Venturing Insights, 13,* e00157. https://doi.org/10.1016/j.jbvi.2020.e00157

Chapter 16
Climate Justice, Renewable Energy, and Food Security in Sub-Saharan Africa: A Bibliometric Analysis of Emerging Research Trends

Joy Nneamaka Obi and Emmanuel Ojo

Abstract This study provides a comprehensive bibliometric analysis of research integrating climate justice and renewable energy in food security literature for sub-Saharan Africa from 2009 to 2024. Analysing 39 publications, we examine the volume, growth trajectory, dominant themes, and collaboration networks within the field. Results indicate a significant focus on adaptive agriculture practices, equitable energy access, and climate-resilient food systems, highlighting the critical need for policies and approaches prioritising equity, inclusion, and resilience. The United States, the United Kingdom, and Brazil have emerged as key knowledge centres, with collaboration patterns revealing global and regional research engagement. Four overarching thematic domains are identified: (1) Food Security and Climate Change Adaptation; (2) Renewable Energy and Sustainable Food Systems; (3) Climate Justice and Equitable Food Policies; and (4) Energy Justice and Equitable Access. This study offers valuable insights for policymakers and researchers, emphasising the need for contextualised climate education and gender-sensitive policies. Our findings underscore the importance of mixed-methods research designs and region-specific adaptation strategies in addressing the complex challenges posed by climate change to food security in sub-Saharan Africa.

Keywords Climate justice · Renewable energy · Food security · Bibliometric analysis

---

J. N. Obi (✉) · E. Ojo
University of Witwatersrand Johannesburg, Johannesburg, South Africa
e-mail: joy.Obi@wits.ac.za

E. Ojo
e-mail: emmanuel.Ojo@wits.ac.za

R. U. Onyeneke et al. (eds.), *Energy Transition, Climate Action and Sustainable Agriculture*, https://doi.org/10.1007/978-3-031-83165-2_16

16.1 Introduction

Sub-Saharan Africa's food security, a cornerstone of its socio-economic stability, faces severe threats from climate change. The region is particularly vulnerable to climate hazards, experiencing reduced crop yields, disrupted rain patterns, and more frequent extreme weather events (Obi et al., 2021, 2023) This vulnerability is further exacerbated by the region's reliance on rain-fed agriculture and limited adaptive capacity (Munonye et al., 2023). Recent studies have highlighted the interconnected challenges of food security, climate justice, and energy transition as critical concerns for sustainable development in developing nations (Neupane et al., 2022). This paper presents a rigorous bibliometric analysis exploring trends and networks in research integrating climate justice and renewable energy concepts within food security literature in sub-Saharan Africa from 2009 to 2024. Bibliometric analysis, a systematic approach to examining extensive literature, allows for the deconstruction of field developments, identification of collaboration patterns, and highlighting of emerging trends (Donthu et al., 2021).

The overarching question guiding this research is: 'What is the growing body of research on integrating climate justice and renewable energy in food security in Sub-Saharan Africa in the academic discourse between 2009 and 2024, and what knowledge may be gained to reimagine research and adaptation strategies to address climate change impacts in the sub-Saharan African region?'.

This study addresses four specific research questions: (I) What are the dominant themes and trends in research linking food security, climate justice, and energy transition from 2009 to 2024? (II) How do publication patterns, citation impacts, and collaborative networks reflect the integration of food security, climate justice, and energy transition research? (III) How have environmental and energy justice concepts been integrated into food security and climate change research? (IV) How are contemporary research developments impacting policies concerning climate justice, energy access, and food security, and what critical research gap exists that must be filled to enhance their interconnectedness? This bibliometric analysis aims to inform policy, practice, and future research by examining trends, patterns, and gaps. It provides a comprehensive picture of the current research on integrating climate justice and renewable energy in food security, which is valuable for prioritising adaptation and mitigation strategies. The study contributes to scholarly discourse on resilient and equitable environments by offering a critical understanding of sustainability and potentially redefining approaches to addressing climate change impacts on food security in sub-Saharan Africa.

The following research objectives will be achieved in the study:

i. analyse the volume, growth, and patterns of literature on integrating climate justice and renewable energy in food security.
ii. identify the top contributing authors, institutions, and countries in the field of integrating climate justice and renewable energy in food security research.
iii. examine the nature and dynamics of collaboration among leading countries and institutions.

iv. identify influential journals that support all research questions by publishing significant work on integrating climate justice and renewable energy in food security.
v. explore the dominant themes and their interconnections in the literature on integrating climate justice and renewable energy in food security.

Climate Justice and Renewable Energy in Food Security: An Environmental and Energy Justice Perspective.

This research directly addresses the conference theme "Climate Justice and Just Energy Transition in Developing Countries" by examining the integration of climate justice and renewable energy concepts within food security research. Our bibliometric analysis of emerging trends aligns with multiple Sustainable Development Goals, including Goal 2 (Zero Hunger), Goal 7 (Affordable and Clean Energy), and Goal 13 (Climate Action). By exploring how these interconnected challenges are being addressed in academic literature, we provide a comprehensive overview of current research directions and identify critical gaps. This transdisciplinary approach is essential for developing countries, particularly in sub-Saharan Africa, where food security, climate justice, and energy transition are inextricably linked. Our study offers valuable insights for researchers, policymakers, and practitioners working towards sustainable and equitable development in the region.

This section untangles the complex web that climate change has woven around food security, particularly within the sub-Saharan African context. Drawing upon theoretical frameworks such as Fricker's epistemic injustice, Fraser's social justice, Energy justice and Sustainability frameworks, we examine major climate change crises from 2009 to 2024 and their impact on food security in sub-Saharan Africa (Abbas et al., 2024; Musara et al., 2021). Climate justice emphasises the need for equitable solutions that consider the disproportionate impacts of climate change on marginalised communities, which often experience environmental injustices. Research indicates that these vulnerable populations are significantly affected by climate variability, threatening their livelihoods and food security (Obi et al., 2021). Renewable energy emerges as a crucial factor in enhancing food security by providing sustainable power for agricultural practices and improving resilience against climate impacts (Neupane et al., 2022).

The energy justice framework underscores the importance of equitable access to renewable energy resources, advocating for the inclusion of marginalised communities in energy decision-making processes (Dorji et al., 2023; Taliep, 2022). Aligning these themes with the Sustainable Development Goals highlights the importance of achieving food security (Goal 2), ensuring access to clean energy (Goal 7), and taking Climate Action (Goal 13). The incorporation of climate justice and renewable energy considerations into food security research serves to tackle systemic disparities while simultaneously fostering sustainable practices that bolster resilience against the impacts of climate change. By examining these issues through the lenses of environmental injustice, sustainability, and energy justice, researchers can inform policies and practices that prioritise the rights and needs of marginalised communities, ultimately contributing to sustainable development and resilience in sub-Saharan Africa (Agbehadji et al., 2023; Cairney et al., 2023; Dorji et al., 2023).

16.2 Methodology

16.2.1 Dataset Collection

The data for this paper was collected through Scopus, a comprehensive and authoritative database for peer-reviewed papers (Hallinger & Nguyen, 2020). The search strategy was informed by the Springer call for papers on "Climate justice and just energy transition conference". The call spotlights the critical role of just Energy transition in advancing equitable climate action focusing on upholding justice in energy reforms, particularly for developing nations. The search string used was:

("food security" OR "climate justice" OR "just energy transition") AND ("environmental justice" OR "energy justice") AND ("energy transition" OR "renewable energy") AND ("climate change" OR "climate adaptation").

16.2.2 Dataset Pre-processing and Refinement

The one-string search strategy used was conducted as of 6 August 2024. To ensure relevance and manageability, a systematic pre-processing protocol was employed, resulting in 40 documents pertinent to the research scope.

Filtration criteria included: temporal scope (2009–2024), document types (peer-reviewed articles, book chapters, reviews, and editorial books), and keywords (9 terms reflecting multidimensional aspects of food security, climate justice, renewable energy transition and climate change). The Scopus was confined to English language publications.

Our bibliometric analysis sought to deliver a detailed overview of research output in the specified domain. In crafting our search strategy, we incorporated all relevant geographical areas, including sub-Saharan Africa. However, the databases we consulted (Scopus database) indicated a significant lack of published works from this region, leading to an underrepresentation of articles from sub-Saharan Africa in our results. This limitation was not a result of intentional exclusion but rather a reflection of the scarcity of pertinent publications in the field (Fig. 16.1).

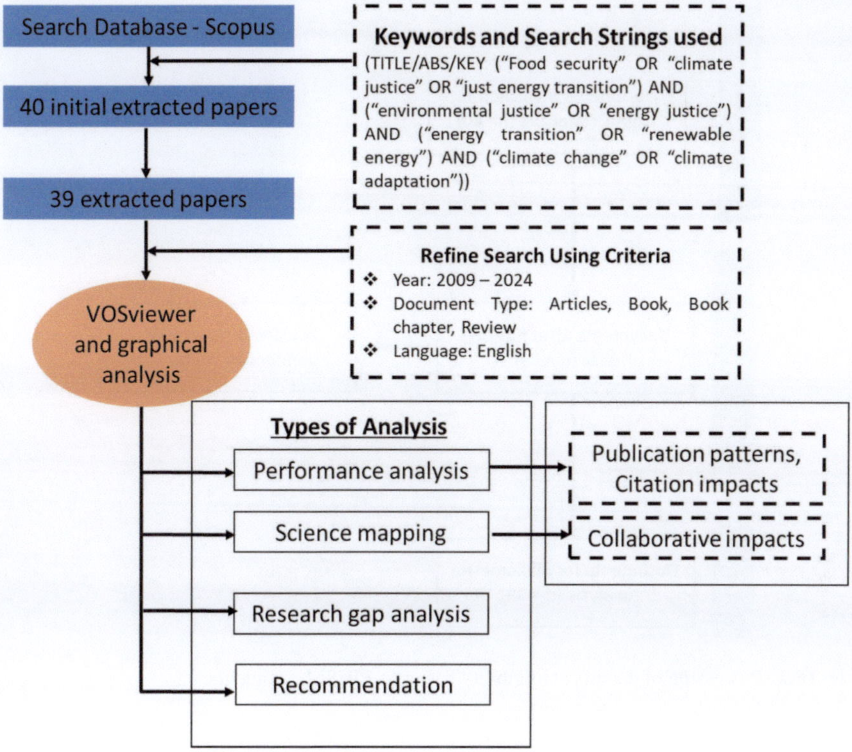

Fig. 16.1 An outline of the research methodology and summary of the research framework

16.2.3 Data Analysis Strategies

The article's analysis strategy comprised two main components:

I. Performance Analysis: Using Scopus, we conducted an initial descriptive analysis of the 39 documents. This included examining publication volumes, citation frequencies, and identifying highly cited articles, authors, journals, and influential institutions and countries.
II. Science Mapping: We employed VOSviewer software for advanced bibliometric mapping and network analysis. VOSviewer's functionality is particularly suited for creating easily interpretable bibliometric maps, with a focus on clear graphical representation (van Eck & Waltman, 2010). This analysis included:

- Creation of visual network illustrations showing connections among authors, institutions, countries, and frequently used keywords
- Citation analysis and bibliographic coupling to explore the interplay and structural linkages of research components
- Cluster analysis to identify key themes within the research

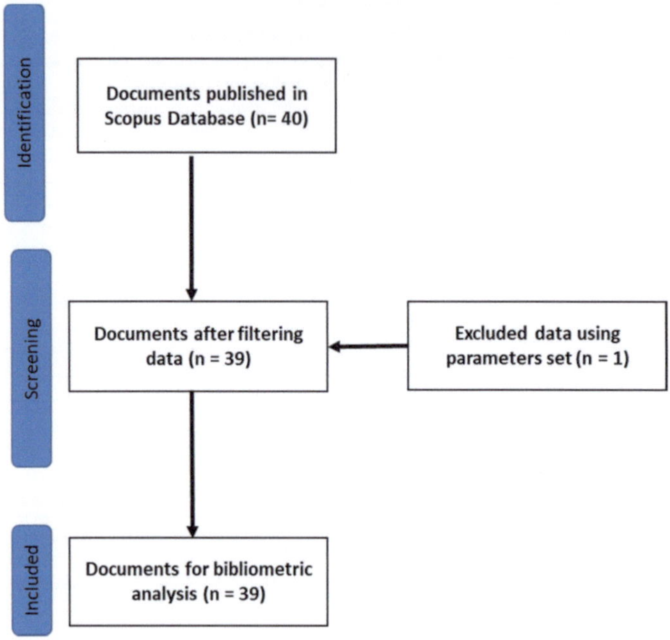

Fig. 16.2 Processing of document eligibility based on PRISMA guideline

The results of our data collection and refinement process are visually represented in Fig. 16.2, which follows the PRISMA guidelines for systematic reviews.

Through this methodological approach, the article aimed to provide a comprehensive overview of the climate change and agriculture research landscape in sub-Saharan Africa, identifying trends, patterns, and gaps to inform future research, policy, and practice (Table 16.1).

16.3 Results and Discussion

This section presents the results of our bibliometric analysis and contextualised the findings within the evolving academic dialogue on integrating climate justice and renewable energy in food security research from 2009 to 2024. It explores gleaned insights, with a focus on network trends, towards re-envisioning research and strategic policies for climate justice and transitioning to renewable energy to impact food security within the sub-Saharan African context.

Table 16.1 Bridging analysis and insight in integrating climate justice and renewable energy in food security research

Section	Title	Research questions	Analysis type
3.1	Research Landscape and Publication Trends (2009–2024)	RQI & RQII: Volume, growth, and patterns of literature on integrating climate justice and renewable energy in food security; Types of publications	Performance analysis
3.2	Key Contributors: Authors, Institutions, and Countries	RQIII: Top contributing authors, institutions, and countries in integrating climate justice and renewable energy in food security research	Performance Analysis
3.3	Collaborative Networks: Partnerships among Leading Countries and Institutions	RQIII: Nature and dynamics of collaborations among leading countries and institutions	Science Mapping
3.4	Influential Journals Shaping the Discourse	(Supports all research questions by identifying influential journals)	Performance Analysis
3.5	Thematic Landscape: Dominant Themes and Interconnections	RQIV: Dominant themes and their interconnections in Integrating Climate Justice and Renewable energy in food security	Science Mapping

16.3.1 Research Landscape and Publication Trends (2009–2024)

The analysis of publication volume and growth trajectory between 2009 and 2024 offers valuable insights into the evolution of climate justice and renewable energy in food security research in sub-Saharan Africa. Figure 16.3 illustrates this growth pattern, revealing distinct phases of development. The study underscores the growing significance of food security, especially considering climate change's effects on agricultural practices and food supply systems. Climate justice tackles the moral and social ramifications of climate change, highlighting the necessity of just solutions that consider the most vulnerable groups impacted by environmental disruptions. Research investigates the link between energy policy and their consequences on agricultural systems and climate resilience. The transition to sustainable energy sources is a significant concern. The trends over time from 2009 to 2014 reveals that early research mostly concentrated on food security alone and interconnectedness between water-food-energy with little integration of the perspectives of energy transition and climate justice. Between 2015 and 2019, there was a discernible surge in multidisciplinary research, with academics starting to link food security with climate justice and energy polices. From 2020 to 20,224, the trends point to a surge in in-depth research that tackles the themes' (food security, climate justice and energy transition) interconnection from an inclusive perspective, demonstrating an increasing recognition of their interdependencies.

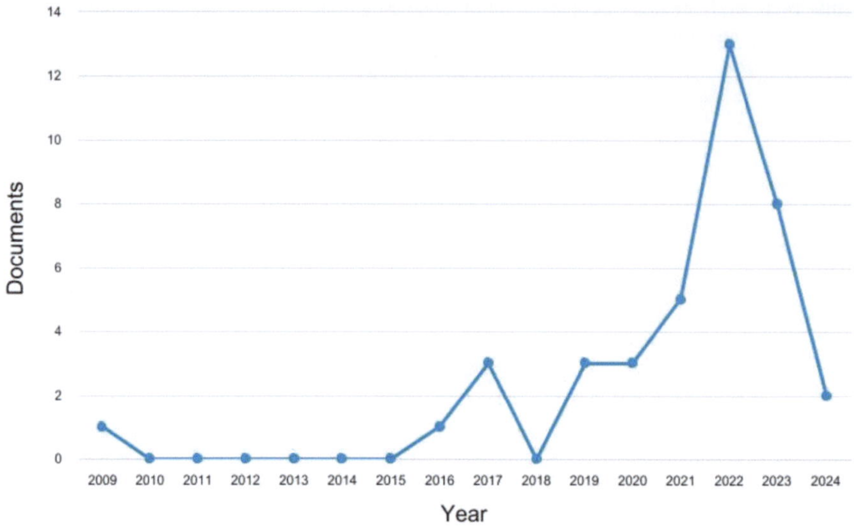

Fig. 16.3 Growth trajectory of publication volume from 2009 to 2024

Regarding publication types, journal articles consistently dominated the landscape throughout the study period. This trend is clearly depicted in Fig. 16.4, which illustrates the distribution of the publication types from 2009 to 2024. It accounted for 76.9% of all publications. This indicates that the primary channel for disseminating research findings in this subject is through peer-reviewed journal publications, underscoring the significance of thorough academic scrutiny and validation in the discourse. Reviews accounted for 7.7% of all publications. This percentage shows that some level of significant effort has been made to synthesize literature, point out gaps and explore novel paths of research. Review articles are essential for consolidating knowledge, evaluate the strength of the field and direction of future research projects. Book Chapter and books accounted for a comparatively small percentage 7.7% each. This reveals they are not dominant forms of publication, but their existence indicates that researchers are incorporating their findings into more general academic discourses, educational contexts, or policy related publications.

16.3.2 Global Funders of Climate Justice and Renewable Energy in Food Security Research

This section untangles the global funders of climate justice, renewable energy, and food security research. as depicted in Fig. 16.5. The largest funder of research related to integrating climate justice, and renewable energy justice in food security are Horizon 2020 Programme and the European Union. This suggests that they

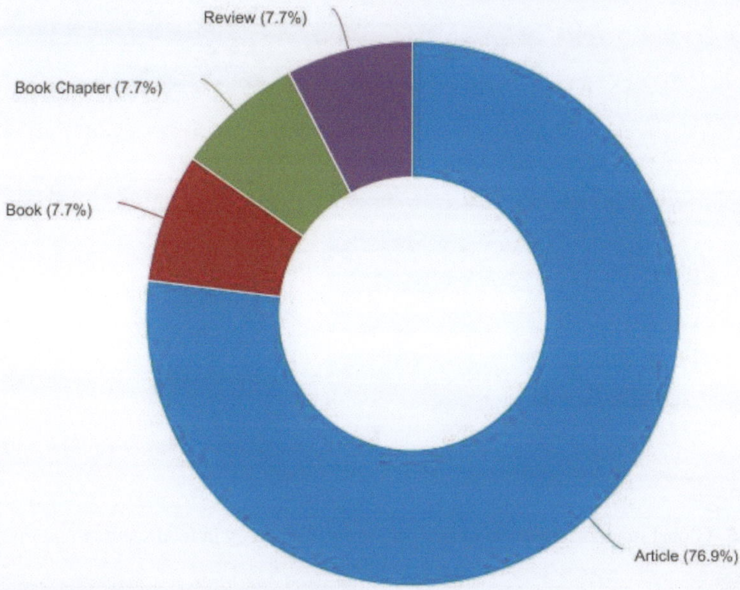

Fig. 16.4 Trend in publication types and growth

have been a major proponent of supporting research effects aimed at addressing the nexus between climate justice, renewable energy, and food security; Their providing substantial funding has enabled researchers to investigate these critical issues and develop solutions that promote equitable access to food security. The leading funder's investment in this field underscores the importance of understanding the complex relationship between climate change, renewable energy, and food security, particularly from a justice-oriented perspective which has likely facilitated the advancement of knowledge and the development of innovative approaches to combat the challenges of climate injustice, use of non-renewable energy sources and food insecurity on a global scale.

16.3.3 Key Players of Top Contributing Authors, Institutions, and Countries in Integrating Climate Justice and Renewable Energy in Food Security

This section incorporates a table and the overlay visualization to present the major contributors, affiliations, and geographical trends in the literature, comparing sub-Saharan Africa with global contributions. By Integrating VOSviewer for bibliometric analysis, the minimal number of documents and citation threshold for authors to be included in the top list was established as 1 and 40, respectively. Only 12 of the

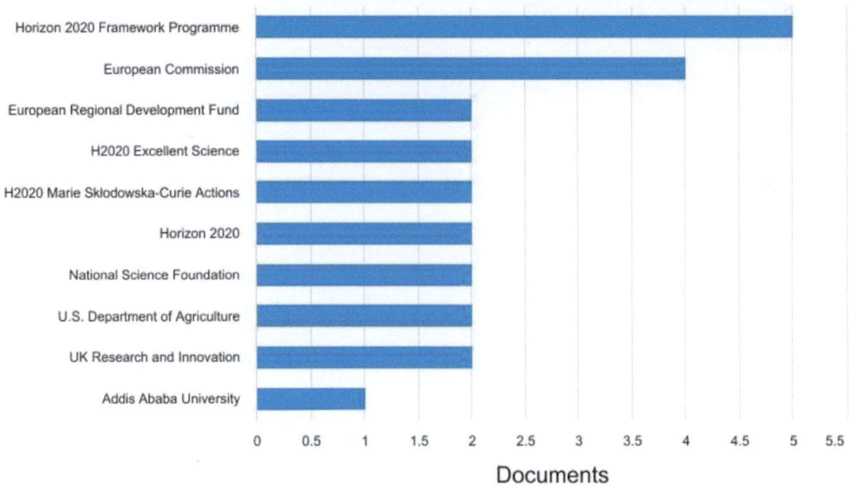

Fig. 16.5 Global funders of climate justice and renewable energy in food security research

101 authors fulfilled these criteria. See Table 16.2 and Fig. 16.6, for the overlay visualisation.

16.3.4 Collaboration Dynamics: Nature and Evolution of Partnership Among Leading Countries and Institutions (2009–2024)

This section presents the collaboration patterns over the period 2009–2024 amongst the top ten countries across the world to elucidate the landscape of global scholarly collaboration. To do this, a rigorous selection methodology was implemented. The criteria for inclusion in this article were determined by two primary quantitative thresholds: a minimum of 3 documents and at least 24 citations per country. These benchmarks were meticulously selected to filter for substantial scholarly output and impact, ensuring that only the most prolific and influential research landscapes were considered. Out of a pool of 30 countries, 11 countries met these stringent parameters, thus forming the cohort for further analysis. The United States with 16 documents and 865 citations, not only surpassed the set thresholds but also demonstrated significant international research engagement, as evidenced by its robust total link strength of 2515. This position underscores the United States' pivotal role in the global knowledge exchange network. See Table 16.3 and Figs. 16.7, 16.8.

Table 16.2 Top authors

Full name	Affiliation	Number of publications	Total citations	h-Index
Barry, John	Queen's University Belfast, Northern Ireland, United Kingdom	1	454	44
Healy, Noel	Salem State University, United States	1	454	13
Brown, Benjamin	University of Edinburg, United Kingdom	1	82	21
Spiegel, Samuel J	University of Cambridge, United Kingdom	1	82	3
Heffron, Raphael j	University of West Indies, St. Augustine, Trinidad and Tobago	1	58	33
Sokolowski, Maciej M	University of Warsaw, Poland, Meiji University, Tokyo, Japan	1	58	13
Baker, Shalandah	University of Hawaii, United States	1	46	6
Campney, Alicia	York University, Canada	1	44	1
Hoicka, Christina E	York University, Canada	1	44	16
Savic, Katarina	York University Canada	1	44	2
Fairchild, Dennis	Emerald Cities Collaboration, United States	1	44	3
Weinrub, Al	Local Clean Energy Alliance (LCEA), United States	1	44	2

16.3.5 Influential Voices: A Bibliometric Analysis of Top Journals Shaping the Discourse on Integrating Climate Justice and Renewable Energy in Food Security Research

The bibliometric analysis of top academic sources by documents and citations provided valuable insights into the influential journals shaping the discourse around re-imagining research on the integration of climate justice and renewable energy in food security.

The criteria for selecting the top journals were based on the total link strength, which measures the citation connections between sources. Table 16.4 shows that 8 sources were chosen from a dataset of 35 academic sources, focusing on those with the greatest total strength. Figure 16.9 further refines the selection by setting minimum thresholds of 2 documents and 20 citations for a source to be included. The resulting list, presented in Fig. 16.9, showcases a diverse range of journals across disciplines such as energy policy, proceedings of the National Academy of Sciences

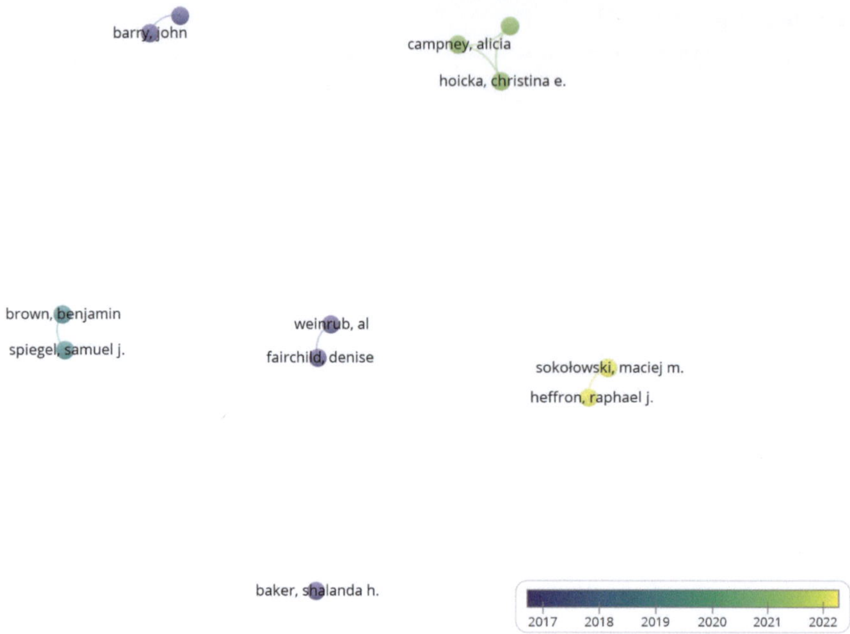

Fig. 16.6 Overlay visualisation of the top 12 authors

Table 16.3 Collaboration dynamics: nature and evolution of partnership among leading countries and institutions (2009–2024)

Country	Citations	Documents	Total link strength
United States	865	16	2515
United Kingdom	774	6	2626
Brazil	260	4	2514
Japan	258	3	2412
Canada	246	5	2050
Germany	243	4	2304
Netherlands	243	5	2216
Spain	241	3	2267
Australia	235	4	2342
South Africa	214	3	2300
Austria	214	3	2200

of the United States, energy research and social sciences, natural resources journal, global transitions, and ecological economics.

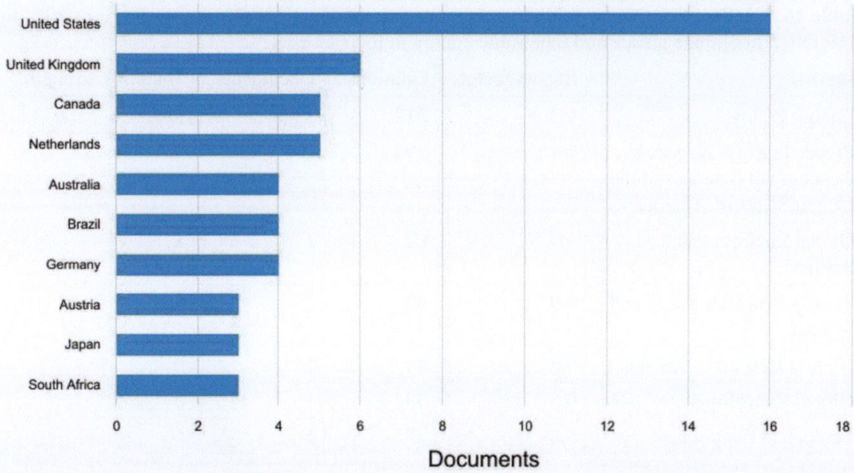

Fig. 16.7 Histogram of global research on climate justice and renewable energy in food security research

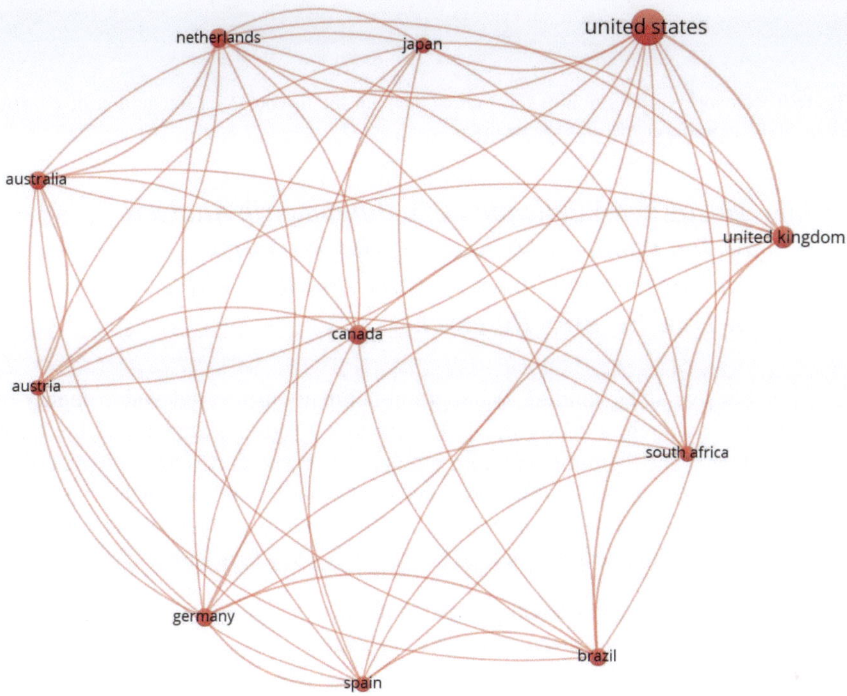

Fig. 16.8 Network visualisation of global research collaboration

Table 16.4 Influential voices: a bibliometric analysis of top journal shaping the discourse on integration of climate justice and renewable energy in food security

Journal	Impact factor	Citations	Documents	Total link strength
Energy Policy	9.3	512	2	5
Proceedings Of National Academy of Sciences of the United States of America	9.4	193	1	1
Global Environmental Politics	3.9	82	1	8
Energy Research and Social Science	6.9	69	2	4
Natural Resources Journal	0.8	46	1	0
Global Transitions	0.6	33	1	0
Ecological Economics	6.6	27	1	6

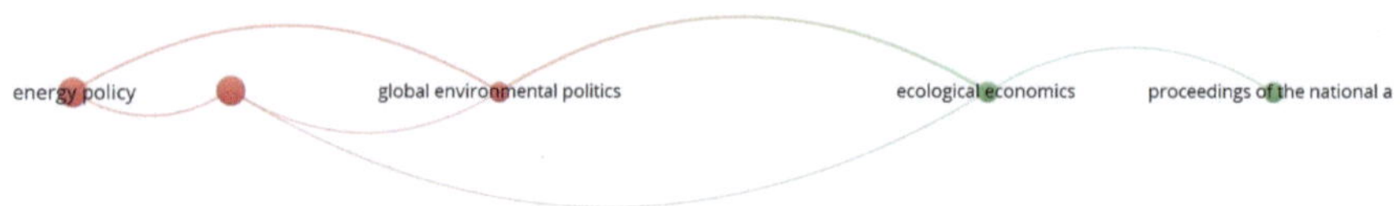

Fig. 16.9 Network visualisation of top journal shaping the discourse on integration of climate justice and renewable energy in food security

16.3.6 Most Cited Publications: A Bibliometric Analysis of Top Publications Shaping the Discourse on Integrating Climate Justice and Renewable Energy in Food Security Research

A comprehensive understanding of integrating climate justice, renewable energy in food security research is very crucial. To this end, a rigorous analysis of the most cited documents and their corresponding focus area was undertaken. From a pool of 39 extracted documents, 10 publications garnered over 25 citations each, as detailed in Table 16.5 a meticulous frequency count and thematic analysis of the keywords within these influential articles yielded the following key focus area:

- Food Security and Climate Change Adaptation
- Renewable energy and sustainable Food Systems
- Climate Justice and Equitable Food Policies, and
- Energy Justice and Equitable Access

A notable trend in the reviewed literature is the predominance of qualitative research methodologies. This underscores the fact that the topic is still in its initial phase of research, hence qualitative methodology serves as an exploratory way to

Table 16.5 Top cited publications (>20)

Title	Cited	Type of study	Method	Data source	Keywords
Mexican energy reform, climate change, and energy justice in indigenous communities	46	Qualitative	Review	Literature review	Climate change, Indigenous community, Environmental justice
Global governance for climate justice: A cross-national analysis of CO2 emissions	33	Quantitative	Observation	Secondary data	Carbon dioxide. Climate justice, Global governance
Defining and conceptualising energy policy failure: The when, where, why, and how	58	Qualitative	Review	Literature review	Climate change, Climate-Energy Policy, Energy justice, Energy policy failure, Just transition, Sustainable development
Ten facts about land systems for sustainability	193	Qualitative	Review	Literature review	Governance, Land use, Social-Ecological systems, Social change, Sustainability, renewable Energy, Environmental justice
Coal, Climate justice, and the cultural politics of energy transition	82	Qualitative	Review	Literature review	Climate change, Climate classification, Coal mine, Coal Fired power plant, Environmental justice, Fossil fuel
Beyond promises: Realities of Climate finance justice and energy transitions in Asia and the Pacific	25	Quantitative	Key Informants Participants	Interview	Climate finance, Climate justice, Energy justice, Energy transition

(continued)

Table 16.5 (continued)

Title	Cited	Type of study	Method	Data source	Keywords
Reconciliation through renewable energy? A survey of Indigenous communities, involvement, and peoples in Canada	44	Quantitative	Survey	Participatory Appraisal and Interview	Community energy, Energy justice, Indigenous people, Reconciliation, Renewable energy
How social movements contribute to staying within the global carbon budget: Evidence from a qualitative meta-analysis of case studies	27	Qualitative	Case Study	Exploratory approach	Carbon budget, Carbon emissions, Climate change mitigation, Environmental justice, Meta-analysis, Social movements
Politicizing energy justice and energy system transitions: Fossil fuel divestment and a "just transition"	454	Qualitative	Review	Literature review	Climate change. Divestment, Energy justice, Energy politics, Fossil fuels, just transition
Energy democracy: Advancing equity in clean energy solutions	44	Qualitative	Review	Literature	Centralized renewable energy, clean energy, climate activism, climate justice

understand the field. However, overreliance on the method, often manifested in the form literature review as observed within the identified focus areas, creates a potential for narrow and incomplete understanding of the complex phenomenon of climate change. While qualitative research undoubtedly offers insights through its ability to identify patterns, trends, and correlations, it is inherently limited in its capacity to delve into the depth and richness of human experiences and perceptions to the nexus climate justice, renewable energy and food security. To fully comprehend the multifaceted nature of this global challenge, a more nuanced approach is imperative.

A mixed-methods research design, incorporating both quantitative and qualitative methodologies, emerges as the optimal framework for investigating climate change. By synergistically combining the strengths of these complementary approaches, researchers can achieve a more holistic and comprehensive understanding of the issue, thereby informing more effective and equitable policy development and intervention strategies.

16.3.7 Thematic Evolution: Dominant Themes and Interconnections in Climate Justice and Renewable Energy in Food Security Research

The preceding sections provided a comprehensive review of the research landscape, the key authors in the field, and the dynamics of collaboration in climate justice and renewable energy in food security research covering the years 2009 to 2024. Building on this groundwork, this section of this bibliometric review, infirmed by VOSviewer data, critically examines scholarly discourses across two decades, unearthing the predominant themes and their interconnections, through analysing keywords, their occurrences, and total link strengths.

The VOSviewer software was used to analyse 40 publications generated by Scopus, revealing 414 keywords. To identify the top 15 keywords, a threshold was set in VOSviewer requiring each keyword to have a minimum occurrence of 5. Based on this criterion, VOSviewer calculated the total strength of the co-occurrence links between these top 22 keywords and other keywords. This in-depth analysis, performed across the entire volume of publications, resulted in the data presented in Table 16.6 and illustrated by Figs. 16.10, 16.11 and 16.12 to present a network graph of keyword co-occurrence to visualise the interconnections of the keywords based on the strength of the co-occurrence.

The keyword table below provides a solid starting point for thematic analysis in this section. This helps to understand which themes are most important by looking at how often they appear and how strongly they are connected to other keywords. The keyword, "climate change," stands out above the rest. This suggests that people are

Table 16.6 Quantitative analysis of keyword frequency and interrelationship in climate justice and renewable energy in food security research (2009–2004)

Keyword	Occurrences	Total link strength
Climate change	24	94
Energy justice	19	70
Environmental justice	17	61
Energy transitions	8	52
Energy transition	11	46
Climate justice	10	42
Energy justices	6	42
Energy	6	40
Energy Policy	8	40
Renewable energy	9	40
Just transition	9	37
Sustainable development	6	36
Carbon	5	33
Social justice	5	30
Sustainability	5	23

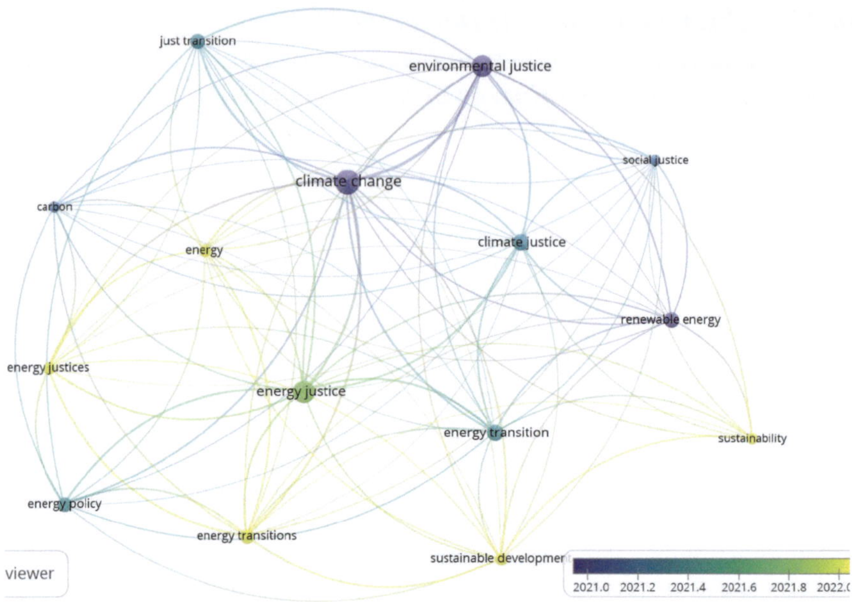

Fig. 16.10 Density visualisation of selected keywords

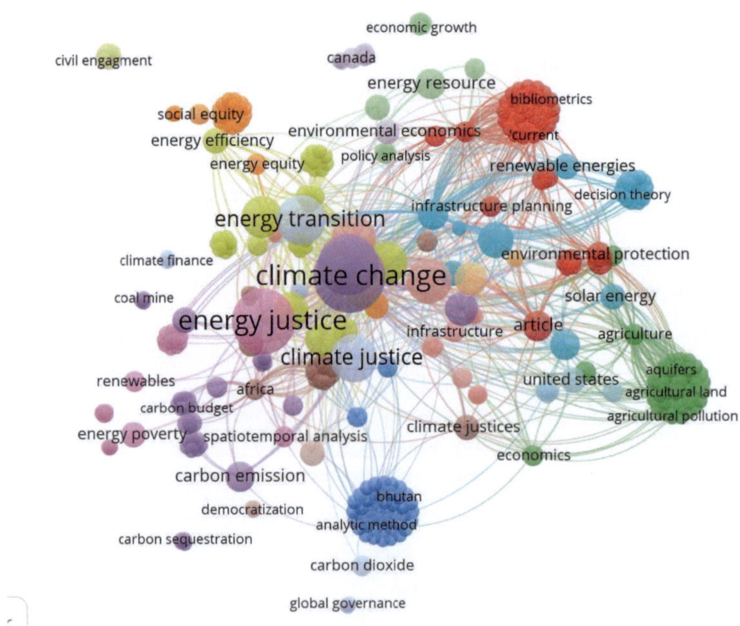

Fig. 16.11 Network visualisation of all keywords

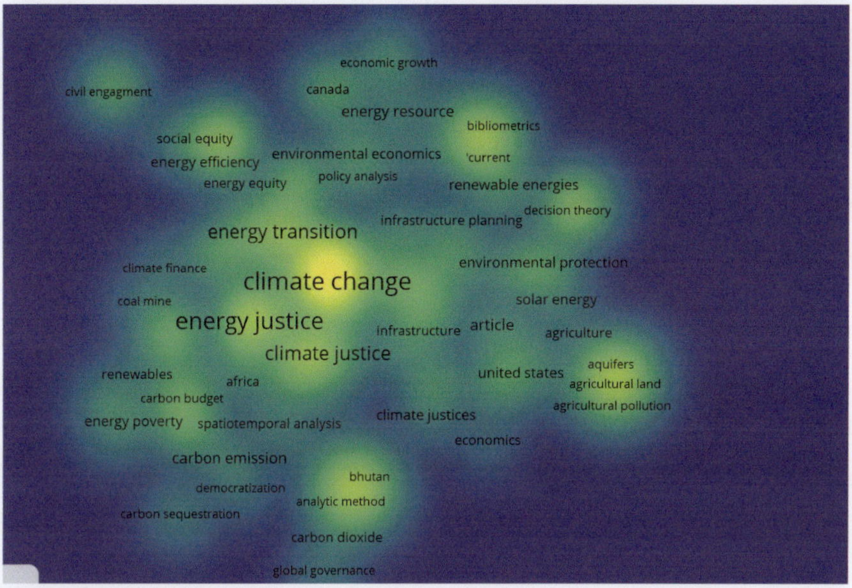

Fig. 16.12 Density visualization of all keywords

at the heart of climate change research and publications, especially when the impact on food security is considered. It is evident from this analysis that climate change has put a spotlight on agriculture, which affects food security in sub-Saharan Africa.

By examining the keyword table and the network graph of keyword co-occurrence, an intricate web of thematic priorities and their dynamic interplay with the keyword "climate change" serving at the core reveals four overarching thematic domains: (I) Food Security and Climate Change Adaptation; (II) Renewable energy and sustainable Food Systems; (III) Climate Justice and Equitable Food Policies, and (IV) Energy Justice and Equitable Access- the study effectively highlighted the multifaceted nature of these issues. Food security and climate change are concerned with how food security is affected by climate change and approaches to establishing sustainable adaptation policies. To advance sustainability and improve food security, renewable energy and suitable food systems should investigate how renewable energy might be incorporated into food systems. In the context of food, climate justice and equitable food prices highlight the necessity of fair policies that address the disproportionate effects of climate change on marginalized communities. The idea of "energy justice" promotes a fair distribution of energy resources and involves marginalised populations in the process of making energy-related decisions. The interplay between these themes allows for a comprehensive understanding of how climate change impacts food security, the role of renewable energy in improving agricultural practices, and the need for equitable policies that address social injustices. This facilitates the identification of research gaps and emerging trends, guiding future studies and policy development. Overall, it underscores the crucial integrated approach to addressing the complex interactions between climate justice, food security, and renewable energy, ultimately promoting resilience in vulnerable communities and development goals.

Table 16.6 Theme 1: Food security and climate change

Keyword	Occurrences	Total link strength
Climate change	24	94
Carbon	5	33
Sustainability	5	23

Table 16.6 Theme 2: Renewable energy and sustainable food system

Keyword	Occurrences	Total link strength
Energy transitions	8	52
Energy transition	11	46
Renewable energy	9	40
Energy policy	8	40

Table 16.6 Theme 3: Climate justice and equitable food policies

Keyword	Occurrences	Total link strength
Climate justice	10	42
Environmental justice	17	61
Social justice	5	30
Just transition	9	37

Table 16.6 Theme 4: Energy justice and equitable access

Keyword	Occurrences	Total link strength
Energy justice	19	70
Energy justices	6	42
Energy	6	40

The intersection of climate justice, renewable energy, and food security has attracted significant scholarly interest in sub-Saharan Africa with climate change, energy provision, and food production challenges. Over the past decade, especially since 2015, there has been a discernible increase in research focused on these interconnected themes. However, much of this scholarship is unevenly distributed, with certain regions receiving disproportionate attention, which creates a geographical imbalance that limits a thorough understanding of the diverse socio-economic and environmental landscapes across sub-Saharan Africa. This disparity in research focus impedes the development of effective and inclusive policy frameworks, emphasizing the urgent need for more context-specific investigations that can address the unique challenges faced by various regions within the continent. By extending research efforts to include less-studied areas, researchers can provide policymakers with a clearer understanding of localized needs, thereby facilitating the creation of sustainable approaches that address the socio-environmental conditions of sub-Saharan Africa.

The role of renewable energy in bolstering food security in sub-Saharan Africa has become increasingly essential, as reliable energy access can dramatically alter agricultural practices. Solar-powered irrigation systems, for instance, have proven to be highly beneficial in increasing crop yields and improving the management of water resources. In regions where water scarcity is a critical impediment to agricultural productivity, these systems provide a means for consistent irrigation, allowing farmers to grow crops more sustainably. By enhancing water utilization, solar-powered irrigation not only elevates agricultural output but also conserves crucial resources, thereby promoting more resilient farming practices across the region. Research by Birhanu et al. (2023) highlights that solar irrigation systems have significantly improved agricultural productivity and resilience in areas with severe environmental challenges. Furthermore, solar dryers are playing an essential role in minimizing post-harvest losses. For example, in Nigeria, the use of solar dryers has effectively preserved tomatoes by reducing their moisture content, enhances their shelf life and marketability. The study by Owureku-Asare et al. (2022) emphasizes the positive impact of these dryers on food availability and quality, demonstrating how renewable energy technologies can substantially enhance food security in regions where post-harvest losses are a persistent issue.

The potential of renewable energy to bolster food security is evident; however, the principle of climate justice is inadequately explored in this context. Climate justice emphasizes the fair distribution of the benefits and challenges associated with climate policies, underscoring the importance of making renewable energy solutions available to all, particularly to marginalized and low-income communities that are often the most susceptible to climate change. The substantial upfront costs of renewable energy technologies can hinder adoption among many communities. To counteract this barrier, innovative financing models, such as pay-as-you-go systems, have emerged in several countries to enhance the affordability of renewable energy. For example, in Kenya, pay-as-you-go solar home systems have enabled households to obtain solar energy through small, manageable payments. Adwek et al. (2020) illustrate that these financing models improve the affordability of renewable energy and empower communities by supplying them with reliable and sustainable energy sources. Such initiatives are vital for fostering climate justice, as they ensure that the benefits of renewable energy are accessible to those disproportionately affected by climate change. Integrating renewable energy with climate-smart agricultural practices can create synergies that improve resilience to climate impacts and enhance food security.

Climate-smart agriculture refers to practices designed to enhance agricultural productivity while minimizing greenhouse gas emissions, complementing renewable energy efforts. One notable example is agroforestry, which integrates agricultural activities with tree cultivation. This method not only improves soil fertility and provides shade but also facilitates carbon sequestration, making it particularly suitable for areas striving to achieve a balance between environmental sustainability and agricultural output. The finding of Muthuri et al. (2023) in East Africa demonstrates that agroforestry systems have significantly increased crop yields and provided farmers with supplementary income through the commercialization of tree products. By merging climate-smart strategies such as agroforestry with renewable energy technologies, agricultural systems in sub-Saharan Africa can enhance their resilience to climate-related challenges while promoting sustainable development. Renewable energy availability enhances various agricultural operations beyond irrigation, including food processing and storage. For example, solar-powered food processing facilities offer a dependable energy source for drying, canning, and packaging agricultural products, thereby reducing post-harvest losses and extending the shelf life of crops. This capability is especially crucial in remote regions with limited access to traditional energy sources, where renewable energy can revolutionize local food systems and foster economic advancement.

The successful integration of climate justice into food security policies necessitates the creation of comprehensive frameworks that consider the socio-political factors associated with climate change. It is vital for marginalized communities, who are frequently the most susceptible to the effects of climate change, to play a significant role in the decision-making processes surrounding renewable energy projects. Utilizing participatory approaches that involve these communities from the outset enhances their sense of ownership but also ensures that the initiatives align with local needs and priorities. A case study from Ghana illustrates this participatory model, where local farmers were involved in the design and implementation of solar-powered irrigation systems, resulting in increased adoption and improved agricultural results. Such frameworks promote equity by enabling communities to influence the solutions that directly affect them, addressing inequalities in resource distribution, and ensuring that climate solutions are adapted to the specific needs of different areas. By integrating the voices of marginalized populations into these frameworks, we can advance climate justice and ensure that the socio-political aspects of climate change are adequately represented in policy development.

Global collaboration plays a vital role in the progress of renewable energy and food security; however, existing research partnerships often highlight significant disparities in the creation and accessibility of knowledge. Countries such as the United States and the United Kingdom are key players in global research networks, and institutions in sub-Saharan Africa tend to be underrepresented in these collaborative efforts. This disparity raises important questions about the fairness of research benefits and the level of engagement of local stakeholders in developing solutions. To rectify these imbalances, it is crucial to encourage more inclusive, locally driven research initiatives that empower African researchers and institutions to take on leadership roles. Organizations like the African Centre for Technology Studies (ACTS) have been pivotal by providing essential training and resources and enhancing research capacity in the region (Chapman et al., 2022). The Renewable Energy for Food Security (REFS) project is a notable example of a successful collaboration between African and European institutions, illustrating the potential of international partnerships to create contextually relevant solutions (IRENA, 2021). Such collaborations enable the amalgamation of diverse expertise with solutions that resonate with local contexts. By establishing global partnerships, capacity building, and knowledge exchange, sub-Saharan Africa can reap the benefits of international research and make substantial contributions. Strengthening research networks and empowering local institutions will enable the region to cultivate the resilience necessary to tackle its distinct challenges in food security and renewable energy.

In summary, renewable energy is essential for improving food security in sub-Saharan Africa and facilitating agricultural activities such as irrigation, food processing, and storage. Although advancements in renewable energy technologies have yielded encouraging outcomes, the incorporation of climate justice into food security policies is an area that necessitates further attention. It is vital to overcome the financial obstacles that hinder marginalized communities from adopting renewable energy, thereby promoting equitable access and ensuring that the advantages of these technologies are broadly shared. Climate-smart agricultural practices complement renewable energy efforts, enhancing resilience and sustainability within systems. To fully realize the potential of renewable energy solutions in sub-Saharan Africa, it is imperative to establish inclusive frameworks that consider the socio-political dimensions of climate justice. Global cooperation and the empowerment of local researchers are essential for progressing this initiative. By nurturing inclusive partnerships and endorsing locally driven projects, we can ensure that strategies for renewable energy and food security effectively meet the unique socio-economic and environmental challenges and ultimately foster a more sustainable and resilient future for the region.

16.4 Conclusion, Recommendation and Limitations

This bibliometric analysis has provided valuable insights into the evolving research landscape on climate justice, renewable energy, and food security in sub-Saharan Africa. This study highlights a growing body of literature addressing these interconnected issues, reflecting an increased awareness of the need for holistic approaches to sustainable development. However, it also revealed significant gaps and opportunities for future research and policy development. The predominance of qualitative research methodologies valuable for exploratory studies, suggests a need for more diverse and robust research designs. Mixed method approaches, incorporating quantitative and qualitative elements, could provide a more comprehensive understanding of the complex dynamics at play. This would enable researchers to identify patterns and trends and delve deeper into the lived experiences and perceptions of communities affected by climate change and food insecurity.

Based on our findings, we propose two critical recommendations. First, there is an urgent need to foster greater collaboration among global research institutions, African universities, and think tanks. This would help build local research capacity and ensure African perspectives are central to climate justice and food security. Second, we recommend developing interdisciplinary research programmes that explicitly link climate justice, renewable energy, and food security. Such programmes should prioritise the involvement of local communities and policymakers to ensure that research outputs are relevant, actionable, and aligned with the needs of those most affected by these challenges.

One limitation of this review is the inherent nature of bibliometric analyses, which focus on quantitative metrics and may not fully capture the depth of qualitative insights into climate justice and renewable energy in food security. Additionally, the reliance on documents indexed within a specific database might exclude relevant literature not captured by the search criteria, potentially limiting the scope of analysis. Finally, a significant limitation of our research is the insufficient representation of sub-Saharan Africa within our bibliometric analysis. Although we did not deliberately omit articles from this area, the lack of pertinent publications in the databases employed (Scopus Core Collection) led to their minimal inclusion. As a result, our findings may not comprehensively reflect the burgeoning research contributions originating in sub-Saharan Africa. Future investigations should explore alternative databases or additional resources to enhance the representation of research output from this region.

Acknowledgements We express our sincere appreciation to the School of Education at the University of Witwatersrand for providing us with essential platforms to produce high-quality research output.

Authors' Contributions **Obi Joy Nneamaka:** Conceptualization, methodology, formal analysis, data curation, writing-original draft preparation, writing-review, and editing. **Emmanuel Ojo:** Conceptualization, methodology, writing-original draft preparation (part of the introduction), writing-review, and editing.

Funding Not applicable.

Data Availability Available based on request by qualified users.

Ethics Approval This article does not contain any studies with human or animal participants performed by any of the authors.

Conflict of Interest Authors declare no conflicts of interest.

References

Abbas, S., Rao, A. R. B. N., Khattak, F., & Ahmad, Y. (2024). Revolutionising sustainability using a new triplet: A system dynamic model. *Ecological Modelling*, *492*. https://doi.org/10.1016/j.ecolmodel.2024.110720

Adwek, G., Boxiong, S., Ndolo, P. O., Siagi, Z. O., Chepsaigutt, C., Kemunto, C. M., Arowo, M., Shimmon, J., Simiyu, P., & Yabo, A. C. (2020). The solar energy access in Kenya: A review focusing on Pay-As-You-Go solar home system. *Environment, Development and Sustainability, 22*(5), 3897–3938. https://doi.org/10.1007/s10668-019-00372-x

Agbehadji, I. E., Schütte, S., Masinde, M., Botai, J., & Mabhaudhi, T. (2023). Climate risks resilience development: A bibliometric analysis of climate-related early warning systems in Southern Africa. *Climate, 12*(1), 3. https://doi.org/10.3390/cli12010003

Birhanu, B. Z., Sanogo, K., Traore, S. S., Thai, M., & Kizito, F. (2023). Solar-based irrigation systems as a game changer to improve agricultural practices in sub-Sahara Africa: A case study from Mali. *Frontiers in Sustainable Food Systems, 7*, 1085335. https://doi.org/10.3389/fsufs.2023.1085335

Cairney, P., Timonina, I., & Stephan, H. (2023). How can policy and policymaking foster climate justice? A qualitative systematic review. *Open Research Europe, 3*, 51. https://doi.org/10.12688/openreseurope.15719.2

Chapman, C. A., Abernathy, K., Chapman, L. J., Downs, C., Effiom, E. O., Gogarten, J. F., Golooba, M., Kalbitzer, U., Lawes, M. J., Mekonnen, A., Omeja, P., Razafindratsima, O., Sheil, D., Tabor, G. M., Tumwesigye, C., & Sarkar, D. (2022). The future of sub-Saharan Africa's biodiversity in the face of climate and societal change. *Frontiers in Ecology and Evolution, 10*, 790552. https://doi.org/10.3389/fevo.2022.790552

Donthu, N., Kumar, S., Mukherjee, D., Pandey, N., & Lim, W. M. (2021). How to conduct a bibliometric analysis: An overview and guidelines. *Journal of Business Research, 133*, 285–296. https://doi.org/10.1016/j.jbusres.2021.04.070

Dorji, T., Morrison-Saunders, A., & Blake, D. (2023). Understanding how community wellbeing is affected by climate change: Evidence from a systematic literature review. *Environmental Management, 72*(3), 568–586. https://doi.org/10.1007/s00267-023-01833-w

Hallinger, P., & Nguyen, V.-T. (2020). Mapping the landscape and structure of research on education for sustainable development: A bibliometric review. *Sustainability, 12*(5), 1947. https://doi.org/10.3390/su12051947

IRENA (International Renewable Energy Agency). (2021). *Renewable energy for agri-food systems*. International Renewable Energy Agency, Abu Dhabi. https://www.irena.org/-/media/Files/IRENA/Agency/Publication/2021/Nov/IRENA_FAO_Renewables_Agrifood_2021.pdf

Munonye, J., Matthew, D., Olaolu, M., Onyeneke, R., Obi, J., Amadi, M., Ibrahim, S., Izuogu, C., & Njoku, L. (2023). The impact of food security and poverty status on rural household farmers in Ikwo, Ebonyi State, Nigeria. *African Journal of Food, Agriculture, Nutrition and Development, 23*(121), 23637–23656. https://doi.org/10.18697/ajfand.121.22670

Musara, E., Grant, C., & Vorster, J.-A. (2021). Inclusion as social justice: Nancy Fraser's Theory in the South African Context. In C. A. Mullen (Ed.), *Handbook of Social Justice Interventions in Education* (pp. 1–20). Springer International Publishing. https://doi.org/10.1007/978-3-030-29553-0_107-1

Muthuri, C. W., Kuyah, S., Njenga, M., Kuria, A., Öborn, I., & Van Noordwijk, M. (2023). Agroforestry's contribution to livelihoods and carbon sequestration in East Africa: A systematic review. *Trees, Forests and People, 14*, 100432. https://doi.org/10.1016/j.tfp.2023.100432

Neupane, N., Chaudhary, P., Rijal, Y., Ghimire, B., & Bhandari, R. (2022). The role of renewable energy in achieving water, energy, and food security under climate change constraints in South Asia. *Frontiers in Sustainable Food Systems, 6*, 1016093. https://doi.org/10.3389/fsufs.2022.1016093

Obi, J., Enete, A., & Munonye, J. (2021). Farm households' levels of vulnerability to extreme weather events in South-Eastern Nigeria. *African Journal of Food, Agriculture, Nutrition and Development*, *21*(02), 17450–17463. https://doi.org/10.18697/ajfand.97.19410

Obi, J., Ifejirika, C., Itam, K., Enete, A., Munonye, J., Osuji, E., Oyoboh, D., Jimmy, S., Egwu, C., Nwachukwu, C., Obetta, A., Nwofoke, C., & Odoh, N. (2023). An assessment of the performance of emergency management agency in the natural hazards management among farm households in the southeast zone, Nigeria. *Acta Agriculturae Slovenica, 119*(2). https://doi.org/10.14720/aas.2023.119.2.12480

Owureku-Asare, M., Oduro, I., Saalia, F. K., Tortoe, C., Ampah, J., & Ambrose, K. (2022). Drying characteristics and microbiological quality assessment of solar-dried tomato. *International Journal of Food Science, 2022*, 1–13. https://doi.org/10.1155/2022/2352327

Taliep, N. (2022). A community engagement model for an inclusive just energy transition in the South. *Social and Health Sciences*, *20*(1 and 2). https://doi.org/10.25159/2957-3645/11411

van Eck, N. J., & Waltman, L. (2010). Software survey: VOSviewer, a computer program for bibliometric mapping. *Scientometrics*, *84*(2), 523–538. Scopus. https://doi.org/10.1007/s11192-009-0146-3

Chapter 17
Accounting for Climate Change and Value of Listed Oil and Gas Firms in Nigeria

Silvia Nwakaego Onyekachi

Abstract The stakeholders' theory presupposes that balancing the interests of all stakeholders with respect to corporate reporting offers a better strategy for driving a successful organization. This study attempted to align with stakeholders' theory as it evaluates the influence of climate change accounting on corporate value of firms in the oil and gas sector of Nigeria. Secondary data were collected from the financial reports of eight listed firms in Nigerian Exchange Group from 2013 to 2022 and content analysis technique was employed to obtain data for climate change disclosures with the aid of Global Reporting Initiative on emissions reporting. The autoregressive distributed lag method was employed in the testing of hypothesis and findings show that climate change disclosure has significant and positive effect on the firm's value at the long run. Thus, the study concluded that climate change disclosure is an important factor that addresses the interests of other non-financial stakeholders to oil and gas companies in Nigeria. It therefore recommended that companies in the sector should pay more attention to preventive and remedial environmentally responsible activities and disclose them in their financial reports.

Keywords Climate change accounting · Firm value · Carbon emissions · Stakeholders' theory · Corporate reporting

17.1 Introduction

Business firms are always on the quest to grow their value through consistent profit-making which underscores the central motive for venturing into business. As they manage myriads of threats and opportunities unleashed from the business environment, a constant check must be made on the growth of the firm's value to ensure they continue to remain relevant to the stockholders. Thus, one can safely regard growth in

---

S. N. Onyekachi (✉)
Department of Accounting, Michael Okpara University of Agriculture, Umudike, Abia State, Nigeria
e-mail: onyekachi.silvia@mouau.edu.ng

R. U. Onyeneke et al. (eds.), *Energy Transition, Climate Action and Sustainable Agriculture*, https://doi.org/10.1007/978-3-031-83165-2_17

shareholders' value as one of the major concerns of an average and rational investor. As the calls for stakeholders' perspectives to be considered in the reporting of business activity which on its own also connotes engagement of activities that portray business firms in the light of being responsible for their community, environment and employees; it poses the question of its relevance to the firm's value, in other words, the going-concern status of the reporting entities.

Firms' value is the prosperity of firms which is dependent on the prices of stock in the market (Fatma & Chouaibi, 2021). According to Islam et al. (2014), there is a positive link between growth in the earnings of companies and stock prices which leads to overall increased value for the shareholders, thus suggesting that consistent profitability of the business establishes its relevance to the shareholder in terms of firms' value.

Climate change on the other hand, is the long-term variations in the temperature and weather of the earth which can be attributed to natural causes or negative human activities. The World Health Organization Fact Sheet (2016) described climate change as change in climate overtime as a result of variability in natural causes or human activity. This change is critical to businesses and every occupant of the planet—earth because it affects the way living organisms breath, eat, move and survive within the earth.

The issue of climate change accounting is linked to the stakeholders' theory and calls for everyone to become environmentally conscious and friendly in carrying out judicious business or personal activities to enhance the planet's sustainability in the future years. Accounting as a discipline being a social function that continues to harness issues from the political, cultural, economic and environmental fronts (Nwankwo & Orji-Okafor, 2023), demands that accountants of legitimate organizations should spearhead the proper accounting and reporting of climate change related activities undertaken by their respective firms during the period under review so as to align with the expectations of other parties having a stake in the business including some investors who have become passionate about preserving the planet.

However, the primary motive of business which entails making profits and maximizing shareholders' value needs to be viewed in the light of remaining environmentally responsible as any action that is mutually exclusive with value increment for the business investor is inimical to the central objective of the venture itself. As expected of every academic phenomena, several research thrusts have been initiated to address possible linkages between climate change accounting and financial performance of the organization with the view to investigate the nature and direction of the relations between climate change accounting and profit-making efficiency of organizations in several sectors of the Nigerian economy (Egbunike et al., 2021; Nwankwo & Orji-Okafor, 2023; Ofurum & Mmadubuobi, 2023). But this study aims to evaluate the climate change accounting with respect to the shareholders' value of oil and gas companies in Nigeria while taking into consideration the fact that the firm value is more substantial than profits for the firm and its investors.

Accounting for climate change is especially critical in the Nigerian oil and gas business environment given its status as a developing market. Adu et al. (2024)

described the Sub-Saharan African region (Nigeria inclusive) as one that is bedevilled by highly bureaucratic and corrupt government with little or no accountability while having weak institutional framework and regulation. The implication of this systemic disadvantage is that effective accounting and reporting mechanisms may not be upheld except it attracts sanctions to the reporting organizations. Thus, the reporting firms who may be driven by pure profit-making motives are bound to relegate climate change initiatives to the background leading to environmental disasters and hazards. In the Nigerian oil and gas sector, carbon emissions have been identified as hazardous implication of oil production (Nwankwo & Orji-Okafor, 2023), requiring intentional management and replenishment initiatives in terms of investments as a remedial action. The importance of this study is linked to the need for firms within the oil and gas sector to align their climate change accounting disclosures to their value and growth to decipher the critical nature of other stakeholders to the legitimacy of their businesses. The specific objective of the study therefore entails to evaluate the effect of climate change accounting on the value of oil and gas firms in Nigeria. The study also hypothesizes as that climate change accounting has no significant effect on the value of oil and gas firms in Nigeria. This research will be of particular interest to the accounting body of knowledge in Nigeria especially as it provides an empirical literature basis for the justification of proper disclosure in respect of climate change and other environmental factors. The paper also adds to literature in providing evidence on the critical role that credible reporting can have on the perception of firms by stakeholders.

17.2 Literature Review

17.2.1 Conceptual Review

17.2.1.1 Climate Change Accounting

According to Agbo and Egbunike (2024), accounting for climate change includes making full reports on the risks and opportunities available to the company due to changes in the in the climate and environmental factors and sharing the company's plans and strategies towards the management of the critical climate change issues such as energy consumption, carbon and greenhouse gas emissions. The need to account for climate change activities in the financial reports of companies is linked to the quest for business investors to put efforts in giving back to the society where it draws it resources and gains from. Again, the climate change risks impact negatively on business results (Giang et al., 2021) and the accounting profession should take a lead in the supporting role for climate change adaptation (Linnenluecke et al., 2015). Furthermore, these firms need to ensure that earth's space is made conducive and habitable for the future generations via proper disclosure of their plans and targets with respect to the replenishment. One of the widely accepted procedures to ensure that firms are involved in environmental preservative practices and reduce global emissions is through disclosure (Zhang et al., 2022; Agbo & Egbunike, 2024).

Moyo and Wingard (2015) submit that climate change may not impacts directly on business environment but have a way of stressing the system to affect the firm's financial position with the passage of time. Though it appears that accounting for climate change is due to pressures faced by the reporting firms from other stakeholders rather than being an issue of regulatory provision, Gardener (2022) opined that firms in United Kingdom have received guidance through the factsheet of November 2021 on how to report or disclose climate change related matters. This suggests that the accounting profession is bracing up to this challenge in providing procedure and defining its position for accountability in climate change disclosure. However, the role of climate change accounting in adding credibility and acceptability to the firms' business and improving the firm's value remains an issue of academic debate as Doobee et al. (2024) rightly captured that three schools of thoughts abound on the subject matter; the one that suggests that there is a link between environmental disclosures and performance, the one that suggests a negative connection and the school of thought that suggests no link between the variables. Thus, creating the need for more empirical submissions on climate change accounting and firm's value especially in Nigeria as an oil producing and developing nation.

17.2.1.2 Firm Value

Firm value, also called corporate value represents the totality of the amount that the stockholders' holding is worth in the market. It is the present worth of all the future inflow a business is expected to earn from its activities (Endri & Fathony, 2020). Corporate value is therefore measured from the perspective of the market value where the forces of demand and supply, inform the worth of the shareholders' claim instead of the purchase price as recorded in the books. Thus, the maximization of shareholders' wealth entails raising the corporate value. Firm value can be measured using the Tobin's Q as applied by Endri and Fathony (2020), where Tobin's Q is calculated as market value of equity plus total liabilities all divided by the total assets. Some research thrusts attempting to capture factors that directly impacts on firm's value identified consistent and totality of corporate performance. Selvam et al. (2016) purposing to explore the determinants of corporate performance through a subjective model, highlighted possible variables that can predict the performance of companies to include employee satisfaction, customer satisfaction, environmental audit, profitability, market value, growth, corporate governance, environmental and social performance. According to the study, market value, profitability and firm growth complement each other; thus, though market value which reflects the present value of shareholders' net-worth can be predicted by the profitability levels of firm accumulated over a period of time, it can also be linked back to the performance of firms through market anomalies that put the advantaged firms in a good light so that they can profit from the advantage of a better firm value. Hence, Selvam et al. (2016) suggested that a multi-directional relationship could subsists between firm performance and values as each of the variables is capable of eliciting statistical response firm the other. But the study did not carry out a statistical analysis to buttress it theoretical linkages even after making clear definitions of each of the variables described as determinants of firm performance rather it suggests that future studies should consider filling this gap. To provide a theoretical backing, the study argues that the position of the stakeholders' theory on what the true objective of a firm should be, demands that the interests of all the stakeholders be harmonized to arrive at the purpose of the firm; and further emphasized that since contemporary business firms are already making attempting to respect all the various interests of stakeholders by adopting social and environmental strategies as well as elaborate financial reporting efforts, empirical studies should link the enhanced firm values to its performance. Valentin (2012) studied the determinants of corporate performance with a particular emphasis on the need to explore a multi-dimensional model which does not only rely on historical facts to predict variations on the corporate performance. The study attempts to measure performance using market value approach which it defined as Tobin's q ratio. Employing the use of regression analysis on the secondary data collected the study holds that market value performance of the firm can be viewed as another measure of corporate performance. So, firm's value is of utmost relevance to the central objective of a business firm and climate change accounting issues that are likely threaten a firm's performance can be detrimental to its value.

17.2.2 Theoretical Framework

The stakeholders' theory according to the position of Harrison et al. (2015), is a practical theory which provides an effective and ethical procedure for managements to guide the activities of their organizations. They opined that stakeholders' theory promotes achievement of better financial performance for organizations that manage the interests of all the stakeholders and balance their activities to promote better interactions. This is the crux of stakeholders' theory. In financial reporting, stakeholders' theory is the foundational outlook that seeks to prioritize the interests of all stakeholders who are linked to the business of the firm (Agbo & Egbunike, 2024). Thus, the theory presupposes that better performance, and corporate value is linked to the information provided for all stakeholders, financial and non-financial stakeholders including the host communities, government and other key players whose basis for choice of patronage is dependent on environmental accountability of firms. Thus, in line with stakeholders' theory for financial reporting, this study links climate change disclosure to corporate value while controlling for firm size and leverage of the selected firms.

17.2.3 *Empirical Review*

17.2.3.1 Climate Change Accounting and Financial Performance

Though accounting for climate change is still a contemporary accounting research issue yet some studies have linked proper disclosure and investments to replenish the climate to the acceptability of the reporting firms which have serious implication for their positive financial performance. For example, Nwankwo and Orji-Okafor (2023) found that accounting for climate change has significant and positive effect on return on investment of oil and gas companies in Nigeria and South Africa. Though the study could be said to have applied the appropriate empirical models and tools for data collection and analyses yet the combination of firms of both countries obscures the individualities and economic differences they could possess. Even though both countries are in the same continent and could possess similar institutional culture, it is still pertinent to paint a separate picture of climate change disclosure and accounting practices of both countries for better recommendations and lessons to be taken from the study. Afolabi et al. (2024) submitted that pollution, habitat destruction and greenhouse gas emissions which was captured as wastes are the elements of climate change that should be treated in the financial reports as environmental issues. So, the study utilized the Driscoll-Kray standard errors as the analytical tool and data from eight listed oil and gas firms to show that waste management disclosure has a significant impact on the financial performance of the selected companies. Though the study combined employee health safety and community development with waste management disclosure to arrive at its model for environmental accounting disclosure, the individual result for waste management disclosure reveals the critical nature of accounting for gas emissions and other wastes that result into climate change to financial performance of firms in the Nigerian oil and gas sector. Moyo and Wingard (2015) established that there is positive and significant correlation between climate change and financial performance of seventy listed companies in the Johannesburg Stock Exchange. The study still recommended amongst others that a consistent measures of climate change performance to be created by the accounting profession to enable comparisons among companies. Beredugo and Mefor (2012) utilized primary data from residents of riverside areas of Nigeria where oil exploration is densely conducted and found that environmental accounting and reporting has significant correlation with sustainable development in Nigeria. Bassey et al. (2013) found that environmentally friendly firms are more likely to make full and significant disclosure of their environmental activities on the financial reports whereas the study called for uniformity in the disclosure pattern of firms to enhance peer review and comparisons overtime. Agbo and Egbunike (2024) opined that climate change related disclosure of quoted oil and gas firms in Nigeria has a positive and statistically significant relationship with their financial performance which was captured using returns on assets. Egbunike et al. (2021) found that accounting for carbon management, specifically greenhouse gas emissions disclosure has significant effect on Tobin's Q of manufacturing firms. Eze and Akaegbobi (2023) found that carbon emissions management has a strong statistically significant effect on the financial performance of oil and gas firms in Nigeria. Aniefor et al. (2023) studied a sample of seven listed companies on the Nigerian Exchange Group and found that both emissions disclosure and environmental disclosure index have significant effects on return on equity of oil and gas firms in Nigeria. Etim et al. (2024) had a similar result using a desktop approach to review existing literature on green accounting and financial performance of listed oil and gas firms in Nigeria. However, Güneysu and Atasel (2022) had a different result as they reported that carbon emissions and financial performance measures; return on assets and earnings per share are negatively correlated.

17.2.3.2 Summary of Research Gaps

Some empirical literature consulted during this study measured carbon emission as the quantity of carbon emissions reported by companies (Güneysu & Atasel, 2022) while majority utilized the GRI index of climate change disclosures for capturing carbon and greenhouse gas emissions accounting (Aniefor et al., 2023; Egbunike et al., 2021). It is also obvious that existing literature have invested great degree of research efforts in linking climate change accounting to financial performance of firms in various sectors of the economy. This created the need for more empirical evidence on the nexus between climate change accounting and corporate value which is one of critical concerns of an average stockholder in the Nigerian oil and gas sector.

17.3 Methodology

17.3.1 Data Source and Description of Variables

The research design adopted for this study is *ex-post facto* design as it sought to decipher the link between two variables of interest: climate change accounting and firm's value. The study purposively selected the firms in the oil and gas sector of the Nigeria Exchange Group which is the interest area of this research. It is essential to note that Nigeria is one of the major oil exporting countries of the world and is largely dependent on the oil revenues hence, the oil and gas sector is an important part of the economy which conversely puts the residents of the oil drilling area in jeopardy through its effects on the environment. The population of the study consists of eight listed firms in the oil and gas sector by Nigerian Exchange Group as of 20th August 2024, which were all selected for sampling since the number is manageable. The listed oil and gas firms include Capital Oil Plc, ConOil Plc, Eterna PLc, Japaul Gold & Ventures Plc, MRS Oil Plc, OANDO Plc, Seplat Energy Plc and TotalEnergies Plc.

The data is secondary in nature and were sourced from the downloaded financial reports of each selected company from 2013 to 2022, using content analyses technique.

17.3.2 Analytical Technique

Autoregressive distributed lag (ARDL) was applied for analyses of data because it is the most appropriate for variables with mixed order of integration while efficiently estimating the long-run and short-run relationship between the variables of research interest. The study adapted the following model from Agbo and Egbunike (2024) to estimate the possible effect that climate change accounting has on firm's value.

$$FV_{it} = \beta_0 + \beta_1 CCA_{it} + \beta_2 FS_{it} + \beta_3 LEV_{it} + \mu_{it} \tag{17.1}$$

where, i represents the number of companies and t, number of years.

FV is firm value, Tobin's Q was used as a proxy for firm size.

CCA is Climate change accounting rated with GRI 305 Emissions.

LEV is leverage ratio applied as a control variable in this research.

FS is firm size applied as a control variable in this model. It is calculated as natural log of total assets.

The Hausman test was applied to select the most appropriate ARDL outcome between pooled mean group, mean group and dynamic fixed effects estimations. The pooled mean group of ARDL is advantageous to the aim of this research as it can estimate both long-run and short-run relationships as well as the speed of adjustment to long-run equilibrium. However, the mean group estimation has less restrictive procedures and is efficient in performing diversity of parameters (Pesaran et al., 2001). The dynamic fixed effect estimator is efficient in limiting the vector coefficient and achieving consistency for long-run and short-run estimations (Shaari et al., 2020). So, the hausman test serves to provide a basis for selecting the most suitable estimator for explaining the relationship between climate change accounting disclosures and corporate value of the selected firms in this study.

The long-run model for both pooled mean group and dynamic fixed effect estimators as adapted from Shaari et al. (2020) are as follows:

$$\begin{aligned} \mathrm{FV}_{it} = \alpha &+ \sum_{j=1}^{p} \lambda_{ij}\mathrm{FV}_{i,t-j} + \sum_{j=0}^{q} \delta_{1ij}\mathrm{CCA}_{i,t-j} \\ &+ \sum_{j=0}^{r} \delta_{2ij}\mathrm{FS}_{i,t-j} + \sum_{j=0}^{s} \delta_{3ij}\mathrm{LEV}_{i,t-j} + \varepsilon_{it} \end{aligned} \tag{17.2}$$

The mean group long-run model is as follows:

$$FV_{it} = \theta_i + \beta_{0i}FV_{i,t-1} + \beta_{1i}CCA_{i,t-1} + \beta_{2i}FS_{i,t-1} + \beta_{3i}LEV_{i,t-1} + \varepsilon_{it} \quad (17.3)$$

The short-run model for all the three estimators are as follows:

$$\begin{aligned}\Delta FV_{it} = {} & \alpha_i + \phi_i\big(FV_{i,t-1} - \lambda_1 CCA_{i,t-1}\lambda_2 FS_{i,t-1} - \lambda_3 LEV_{i,t-1}\big) \\ & + \sum_{j=1}^{p} \lambda_{ij}\Delta FV_{i,t-j} + \sum_{j=0}^{q} \delta_{1ij}\Delta CCA_{i,t-j} + \sum_{j=0}^{r} \delta_{2ij}\Delta FS_{i,t-j} \\ & + \sum_{j=0}^{s} \delta_{3ij}\Delta LEV_{i,t-j} + \varepsilon_{it} \quad (17.4)\end{aligned}$$

ere; λ_i are long-run paramers; ϕ_i is the parameter for error-correction term; α_i company specific effect; ε_{it} is the error term (Eqs. 17.1, 17.2, 17.3 and 17.4).

17.4 Results and Discussion

The study conducted some preliminary tests to enable proper description of the data collected and inform the selection appropriate analytical technique for testing the formulated hypothesis. The result obtained for descriptive statistics is shown on Table 17.2.

Table 17.1 Specification of research variables

Variable	Source	Description
Climate change accounting	Financial reports of each selected company (2013–2022) and GRI 305 Emissions (2016)	Raw scores of each item contained in Global Reporting Index (GRI 305 emissions) total of 21 items. A company is scored 1 for each item identified in its financial report of a given year or 0 if the information was not disclosed
Firm value	Financial statement of each selected company (2013–2022)	Firm value was measured using Tobin's Q of the selected companies. Tobin's Q is calculated as market value of firm's stock in Naira divided by its Net Assets
Firm size	Financial statement of each selected company (2013–2022)	Calculated as natural log of total assets. It served as control variable in the equation
Leverage	Financial statement of each selected company (2013–2022)	Leverage represents the level at which borrowing of the companies at a time is covered by investments in its equity. This is calculated as the total debt in Naira divided by total equity value

Source Research review 2024

Table 17.2 Results of descriptive statistics of variables

	CCA	Firm value	LEVERAGE	Firm size
Mean	11.83750	0.979500	0.412850	9.876022
Median	12.00000	0.780000	0.421439	9.898099
Maximum	20.00000	6.290000	0.476498	11.88440
Minimum	5.000000	0.140000	0.289121	8.417862
Std. Dev.	3.953668	0.812195	0.040901	0.765699
Skewness	0.321814	4.250703	−0.860016	0.388137
Kurtosis	2.136214	25.98938	3.860160	3.219495
Jarque–Bera	3.867940	2002.619	12.32794	2.169263
Probability	0.144573	0.000000	0.002104	0.338026
Sum	947.0000	78.36000	33.02801	790.0818
Sum Sq. Dev.	1234.887	52.11318	0.132160	46.31730
Observations	80	80	80	80

Source Eviews 13 output

The outcome on Table 17.2 indicates that climate change accounting has a mean value of 11.84 as against the minimum and maximum values of 5 and 20 respectively. It implies that the practice of climate change disclosure is well spread even though the disclosures made in some years are very poor as against the GRI standard for sufficient disclosure. The Tobin's Q which is the firm's value indicate that for some years covered by the study, the market value of the selected firms dipped and for other years, it bloomed however the median value suggested that the midpoint value is 0.78 implying that over half the entire observations for firm's value is lower than 1. The leverage and firm size have lower variability which indicates that most of the firms may have comparable outcomes for leverage and firm size. The Jarque–Bera probability suggests that only climate change accounting and firm size are normally distributed since their outcomes are well above 5% benchmark implying that null hypothesis which proposed that the series are normally distributed should be accepted but for leverage and firm value, the study rejects the null hypothesis and concludes that the series are not normally distributed.

The study also conducted a stationarity test for each of the variables to ascertain their order of integration. However, prior to conducting of the unit root test, cross-section dependence tests were taken for each variable to decipher the appropriate generation of unit root method to use for the panel data. The results obtained are summarized and presented on Table 17.3.

The result for cross-section dependence indicate that firm value and leverage should be tested for stationarity using one of the first-generation methods and the study applied the Levin, Lu and Chu t whereas the climate change accounting and firm size were tested with Im, Pesaran & Shin W which is a second-generation unit root test method. The unit root outcome also shows that the variables are integrated on a mixed order. Thus, autoregressive distributed lag estimation was adopted for

Table 17.3 Cross-section dependence and unit root tests results

Test criteria	Climate change accounting	Firm value	Leverage	Firm size
Breusch-Pagan LM	0.0000	0.4200	0.0000	0.0000
Pesaran scaled LM	0.0000	0.9091	0.0000	0.0000
Bias-corrected scaled LM	0.0000	0.7412	0.0000	0.0000
Pesaran CD	0.0000	0.8801	0.0592	0.0000
Decision on unit root generation	2nd generation	1st generation	1st generation	2nd generation
Method of unit root	Im,Pesaran & Shin W	LLC	LLC	Im,Pesaran & Shin W
Unit root result	1(1)	1(0)	1(0)	1(1)

Source Research analyses output extracted from Eviews 13

analysing the influence of climate change accounting on firm's value since the criteria for using ordinary least square regression method was not met.

The study also conducted the correlation analysis of the variables to check the correlation coefficient of the variables for cases of multicollinearity as high correlations among the independent variables suggests the existence of multicollinearity. The results obtained are contained in Table 17.4.

The results on Table 17.4 show that the variables are not highly correlated with climate change accounting indices. The highest correlation coefficient is between leverage and firm value and the outcome is 26%. This indicates that the variables are not highly correlated and implies that multicollinearity is not an impairment factor to the variable's estimation. The results also show that climate change accounting is positively correlated with firm size, and value but it is negatively correlated with leverage.

Table 17.4 Multicollinearity estimation of variables

Correlation analysis: ordinary

Observations	CCA	FIRMSIZE	LEVERAGE	FIRM VALUE
FIRMSIZE	0.037071	1.000000		
LEVERAGE	−0.038538	0.157217	1.000000	
FIRM VALUE	0.037483	0.236554	0.262603	1.000000

Source Extracted from Eviews 13 output

Table 17.5 Short-run panel autoregressive distributed lag estimation of climate change accounting on firms' value

Test statistics	Coefficient	T-statistics	P-values
Cointegrating equation	−0.8958	−6.3952	0.0000
Short run coefficients:			
Climate change accounting	0.1945	1.3617	0.1779
Firm size	0.8701	2.0674	0.0426
Leverage	0.2722	0.0877	0.9304
Constant	−0.5315	−4.0860	0.0001

Source Research Analyses Output extracted from Eviews 13

17.4.1 Empirical Result

Restatement of hypothesis.

Climate change accounting has no significant effect on the value of oil and gas firms in Nigeria.

The results in Table 17.5 show the short run ARDL estimation for variables. The short-run outcome indicates that climate change accounting has no significant effect on firm's value as is evident in the values of the t-statistics of 1.36 which is below 2 and probability value of above 5% statistical significance. The same applies for leverage, which is one of the control variables for the equation as its probability value is greater than 5%. But firm size has a positive coefficient of 0.87 and a corresponding probability value of less than 5% thus the study posits that firm value associates positively with variations in firms' value. Considering the model's speed of adjustment, the cointegrating equation of −0.8958 is significant at a 1% statistical level implying that the variables maybe significantly associated in the long run. The significant level of the cointegrating equation outcome also shows that even though climate change disclosures of the selected firms do not influence their corporate values in the short run, it is likely to be resolved at the long run to influence firms' values significantly.

Furthermore, ARDL pooled mean group estimator was adopted for analysing the model based on the outcome of the Hausman test on Table 17.6. The null hypothesis of the Hausman test presupposes that pooled mean group is better than mean group estimation when the probability value of the estimation is greater than 5% thus in consideration of the probability of 0.33 approximately, the pooled mean group estimation was adopted for explaining the relationship between climate change accounting and corporate value of oil and gas firms in Nigeria.

The long-run results on Table 17.6 indicate that climate change accounting has significant effect on the values of oil and gas firms in Nigeria. This was inferred from the t-statistics value of 5.16 which is well above 2; and the probability value of 0.00 which is below the 5% statistical significance. The implication of the outcomes is that the null hypothesis of this study which proposes that climate change accounting has no significant effect on firm value of listed oil and gas firms in Nigeria should be rejected in favour of its alternative position.

Table 17.6 Long-run panel autoregressive distributed lag estimation of climate change accounting and firm's value

Variables	Pooled mean group		Mean group		Dynamic fixed effect	
	Coefficient	Probability	Coefficient	Probability	Coefficient	Probability
CCA	0.0055	0.0000	−0.2718	0.2539	−0.005817	0.4677
FS	0.0210	0.4306	0.5897	0.5838	−0.078057	0.7318
Leverage	2.4823	0.0006	−2.7806	0.5860	6.319117	0.3162
Hausman test			1.1097	0.3311	1.2865	0.7323

Source Research analyses output extracted from Eviews 13

The results of this study align with the findings of Agbo and Egbunike (2024) though their study specifically assessed the influence of climate change accounting on financial performance of listed oil and gas firms in Nigeria using returns on assets as a proxy for financial performance. It goes to strengthen the argument that climate change practices and disclosures especially by oil and gas firms whose business activities obviously impacts on the environmental safety is gradually becoming a critical topic in the Nigerian stock market. Asides, the friendly disposition of the immediate host communities where these firms operate, the fines and other risks they tend to avoid by being environmentally conscious and accountable is beneficial to their net worth overtime. Furthermore, the results obtained by Adu et al. (2024) for banks drawn from sixteen Sub-Saharan African countries also show that climate change initiatives and disclosure improves the performance of the selected banks. Other studies that obtained similar results for carbon emissions disclosure are Egbunike et al., (2021), Eze and Akaegbobi (2023), Aniefor et al. (2023) and Etim et al. (2024). The study also notes that this finding opposes the position adopted by Güneysu and Atasel (2022) as they found that total carbon emissions have no significant relationship with firms' value. It is important to consider that the firms they studied are non-financial firms in Turkey; the study also utilized total carbon emissions as data instead of disclosures as applied in this work. So, their conclusion in other words is that total emissions do not correlate significantly with thc firm's value. The same applies for Giang et al. (2021) who concluded that climate change on its own impacts negatively and significantly on the financial performance firms in Vietnam.

The long-run results of the pooled mean group estimation also shows that leverage as a control variable has a significant association with the values of the firms in the oil and gas industry. This is based on the pooled coefficient of 2.48 which is significant at the level of 1%. This outcome suggests that the leverage of the firms within the Nigerian oil and gas sector maintain a leverage position that interacts positively with variation in their corporate value. It also implies that these firms are properly leveraged in a way that promotes their image and consequently, their market value. According to the research made by Farooq and Masood (2016), suitably leveraged cement firms in Pakistan were found to have their leverage ratios align significantly and positively with corporate values thus implying that companies who maintain their capital mix are likely to grow their market value through the going-concern boost that is associated with it. But Ibrahim and Isiaka (2020) found that financial leverage of some selected quoted firms in Nigeria has significant and negative effect on the firms' value. This submission does not align with the findings of this research however its is should be noted that the present study was restricted to oil and gas firms in Nigeria and also the time frame of both studies do not exactly match. Pratt et al. (2023) opined that negative links between leverage and firms' value at any time may just be a temporary occurrence which is bound to resolve at the long run because their findings indicate that financial leverage generally promotes corporate value.

17.5 Conclusion and Recommendations

The study concludes based on the results obtained during this study, that climate change disclosure is essential to the maximization of firm value of companies in the oil and gas sector of Nigeria. Thus, it is recommended that firms in the oil and gas sector of Nigeria should be deliberate in their climate change remediation and preventive practices as well as being consistent in the reporting of same. The study also suggests that climate change practices disclosures should be mandatory for companies operating in the oil and gas sector of Nigeria. Also, the accounting body and stakeholders in Nigeria from academic, regulatory and professional fronts should articulate specific standards or procedure to guide accounting for climate change preventive and remedial activities by firms. It is important that companies who have demonstrated responsibility for climate preservation should report their affairs in a uniform manner to make the reports comparable among their peers or several years. Climate change issues and remedial actions especially for oil and gas companies whose ordinary business activities partly jeopardize or impact negatively on the atmospheric condition of the Nigerian environment should be mandated to disclose their actions in that respect to other parties who have interest in the companies, in line with the stakeholders' theory.

Some of the limitations faced includes availability of electricity and costs of data for assessing the websites of the selected companies to download their financial reports and content analyses of each financial report to compile the scores of climate change disclosures. However, the researcher was able to wade through the challenges and conclude the research exercise.

Acknowledgements I recognize the unquantifiable efforts and patience of the reviewers who painstakingly went through this paper and made invaluable recommendations and constructive criticisms. I appreciate Professor U. Uwazie for encouraging towards academic research productivity and Mr. Churchill for availing me of the software utilized for the analytical procedure of this research.

Funding No funding was received to assist with the preparation or for conducting this research.

Data Availability Statement The datasets used or analysed during the current study are available from the corresponding author on reasonable requests.

Ethical Clearance The researcher has conducted a study which does not involve eliciting information from persons. The financial reports and Global reporting initiatives index applied are information readily available on the internet. Also, the works consulted during this research were duly cited and referenced. So, there is no known ethical standards neglected and no application was made for ethical clearance.

Conflict of Interest I, Silvia Nwakaego Onyekachi therefore, declare that I have no known competing financial interests or personal relationships that could have influenced the work reported in this paper.

References

Adu, D. A., Abedin, M. Z., Saa, V. Y. & Boateng, F. (2024). Bank sustainability, climate change initiatives and financial performance: The role of corporate governance. *International Review of Financial Analysis,* 95(B). https://doi.org/10.1016/j.irfa.2024.103438

Afolabi, A. R., Amosun, O. O., Okunade, S. O. & Agbor, M. N. (2024). Environmental accounting disclosures and financial performance of listed oil and gas companies in Nigeria: An application of Driscoll-Kraay standard errors approach. MPDI Preprints. https://doi.org/10.20944/preprints 202403.1553.v1

Agbo, E., & Egbunike, C. F. (2024). Climate change disclosure and firm performance of quoted oil and gas firms in Nigeria. *Annals of Management and Organization Research, 5*(3), 163–173.

Aniefor, S. J., Onatuyeh, E. A., & Orife, C. O. (2023). Carbon emissions and financial performance: A study of oil and gas firms in Nigeria. *Journal of Xidian University, 17*(11), 382–396.

Bassey, B. E., Effiok, S. O., & Eton, O. E. (2013). The impact of environmental accounting and reporting on organizational performance of selected oil and gas companies in Niger Delta region of Nigeria. *Research Journal of Finance and Accounting, 4*(3), 57–73.

Beredugo, S. B., & Mefor, I. P. (2012). The impact of environmental accounting and reporting on sustainable development in Nigeria. *Research Journal of Finance and Accounting, 3*(7), 55–63.

Doobee, L. P., Ironkwe, U. I., & Nwaiwu, J. N. (2024). Green accounting practice and listed oil and gas companies performance metric in Nigeria. *Social Science and Humanities Research, 7*(6), 1–20.

Egbunike, P., Afodigbueokwu, A. I., & Hilary, E. (2021). Carbon management accounting and performance of quoted consumer goods manufacturing firms in Nigeria. *International Journal of Advances in Engineering and Management, 3*(5), 846–860.

Endri, E., & Fathony, M. (2020). Determinants of firm's value: Evidence from financial industry. *Management Science Letters, 10*, 111–120.

Etim, E. O., Umo, U. P., Udoh, O. A., & Edet, J. P. (2024). An exploratory research on effect of green accounting on financial performance of oil and gas companies in Nigeria. *International Journal of Economics and Financial Management, 9*(1), 81–95.

Eze, M. N., & Akaegbobi, G. N. (2023). Carbon emissions management and performance of selected oil and gas firms in Nigeria. *Journal of Management Sciences Research, 1*(1), 35–50.

Fatma, H. B., & Chouaibi, J. (2021). Corporate governance and firm value: A study on European financial institutions. *International Journal of Productivity and Performance Management*. https://doi.org/10.1108/IJPPM-05-2021-0306

Farooq, M. A., & Masood, A. (2016). Impact of financial leverage on value of firms: Evidence from cement sector of Pakistan. *Research Journal of Finance and Accounting, 7*(9), 73–77.

Gardner, L. (2022). *How climate change is impacting financial statements.* https://www.grantthornton.co.uk/insights/how-climate-change-is-impacting-financial-statements/

Giang, N. T., Hanh, T. M., Hien, P. T., Trinh, N. T., Huyen, N. T. K., & Trang, V. H. (2021). The impacts of climate change risks on financial performance: Evidence from listed manufacturing firms in Vietnam. *Advances in Economics, Business and Management Research, 196*, 581–595.

GRI 305 Publication (2016). Global reporting initiative: 305. https://www.globalreporting.org/standards/media/1012/gri-305-emissions-2016.pd

Güneysu, Y., & Atasel, O. Y. (2022). Karbon emisyonları ile finansal performans arasındaki ilişkinin incelenmesi: BIST100 endeksinde bir araştırma. *Fırat Üniversitesi Sosyal Bilimler Dergisi, 32*, 3(1183–1193).

Harrison, J. S., Freeman, R. E., & Abreu, M. C. S. (2015). Stakeholder theory as an ethical approach to effective management: Applying the theory to multiple contexts. *Review of Business Management, 17*(55), 858–869.

Ibrahim, U. A., & Isiaka, A. (2020). Effect of financial leverage on firm value: Evidence from selected firms quoted on the Nigerian stock exchange. *European Journal of Business and Management, 12*(3), 124–135.

Islam, M. R., Khan, T. R., Choudhury, T. T., & Adnan, A. M. (2014). How earning per share (EPS) affects share price and firm value. *European Journal of Business and Management, 6*(17), 97–108.

Linnenluecke, M., Birt, J., & Griffiths, A. (2015). The role of accounting in supporting adaptation to climate change. *Accounting and Finance, 55*(3), 607–625.

Moyo, M., & Wingard, H. C. (2015). An assessment of the impact of climate change on the financial performance of South African companies. *Journal of Governance and Regulation, 4*(2), 49–62.

Nwankwo, K. N., & Orji-Okafor, T. G. (2023). Accounting for climate change and operational performance of quoted oil and gas firms in Nigeria and South Africa. *Journal of Global Accounting, 9*(3), 270–287.

Ofurum, C. D., & Mmadubuobi, L. C. (2023). Effect of carbon emissions and financial performance of Nigerian oil and gas firms. *International Journal of Research Publication and Reviews, 4*(12), 4777–4786.

Pesaran, M. H., Shin, Y., & Smith, R. (2001). Bounds testing approaches to the analysis of level relationships. *Journal of Applied Econometrics, 16*, 289–326.

Pratt, W. R., Barboza, G. A. & Brigida, M. (2023). Leverage and firm value. *Economic Notes, 52*(2). https://doi.org/10.1111/ecno.12218

Selvam, M., Gayathri, J., Vasanth, V., Lingaraja, K., & Marxiaoli, S. (2016). Determinants of firm performance: A subjective model. *International Journal of Social Science Studies, 4*(7), 90–100.

Shaari, M. S., Abidin, N. Z., & Karim, Z. A. (2020). The impact of renewable energy consumption and economic growth on CO_2 emissions: New evidence using panel ARDL study of selected countries. *International Journal of Energy Economics and Policy, 10*(6), 617–623.

Valentin, C. (2012). *Determinants of corporate financial performance.* www.dafi.asce.ro/revista/6/costeavalentin.pdf

World Health Organization Fact Sheet. (2016). *Protecting health from climate change.* Retrieved on 22nd August 2024 from https://www.who.int/docs/default-source/wpro---documents/hae---regional-forum-(2016)/climatechange-factsheet-rfhe.pdf?sfvrsn=75d570fd_2

Zhang, Y., Wang, C., & Cai, W. (2022). Carbon pricing policy, revenue recycling schemes, and income inequality: A multi-regional dynamic CGE assessment for China. *Resources, Conservation and Recycling, 181*, 106246.

Chapter 18
Climate Change and Agriculture: Mitigation, Adaptation and Resilience in Abakaliki, Ebonyi State

Jephter Ebuka Orji, Happiness Chinyereugo Orji, Chibuzo Uzoma Izuogu, Chidinma Sussan Iroegbu, and Eyibio Nkpouto

Abstract Undoubtedly, climate change is a global issue that impacts all aspects of human life today. The review evaluated the effect of climate change on agricultural production, mitigation, adaptation and resilience in Ebonyi state. Previous studies on climate change and agriculture concentrated on adaptation alone, mitigation alone, some on mitigation and adaptation, while others are on adaptation and resilience and are published more in international journals. However, there are very few studies on climate change and agriculture: mitigation, adaptation and resilience in Abakaliki, Ebonyi state, hence the need for the study. Reviewing the mitigation, adaptation, and resilience of climate change and agriculture in Abakaliki, Ebonyi state, was the goal of the study. From the initial search of 100 journal papers, only 51 met the inclusion criteria using PRISMA procedures. 51 research papers that included information on agriculture and climate change, mitigation, adaptation, and resilience, and were published in local and international journals between 2013 and 2023, were systematically reviewed. The review identified the effects of climate change to include that amount of CO_2 and other greenhouse gases in the atmosphere has increased globally, which has increased the average world temperature and caused unpredictable weather patterns. Climate change has also shown several effects on

J. E. Orji (✉) · C. S. Iroegbu
Department of Soil Science, Alex Ekwueme Federal University Ndufu Alike, Ikwo, Ebonyi State, Nigeria
e-mail: orji.jephter@funai.edu.ng

C. S. Iroegbu
e-mail: susan.iroegbu@funai.edu.ng

H. C. Orji
Department of Agricultural Economics and Extension, Ebonyi State University, Abakaliki, Ebonyi State, Nigeria

C. U. Izuogu
Department of Agricultural Extension, Alex Ekwueme Federal University Ndufu Alike, Ikwo, Ebonyi State, Nigeria

E. Nkpouto
Department of Soil Science, University of Nigeria, Nsukka, Nigeria

R. U. Onyeneke et al. (eds.), *Energy Transition, Climate Action and Sustainable Agriculture*, https://doi.org/10.1007/978-3-031-83165-2_18

agriculture and food production. Through changes in agro-ecological conditions, it directly affects food production. It also influences the increase and distribution of earnings, which in turn determines the demand for agricultural products. Given the blatantly evident detrimental effects of climate change on agriculture, mitigation, adaptation and resilient measures are required. The results also identified that most of the papers were published in international journals. The review observed the need for more information on adaptation and resilience in the study area since the effect of climate change is persisting and should be published locally.

Keywords Climate change · Agriculture · Mitigation · Adaptation · Resilience

18.1 Introduction

One of the greatest challenges to humanity right now is climate change (Orji & Iroegbu, 2021). Since the late nineteenth century, the earth's temperature has already increased by 1.1 °C. Every part of the world is already feeling the effects of this, with heat waves and droughts becoming more common and intense, altered rainfall patterns, flooding, ice melting, and habitat disruption. Sea level rise is one of the climate's effects that will take years or even millennia to fully manifest. This climate change results in erratic weather patterns and the impacts are already being felt globally especially in the tropics (Bandyopadhyay, 2013; Mbah et al., 2016). The major driving force of the observed changing climate according to researchers has been identified as global warming (Olayinka, 2018).

Global warming is attributed to increased concentrations in the atmosphere of greenhouse gases (GHGs) such as CO_2, CH_4 and N_2O (IPCC, 2014; Orji & Iroegbu, 2021). The earth's mean surface temperature has risen by around 0.8 °C (1.4 °F) since the early twentieth century, with almost two thirds of the increase occurring since1990 (Auwal et al., 2018). Bandyopadhyay (2013) observed that global warming is the major cause of changes in the various climatic events of the global environment. According to Olayinka (2018), global warming can be defined as increase in the earth's global average temperature resulting from human activities such as burning of fossil duel, animal husbandry and changing land use. However, there are some gases in the atmosphere which acts a blanket preventing the escape of the infra-ray radiation by absorbing and emitting it back thereby warming the earth's surface and troposphere, such gases are called greenhouse gases (GHGs) which include Carbon (iv) Oxide (CO_2), Methane (CH_4), Nitrogen (i) Oxide (N_2O) and Chlorofluoro carbons (CFCs) (IPCC, 2014; Orji & Iroegbu, 2021).

Human activities especially combustion of fossil fuels, a primary source of energy at present, and changes in land use are major contributors to the increase of CO_2 concentrations in the atmosphere (Lal, 2015; IPCC, 2021). Previous studies have shown that there has been increase in global average air and ocean temperatures leading to melting of permafrost and increase in sea levels as a result of global warming. Since the industrial revolution, CO_2 concentration increased from 270 ppm to more than 360 ppm. Currently, there has been an increase of between 0.4 and 0.8 °C in the earth's global air temperature (Olayinka, 2018) and this year 2024 has been observed to be hottest year so far globally.

Agriculture which is the hub of food production in the whole is seriously under threat due to climate change. With its significant impact on livelihoods and economies throughout the continent, agriculture is a particularly significant sector in Africa. According to Leisner (2020), the average contribution of agriculture to GDP in Sub-Saharan Africa is fifteen percent. Since 70% of Africans depend on rain-fed agriculture for a living, the continent is especially vulnerable to climate change due to its geographical location. 80% of the farmed land in Sub-Saharan Africa is made up of smallholder farms (IPCC, 2021; Smith et al., 2014). A major threat to agricultural output and food access is predicted by the Intergovernmental Panel on Climate Change (IPCC, 2014) as a as a result of climate variability and change in climate.

Unfortunately, the agricultural sector has been found as a major land user and consumer of fossil fuels, agriculture also directly contributes to greenhouse gas emissions through practices like raising livestock and producing rice. According to the Intergovernmental Panel on Climate Change, land use, agriculture, and fossil fuels have been the three main causes of the increase in greenhouse gases observed over the past 250 years (CCA, 2021). The agricultural food system is responsible for a significant amount of the greenhouse gas emissions that are produced (CCA, 2021).

Mitigation in this regard is defined as actions that lower greenhouse gas (GHG) emissions or improve carbon sinks' ability to absorb GHGs from the atmosphere. Acts that can improve natural sinks, like reforestation, are also included in the mitigation of climate change, afforestation, farming system and other appropriate agricultural management practices that could increase carbon sequestration (Auwal et al., 2018; Chris-Emenyonu et al., 2020). Because of extreme weather conditions including intense rainfall and high soil temperatures, tropical soils are naturally low in fertility, have low levels of organic carbon, are rapidly deforested, and degrade physically more quickly than other soil types (Are et al., 2018). Adaptation to climate change is the process of adjusting to the effects of climate change and resilience is the tendency for a system to envisage, curtail, tolerate and survive the outcome of incidences that are hazardous to its existence in such an efficient and timely manner by utilizing sustainable execution of preservation, advancement, and restoration of its essential structure and functions. Resilience strategies may be evaluated in terms of buffer capacity, self-organization, and adaptive capacity (Izuogu et al., 2021).

The probable consequences of climate change include variation in temperature regimes, rainfall pattern, agricultural yields, sea level rise, regular occurrence of extreme weather events like heat waves, droughts, heavy intensity rainfall, cyclones, hurricanes, melting of glaciers or ice, forest fire, flooding change in ecosystem function, shifts in species distribution, species extinctions desert encroachment, rise in sea level etc. (Magesa et al., 2023). By 2050, greenhouse gas emissions are expected to increase by 52% (CCA, 2021). Events including heat waves, droughts, storms, and floods will cause changes in regional weather patterns (CCA, 2021). Climate change affects water availability, water is increasingly scarce. 90 percent of fresh water supplies will disappear by 2030 if the current trends continue without mitigation.

In addition to decreasing stability and resilience, fragmentation, extinction of species, and degradation of ecosystem quality are all consequences of climate change. It also affects biodiversity and renewable natural resource depletion. Higher temperatures have an impact on many agricultural systems since certain plants are vulnerable to them at crucial times like flowering and seed development (Auwal et al., 2018; Tawakalitu et al., 2022). Numerous pests and diseases that affect humans, animals, and crops are susceptible to variations in the climate, particularly in temperature. Certain illnesses and pests may spread to previously undiscovered places (Brandyopadhy, 2013). With the level of the negative impacts of change in the climate globally on agriculture, proactive measures for mitigation, adaptation and resilience must be discovered and adopted globally especially in Nigeria and Abakaliki, Ebonyi state in particular to reduce and overcome the impacts.

Numerous existing studies on climate change and agriculture have been observed to focus on either mitigation alone, or adaptation alone, or mitigation and adaptation or adaptation and resilience but studies on climate change and agriculture: mitigation, adaptation and resilience in Abakaliki, Ebonyi state is very scanty hence the need for the review study. The review study seeks to create more awareness through published articles in the study area which will help the people living the study area to be well informed of the effects of climate change, the mitigation, adaptation and resilience practices. This paper aimed at reviewing climate change and agriculture: mitigation, adaptation and resilience in Abakaliki, Ebonyi state.

18.2 Methodology

The review study adopted systematic review. The review was conducted using existing literature through the implementation of a qualitative approach that provided information on agriculture and climate change, mitigation, adaptation and resilience in both international and local papers. Secondary sources of information for the study included publications from online open access journals such as Scopus, Web of science, Google scholar and publications by universities. In a bid to harness existing and relevant data for the study, a literature study was conducted with the Preferred Reporting Items for Systematic Reviews and Meta Analysis (PRISMA) approach as was previously used by Hackfort (2021). The statement of inclusion and exclusion criteria was used in determining the eligibility criteria of research articles.

18.2.1 Criteria for Inclusion and Exclusion

Article inclusion and exclusion criteria were set out in the research using six (6) steps as follows:

i. The area of knowledge for the article must be related to and focused on agriculture and climate change, mitigation, adaptation and resilience published internationally and local(Nigeria).
ii. Articles must be published in English language;
iii. Articles must have been published between 2013 and 2023, this implied that articles that were published earlier on were excluded;
iv. Articles must be of empirical data;
v. All non-peer reviewed articles were excluded;
vi. Full text must be available.

During the literature search, a total of 100 studies were obtained but our selection criteria gave rise to a final sample of 51 final studies reviewed. Therefore, a total of 51articles were used for the review. Of the 45 papers included in the review, 20 were on agriculture and climate change in international and local papers, 25 were on mitigation and adaptation in international and local, while 6 were on adaptation and resilience in international and national papers. Having set our inclusion and exclusion criteria for the content analysis, as earlier mentioned, the PRISMA method was chosen to be sure of the scientific standard with guaranteed transparency. The acceptable databases were identified and were used for the search. The sources used included web of science, Scopus, google scholar and university-based papers. The key words that were used in the search covered titles, keywords, abstracts, climatic change, agriculture, mitigation, effect of climate change, adaptation and resilience. We extracted 100 articles from the databases which contained 28 from Scopus, 9 from Web of Science. 18 from University journals and 45 from Google scholar. After reading through the papers and subjecting them into the evaluation processes of inclusion and exclusion using the PRISMA guideline, 51 papers were found to be included and were used for the study (Fig. 18.1).

18.2.2 Effects of Climate Change

The review study revealed the following effects of climate change: changes in temperature regimes, patterns of rainfall, agricultural yields, rising sea levels, frequent occurrences of extreme weather events such as heat waves, droughts, intense rainfall, cyclones, hurricanes, melting of glaciers or ice, forest fires, flooding, altered ecosystem functions, changes in species distribution, extinctions of species, encroachment of deserts, rising sea levels, etc. (Sindhu, 2011; Smith et al., 2014). By 2050, it is anticipated that greenhouse gas emissions will increase by 52% (Orji & Iroegbu, 2021). Heat waves, droughts, storms, and floods are examples of phenomena that can alter regional weather patterns (Smith et al., 2014). Water is becoming more and more scarce due to climate change. 90 percent of fresh water supplies will disappear by 2030 if the current trends continue without mitigation.

The loss of biodiversity and renewable natural resources, as well as decreased stability and resilience, fragmentation, extinction of species, and degradation of

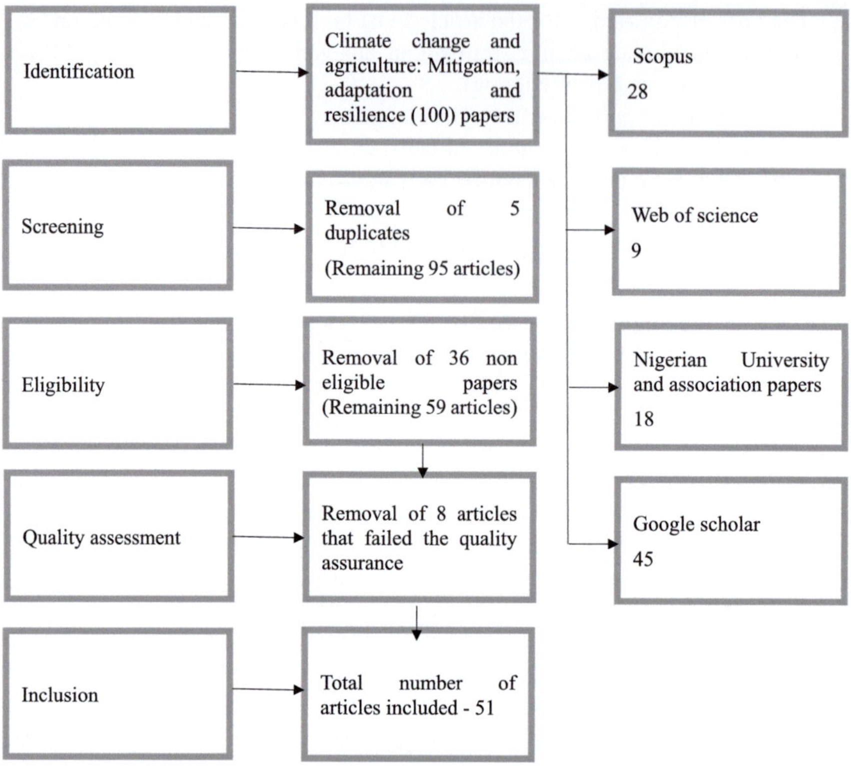

Fig. 18.1 PRISMA protocol search flow research

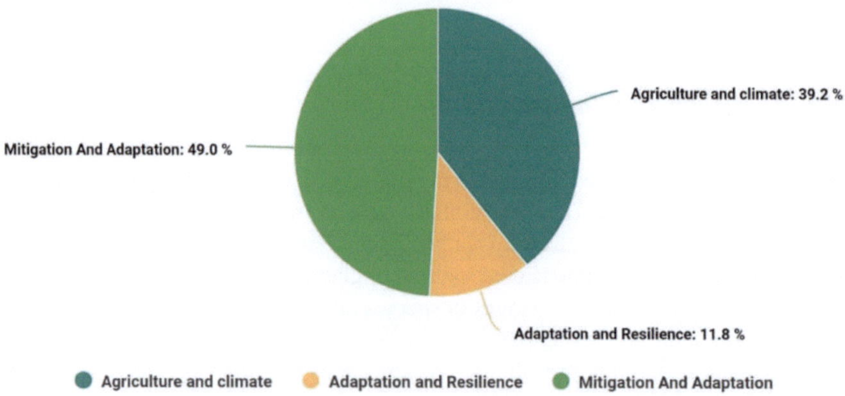

Fig. 18.2 Article distribution according to the aspect or area of coverage using pie chart

ecosystem quality, are all consequences of climate change. Higher temperatures will have an impact on many agricultural systems because certain plants are sensitive to them at crucial times like flowering and seed development (Auwal et al., 2018). Numerous pests and diseases that affect humans, animals, and crops are susceptible to variations in the climate, particularly in temperature. Certain pests and illnesses could spread to previously undiscovered locations (Brandyopadhy, 2013).

Ajetomobi (2016) investigated variations in extreme weather, with a focus on variations in extreme temperature and extreme rainfall, as part of his research on the development of food crops and climate change. Regarding high temperatures, he concludes that while groundnut, cowpea, and melon yields are positively impacted, the average yields of cassava, cocoyam, cotton, millet rice, sorghum, yam, and maize are severely impacted. Ajetomobi concludes that there is a negative correlation between average yields for rice, sorghum, and cassava in relation to extreme rainfall, but a positive correlation for the other crops. On the other hand, the effects of total rainfall throughout each crop growth season are negligible for sorghum, yam, maize, and melon, and negative for cotton and millet.

According to Rani and Maeswari (2015), there are two main ways that climate change affects agriculture: it affects the geographical boundaries of the regions that produce different kinds of crops and livestock, as well as the potential yields of those crops and livestock. Climate change's unexpected repercussions may negatively impact Sub-Saharan Africa's natural and human systems in several ways (Serdeczny et al., 2016). The high mortality rate in fisheries and livestock is one of the depressing effects of climate change in Nigeria's agriculture sector (Cervigni et al., 2013; Chukwunonso et al., 2020), leaching (Abdullahi, 2018; Odewumi et al., 2013; Nwalem et al., 2019), floods (Iheoma, 2015; Ikehi, 2014), wilting of seedlings after germination and transplanting (Babatolu & Akinnubi, 2016), and drought (Abdullahi, 2018; Abubakar & Yamusa, 2013; Chukwunonso et al., 2020; Odewumi et al., 2013). The Length of Growing Period (LGP) is a significant feature of crop productivity that has been impacted by climate variability. Even though LGP may rise as a result of global warming, higher day and nighttime temperatures may promote the physiological development of crops, resulting in earlier crop maturity and lower agricultural output. Rainfall variability is predicted to grow as a result of climate change on a global scale. Because of the increased atmospheric requirements brought on by the higher temperatures, there will be less rainfall.

As this determines the planting season, the evaluated publications also expressed worry regarding abnormalities in the late onset, frequency, and intensity of rainfall. The farmers said that as a result, they frequently utilize irrigation at this early stage. Numerous farmers conveyed their apprehension regarding the threat presented by the growing frequency of erosion and flooding, noting that these occurrences have previously carried away their standing vegetable crops and topsoil, diminishing the fertility of the soil. The farmers reported that they had to pay more to apply fertilizer and manure to restore soil fertility.

18.3 Results

18.3.1 *Distribution of Articles According to the Area Covered and Category of the Journal Papers*

The selected journal papers were distributed according to the area of coverage; the papers that covered the area of agriculture and climate change were 20 (32.9%), 25 (49.02%) papers both international and local covered the area of adaptation and mitigation of climate change effects whereas 6 (11.8%) international and local papers covered the area of adaptation and resilience of the effects of climate change (Figs. 18.2 and 18.3) with the highest value at the papers on adaptation and mitigation. This agrees with the findings of Abbass et al. (2022) and IPCC (2021) that many mitigation and adaptation strategies have been developed.

Figures 18.3 and 18.4 show the distribution of the journal papers according to the area of coverage and location of publication. 15 papers on climate change and agriculture were published internationally which represents 29.4% of the articles whereas 5 papers representing 9.8% on climate change and agriculture were published locally. Similarly, 18 journal articles representing 35.3% and 7 journal papers representing 13.7% on climate change mitigation adaptation were published internationally and locally respectively. 4 and 2 papers representing 7.9 and 3.9% on adaptation and resilience were published internationally and locally respectively (Figs. 18.3 and 18.4). The results showed that there are more papers on climate change and agriculture, mitigation, adaptation and resilience published internationally than locally. The results further showed that papers on mitigation and adaptation recorded highest value whereas the lowest was recorded by the papers on adaptation and resilience (Figs. 18.1, 18.2, 18.3 and 18.4). This shows that more attention in terms of research and publication is needed in adaptation and resilience both locally and internationally but more locally for the sake of the rural people in the study area (Fig. 18.5).

18.3.2 *Mitigation*

Reducing the effect of climate change on our world is known as mitigation. To lower the atmospheric concentration of greenhouse gases, human intervention is required (Sigh, 2016). The IPCC (2014; 2021) defines mitigation as actions taken to lower greenhouse gas (GHG) emissions or improve carbon sinks' ability to absorb GHGs from the atmosphere (Auwal et al., 2018). Moreover, actions to improve natural sinks—like reforestation—and enhanced soil carbon sequestration through sensible agricultural management are included in the mitigation of climate change. Some of the mitigation strategies/practices according to the reviewed literatures include afforestation, proper conservation of seeds, reduced use of fossil fuel for energy generation, avoidance of deforestation, minimum tillage operation, etc. (Abbass et al., 2022; Okoroh et al., 2016).This is a necessary step toward achieving net-zero and

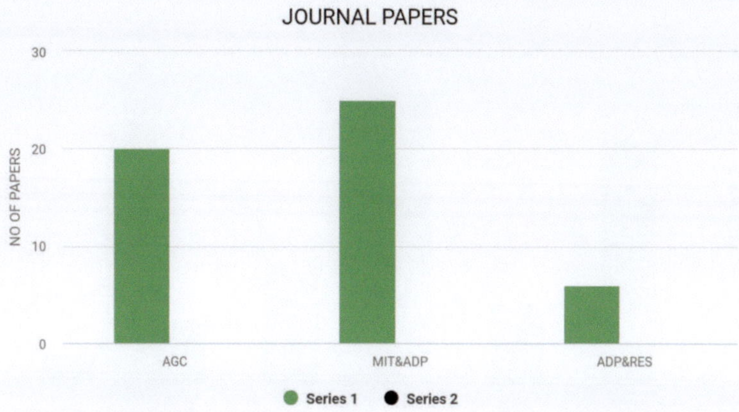

Fig. 18.3 Article distribution according to the aspect or area of coverage using bar chart

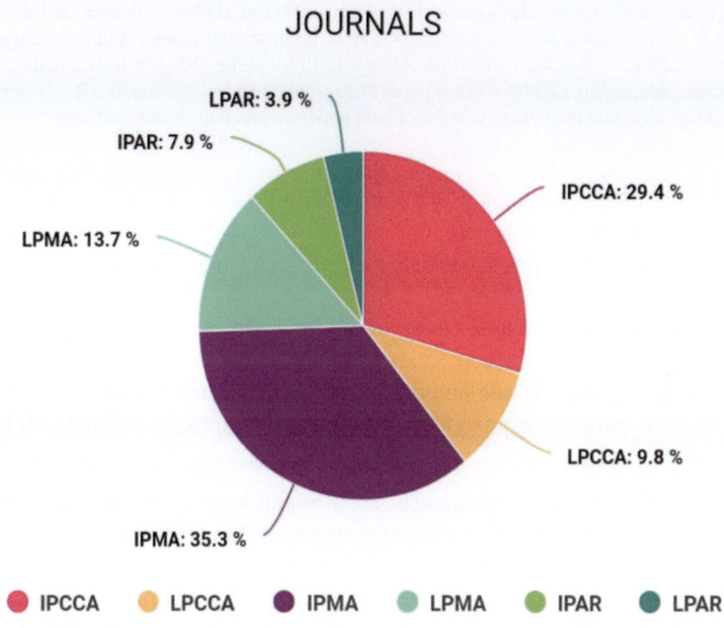

Fig. 18.4 Article distribution according to the location of the paper (either international or local) using pie chart

limiting the rise in global temperature to 1.5 °C. Global mitigation strategies which includes all actions that remove or reduce greenhouse gas emissions suggested by the reviewers for the farmers in the study area to also adopt include the following:

- the implementation of low-carbon energy sources;

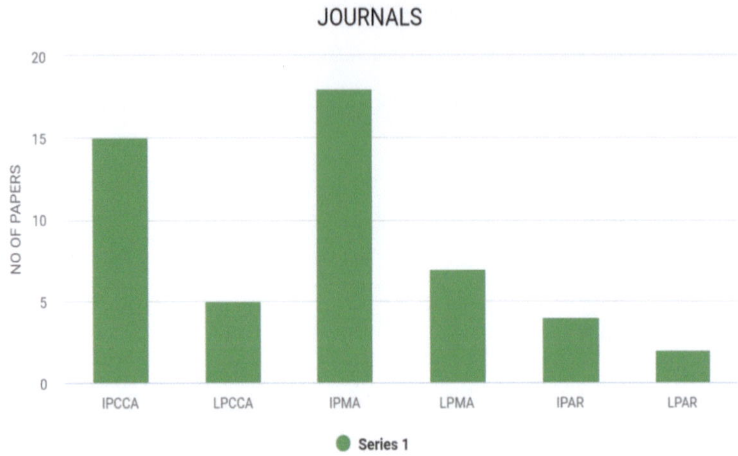

Fig. 18.5 Article distribution according to the location of the paper (either international or local) using pie chart. AGC = Agriculture and climate change, MIT &ADP = Mitigation and adaptation, ADP&RES = Adaptation and resilience, 1PCCA = International paper on climate change and agriculture a, LPCCA = Local paper climate change and agriculture, IPMA = International paper on mitigation and adaptation, LPMA = Local paper on mitigation and adaptation, IPAR = International paper on adaptation and resilience, LPAR = Local paper on adaptation and resilience

- the utilization of developing technologies to mitigate or eliminate carbon dioxide emissions from the atmosphere;
- the preservation of naturally occurring carbon sinks, like soil, forests, and oceans; and
- the modification of human behavior to minimize our collective carbon footprint, like utilizing locally produced goods and altering our modes of transportation.
- by accelerating the rate and amount of carbon sequestered in biomass, planting new trees, restoring degraded forests, and improving existing forests all help to mitigate climate change.

Low-carbon energy sources should be used; emerging technologies should be used to reduce or completely stop carbon dioxide emissions from the atmosphere; naturally occurring carbon sinks like soil, forests, and oceans should be preserved; and human behavior should be changed to reduce the overall carbon footprint of society, such as buying locally made goods and changing our ways of transportation. Planting new trees, reforesting damaged areas, and enhancing already-existing forests all contribute to mitigating climate change by increasing the pace and amount of carbon stored in biomass (Auwal et al., 2018; IPCC, 2021).

18.3.3 Adaptation

Adjusting to the effects of change on our world is the definition of adaptation to climate change (McNamara and Buggy (2017). It entails being aware of the negative effects of climate change and acting to mitigate, adapt to, and prepare for such effects. It may also mean benefiting from the effects of climate change (EU, 2021; Auwal et al., 2018). The literatures reviewed showed farmers in Nigeria, southeast and Ebonyi state have adopted some adaptive measures which include planting of trees or reforestation, changing plant dates of crops, cooling livestock pens regularly, use of irrigation method, adopting heat tolerant species of animals, use of resistant varieties, using different crop varieties, adopting soil conservation approach, expansion of rain water harvesting, barriers and hedges along contours to the soil erosion and altering the timing or location of cropping activities (Oselebe et al., 2016; Castells-Quintana et al., 2018).

The following different adaptations strategies are summarized by (Auwal et al., 2018; Bandyopadhyay, 2013; Lal, 2015); they are developing improved varieties, adjusting crop sowing time, managing pests and diseases, water management, land shaping, use of bio-fertilizer, livestock systems for the future, aquaculture management, conservation of forests, soil conservation, etc.

The impacts of climate change on agriculture are already being observed in various regions of the country, according to a study of smallholder farms in the regions (Malhi et al., 2022). As a result, some farmers there have introduced new technologies and seed varieties, and intercropping, crop, soil, land, water, and livestock management systems have changed (EU, 2022; Karimi et al., 2022; Ogisi & Begho, 2023). Some recommendations for strategies to adapt include doing away with supply chain and home food waste, promoting varied and vegetable-rich diets, and ensuring that food is accessible to everyone worldwide (SAPEA, 2020; Wiréhn, 2018). It is widely acknowledged by scholars that agricultural innovation plays a crucial role in mitigating the possible consequences of climate change (Lal, 2020). According to Orji and Iroegbu (2021), Tajudeen et al., (2022) and Leisner (2020), this entails enhanced soil management, water-saving technology, crop compatibility, crop rotation, introduction of diverse crop varieties, and support for community-based adaptation measures.

Other adaptation actions practiced in other places and across the globe that farmers in Ebonyi state should also consider include:

- building barriers against sea level rise;
- acting to safeguard infrastructure and buildings against weather events with abrupt or gradual onset;
- adjusting crop types for food production and farming to correspond with conditions that are wetter, colder, or warmer; and
- modifying behavior to mitigate the effects of climate change, such as cutting back on food waste;
- cultivating crops resistant to drought
- Water-saving technology should be used,

- satellites should be used to monitor the environment,
- adaptation science and climate analytics should advance quickly to provide cutting edge climate data, and
- digital tools should be scaled up to increase human capacity for adaption.

To gain a better understanding of the problems and the most effective ways to solve them, government and international research and investments in agricultural production and infrastructure are required. As well as funding, insurance, and safety nets for vulnerable people, current government policies and programs are intended to provide environmentally conscious government subsidies, educational campaigns, and financial incentives (Arya, 2021*;* EU, 2022). Additionally, giving precise weather forecasts and early warning systems to underdeveloped or isolated places will enable better planning; by utilizing and sharing the existing technology, the international community can address and mitigate the global issue of climate change (Magesa et al., 2023; Singh, 2016*). The review study also suggested* adaptation policy cycle and support provided by the UN climate change regime (UNFCCC, 2021) to be adopted by the stakeholders in agriculture and farmers in Abakaliki, Ebonyi state.

18.3.4 Resilience

The capacity of a system to take in, tolerate, and recover from a negative occurrence is known as resilience. Municipalities are positioned to withstand natural catastrophes and other hazardous effects of climate change through a combination of policies, infrastructure, services, transportation, energy infrastructure, and planning *(*Kahiluoto et al., 2019*)*. Being resilient to climate change may also entail anticipating its effects and developing the capacity to bounce back swiftly from them. Changes in climate resilience is the capability and aptitude to anticipate and mitigate the negative effects of climate change (Nelson, 2011; Onyeneke et al., 2019). Although it is linked to adaptability, it is not the same. According to Onyeneke et al. (2019), resilience is the ability to prepare for the effects of climate change, especially the ability to recover from hazardous climate-related events. Adaptation is the process of "adjusting to the actual or expected climate and its effects" (Acevedo et al., 2020; EU, 2021; Pierre et al., 2022). Practices that increase climate resilience which should be taken into consideration according to the reviewed literatures include:

- creating early warning systems to forecast and track dangerous events, such as extreme weather;
- expanding green spaces in cities and rural areas to boost their ability to absorb floods;
- planting trees to help mitigate excessive heat in cities and other areas where there are no trees;
- continually provide electricity, food, water, shelter, medical care, and transportation, among other necessities;
- preserving the functionality of contemporary infrastructure and systems;

- preserving the livability and comfort of systems that facilitate contemporary lifestyles
- supporting populations that may be particularly vulnerable, such as the elderly and low-income earners both in the cities and rural areas;
- considering the environment and natural forces when planning land-use systems and infrastructure especially in the cities;
- increasing safety through enhanced emergency preparedness and strong coordination between pertinent agencies and first responders;
- providing a variety of transportation and access options;
- and constructing sustainable, flexible, and durable systems and structures, for example buildings.

Building stronger, more durable systems and promoting sustainable development are made possible by resilience. It stimulates investment in infrastructure and jobs, which can significantly boost economies. Resilient community planning helps settlements plan, adapt, and prosper in the face of climate change. As a result, concepts related to resilience are making their way into various agencies' policies and designs. Planning for resilience can also stop fatalities, severe financial losses, and displacement in the areas most susceptible to natural catastrophes (Donald, 2011; IPCC, 2021).

As part of resilience approach, the 2024 NIMET predictions have been serving as a reliable guide. The prediction indicates that the onset dates of 2024 rainy season in the inland states of the south are likely to occur between March and April and in the central states, it is expected to be in May. The states in the north are likely to have rainy season onset dates between June and July and some states like Ekiti, Oyo, Ondo, Edo, Delta, Anambra, Enugu and Ebonyi are likely to experience delayed onset (NIMET report, 2024). It was predicted that some states will experience shorter, longer and normal length of rainfall season. The predicted day and nighttime temperatures and departure from normal for the five critical months Jan–May. The prediction shows that most part of the country are likely to experience warmer than normal day time temperature. Due to the climate change, the daytime temperature experience across Nigeria in March 2024 are predicted to range between 31 and 40 °C. The predictions were confirmed to be valid through experienced weather changes and the meteorological observations and reports nationwide.

18.4 Conclusions, Recommendations and Limitations

Climate change and its effects are a reality globally especially in the tropical regions. Today, the impacts are obvious in agriculture, weather conditions, health, social, economy and others. It is, therefore, necessary to adopt measures and practices that will reduce and control the activities giving rise to GHGs. There is urgent need for greater attention towards environmental remediation and application of technology solutions as well as mitigation approach that may be applied. The impact of climate change is obvious especially this year 2024, the day and nighttime temperature will continue to rise as a result of already accumulated GHGs in the atmosphere. Therefore, as we adopt the mitigation practices, there is urgent need for adaptation strategies and resilient approaches globally as suggested in the work to cope with challenges of climate change. The review study therefore recommends that experts of climate change should engage more on studies on mitigation, adaptation and resilience as a holistic approach to battle against climate change even at the global scale.

The study also recommends steady research on climate change, adaptation and resilience should be encouraged and funded by government and non-governmental organizations so that findings will be published especially in the local journals and publish with the local dialects.

Acknowledgements We wish to acknowledge the encouragement and contribution of Prof Chinedum Uzoma Nwajiuba and his family towards knowledge. We also appreciate Dr Robert U. Onyeneke for his encouragement and mentorship.

Funding No funds, grants, or other support was received.

Data Availability Statement The data set used during the current study are available from the corresponding author on reasonable request.

Ethical Clearance Not applicable.

Conflict of Interest The authors have declared no conflict of interest.

References

Abbass, K., Qasim, M. Z., Song, H., Murshed, M., Mahmood, H., & Younis, I. (2022). A review of the global climate change impacts, adaptation, and sustainable mitigation measures. *Environmental Science Pollution Resource, 29*, 42539–42559.

Abdullahi, H. G. (2018). *Drought and drought mitigation in Yobe State, Nigeria (Doctoral dissertation)*. University of Wolverhampton.

Abubakar, I. U., & Yamusa, M. A. (2013). Recurrence of drought in Nigeria: Causes, effects and mitigation. *International Journal of Agriculture Food Science Technology, 4*(3), 169–180. http://www.ripublication.com/ijafst.htm

Ajetomobi, J. (2016). Effects of weather extremes on crop yields in Nigeria. *African Journal of Food, Agriculture, Nutrition and Development, 16*(4), 1116–1184.

Acevedo, M., Pixley, K., Zinyengere, N., Meng, S., Tufan, H., Cichy, K., ... & Porciello, J. (2020). A scoping review of adoption of climate-resilient crops by small-scale producers in low- and middle-income countries. *Nature Plants, 6*, 1231–1241.

Arya, J. P. (2021). *Climate change and agriculture in South Asia: Adaptation options in smallholder production systems*. Springer. Retrieved from https://link.springer.com

Auwal, M., Adamu, U. K., Adam, I. A., Dambatta, M. A., Dawaki, K. D., & Bashir, A. Y. (2018). Review on effect of climate and population growth on agricultural production and food security. In *Proceedings of the 42nd Annual Conference of SSN* (pp. 466–474).

Babatolu, J. S., & Akinnubi, R. T. (2016). Smallholder farmers' perception of climate change and variability impact and their adaptation strategies in the Upper and Lower Niger River Basin

Development Authority areas, Nigeria. *Journal of Petroleum & Environmental Biotechnology, 7*, 279. https://doi.org/10.4172/2157-7463.100027

Bandyopadhyay, B. K. (2013). Climate change impacts and adaptation for coastal areas: Issues and options. *Journal of the Indian Society of Soil Science, 61*(Supplement), 598–5105.

Castells-Quintana, D., Lopez-Uribe, M. del P., & McDermott, T. K. (2018). Adaptation to climate change: A review through a development economics lens. *World Development, 104*, 183–196.

Cervigni, R., Valentini, R., & Santini, M. (2013). *Toward climate-resilient development in Nigeria*. The World Bank. https://doi.org/10.1596/978-0-8213-9923-1

Chris-Emenyonu, C. M., Onweremadu, E. U., Njoku, J. D., Ahukaemere, C. M., & Aririguzo, B. N. (2020). Soil carbon sequestration in forest soils in relation to parent material and soil depth in south-eastern Nigeria. *American Journal of Climate Change, 9*, 425–437. https://doi.org/10.4236/ajcc.2020.94025

Climate change and agriculture | Union of Concerned Scientists. (2021). Retrieved from www.ucsusa.org

European Union. (2021). *European missions on adaptation to climate change: A climate resilient Europe*. Retrieved from https://ec.europa.eu/climate-adaptation

European Union. (2022). *Climate resilience and adaptation to climate change. Science for Climate Action*. Retrieved from https://ec.europa.eu/climate-adaptation

Grigorieva, E., Livenets, A., & Stelmakh, E. (2023). Adaptation of agriculture to climate change: A scoping review. *Climate, 11*(10), 202. https://doi.org/10.3390/cli11100202

Hackfort, S. (2021). Patterns of inequalities in digital agriculture: A systematic literature review. *Sustainability, 13*, 12345. https://doi.org/10.3390/su132212345

Iheoma, C. G. (2015). Impact of climate change on agricultural production and sustainability in Nigeria. *Asian Journal of Agricultural Extension, Economics & Sociology, 4*(1), 29–41. https://doi.org/10.9734/AJAEES/2015/11518

Ikehi, M. E. (2014). *Perceived impacts of climate change on agricultural production in Niger Delta region of Nigeria (Master's thesis)*. University of Nigeria.

Intergovernmental Panel on Climate Change (IPCC). (2021). *Sixth assessment report: Working group I report on climate change*. Retrieved from https://www.ipcc.ch/report/ar6/wg1/

IPCC. (2014). Climate change 2014: Impacts, adaptation, and vulnerability. In *Contribution of working group II to the fifth assessment report of the Intergovernmental Panel on Climate Change*. Cambridge University Press.

Izuogu, C. U., Onyeneke, R. U., Njoku, L. C., Azuamairo, G. C., & Atasie, M. C. (2021). Repositioning Nigeria's agricultural extension system towards building climate change resilience. *Sarhad Journal of Agriculture, 37*(1), 180–189.

Kahiluoto, H., Kaseva, J., Balek, J., Olesen, J. E., Ruiz-Ramos, M., & Gobin, A. (2019). Decline in climate resilience of European wheat. *Proceedings of the National Academy of Sciences of the United States of America, 116*(1), 123–128. https://doi.org/10.1073/pnas.1804387115

Karimi, V., Valizadeh, N., Rahmani, S., Bijani, M., & Karimi, M. (2022). Beyond climate change: Impacts, adaptation strategies, and influencing factors. In S. A. Bandh (Ed.), *Climate change* (pp. 49–70). Springer.

Lal, R. (2015). Restoring soil quality to mitigate soil degradation. *Sustainability, 7*, 5875–5895. https://doi.org/10.3390/su705587

Lal, R. (2020). Advancing climate change mitigation in agriculture while meeting global sustainable development goals. In *Soil and water conservation: a celebration of 75 years*. Soil and Water Conservation Society.

Leisner, C. P. (2020). Review: Climate change impacts on food security—Focus on perennial cropping systems and nutritional value. *Plant Science, 293*, 110412. https://doi.org/10.1016/j.plantsci.2020.110412

Magesa, B. A., Mohan, G., Matsuda, H., Melts, I., Kefi, M., & Fukushi, K. (2023). Understanding the farmers' choices and adoption of adaptation strategies, and plans to climate change impact in Africa: A systematic review. *Climate Services, 30*, 100362. https://doi.org/10.1016/j.cliser.2023.100362

Malhi, G. S., Kaur, M., & Kaushik, P. (2022). Impact of climate change on agriculture and its mitigation strategies: A review. *Sustainability, 13*, 1318. https://doi.org/10.3390/su13031318

Mbah, E. N., Ezeano, C. I., & Saror, S. F. (2016). Analysis of climate change effects among rice farmers in Benue State Nigeria. *Current Research in Agricultural Sciences, 3*(1), 7–15. https://doi.org/10.9734/AJAEES/2016/22512

McNamara, K. E., & Buggy, L. (2017). Community-based climate change adaptation: A review of academic literature. *Local Environment, 22*, 443–460.

Nelson, D. R. (2011). Adaptation and resilience: Responding to a changing climate. *Wiley Interdisciplinary Reviews: Climate Change, 2*(1), 113–120. https://doi.org/10.1002/wcc.91

NIMET report. (2024). *2024 seasonal climate prediction: Rainfall and temperature predictions*. Nigerian Meteorological Agency.

Nwalem, M. P., Asogwa, B. C., & Aye, G. C. (2019). Analysis of climate change manifestations among sesame farmers in Benue State, Nigeria. *Journal of Scientific Research and Reports, 22*(2), 1–7. https://doi.org/10.9734/JSRR/2019/31433

Odewumi, S. G., Awoyemi, O. K., Iwara, A. I., & Ogundele, F. O. (2013). Farmer's perception on the effect of climate change and variation on urban agriculture in Ibadan Metropolis, South-western Nigeria. *Journal of Geography and Regional Planning, 6*(6), 209–217. http://www.academicjournals.org/JGRP https://doi.org/10.5897/JGRP2013.0370

Ogisi, O. D., & Begho, T. (2023). Adoption of climate-smart agricultural practices in sub-Saharan Africa: A review of the progress, barriers, gender differences and recommendations. *Farming Systems, 1*, 100019.

Okoroh, J. P., Olaolu, M. O., & Igbokwe, E. M. (2016). Climate change mitigation and adaptation strategies used by farmers in Imo state. *Nigerian Journal of Agricultural Extension, 20*(2). https://doi.org/10.4314/jaev.v20i2.10

Olayiuka, A. (2018). Global warming and climate change: Mitigation through soil organic matter management. *Nigerian Journal of Soil Science, 28*(1), 83–90.

Onyeneke, R. U., Nwajiuba, C. A., Emenekwe, C. C., Nwajiuba, A., Onyeneke, C. J., Ohalete, P., & Uwazie, U. I. (2019). Climate change adaptation in Nigerian agricultural sector: A systematic review and resilience check of adaptation measures. *AIMS Agriculture and Food, 4*, 967–1006. https://doi.org/10.3934/agrfood.2019.4.967

Oselebe, H. O., Nnamani, V. C., Efisue, A., Onu, D., Eze, S. O., & Ogunji, J. O. (2016). Perceptions of climate change and variability, impacts and adaptation strategies by rice farmers in South-east Nigeria. *Our National Journal*. http://nepjol.info/index.php/ON https://doi.org/10.3126/on.v16i2

Orji, J. E., & Iroegbu, C. S. (2021). Mitigation of climate change through soil (terrestrial) carbon sequestration and organic matter management on tropical soils. In *Proceedings of the 54th annual conference of the agricultural society of Nigeria* (pp. 1109–1117). Alex Ekwueme Federal University Ndufu Alike, Ebonyi State

Rani, B. R., & Maeswari, K. S. (2015). Training need assessment on vulnerability and enhancing adaptive capacity of communities on climate change. *International Journal of Agricultural Science and Resource, 5*(3), 61–68.

Science Advice for Policy by European Academies. (2020). *A sustainable food system for the European Union*. SAPEA. https://doi.org/10.26356/sustainablefood

Serdeczny, O., Adams, S., Baarsch, F., Coumou, D., Robinson, A., Hare, W., & Reinhardt, M. (2016). Climate change impacts in Sub-Saharan Africa: From physical changes to their social repercussions. *Regional Environmental Change*. https://doi.org/10.1007/s10113-015-0910-2

Sindhu, J. S. (2011). Potential impacts of climate change on agriculture. *Indian Journal of Science and Technology, 4*(3), 348–353. https://doi.org/10.17485/ijst/2011/v4i3.32

Singh, S. K. (2016). Climate change: Impact on Indian agriculture and its mitigation. *Journal of Basic and Applied Engineering Research, 3*(10), 857–859.

Smith, K. R., Woodward, A., Campbell-Lendrum, D., Chadee, D. D., Honda, Y., Liu, Q., & Sauerborn, R. (2014). Climate change impacts, adaptation, and vulnerability. Part A: Global and sectoral aspects. In *Contribution of working group II to the fifth assessment report of the intergovernmental panel on climate change* (pp. 736–770). Intergovernmental Panel on Climate Change.

Tajudeen, T. T., Omotayo, A., Ogundele, F. O., & Rathbun, L. C. (2022). The effect of climate change on food crop production in Lagos State. *Nigeria. Foods, 11*(24), 3987. https://doi.org/10.3390/foods11243987

United Nations Framework Convention on Climate Change. (2021).

Wiréhn, L. (2018). Nordic agriculture under climate change: A systematic review of challenges, opportunities and adaptation strategies for crop production. *Land Use Policy, 77*, 63–74. https://doi.org/10.1016/j.landusepol.2018.05.041

Chapter 19
Perception of Climate Change in Ohaozara and Abakaliki LGAs of Ebonyi State, Nigeria: An Assessment of the Health and Agricultural Sectors

Iyke Uwazie Uwazie, Kelechi Clara Anyanwu, and John Joy Chukwu

Abstract This study conducted an empirical analysis of the perception of climate change in the Ohaozara and Abakaliki local government areas of Ebonyi State, Nigeria, focusing on its impact on the health and agriculture sectors. The motivation stemmed from the increasing vulnerability of these sectors to climatic changes, prompting the need to understand local perceptions and the associated effects. Data were collected through a survey of 405 household heads, with 202 respondents from Ohaozara and 203 from Abakaliki. Descriptive statistics were used to present respondents' perspectives on climate change, with 80% acknowledging its occurrence, while 20% did not perceive any changes. To assess the impact of climate change, binary logistic regression was applied after controlling for collinearity using a contingency coefficient test. The results revealed a significant positive relationship between climate change and both the agricultural and health sectors, highlighting climate change as a key factor affecting crop productivity and public health in the region. The study concludes that climate change has a considerable impact on agriculture and health, calling for targeted interventions. It recommends government support in the form of loans, grants, and irrigation systems for farmers, along with financial backing for healthcare facilities and equipment. Additionally, educating households about climate change is critical for effective mitigation. This research underscores the importance of adaptive policies to safeguard vulnerable communities in Nigeria from the adverse effects of climate change.

Keywords Perception · Climatic change · Health · Agricultural sectors · Ohaozara · Abakaliki

---

I. U. Uwazie (✉)
Department of Economics, College of Management Sciences, Michael Okpara University of Agriculture, Umudike, Umuahia, Abia State, Nigeria
e-mail: ui.uwazie@yahoo.com

K. C. Anyanwu · J. J. Chukwu
Department of Economics and Development Studies, Alex Ekwueme Federal University Ndufu Alike, Ikwo, Ebonyi State, Nigeria

R. U. Onyeneke et al. (eds.), *Energy Transition, Climate Action and Sustainable Agriculture*, https://doi.org/10.1007/978-3-031-83165-2_19

19.1 Introduction

Undoubtedly, due to human-induced changes to the natural world, Nigeria and Africa are seeing an increasing amount of weather unpredictability (Idowu & Ayoola, 2010). Climate change is a worldwide issue that endangers the ability to maintain long-term development, secure access to food, support agriculture, protect human health, and ensure water supply (Brown & Crawford, 2008).

The 2001 report by the Intergovernmental Panel on Climate Change (IPCC) defines climate change as changes in climatic patterns that are caused by human activity, namely the increase in greenhouse gases in the Earth's atmosphere. Climate change is caused by both natural and human factors. Natural factors include changes in solar output, natural aerosol emissions, and fluctuations in the earth's orbital characteristics.

Human factors include the burning of fossil fuels, industrial activities, production of cement, changes in land use, deforestation, and growing crops (IPCC, 2001, 2007; Van der Werf et al., 2009).

Climate change affects agriculture in that most crops in Nigeria that are rain-fed are affected by climate change. Crop production is affected by climate change with evidence of unstable rainfall and variations in the pattern or structure of rainfall which leads to a decline of 2.5% on crop harvest, devastated farmland and floods which causes migration of pest and disease, while high temperature suffocates crops (Iranica, 2010).

Nigeria livestock production is also affected by climate variations, this has caused increases in parasites and epidemic diseases. It has also led to Scarce grazing land and water resources as a result of minimal precipitation and elevated temperatures. (this is the case of Northern Nigeria) (URT, 2005; DILAPS, 2007) this invariably irises deaths of livestock which have been reported almost each year (Shemsanga et al., 2010; Vanda et al., 2008; URT, 2003). Climate change has been the major cause of reduced livestock productivity through delayed reproductive cycles, reduced income for sellers, reduced meats for consumption, reduction in investment profit by 20% (Iranica, 2011).

Another sector of the economy that is seriously challenged is the health sector. Although Nigeria is involved in various activities that promote health and health outcomes, the country faces fundamental challenges that limit growth and development of the sector. The World Health Organization (WHO) estimates about 150,000 deaths annually (especially in African countries) attributed to climate change (Nkoma & Nyong, 2006) Similarly, Opele and Ikenweiwe (2010) observed that climate change leads to increase in respiratory disease due to increase in the level of pollutant malaria, skin ailment, heat stroke, portable water shortage due to foods etc.

In relation to the issue at hand, previous administrations have endeavoured to establish a secure and conducive environment for the well-being of its populace (Marmot & Allen, 2012). Addressing the impact of climate change enhances the functioning of the ecosystem, as 75% of the population is engaged in agricultural activities and their well-being is crucial for sustained productivity (Ikpi & Akinyosoye,

2013). Prior to this period, the climate system occurs in its stable state in the sense that the temperature was at its bearable rate and rainfall also and things were at its natural state before the economy evolved slowly paving way for an industrialized economy. These industrial activities brought about the emission of harmful substances as production was being carried out causing a depletion of the ozone layer of the earth core and slowly giving way to global warming and then climate change became a major concern. Overtime Nigeria has not been able to curb the role of change in the climate system despite experiencing periods of economic growth. Developing nations often bear the brunt of climate change impacts resulting from the actions of wealthy nations. This is mostly owing to the limited capacity of developing countries to mitigate these consequences, which is a consequence of poverty and inadequate technological advancement. If the developed nations promptly adopt substantial emission reductions and decrease their water pollution levels, while the developing countries adopt clean technology and reduce deforestation rates while enhancing afforestation efforts, the rate of carbon sinks will improve, and depletion of them will decrease.

By improving the quality of the atmosphere and allowing the ozone layer to naturally regenerate, the rate of global warming will be significantly decreased, ultimately eliminating its impact on individuals and the ecosystem over time (Peter Akpodiogaga, 2010). For any nation to undergo economic growth (which progressively leads to economic development) it should be productive in the sense that productive activities (industrial activities) take place, and these activities emits gaseous substances to the air causing a depletion of the natural state of greenhouse gas and causing global warming in the economy and overtime it causes a change in the climate system. For example, in 2012 there was an occurrence of a massive flood experience in the southern region of Nigeria which caused the destruction of not only individual's properties but also the lives of many citizens.

We noticed that most times weather forecast for some days are not always accurate, the reason is because overtime the climate system is affected by human activities that alters' the state of atmospheric balance. The rise in temperature in the Sahara Desert due to climate change has resulted in a decrease in agricultural productivity, prompting the Fulani herdsmen to migrate from the northeast to the greener plateau and middle belt regions in search of better grazing lands. This movement has led to conflicts with crop farmers, resulting in numerous fatalities (Beyioku, 2016). Additionally, it is observed that a shift in the climate system led to the conversion of the forest in Oyo state into grassland in the southern region of Nigeria. The coastal region of Nigeria is being threatened by the increasing sea levels. Due to its significance as a major source of the nation's oil wealth, the Niger Delta is highly susceptible to floods and has also suffered from severe oil contamination, in addition to the dangers posed by rising sea levels.

Climate change occurs within the framework of developmental stress, poverty, food scarcity, drought, outbreaks of infectious diseases, land degradation, and shifts in the environment (Brown & Crawford, 2008). The healthcare industry and agriculture sector are adversely affected by climate change, primarily due to the negligence of competent authorities towards its implications.

This study aims to investigate the perception of climate change and its impact on the agriculture and health sectors in Ohaozara and Abakaliki Local Government Areas (LGAs) of Ebonyi State, Nigeria. The research objectives are twofold: first, to assess the local perception of climate change; and second, to establish the relationship between this perception and its effects on agriculture and health. To achieve these objectives, the study formulated two research questions:

(1) Does climate change significantly affect the agricultural sector in Ohaozara and Abakaliki LGAs?
(2) Does climate change significantly affect the health sector in these areas? Corresponding null hypotheses were tested:

$H0_1$—Climate change does not have a significant effect on the agricultural sector, and .

$H0_2$—Climate change does not have a significant effect on the health sector.

19.2 Methodology

19.2.1 Population and Area of Study

Basically two (2) Local Government Areas, namely Ohaozara and Abakaliki LGA which are in Ebonyi state are the focus of this study. Ebonyi state, one of the 36 states in Nigeria comprises of 13 local government areas namely Abakaliki, Ebonyi, Ohaukwu, Ezza South, Afikpo north, Ezzasouth, Afikpo south, Ikwo, Ivo, Ishielu,Onicha and Ohaozara. But this study focuses on Ohaozara and Abakaliki LGA. Ohaozara which has a land mass of about 312km^2, density of 628.9 km^2 and comprises of an estimated population of 196,200 (NPC, 2016); Abakaliki which has a land mass of about 584km, density of 339.2 km^2, with the estimated population of 198,100 (NPC, 2016) Abakaliki lies at the intersection of Enugu, Afikpo and Ogoja roads. They both fall within the Imo-Cross River basin province of south-eastern Nigeria. Both LGAs are made up of both indigenes and non-indigenes of Ebonyi state. These include both the employed and non-employed in the locality, students, civil servants, artisans, public servants, health workers, farmers, traders and so on.

The population of this study would be selected households in Abakaliki and Ohaozara LGA of Ebonyi state. The estimated average person per household in Abakaliki and Ohaozara is five (5) persons (NBS, 2006). We divide the total population of the people of Ohaozara and Abakaliki LGA by the number of average persons per household, this result to the total of 29,725 for Ohaozara and 39,620 for Abakaliki respectively.

19.2.2 Sample Size Determination

Following, Ojide (2016), the sample size determination is expressed as follows

$$\boldsymbol{n} = \frac{\boldsymbol{N}}{1 + \boldsymbol{N}(\boldsymbol{e}^2)}1 \tag{19.1}$$

where: n is the sample size, and N represents the estimated portion of the target population with a specific attribute. e represents the margin of error.

Thus, for Ohaozara

$$n = \frac{29725}{1 + 29725(0.07^2)} \tag{19.2}$$

$$n = 202$$

For Abakaliki

$$\frac{39620}{1 + 39620(0.07^2)} \tag{19.3}$$

Pn = 203.

The figure above will be used as the sample size for this research work and 0.07 (70%) is the sampling error due to time and money constraint.

19.2.3 Sampling Techniques and Instrument of the Study

A probability sampling method was employed in this research, ensuring that every unit in the population had an equal or greater than zero chance of being selected (Lohr, 1999). This approach helped to minimize bias due to sampling error. The instrument used for data collection was a well-structured questionnaire, which included both open-ended and close-ended questions. The questionnaire was designed to gather information on the respondents' demographic and socio-economic characteristics, as well as their perspectives on agriculture, health, and climate change.

The questionnaires were administered to the respondents by trained research assistants in 2024. To ensure accuracy and prevent misinterpretation, the research assistants explained each question in detail before the respondents completed the questionnaire. After completion, the questionnaires were retrieved and used for data analysis.

19.3 3 Method of Data Analysis

This study utilized cross-sectional data. This entails quantifying several factors within the chosen sample at a given timeframe. This objective will be accomplished by implementing a meticulously designed questionnaire that will be distributed to the participants. This research study will utilize a variety of analytical tools.

Descriptive statistical analysis was employed to examine the socio-economic characteristics of the households, and data processing will be conducted using SPSS (Statistical Package for Social Science). The econometric method of logit regression will be used to analyze the data obtained (Mohammed et al., 2017; Toan et al., 2014). The method was chosen because of its ease interpretation and effectiveness in result.

19.3.1 Model Specification

In this paper, the dependent variable in this study is both dummy variables which make binary outcomes (zero and one) applicable and quantitative variable with no dummy. Non-linear models are more appropriate because binary outcomes are generating non-linear. Perception is quantified using a dummy variable in the model. A value of 1 is assigned to respondents who perceive climate change, while a value of 0 is assigned to respondents who do not perceive climate change. Based on the research conducted by Mohammed et al. (2017), a logistic regression model was utilized to examine the relevant variables that are associated with the perception of climate change among respondents in Ohaozara and Abakaliki. Furthermore, the researchers employed descriptive and analytical statistical techniques, similar to those utilized by Toan et al. (2014). The data obtained will be summarized using basic measures such as mean and percentage.

The primary aim of this study is to examine the perception of climate change. This will be accomplished with descriptive analysis, with mean and percentage being employed to present the data findings.

The second purpose is to establish a correlation between the perception of climate change impacts and the agriculture and health sectors in Ohaozara and Abakaliki LGA, respectively. The analysis will employ a logistic regression technique, specifically binary logistic regression, to gain a deeper insight into how people perceive the impact of climate change.

The Logit model, as derived from Gujarati (2013), is defined as:

$$p_i = \frac{1}{1 + e^{-zi}} = \frac{e^z}{1 + e^z} \tag{19.4}$$

where Zi $= \beta 1 + \beta 2 \mathrm{X}i + \beta 3 \mathrm{X}j$

Equation 19.2 denotes the cumulative stochastic distribution functions. It can be confirmed that when the value of Pz varies from 0 to 1, the range of Zi extends from negative infinity to positive infinity. Additionally, Pz exhibits linearity with respect to parameters, but non-linearity when it is associated with Zi. If *P*i, the probability that respondents perceive climate change is 1 and for respondents who do not perceive climate change effect is (1-Pz), Thus, the probability of the perception is expressed as

$$1 - Pi = \frac{1}{e^{z} + 1} \tag{19.5}$$

We rewrite the equation as

$$\frac{p}{1 + Pi} = \frac{e^{z}}{e^{z} + 1} x \frac{e^{z} + 1}{1} = e^{Z1} \tag{19.6}$$

$\frac{p}{1-pi}$ This is the odds ratio in favor of a perception of climate change.
Applying the natural logarithm function to Eq. (19.4), we obtain

$$\mathrm{Li} = \mathrm{Ln}\left(\frac{\mathrm{Pi}}{1 - \mathrm{Pi}}\right) = \mathrm{Zi} \tag{19.7}$$

Zi maintains its definition as stated previously. Zi is a function that depends on N explanatory variables. In this study, the logit model will be used to run two regressions, namely: for agriculture and for health.

$$Li(healthstatus) = in\frac{Pi}{1 - Pi} = \beta 1 + \beta 2Xi + \beta 3Xj + ui \tag{19.8}$$

$$Li(Cropproduction) = in\frac{Pi}{1 - Pi} = \beta 1 + \beta 2Xi + \beta 3Xj + ui \tag{19.9}$$

The models are explicitly stated as:

$$\begin{aligned} HealthStatus = {} & \beta 1 + \beta 2(ageyrs) + \beta 3(gender) \\ & + \beta 4(HHincome) + \beta 5(Eduyrs) \\ & + \beta 6(HExp) + \beta 7(AvaiHF) \\ & + \beta 8(Ra\inf all) + \beta 9(Temp) + \mu i \end{aligned} \tag{19.10}$$

For the above parameters, where pi is the intercept

β_1 = Age in years; β_2 = Gender of Household heads; β_3 = Household income;

β_4 = Education of household heads in years; β_5 = Household expenditure in numbers;

β_6 = Availability of health facilities; β_7 = Rainfall perceived; β_8 = Temperature perceived.

ui = Error term

$$\begin{aligned} CropProduction = {} & \beta 1 + \beta 2(ageyrs) + \beta 3(gender) \\ & + \beta 4(HHsize) + \beta 5(HHincome) \\ & + \beta 6(Eduyrs) + \beta 7(Rainfall) \\ & + \beta 8(Temperature) + \beta 9(Salinitylevel) + ui \end{aligned} \tag{19.11}$$

For the above parameters, where Pi is the intercept; β_1 = Age in years

β_2 = Gender of Household heads; β_3 = Household Income; β_4 = Household Size.

β_5 = Education of household heads in years; β_6 = Rainfall perceived; β_7 = Temperature perceived; β_8—Salinity level; ui = Error term.

Equations (19.8) and (19.9) are logistic function, showing the log of odds that a household perceive climate change given the socio-economic, health and agriculture determinant.

Where: Perception of climate change (regressand) = 1 for households that perceive climate change and 0 if otherwise. The analysis of climate change perception will involve the use of descriptive statistics to examine socio-economic variables, as well as the effects of climate change on health and agriculture. The binary logistic regression method will be used to assess the perception of the impact of climate change on health and agriculture, using health status and crop production as proxies (Uddin et al., 2017).

Apriori Expectations.

19.3.2 Variable Specification

For the first objective, which addresses how the agriculture and health sectors perceive climate change, the dependent variable is a dummy variable. Based on a thorough examination of existing literature, researchers utilized the perception of a reduction or augmentation in temperature and/or rainfall as indicators to assess climate change (Wolfgang et al., 2017). The study utilized independent factors such as socio-economic variables (IPCC, 2007), agricultural determinants (Mohammed et al., 2017), and health determinants (Aminul et al., 2012).

The IPCC (2012) states that the key factors influencing the impact of climate change on health are variations in climate, including both short-term and long-term changes in temperature, rainfall, and precipitation. The study conducted by Chadwick et al. (2012) examines the accessibility of public health services and the strategies implemented to safeguard the well-being of individuals.

The gender of the household heads will serve as a proxy variable in the model. It will be utilized to ascertain the perception of climate change, as well as the perception of its impact on agriculture and health. Most of the literature examined utilized gender as a factor in determining the perception of climate change (Mohammed et al., 2017; Toan et al., 2014).

The age of household heads is considered an independent variable in the model. It will be utilized to ascertain the perception of climate change, as well as its effects on agriculture and health. Most of the literature analyzed used age in years as a factor (Mohammed et al., 2017). However, it was later excluded from their research due to its high correlation with three other variables.

Household income is a vital socio-economic variable utilized as an independent variable in this study. The purpose of this study is to evaluate the influence of household income on people's perception of climate change's effects on agriculture and health. Previous research, such as the work of Adebayo Oluwole in 2017, has also examined the role of household income in this context.

The variable "education years" will serve as an independent variable in this research. It will be utilized in the model to ascertain the perception of climate change and its effects on health and agriculture. Edu years depicts if the household head if educated will be more likely to have good perception or otherwise. Most literatures capture Eduyrs such as, Abetibowo and Hessan (2005) and Adebayo Oluwo (2017) among others.

Rainfall is an independent variable because it is a basic proxy for perception of climate change and an increase or decrease might have impacts on the health statues of individuals or their agriculture (crop produce). Research conducted by, Mohammed et al. (2017), Toan et al. (2014). Cecilia and Kafula (2017) among others suffice.

Temperature is another determinant of climate change, and it is used in this research work as an independent variable to access its impact on agriculture and health status of respondents. See also literature such as Abetibowo and Hessan (2005), Cecilia and Kafula (2017), Aminu et al. (2012).

Availability of health facilities is an explanatory variable, and it is used in this model to determine perception of climate change and its impact of health, i.e. if there are available health facilities to accommodate those persons who have disease\health issues as a result of changes in climate. Md Arninul et al. (2012) used availability of health facilities in their literature.

Salinity level is an explanatory variable and basically will be used to test for the second objective, the impact of climate change on agriculture being that salinity is a key to finding out the state of the soil for crop productivity and used to find out the neatness of water found in the locality for individual's intake. See literatures such as Leweri, Chisanga (2017), Mckinsey and Evenson (1998).

Farm size is an independent variable to determine the perception of climate change on agriculture (crop productivity) which will be used to test for the second objective of the research work. Research works such as Mendelsohn and Dinar (2003), Apata (2011).

Crop productivity is a dependent variable for the second objective which is to determine if there's an impact of climate change on crop productivity because agricultural activities are vast. To enable accuracy in data received from respondents, crop productivity was deemed preferable. Literatures such as Adejuwon (2004), Lortyom et al. (2017), Mendelsohn and Dinar (2003) among other research work.

Health status is a dependent variable in the second objective which is to determine if there is really an Effects of global warming on the well-being of participants. Studies show that factors such as extreme temperature or rainfall causes discomfort in humans (Larsen, 2003), Toan et al. (2014) among other research works.

19.3.3 Estimation Techniques and Procedure

The choice of the techniques suitable to be used in this paper hinges on the research problems and objectives of any study (Koutsoyannis, 1977). A well-structured questionnaire which captures key answers regarding the study was shared with 202 households for Ohaozara LGA and 203 households for Abakaliki LGA. A Logit regression econometric model was employed for data analysis to examine the factors influencing the influence of climate change on agricultural and health status. Descriptive statistics were utilized to reflect the perception of climate change.

19.4 Results and Discussion

19.4.1 Demographic Characteristics of the Respondents in Both Local Governments

The demographic characteristics of respondents in Ohaozara and Abakaliki LGA are displayed in Tables 19.1 and 19.2 respectively. The questionnaire results revealed that the mean age of the respondents in the research region of Ohaozara was 2.18 (36–45 years) and 2.13 (36–45 years) for respondents in Ohaozara on average, 3.4 (secondary school and NCE/OND) was the mean of education years for respondents in Ohaozara and 3.41 (secondary school and NCE/OND) for Abakaliki LGA (compared to Ohaozara people in Abakaliki seemed to have higher education levels, especially, more of them had first degree and second degree education qualification). Also, the medium sized household number for Ohaozara was 5.75 (21%) and Abakaliki (5.81) that is a home of moderate size consisting of five to six persons. The source of households' income for Ohaozara LGA (mean of 2.4) and (2.43 mean) for Abakaliki LGA where majority of respondents were traders/business owners and/or engage in agricultural activities (although it would be thought of that individuals living in the rural area of Abakaliki would be more engaged in business or salaried employment). The income received by household heads on average from the survey is 2.94 (31,000 —50,000) for individuals in Ohaozara and 2.96 (31,000–50,000) for individuals in Abakaliki, which is slightly above the national average of #43,200 (Trading Economics, 2020). Asekun- Olarinmoye (2013) reported a similar pattern of family income in their study (Table 19.3).

19.4.2 Objective 1

The purpose of this study is to assess how climate change is perceived in relation to the agriculture and health sectors in the Ohaozara and Abakaliki local government areas of Ebonyi, Nigeria. This evaluation is part of the first goal of the study.

Table 19.1 Apriori expectations

Variables	Expected signs	on Expected signs of
	Crop production	Health status
Age	−\+	+
Gender	−\+	−\+
HHincome	+	+
Eduyrs	+	+
Frmsize	+	
HHexp	"	+\−
AHF		+
Rainfall	+\−	+\−
Temperature	−\+	+\−
Salinity level	+	+
AgrExp	+	+

Source Authors Compilation 2024

Table 19.2 Demographic characteristics of study population for Ohaozara ($n = 202$)

Descriptive statistics					
	N	Minimum	Maximum	Mean	Std. deviation
Ageyrs	202	1	6	2.18	1.107
Household size	202	0	18	5.75	3.389
Source of household income	202	0	6	2.45	0.987
Hhinc	202	0	5	2.94	1.584
Eduyrs	202	0	7	3.41	1.419
Valid N (listwise)	202				

Table 19.3 Demographic characteristics of study population for Abakaliki (n = 203)

Descriptive statistics					
	N	Minimum	Maximum	Mean	Std. deviation
1 age years	202	1	6	2.13	1.091
Household size	202	0	18	5.81	3.333
Source of household income	202	0	6	2.43	1.001
Household income	202	0	5	2.96	1.577
Education years	202	0	7	3.41	1.433
Valid N (listwise)	202				

19.4.2.1 Perception of Climate Change and Impact of Climate Change

According to the data shown in Tables 19.8 and 19.9 (included in the Appendix), among the individuals who are aware of climate change in both Local Government Areas (LGAs), the majority of household heads perceive climate change primarily as a rise in temperature, unpredictable rainfall, and flooding or heavy floods. The primary effects of climate change frequently cited by both local government areas (LGAs) were the influence on agricultural yields (28%), the effect on human well-being (25.7% for Ohaozara and 28.2% for Abakaliki), and the impact on the natural surroundings (20.3% for Ohaozara and 19.8% for Abakaliki). During the study, it was observed that the majority of individuals acknowledged the existence of climate change, specifically in the form of fluctuations in temperature and rainfall. Additionally, 35% of respondents associate climate change with occurrences such as flooding or excessive rainfall during the rainy season, as well as soil aridity during the dry season. In addition, we inquired about the respondents' perspectives regarding the primary factor or catalyst behind climate change. 41.6% of household heads in Ohaozara and 45% of household heads in Abakaliki attributed climate change to human activities. In contrast, 32.2% of household heads in Ohaozara and 30% of household heads in Abakaliki believed that climate change is caused by a Supreme Being, while others mentioned alternative causes such as Satan, 'Amadioha', pollution, and natural causes.

Participants were also instructed to contemplate their perspective on climate change and its impact. In Ohaozara, 42.6% of respondents and in Abakaliki, 39.1% of respondents acknowledged the existence of climate change but did not observe any significant impacts. On the other hand, 57.4% of respondents in Ohaozara and 60.4% in Abakaliki reported experiencing severe or very severe effects of climate change. Additionally, they reported that agricultural operations and produce are significantly impacted by climate change, with both Local Government Areas (LGAs) seeing a 45% influence. In Ohaozara, the effect is 18.33%, while in Abakaliki it is 17.8%. Furthermore, 9.9% and 10.9% of respondents in Ohaozara and Abakaliki, respectively, claimed that climate change leads to health dangers and loss of revenue. Regarding respondents' view of government intervention or any government plan to mitigate these effects, 21.3% in Ohaozara and 29.7% in Abakaliki answered "yes" indicating their awareness of government plans, while 66.5% in Ohaozara and 70.3% in Abakaliki responded "no" to the question.

19.4.2.2 Perception of Climate Change on Health

To assess the nexus of climate change on agriculture and health sector in Ohaozara and Abakaliki LGA Ebonyi, Nigeria, a descriptive statistics and tabular representation would be used. For both local government areas there is no significant differences on the response of household heads as regards changes in the patterns of illness in their families as shown in Tables 19.5 and 19.6, household heads were asked how frequent members of their household fall sick, in Ohaozara LGA 53.4% stated that they do not fall sick often and 50% stated that they fall sick often and for Abakaliki 74.8% said they do not fall sick often but 25.2% said they fall sick often.

Also, respondents were asked if they fall sick during the rainy season and what health problems/disease occurs due to rainfall changes as compared to 5 to 10 years ago; for both local government malaria, normal cold, fever and cough had the most occurrence during the raining season, other kind of disease noted by the respondents were pox, skin disease, pneumonia, etc. Respondents also noted a discernible shift in temperature when comparing it to the conditions experienced 5 to 10 years earlier., the most occurring health problem for both local government were headaches, fatigue, dizziness, skin burn/blistering, measles, heatstroke among others (during the course of administering the questionnaire, we were engaged in a conversation with a female household head who is also a nurse and she said "Family members, particularly my kids, are falling worse than they used to be".

Household heads also perceived an increase in their health expenditure after any disaster occurs (59% for both LGA), which increases their health spending compared to 5 to 10 years ago. Being that in Ebonyi state (salt of the nation), one of their natural resources is salt; those salty content in water deteriorate human health, this prompted the question 'do you believe human health is deteriorating due to increase in water salinity'. 65.8% of respondents in both LGA confirmed that water salinity deteriorates human health. In the words on a respondent "we cannot drink the water from the wells, even from the boreholes because it has lots of salty content and can only be used when heated". Respondents were also asked if there was any government facility nearby to enable individuals access health care facilities when needed, 69% of respondents from both local governments noted that there are available health facilities around the locality.

19.4.2.3 Perception of Climate Change on Agriculture

The survey collected responses from heads of households in both Local Government Areas (LGAs) regarding their awareness of the effects of climate change and any observed climate variations within the past 5 to 10 years. Tables 19.11 and 19.12 in the appendix provided data on the perception of climate change impacts on agriculture and agricultural practitioners. The tables presented information on individuals engaged in various agricultural activities such as aquaculture, livestock husbandry, crop plantation, poultry, and sale of farm produce in Ohaozara. This data is in response to the first question regarding their involvement in agricultural activities and the specific types of activities they are engaged in. We discovered that 25% of individuals did not recognize any influence from climate change and were not engaged in any agricultural pursuits, while 31.2% acknowledged the impact of global warming but were not involved in agriculture. The majority of respondents in Ohaozara, approximately 82 individuals, acknowledged the effects of climate change on crop cultivation and poultry farming, accounting for 5.8% of the respondents. However, a smaller group of those involved in agriculture, notably crop cultivation (28 individuals), reported no apparent influence, which was similarly the case for poultry farming (2% of respondents). In Abakaliki LGA, the majority of respondents were engaged in crop plantation, chicken farming, and fish farming. They reported that the influence of climate change was most noticeable on crop plantation (54%), followed by poultry farming (6.5%), and fish farming (3.9%). Respondents observed a correlation between climate change and an elevation in saline levels, affecting both local governments. This was reported by 61% of the respondents. Household heads also observed that climate change has affected their agricultural revenue. This might be attributed to the decline in the value of their agricultural produce when sold, resulting in lower yields compared to 5 to 10 years before. This holds true for both Local Government Areas (LGAs). Household leaders have observed the effects of environmental degradation on their agricultural property, particularly those who possess less than 1 to 2 hectares. Additionally, farmers who grow yam, cassava, and rice have reported that climate change significantly affects their crops.

To fulfill the second objective of the paper, an econometric estimation was conducted to identify and analyze the effect of climate change on the agriculture and health sectors in Ohaozara and Abakaliki LGA, Ebonyi, Nigeria. Prior to doing the data analysis, the contingency coefficients test was utilized on all Tables 19.10, 19.11 and 19.12 to identify and exclude dependent variables that exhibited significant collinearity and strong correlation with each other. Please refer to the appendix for further details. In Ohaozara, there was a significant correlation between education years and household income as well as age years. Additionally, there was a substantial correlation between health spending and household income, as well as the availability of health facilities. Regarding the factors of agriculture, there was a significant association between age and education, household size and age, as well as climate change's impact on farm size and agriculture.

The logistic regression analysis, as shown in Table 19.8, indicates that age in years has a positive but statistically insignificant relationship with the impact of climate change (at a significance level of 5%). This suggests that individuals in their middle age are more likely to have knowledge about the effects of climate change compared to both younger and older individuals. It can be inferred that older individuals, who are older, are better equipped to handle the effects of climate change. We believe that years of expertise come with age, which is why household leaders were chosen. Age is a crucial element that influences the perspective of respondents, as demonstrated in previous research (Toan et al., 2014). However, in this study, the number of years of age was determined to be inconsequential.

The economic implication for the positive sign could also indicate that climate change has more effect on those who are young (0–17) and on those who are old (above 60): being that those age bracket (young and old) depend on the middle age (18–60), it could reduce their productivity which invariably reduce the growth of the economy. There was a positive correlation between gender and the impact of climate change, but it was not statistically significant at the 5% level. The odds ratio indicates a positive correlation between the impact of climate change and gender, with male household heads being more affected than female household heads (1.293). In Nigeria, the male figure bears the burden of greater responsibility than the female gender; the positive relationship could imply that the male gender is being affected more than the female gender of household heads, and when this is the situation, then the productivity of the male gender will diminish over time this will reduce national productivity.

Household income is negatively related to the effect of climate change but significant (10%), the odds ratio (0.865) shows a negative relationship also between the dependent and independent variable. Contrary to the researcher's expectations, the negative sign in this relationship contradicted the assumption that the sign would be positive. The reason behind this assumption was that households with higher income would be more capable of understanding and coping with the impact of climate change compared to those with lower income. Instead, the findings suggested that higher household wealth was associated with a lower likelihood of household heads perceiving the impact of climate change. The economic meaning of a negative relationship is that climate change does not impact on the amount of income received by households. In other words, climate change does not affect the source of household income.

The impact of climate change is not significantly influenced by the number of years of education, as determined by a 5% or 10% level of significance. The researcher hypothesized that there would be a positive correlation between education level and the ability to perceive and cope with the effects of climate change. This assumes that those with higher education are more likely to have a heightened awareness and capacity to address the impact of climate change compared to those who are illiterate. In their research work, Asekun-Olarinmoye et al. (2014) also observed a negative association.

The presence of a health facility was found to have a statistically significant impact at a 5% level (p-value = 0.002). This suggests that places with nearby health services are more likely to effectively address and accurately perceive the impact of climate change. The rationale for this is that health professionals would have the capacity to provide comprehensive information on the impact of climate change and deliver appropriate healthcare services to individuals in their community. The economic consequence of climate change is the potential for significant impacts on health infrastructure and equipment. For instance, excessive rainfall could lead to the flooding of health institutions or their warehouses. In such situations, individuals would be unable to avail themselves of the necessary healthcare services. Bamidele et al. (2013) found that frequent illness had a favorable and significant effect at a 5% level of significance.

Prior to this investigation, the researcher anticipated a positive correlation, based on the reasoning that climate change should have an impact on the health of individuals in the specific area, particularly in the primitive locale. Individuals recognize that climate change impacts their health, as evidenced by an increase in the frequency of illness reported by respondents. The economic ramifications of the positive correlation between climate change and health are significant, potentially resulting in severe consequences and even increased mortality rates. As climate change intensifies, there is a rise in the frequency and severity of hazardous weather events, such as fluctuations in temperature and rainfall.

The impact of rainfall on health was beneficial, although it did not reach statistical significance at the 5 or 10% level. The impact of climate change on health is strongly correlated with rainfall. An increase in rainfall will result in a higher impact on the health of respondents, hence exacerbating the effects of climate change. Rainfall being one of the determinants of climate change shows a positive relationship indicating that climate change has caused severe effects and caused variations in rainfall. Variations such as irregular rainfall (which could lead to changes in agriculture production as some crops produce in seasons), too much rainfall (which could lead to destruction of crop produce thereby reducing farmer's income and their farm land via erosion; it could also cause flooding which might affect water quality leading to health problems of individuals) or little or no rainfall (this will affect crop productivity and human health via water access). This result is in line with the researcher's expectation of a positive relationship and other research works confirmed it (Chershich, 2018; Cecilia et al., 2017; Toan et al., 2014; Haque et al., 2012).

Temperature on health was positive and insignificant at 5% or 10% level. The positive relationship reflects that climate change effect would increase or reduce temperature which has a replicate effect on health; temperature increase causes heat stroke, skin blistering etc. Temperature could also cause water to dry up, affecting human health. The economic implication of this positive relationship has various effects in the economy such as unfavorable weather conditions for individuals (which would affect their productivity), tourism (generally individuals seek comfortability, tourist might reduce their visit because of this weather variability) etc. This positive relationship between temperature and health indicates that the researcher was right about the expectation and other research work (Asekun-Olarinmoye et al., 2014; Fairweather, 2020; Haque et al., 2012; Toan et al., 2014).

According to Table 19.4 results from the logistic regression model, gender is positively and not significantly correlated with the effects of climate change (at the 5% level). This suggests that men are more likely than women to be aware of the effects of climate change, suggesting that male household heads are better equipped to recognize and manage these effects. This finding of positivity aligns with the research conducted by Loan et al. (2014).

The impact of climate change is inversely correlated with household income, and this link is statistically negligible at the 5% and 10% significance levels. Additionally, the odds ratio of 0.916 indicates a negative association between the dependent and independent variables. Contrary to the researcher's initial expectations, the negative sign in this relationship was unexpected. Prior to this research, it was anticipated that the sign would be positive. The reason behind this assumption was that households with higher income would be more capable of coping with and understanding the impact of climate change compared to those with lower income. In other words, the detrimental impact of climate change on agricultural activities should be less pronounced for households with higher income. However, the outcome suggested that household heads with higher wealth were less likely to perceive the impact of climate change.

Table 19.4 Estimates of a binary Logit regression model that examines the impact of climate change on the health of respondents in Ohaozara LGA

Variables	B	S.E	Wald	Df	Sig.	Exp(B)
Gend	0.411	0.372	1.219	1	0.270	1.508
Hhinc	−0.061	0.097	0.394	1	0.530	0.941
Ageyrs	0.055	0.140	0.157	1	0.692	1.057
avaiHF	0.752	0.301	6.256	1	0.012	2.122
Step 1[a] freilIness	0.729	0.369	3.909	1	0.048	2.072
Rfallonhlt	0.042	0.116	0.132	1	0.716	1.043
Temponhlt	0.067	0.064	1.096	1	0.295	1.069
Constant	−1.230	0.680	3.276	1	0.070	0.292

Note [a]Variable(s) entered on step 1: Gend. hhinc, Ageyrs, avaiHF, freilIness, Rfallonhlt, Temponhlt.
Source Authors Compilation, 2023

Table 19.5 Reports the estimates of binary Logit regression model based on the effect of climate change on agriculture of respondents in Ohaozara LGA

Variables	B	S.E	Wald	Df	Sig	Exp(B)
Gend	0.237	0.364	0.424	1	0.515	1.267
1 ill inc	−0.088	0.099	0.787	1	0.375	0.916
Eduyrs	−0.015	0.111	0.018	1	0.893	0.985
Hhsize	0.034	0.044	0.584	1	0.445	1.035
Step l[a] IncreinSL	0.039	0.252	0.024	1	0.877	1.040
Frmsize	0.034	0.134	0.066	1	0.797	1.035
cconagri	0.163	0.107	2.335	1	0.126	1.177
Constant	−0.120	0.621	0.037	1	0.847	0.887

Note [a]Variable(s) entered on step 1: Gend, hhinc, Eduyrs, hhsize, IncreinSL, Frmsize, cconagri

The correlation between education level and the impact of climate change was found to be negative and statistically insignificant at the 5% or 10% significance level. Before obtaining this result, the researcher hypothesized that there would be a positive correlation between education years and the impact of climate change. This is because individuals with higher education are expected to possess a better understanding and ability to address the effects of climate change compared to those who are illiterate. They are also assumed to have the capacity to predict future scenarios and have higher access to information regarding the perceived impact of climate change. Climate change may impact the rate at which children learn, as well as the condition of academic infrastructure such as school buildings and libraries. Previous studies have found a direct correlation between education and the impact of climate change (Mohammed et al., 2017).

Table 19.6 Estimates of binary logit regression model based on the effect of climate change on the health of respondents in Abakaliki LGA

Variables	B	S.F	Wald	Df	Sig.	Exp(B)
Gend	0.475	0.376	1.599	1	0.206	1.608
. tge\rs	0.006	0.145	0.002	1	0.966	1.006
Hhinc	−0.009	0.105	0.007	1	0.934	0.991
Eduyrs	−0.171	0.118	2.104	1	0.147	0.843
Step l[a]avaiHF	0.751	0.301	6.216	1	0.013	2.118
freillness	0.622	0.376	2.737	1	0.098	1.863
Rfallonhlt	0.055	0.117	0.223	1	0.637	1.057
Temponhlt	0.064	0.064	1.009	1	0.315	1.067
Constant	−0.727	0.753	0.933	1	0.334	0.483

Note [a]Variable(s) entered on step 1: Gend, Ageyrs, hhinc, Eduyrs, avaiHF, freillness, Rfallonhlt, Temponhit

The household size had a beneficial effect, but it was not statistically significant at the 5% and 10% level. This suggests that individuals living in households with a greater number of members (resulting in a larger family size) may be better equipped to cope with the impact of climate change and possess a heightened capacity to comprehend the consequences of climate change. The rationale behind this is that having a big family size fosters increased interaction among family members, leading to a heightened awareness of the impact of climate change and the generation of more effective strategies for addressing it, ultimately resulting in improved agricultural output. The economic consequence of having a large house size for agricultural purposes is that it allows for enhanced labor services. Climate change has an impact on labor productivity since it can impair the well-being of humans, thereby diminishing their ability to work. Another study corroborated these findings, however with a documented negative yet statistically significant correlation (Wolfgang et al., 2017).

The salinity level exhibited a positive but statistically insignificant increase at the 5% and 10% significance levels. Prior to this study, the researcher anticipated that Ebonyi state would have high levels of salt in its soil, which could be detrimental to health and could decrease the nutritional value of agricultural products. The economic consequence of a positive correlation is that climate change has an impact on the soil, resulting in a rise in saline levels. This might potentially diminish the nutritional value of food yields, resulting in hunger and potentially leading to mortality. Additionally, it may impact the quality of water sources, hence increasing health issues.

The relationship between farm size and the impact of climate change was found to be statistically non-significant at both the 5 and 10% significance levels. The outcome deviated from the researcher's anticipated positive correlation. One possible explanation is that household heads engaged in agricultural activities possess the potential but may lack adequate education, training in agriculture, and exposure to effective communication, which could hinder their ability to perceive the impact of climate change on their area. The findings indicate that an increase in farm size is associated with a decrease in the likelihood of perceiving the impact of climate change. Previous studies yielded similar findings with small variations. The studies conducted by Udin et al. (2017), Sanog et al. (2012), and Semenza et al. (2008).

The impact of climate change on agriculture was statistically significant at a 10% level. The study involved soliciting evaluations from household leaders regarding the impact of climate change on their agricultural activities. Given that most agricultural crops rely on rainfall for irrigation, the anticipated rise in droughts and floods due to climate change will pose significant challenges to food production, both in terms of quantity and quality. Furthermore, the recent increase in the pricing of food items can exacerbate the problem of food insecurity and the danger of malnutrition, particularly among impoverished urban dwellers. This pertains to significant studies regarding climate change, such as (Bello et al., 2012; Etsia et al., 2002).

In determining the impact of climate change on agriculture, age years and the impact of climate change on agriculture were removed from the model, and farm size was maintained in the logistic regression model. This was done for health determinants, where health expenditure was left out and education years were kept in the model (Table 19.5). Tables 19.9 and 19.10 display the outcomes of the binary logistic regression analysis examining the impact of climatic change, social factors, and determinants of health and agriculture in Abakaliki LGA. Table 9.10 presents the findings of the logistic regression model, indicating that gender has a positive but statistically insignificant relationship with the impact of climate change (at the 5 or 10% significance level). This suggests that males are more likely than females to have knowledge of the effects of climate change. Gender is an important socio-economic factor in many assessments, particularly when considering climate change. This is evident in studies such as the one conducted by Toan et al. in (2014), where the significance of gender is highlighted.

The relationship between age and the impact of climate change was favorable, but it did not reach statistical significance at the 5% or 10% level. The odds ratio indicates a favorable correlation between the impact of climate change and age (1.006). The positive economic implication suggests that climate change has a greater impact on individuals in the age groups of 0–17 and above 60. These age brackets, which rely on the productivity of the middle age group (18–60), may experience reduced productivity, ultimately leading to a decline in economic growth.

The impact of climate change is inversely correlated with household income, and this link is statistically insignificant at the 5% or 10% level. Additionally, the odds ratio of 0.991 indicates a negative association between the dependent and independent variables. Despite the researcher's expectations, the presence of a negative sign in this association was contradictory. Before this research, it was anticipated that the sign would be positive. This was based on the reason that households with higher income would have greater capacity to cope with and comprehend the impact of climate change compared to those with lower income. Instead, the findings suggested that households with higher incomes were less likely to perceive the impact of climate change. The negative association suggests that household income is not influenced by any kind. The economic conclusion is that even though climate change exists, it does not alter or impact an individual's income.

The impact of climate change was found to have a substantial negative correlation with the level of education, with a coefficient of −5%. The researcher hypothesized that there would be a positive correlation between education level and the ability to perceive and cope with the effects of climate change. This is because individuals with higher education are expected to have a better understanding and be more capable of taking appropriate measures to ensure better health outcomes for their households. In their research work, Asekun-Olarinmoye et al. (2014) likewise found a negative association.

The presence of health facilities was found to have a statistically significant impact at a 5% level (p-value = 0.013). This suggests that regions with nearby healthcare services are more likely to effectively address and comprehend the impact of climate change. The rationale behind this is that health professionals would be able to educate household leaders about the impact of climate change and provide appropriate healthcare services to individuals in their community. Previous research has documented this finding (Bamidele et al., 2013).

The occurrence of illness was found to have a positive and statistically significant relationship, with a significant level of 10%. This is the anticipated correlation between the variables, which differs from the findings in Ohaozara LGA where a negative correlation was observed between household frequent illness and the impact of climate change on individual health in that area. The impact of rainfall on health was beneficial, although it did not reach statistical significance at the 5 or 10% level. The impact of climate change on health is directly correlated with rainfall. An increase in rainfall can lead to the proliferation of insects, particularly mosquitoes, which in turn can contribute to the spread of diseases. This, in turn, exacerbates the effects of climate change on human health. This outcome aligns with the researcher's anticipation of a favorable correlation, which has been substantiated by additional scholarly investigations conducted by Toan et al. (2014), Cecilia et al. (2017), and Chershich (2018).

The impact of temperature on health was shown to be positive but statistically inconsequential at the 5% or 10% significance level. The correlation between temperature and health, as influenced by climate change, suggests that climate change can lead to either an increase or a decrease in temperature, which in turn affects the well-being of houses and their inhabitants. The researcher's prediction regarding the existence of a positive link was accurate, as supported by previous studies conducted by Toan et al. (2014), Haque et al. (2012), Asekun-Olarinmoye et al. (2014), and Fairweather et al. (2020).

The logistic regression model, as shown in Table 19.6, indicates that gender has a positive and statistically significant relationship with the impact of climate change (at a significance level of 10%). This suggests that the likelihood of being aware of the effects of climate change is higher among males compared to females. In other words, male household heads are more likely to perceive and effectively address the effects of climate change compared to their female counterparts. This finding of positivity aligns with the research conducted by Uddin et al. (2017).

There is a positive correlation between household income and the impact of climate change, however this relationship is statistically insignificant at the 5 and 10% levels. The researcher anticipated a favorable correlation in this connection before doing the investigation. The underlying justification is that households with higher income are more equipped to comprehend and address the impact of climate change compared to those with lower income. Household heads with greater income are more likely to have greater access to resources, including training on the effects of climate change. Semanza et al. (2008) discovered a favorable correlation between household income and the perception of climate change. Similarly, Akande and Howlader (2015) elucidated that household income significantly influenced the perception of the impact of climate change. Uddin et al. (2017) also identified a similar relationship between household income and perception of climate change.

There was a negative correlation between the level of education and the impact of climate change, which was statistically significant at a 10% level of significance. Before obtaining this result, the researcher hypothesized that the number of years of education would be positively correlated with the impact of climate change. This is because educated household heads engaged in agricultural activities possess greater knowledge, comprehension, and capacity to anticipate and respond to anticipated changes. Previous studies have found a favorable correlation between education and the impact of climate change, as indicated by Mohammed et al. (2017) and Uddin et al. (2017).

The household size had a negative and statistically negligible effect at both the 5% and 10% significance levels. This suggests that individuals residing in homes experiencing an increase in family size are more likely to cope with the impact of climate change and possess a heightened capacity to comprehend the effects of climate change. The negative sign contradicts the anticipated sign of the researcher. The rationale behind this is that having a big family size fosters increased interaction among family members, leading to a heightened awareness of the impact of climate change and the generation of more effective strategies for addressing it, resulting in improved agricultural output. Other studies have also revealed a negative yet significant correlation, as indicated by Wolfgang et al. (2017) and Uddin et al. (2017). The salinity level exhibited a positive but statistically insignificant increase at the 5% and 10% significance levels. Prior to this study, the researcher anticipated a positive correlation between the impact of climate change and the rise in salinity. This is because Ebonyi state is recognized for having high salt levels in its soil, which can be detrimental to health and may also diminish the nutritional value of agricultural products, thereby contributing to food insecurity.

The relationship between farm size and the impact of climate change was found to be positive but not statistically significant at the 5% and 10% levels. The outcome deviated from the researcher's anticipated outcome of a positive correlation. The findings indicate that as farm size increases, the likelihood of seeing the impact of climate change reduces. Increasing the scale of a farm necessitates higher levels of investment and production inputs, such as seeds, fertilizers, and pesticides. These factors impose financial strain on farm budgets. Utilizing these inputs necessitates additional education, experience, and administrative capability, which can potentially impact the view of the household head regarding the impact of climate change. In a study conducted by Udin et al. (2017), it was found that there is a substantial negative relationship between farm size and the impression of climate change. The impact of climate change on agriculture was shown to be statistically significant at a 10% level. The evaluation involved soliciting ratings from household leaders regarding the impact of climate change on their agricultural activities. This pertains to significant studies regarding climate change, such as Etsia et al. (2002) and Bello et al. (2012).

19.5 Conclusion, Policy Recommendations and Limitations

This study investigates the impact of climate change on agriculture and health in the Ohaozara and Abakaliki Local Government Areas (LGAs) of Ebonyi State, Nigeria. The findings reveal that climate change significantly affects both agriculture and health, with various socio-economic factors influencing households' perception of its effects. While the results indicate a positive relationship between gender, household size, and education level and the perception of climate change, the statistical significance varied. For instance, males were found to be more aware of climate change, and the availability of health facilities played a critical role in mitigating the impact of climate change on health. The study also finds that education has a negative correlation with the perception of climate change, which may reflect disparities in the ability to cope with its impacts. Hence, some policy options for governments and policymakers. The government should provide targeted support, including financial aid, irrigation infrastructure, and climate-smart agricultural technologies, to help farmers cope with the adverse effects of climate change. Enhancing access to healthcare facilities, especially in rural areas, is crucial for mitigating the health impacts of climate change. Investments in healthcare infrastructure and public health campaigns focusing on climate change-related diseases should be prioritized. Educational programs that raise awareness about climate change and its impacts should be integrated into both formal and informal education systems. These initiatives can empower communities to adapt to changing environmental conditions. Policymakers should ensure that climate change strategies are gender-sensitive, recognizing the different roles men and women play in agricultural production and resource management.

The study has several limitations. First, the analysis primarily focuses on the agricultural and health sectors, potentially overlooking the broader socio-economic impacts of climate change in the region. Second, the study's reliance on self-reported data through questionnaires may introduce biases, as respondents' perceptions may not fully reflect actual conditions. Lastly, the exclusion of certain variables, such as the exact nature of climatic events (e.g., droughts or floods), limits the comprehensiveness of the analysis.

Future research should incorporate more diverse environmental indicators, such as temperature variability, extreme weather events, and their direct effects on agricultural yields and health outcomes. Additionally, expanding the geographical scope of the study to include more regions in Nigeria could provide a broader understanding of the country's climate change challenges. Further exploration of the role of policy interventions and international climate agreements in shaping local adaptation strategies would also be beneficial.

Acknowledgements The Authors wish to acknowledge the organizers of the maiden climate justice and just energy transition in developing countries conference in Honour of Prof. Chinedum Nwajiuba @60 and to the editor and anonymous peer reviewers of this paper.

Funding This study did not receive any specific funding.

Data Availability Statement The data for this study will be available on request.

Ethical Clearance All ethical clearance was followed.

Conflict of Interest The Authors declares that there is no conflict of interest.

Appendix

See Tables 19.7, 19.8, 19.9, 19.10, 19.11 and 19.12.

Table 19.7 Perception of climate changes and impact of climate change for Ohaozara LGA

		Count	Column N%
Interception of climate change	0	27	13.4
	Flood/heavy food	33	16.3
	Rain	9	4.5
	Heavy rain	26	12.9
	Irregular rainfall	33	16.3
	Increase in temperature	58	28.7
	Decrease in temperature	6	3.0
	Hardness of soil	5	2.5
	Others	5	2.5
Impact due to climate change	0	34	16.8
	Impact on human health	52	25.7
	Impact on crops	57	28.2
	Impact on livestock	10	5.0
	Impact on environment	41	20.3
	Unknown	8	4.0
Effects of climate change on household	0	32	15.8
	Loss of agriculture activities	91	45.0
	Loss of trees/gardens/houses	11	5.4
	Loss of domestic animals	10	5.0
	Loss of income	20	9.9
	Health hazards	37	18.3
	6	1	0.5
Effects of climate change	Not severe	86	42.6
	Severe and very severe	116	57.4
Cause of climate change	0	38	18.8
	Supreme being	65	32.2
	Human activities	84	41.6
	Others	15	7.4
Government plan	0	24	11.9
	Yes	43	21.3
	No	135	66.8

Source Authors Compilation from Respondents Response, 2022

Table 19.8 Perception of climate change and impact of climate change for Abakaliki LGA

		Count	Column N%
Interpretation of climate change	0	24	11.9
	Flood/heavy flood	35	17.3
	Rain	9	4.5
	Heavy rain	29	14.4
	Irregular rainfall	30	14.4
	Increase in temperature	59	29.2
	Decrease in temperature	6	3.0
	Hardness of soil	5	2.5
	Others	5	2.5
Impact of climate change	No	47	23.3
	Yes	155	76.7
Government plan	Yes	60	29.7
	No	142	70.3
Cause of climate change	0	36	17.8
	Supreme being	61	30.2
	Human activities	91	45.0
	Others	14	6.9
Effects of climate change on household Effect of Climate change	0	30	14.9
	Loss of agricultural activities	91	45.0
	Loss of trees/gardens/houses	12	5.9
	Loss of domestic animals	10	5.0
	Loss of income	22	10.9
	Health hazards	36	17.8
	6	1	0.5
	Not severe	79	39.1
	Severe and very severe	122	60.4
	2	1	0.5
Impact due to climate change	0	32	15.8
	Impact on human health	54	26.7
	Impact on crops	57	28.2
	Impact on livestock	10	5.0
	Impact on environment	40	19.8
	Unknown	9	4.5

Source Authors Compilation from Respondents Response, 2022

Table 19.9 Perception of climate change on health for Ohaozara Local Government Area

		Count	Column N%
Frequent illness	Not often	108	53.5
	Very often	92	45.5
	2	2	1.0
Rainfall on health	0	10	5.0
	Malaria	117	57.9
	Pox	5	2.5
	Normal cold/cough/fever	60	29.7
	Skin disease	4	2.0
	Asthma	1	0.5
	Pneumonia	4	2.0
	Others	1	0.5
Temperature on health	0	14	6.9
	Skin burning/blistering	45	22.3
	Heat stroke	12	5.9
	Burning sensation	10	5.0
	Cardiovascular disease	3	1.5
	Measles	16	7.9
	Headaches/fatigue/dizziness	99	49.0
	Hypertension	1	0.5
	Others	2	1.0
Health expenditure	0	27	13.4
	Yes	121	59.9
	No	53	26.2
	3	1	0.5
Health expenditure	0	14	6.9
	5000–20,000	118	58.4
	21,000–30,000	54	26.7
	31,000–40,000	11	5.4
	41,000–50,000	1	0.5
	Above 51,000	3	1.5
	6	1	0.5
Source of drinking water	0	4	2.0
	Rivers	32	15.8
	Supplied water through pipes	95	47.0
	Shallow wells	2	1.0
	Deep wells	16	7.9
	Rain water	27	13.4

(continued)

Table 19.9 (continued)

		Count	Column N%
	Others	26	12.9
Salinity level	0	17	8.4
	Yes	133	65.8
	No	52	25.7
Available health facility	0	11	5.4
	Yes	140	69.3
	No	50	24.8
	3	1	0.5

Source Authors Compilation from Respondents Response, 2022

Table 19.10 Perception of climate change on health for Abakaliki Local Government Area

		Count	Column N%
Frequent illness	Not often	151	74.8
	Often and very often	51	25.2
Rainfall on health	0	9	4.5
	Malaria	122	60.4
	Pox	5	2.5
	Normal cold/cough/fever	54	26.7
	Skin disease	5	2.5
	Asthma	1	0.5
Temperature on health	0	12	5.9
	Skin burn/blistering	44	21.8
	Heat stroke	14	6.9
	Burning sensation	9	4.5
	Cardiovascular disease	3	1.5
	Measles	15	7.4
	Headaches/fatigue/dizziness	102	50.5
	Hypertension	1	0.5
	Others	2	1.0
Health expenditure increase	0	25	12.4
	Yes	120	59.4
	No	56	27.7
	3	1	0.5
Health expenditure	0	14	6.9
	5000–20,000	119	58.9
	21,000–30,000	51	25.2
	31,000–40,000	11	5.4
	41,000–50,000	2	1.0
	Above 51,000	4	2.0
	6	1	0.5
Source of drinking water	0	4	2.0
	Rivers	34	16.8
	Supplied water through pipes	93	46.0
	Shallow Wells	2	1.0
	Deep Wells	16	7.9
	Rainwater	28	13.9
	Others	25	12.4
Available Health Facility	0	11	5.4
	Yes	138	68.3

(continued)

Table 19.10 (continued)

		Count	Column N%
	No	52	25.7
	3	1	0.5
Salinity	0	17	8.4
	Yes	133	65.8
	No	52	25.7

Source Authors Compilation from Respondents Response, 2022

Table 19.11 Perception of climate change on agriculture in Ohaozara LGA

		Impact of climate change			
		No		Yes	
		Count	Column N%	Count	Column N%
Agriculture activities	0	12	25.0	48	31.2
	Fish farming	3	6.2	6	3.9
	Animal husbandry	0	0.0	4	2.6
	Crop plantation	28	58.3	82	53.2
	Poultry	2	4.2	9	5.8
	Sales of farm produce	2	4.2	3	1.9
	Others	1	2.1	2	1.3
Increase in salinity Level	0	9	18.8	5	3.2
	Yes	21	43.8	95	61.7
	No	18	37.5	54	35.1
Agriculture income	0	15	31.2	57	37.0
	5000–20,000	6	12.5	17	11.0
	21,000–30,000	11	22.9	31	20.1
	31,000–40,000	9	18.8	17	11.0
	41,000–50,000	2	4.2	9	5.8
	Above 51,000	5	10.4	23	14.9
Farm size	0	13	27.1	51	33.1
	Less than 1 hectare	11	22.9	34	22.1
	1–2 hectares	14	29.2	38	24.7
	2–4 hectares	6	12.5	19	12.3
	Above 4 hectares	4	8.3	12	7.8

Source Authors Compilation from Respondents Response, 2022

Table 19.12 Perception of climate change impacts on agriculture in Abakaliki LGA

		Impact of climate change			
		No		Yes	
		Count	Column N%	Count	Column N%
Agriculture activities	0	12	25.5	48	31.0
	Fish farming	3	6.4	6	3.9
	Animal husbandry	0	0.0	5	3.2
	Crop plantation	27	57.4	84	54.2
	Poultry	2	4.3	10	6.5
	Sales of farm produce	2	4.3	1	0.6
	Others	1	2.1	1	0.6
Increase in Salinity Level	0	8	17.0	5	3.2
	Yes	8	44.7	95	61.3
	No	21	38.3	55	35.5
Agriculture income	0	15	31.9	58	37.4
	5000–20,000	6	12.8	16	10.3
	21,000–30,000	10	21.3	28	18.1
	31,000–40,000	9	19.1	20	12.9
	41,000–50,000	2	4.3	11	7.1
	Above 51,000	5	10.6	22	14.2
Farm size	0	13	27.7	50	32.3
	Less than 1 hectare	11	23.4	37	23.9
	1–2 hectares	14	29.8	37	23.9
	2–4 hectares	5	10.6	8	11.6
	Above 4 hectares	4	8.5	13	8.4
Crop cultivate	0	13	27.7	53	34.2
	Yam	17	36.2	68	43.9
	Rice	10	21.3	19	12.3
	Maize	3	6.4	5	3.2
	Cash crops	1	2.1	4	2.6
	Others	3	6.4	6	3.9

Source Authors Compilation from Respondents Response, 2022

References

Adejuwon, S. A. (2004). The impacts of climate variability and climate change on crop yield in Nigeria. In: *Paper presented at stakeholders' workshop on assessment of impacts and adaptation to climate change*. Obafemi Awolowo University, Ile-Ife, Nigeria.

Apata, T. G. (2011). Effects of global climate change on Nigerian agriculture: An empirical analysis. *CBN Journal of Applied Statistics, 2*(1), 31–50.

Asekun-Olarinmoye, E., Bamidele, J. O., Odu, O. O., Olugbenga-Bello, A. l., Abodunrin, O., Adebimpe, W. O., Oladele, E., Adeomi, A. A., Adeoye, O., & Ojofeitimi, O. Public perception of climate change and its impact on health and environment in rural southwestern Nigeria. *Res Rep Trop.*

Bamidele, R. (2013). *Conceptualizing the Relationship between Climate Changes and Human Health in Nigeria.*

Brown, O., & Crawford, A. (2008). *Assessing the security implications of climate change for West Africa: Country case studies of Ghana and Burkina Faso.*

Haque, M. A., Yamamoto, S. S., Malik, A. A., & Sauerborn, R. (2012). Households' perception of climate change and human health risks: A community perspective. *Environmental Health, 77*(1), 1–12.

IPCC, Intergovernmental Panel on Climate Change (2001). *Climate change: the scientific basis.* http://www.ipcc.ch/

Marmot, M., Allen, J., Bell, R., & Goldblatt, P. (2012). Building of the global movement for health equity: from Santiago to Rio and beyond. *The Lancet, 379*(9811), 181–188.

Toan, D. T. T., Kien, V. D., Giang, K. B., Minh, H. V., & Wright, P. (2014). Perceptions of climate change and its impact on human health: An integrated quantitative and qualitative approach. *Global Health Action, 7*(1), 23025.

Uddin, M. N., Bokelmann, W., & Dunn, E. S. (2017). Determinants of farmers' perception of climate change: A case study from the Coastal Region of Bangladesh. *American Journal of Climate Change, 5*(1), 151–165.

Chapter 20
Artificial Intelligence and Media Coverage of Climate Change in Nigeria

Okechukwu Christopher Onuegbu, Joseph Oluchukwu Wogu, and Ogonna Wilson Anunike

Abstract Climate change is a developmental issue desiring the commitments of all stakeholders, including the mass media to be mitigated. With the emergence and use of Artificial Intelligence, the media practitioners are expected to explore it. Hence, the need for this paper. It adopted survey design using questionnaires as instruments for data collections and administered them on 500 journalists purposively selected across the country. The objectives were to; ascertain the media practitioners' knowledge level of climate change and artificial intelligence, assess the media use of artificial intelligence in coverage of climate change, and find out the media practitioners' extent of use of artificial intelligence in disseminating climate change hazards in Nigeria. The results of this study showed that media practitioners' knowledge and use of artificial intelligence in disseminating climate mitigation was poor. It recommended increasing training of journalists for better understanding of both climate change and artificial intelligence.

Keywords Artificial intelligence · Climate change · Journalists · Media coverage · Media practitioners · Nigeria

20.1 Introduction

Climate change refers to long-term changes in temperature or weather patterns over a long period of time (United Nations, 2023). These changes occurring in humidity, temperature, sunshine and others at different geographical zones in diverse ways for days, months and years, have adverse effects on humans and the entire ecosystem (Wang et al., 2021; Nasa, 2023; United Nations Framework Convention on Climate Change, UNFCCC, 1992). Basically, there are two major causes of climate change.

O. C. Onuegbu (✉) · J. O. Wogu
Department of Mass Communication, University of Nigeria, Nsukka, Nigeria
e-mail: oonuegbu3@gmail.com

O. W. Anunike
Department of Mass Communication, Federal Polytechnic, Oko, Nigeria

R. U. Onyeneke et al. (eds.), *Energy Transition, Climate Action and Sustainable Agriculture*, https://doi.org/10.1007/978-3-031-83165-2_20

One is natural occurrences such as changes in the sun or solar fluctuations, volcanic emissions, variations in earth's orbit and levels of carbon dioxide (CO_2) (British Geological Survey, 2021).

The second factor or cause of climate change is human or human activities. The key human activities contributing to climate change are greenhouse gas emissions which include carbon dioxide, methane, nitrous oxide, hydrochlorofluorocarbons (HCFCs), hydrofluorocarbons (HFCs) and ozone (National Geographic Society, 2023). Greenhouse gases also contribute to global warming, a long-term warming of the planet's overall temperature (Brown, 2024). Global warming has been ongoing over a century, but humans increase their concentration in the atmosphere while carrying out productions and consumptions, transportation, operating machineries or industrial activities (Islam & Kieu, 2021; Mann, 2024).

The effects of climate change on humans are numerous to mention. From Asia to Africa, North America, South America, Antarctica, Europe, and Oceania continents, there had been reports of earthquake, tsunami, pollution, landslide, fires or floods disasters, droughts, hotter or colder days affecting some people or nations and those being consumed by sea-level rise (Pona et al., 2021). Some have lost their shelter, food, families and other valuables. In 2022, the Intergovernmental Panel on Climate Change (IPCC) report indicated that not less than 40% of the world population are vulnerable to climate change (Haines, 2022).

In the African continent, specifically in Nigeria, evidence of climate change includes the rising temperatures, changes in precipitation patterns, sea-level rise, flooding, erosion, and other ecological challenges (Dunne, 2020). Additionally, extreme weather events, desertification, drought, land degradation, and biodiversity loss are experienced in various parts of the country (Haider, 2019; Okere, 2021). Climate change related disasters have led to loss of 662 lives, a record of 3,174 injuries, displaced 2.5 million people and destroyed 200,000 homes as at 2022 (Jaiyeola, 2023). But the impacts of climate change differ among the six geopolitical zones of Nigeria, namely North Central, Northeast, North West, South East, South South and South West.

The mass media as information purveyor, educator, entertainer and watchdog have numerous roles to play on climate change (Mike et al., 2013). The society entrusted both the mainstream and the new media with agenda setting responsibilities, including surveillance, correlation and cultural transmission (Hanson, 2017). The Sect. 22 of the Nigerian constitution (as amended) even empowered them to always hold the government accountable. Therefore, the media are expected to make the public understand what climate change is and what climate change is not, and climate adaptation and mitigation (Lee & Romero, 2022). This includes in-depth knowledge of climate change terminologies and critical issues such as greenhouse gas (GHG) emission, climate justice, energy and climate opportunities, climate finance, climate action and climate technology to improve public responses towards it (Campbell et al., 2023). This is achievable through investigative reports, features, documentary, personality interviews, among others (UNESCO, 2017).

Experts also believed that the knowledge of Artificial Intelligence (AI) can assist the media to always generate and publicise climate change stories with a view to mitigating the impact so fast. They argued that AI can assist the world in mitigating the impact of greenhouse gas emissions between 5 or 10% (Matias & Brandt, 2023). This includes predicting weather patterns, identifying pollution, mapping the impact of deforestation, tracking icebergs, improving agricultural and environmental impact. Others are waste recycling and analysis, cleaning of oceans from plastic pollutants, predicting climate disasters, energy production, climate education, solar geoengineering (Masterson, 2024; Snow, 2019; UN news, 2023).

Climate change and Artificial Intelligence (AI) are currently raising global concerns. While climate change is impacting negatively on society through loss of lives and properties, AI provides opportunities for the people to carry on more tasks with less effort and time. Thus, there is a growing need for more scholarly research on Artificial Intelligence and media coverage of climate change in Nigeria. Many studies in this domain were conducted outside Nigeria, and focused mostly on how journalists and newsrooms are exploring AI in media coverage and reporting. There is a dearth of study on how AI is helping the media in coverage of climate change in Nigeria, which faces extreme weather conditions, rising sea levels, erratic rainfall patterns, flooding, erosion and other ecological issues. This is a gap in knowledge which the current study will fill, as well as expand the body of scholarly findings on climate change and AI globally. Hence the study focused only on media practitioners' knowledge of climate change and AI, and how they are utilising the knowledge in their professional practices. It was conducted under three research objectives namely, to ascertain the media practitioners' knowledge level of climate change and artificial intelligence in Nigeria; to assess the media use of artificial intelligence in coverage of climate change in Nigeria; and to find out the media practitioners' extent of use of artificial intelligence in disseminating climate change hazards in Nigeria. The study's research questions deduced from the research objectives were: what are the media practitioners' knowledge level of climate change and Artificial Intelligence (AI) in Nigeria? Are the media practitioners' utilising Artificial Intelligence in coverage of climate change in Nigeria? To what extent are media practitioners in Nigeria utilising AI in disseminating climate change hazards?

Research Hypothesis

H1: There is no significant relationship between the media practitioners' knowledge level of Artificial Intelligence and Climate change among genders in Nigeria.

20.2 Literature Review

Artificial Intelligence (AI) is an innovation in the digital world currently making it easier for people to perform different tasks faster. This conglomerate of computer software encompassing machine learning, deep learning and natural language processing, were developed to reason, create ideas, generate contents, among other actions performed by humans (Copeland, 2024). They are of various types but most popular with regards to their abilities to think and behave like humans are reactive machines, limited memory machines, theory of mind and self-aware AI (Joshi, 2019).

Reactive machines AI have no memory to retain and recall past data, limited memory machines AI have memory to retain past events for a short period of time, theory of mind AI were built to hold emotions and intentions like humans, while self-aware AI is being built for applications that would possess super AI capabilities (IBM data & AI team, 2023). All these AI machines have assisted greatly in developing different kinds of AI tools currently serving as virtual assistants, researchers and data banks, chat bots, among others. They include ChatGPT, Google Assistant, OpenAI, DeepAI, Elicit, Midjourney, Gemini, Co-pilot, QuillBot, Civitai and Sora (Willing, 2024).

There are also AI tools good for writing, cyber security and advertising like Article Forge, Grammarly, AI Writer, Writesonic and Copy.ai. Others are Amazon Advertising, Adobe Advertising Cloud, The Trade Desk, AppNexus, MediaMath, Criteo, Taboola, Outbrain, Full Fact and ClaimBuster. Some of them are useful in personalising news feeds, creating myriads of media contents, optimising ad placements and targeting, identifying and fact-checking misinformation, predicting audience behaviour and trends.

Several studies have equally shown that media practitioners especially those in the developed countries are already utilising these AI tools in their reportage (Bechett, 2023; Hurst, 2023). The media outfits such as the Guardian, Reuters, Nucleo, CBC, National Public Radio, the Canadian Broadcasting Corporation, USA Today and Financial Times have even developed AI policies or guidelines, assuring the public that the tools could neither displace human contributions nor impact negatively in their editorial policies (Becker et al., 2023; Caswell, 2023; Cools & Diakopoulos, 2023).

However, there are dearth of scholarships on how media practitioners and their organisations are utilising AI tools for reporting climate change. But there are several AI companies working tirelessly to ensure that climate change is mitigated using AI tools. They include Space Intelligence in Scotland, Eugenie.ai in California, Greyparrot in the United Kingdom, Ocean Cleanup in the Netherlands, and Sipremo in Brazil. Others are Google DeepMind, Climate AI, Carbon Tracker AI and Riskthinking.AI. The United Nations is also utilising AI in mitigating the impact of climate across some communities in Chad, Burundi and Sudan (Snow, 2019).

Nevertheless, in Africa, only about 20% of media companies are utilising AI in the newsroom due to poor knowledge, technical-know-how and insufficient resources to invest in it (Abiola, 2023). In Nigeria specifically, media organisations using AI include Legit.ng, Dubawa and the Cableng. A study by Guanah et al. (2020) also showed that Nigerian media are not giving prominence to AI in their reports.

20.3 Empirical Review

A study by Guanah et al. (2020) showed that Nigerian media are not given adequate attention to AI in their publication as only 64 reports were published by a select newspaper in the year 2019. The study which adopted content analysis and a sample size of 285, found that the media give more priority to political stories than AI. This was supported by Nnamdi and Nwanyanwu (2021) whose survey study also found that journalists in Nigeria have not embraced AI due to power supply issues, inadequate infrastructure and finance to purchase and maintain this equipment.

Bello et al. (2023) in a survey of 376 journalists in Lagos and Kwara States, Nigeria, found that there is a high level of awareness of artificial intelligence journalism among media practitioners, but many were yet to adopt it in their day-to-day professional practice. In a quantitative study, Okocha and Ola-Akuma (2022) also found that although most Nigerian journalists are aware of the idea of robot journalism, they were not yet ready to accept it. But in developed countries many media organisations have adopted AI. Hase et al. (2021) in a panel analysis of ten countries in Global North and Global South found that the news media coverage of climate change varies across countries but is often associated with political, scientific, and (partly) societal focusing events.

20.4 Theoretical Framework

This study was situated under the Technological Determinism (TD) theory and Knowledge, Attitude and Practice (KAP) model. The two were applied to prevent any lapse that could result from using one theory. TD which was propounded by Thorstein Veblen (1857–1929), an American sociologist and economist. The theory believes that technology is a governing force or controls society. It assumes that technology determines social change (Bimber, 1990) but humans are usually at centre stage. That is, technological advancements or innovations change how we think, react, reason or live in a society. Artificial Intelligence is one aspect of technological innovation currently shaping the minds of people because it carries out almost what humans do. It has turned the newsroom, introduced new ways of doing things, including climate action. Thus, Okiyi and Nsude (2019) argued that society is seen as being determined by technology because new technologies like AI transform society at every level. However, Dafoe (2015 explained that engaging with technological determinism enhances the study of technology and our ability to shape our sociotechnical systems. This study was situated under technological determinism because of the three research objectives. The knowledge of TD would help to explain how technologies like AI are currently controlling society.

For the Knowledge Attitude and Practise Model (KAP), media practitioners with the knowledge of climate change and Artificial Intelligence could easily utilise them accordingly. KAP came into existence in the 1950s. Although the model is mostly used in conducting surveys in medicine, it's also useful in research in social sciences and other fields. KAP helps to measure or examine behaviours of people and attitudes when they acquire new knowledge or practise. It is also good at finding out people's feelings, actions and reactions to issues or circumstances of public importance like climate change (Kaliyaperumal, 2004). Climate change and Artificial Intelligence have become a public discourse due to challenges they pose on earth and its inhabitants. The ability and capacity of the media to help mitigate effects of climate change on the living and nonliving creatures is also not in doubt (Andrade et al., 2020).

Hence, knowledge is very important in addressing it. Knowledge is said to be of four categories, namely, scientific or social scientific, local, tacit and self-reflective (Hulme, 2018; Liao et al., 2022). Previous KAP studies have proven that having sufficient knowledge influences the attitudes (behaviours) of professionals like doctors (and journalists) positively especially while discharging their primary responsibilities (Kwak et al., 2022; Rozul et al., 2020). Also, adequate knowledge of climate change is necessary to mitigate the climate change impact on society (Ofori et al., 2023). It can also help to formulate communication strategies on climate change for the general or public use (Atani, 2017). Generally, the Knowledge Attitude and Practise Model (KAP) helps to understand the knowledge of mass media practitioners in Nigeria, their attitudes and practices towards use of AI in the coverage of climate change (issues and challenges, climate adaptability, mitigations and sustainability).

20.5 Methodology

20.5.1 Research Design

This study was conducted between November 13, 2023, to January 13, 2024. It adopted a survey design of descriptive research design and administered questionnaires prepared on Google Form in line with the three research objectives.

20.5.1.1 Study Population

The Study population was 15,000 comprising all the media practitioners registered with the Nigeria Union of Journalists (NUJ). But a sample size of 500 was chosen from the population as recommended by Comrey and Lee (1992) that it was 'very good' for multivariate studies.

20.5.1.2 Study Procedure

The researchers adopted purposive sampling and administered questionnaires on media practitioners registered with the Nigeria Union of Journalists (NUJ) across 37 councils (36 States and the Federal Capital Territory chapters) of the NUJ through their respective WhatsApp groups. To validate the study instrument, the researchers consulted accomplished researchers from the media, Geography and Information Communication Technology (ICT) experts respectively. Test–retest method was used to ascertain the reliability of the research instruments. The data were extracted using the SPSS. The data were analysed using pie charts, histogram, frequency tables and simple percentage.

20.6 Results and Discussion

Table 20.1 shows the respondents' biodata. The percentages of females' respondents were 56% as against the males 44%. Also, the respondents within the ages of 26–37 were more (29%) followed by 38–47 years (25%) and 18–27 years (20%). Others were: 48–57 years (15%) and 58 years and above (11%). Similarly, the respondents' marital status indicates that the Single were more (44%), followed by Married (38%), Widowed (11%) and Divorced (7%). For the response years of professional experiences, those who had practised for 6–10 years were 42%, 0–5 years (20%), 21 years and above (12%), and 16- 20 years (11%). The data also show that 65% of the respondents have first degrees (HND/BSc. /BA), 16% (MSc. /MA), and PhD (12%). The respondents' practising in the New Media (Internet based media organisations) were 52% as against 48% others practising with the mainstream media (radio, televisions, newspapers and other periodicals).

Figure 20.1 shows the respondents' knowledge level of climate change and Artificial Intelligence. The data rated the respondents' knowledge level of climate change as: Excellent-17%, Good-21%, Fair-43%, and Poor-19%. Also, the knowledge level of AI among the respondents are: Excellent- 10%, Good-14%, Fair-18% and Poor-57%.

The Fig. 20.2 above shows the respondents' who have reported climate change with AI tools. According to the data, 14% of the respondents have done that-Yes. However, majority (86%) have not done that before (No).

The Fig. 20.3 describes the respondents' level of satisfaction with the AI tool used in sourcing climate change stories. The data indicate that 4% of the respondents were very dissatisfied over the AI tools used in reportage of climate change. Also, 0% of the respondents were slightly dissatisfied, while 49% of the respondents were neither satisfied nor dissatisfied (neutral). The respondents who were satisfied with AI tools used were 22% but 25% others were very satisfied.

Figure 20.4 indicates the respondents' extent of use of AI tools in disseminating climate mitigation and adaptation strategies. According to the data, the respondents are underutilising AI in disseminating climate change mitigation strategies. Hence, only 15% of the respondents are largely using it. There are also 17% of the respondents who are using AI somehow or somewhat, while 46% others are using it very little. Also, 22% of the respondents have not used it at all (Not at all).

Table 20.1 Respondents biodata

Variables	Frequency	Percentages (%)
Respondents gender		
Male	222	44
Female	278	56
Respondents age		
18–27 years	99	20
26–37 years	147	29
38–47 years	123	25
48–57 years	77	15
58 years and above	54	11
Respondents' marital status		
Single	222	44
Married	188	38
Divorced	34	7
Widowed	56	11
Respondents years of professional experience		
0–5 years	101	20
6–10 years	212	42
11–15 years	74	15
16–20 years	53	11
21 years and above	60	12
Educational qualification		
OND/NCE	34	7
HND/BSc. /BA	323	65
MSc./MA	81	16
PhD	62	12
Medium		
Mainstream media	239	48
New Media	261	52
Region of practising		
North Central	88	18
North East	67	13
North West	54	11
South East	92	18
South South	82	17
South West	117	23
Religion of the respondents		
African traditional religion	87	17
Christianity	149	30

(continued)

Table 20.1 (continued)

Variables	Frequency	Percentages (%)
Islam	174	35
Others	90	18

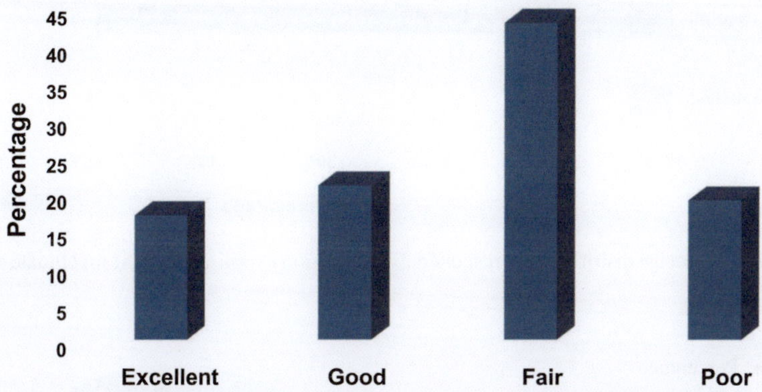

Fig. 20.1 Percentage distribution of the respondents' knowledge level and use of AI

Fig. 20.2 Percentage distribution of respondents' use of AI in reporting climate change

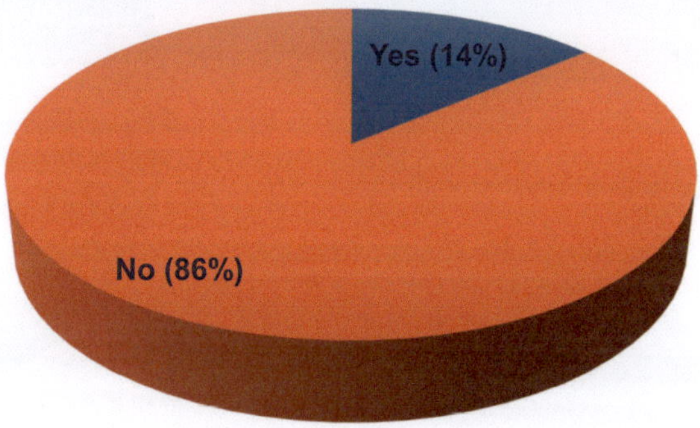

The data in Fig. 20.5 indicate the climate change terminologies reported most by the respondents' using AI. According to the data, most of the respondents have used AI in generating and reporting flood related issues. Hence, 29% of the respondents said they have utilised AI in reporting flooding, while 19% of the respondents have used it for reporting gully erosion related issues. There are also 25% of the respondents who have used AI in reporting about drought more than those utilising it in reporting greenhouse gases which accounted for 15%. The data also showed that only 12% of the respondents' have utilised AI in reporting climate action.

Table 20.2 presents the t-test result for the research hypothesis 1 which clarifies the difference between the media practitioners' knowledge level of AI and climate change among genders in Nigeria. A two-tailed t-test for independent samples (equal

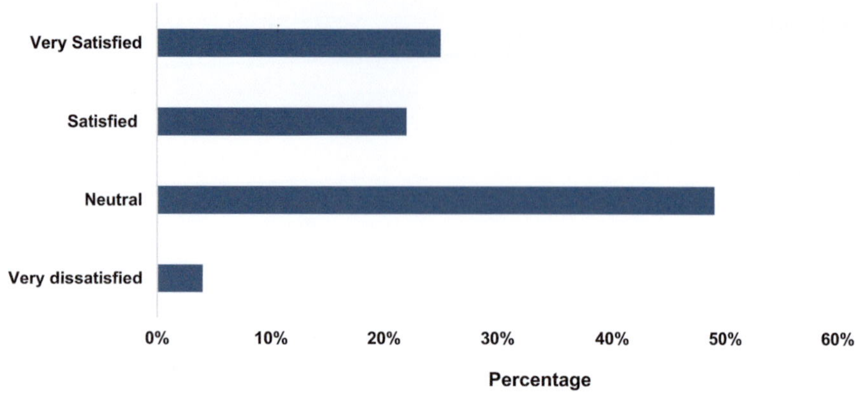

Fig. 20.3 Percentage distribution of respondents satisfaction in utilisation of AI for climate change reportage

Fig. 20.4 Percentage distribution of the respondent's extent of use of AI

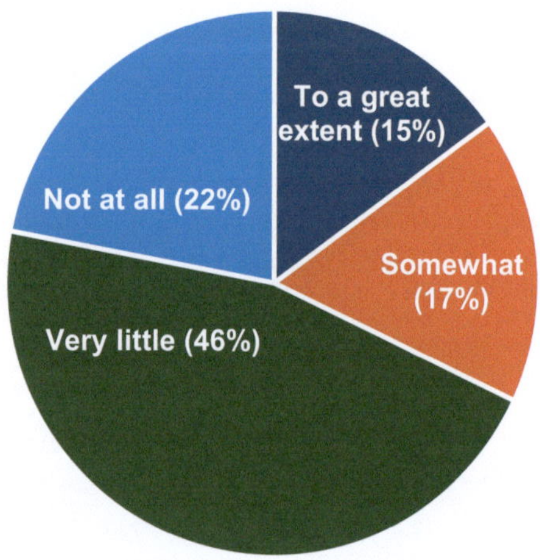

variances assumed) showed that the difference between female and male with respect to the dependent variable (Knowledge level of climate change and Artificial Intelligence) was not statistically significant, $t(4) = -1.13$, $p = 0.32$, 95% confidence interval [-20.7, 8.7]. Thus, the null hypothesis is retained.

20.7 Discussion of Findings

This study was centred on Artificial Intelligence and media coverage of climate change in Nigeria. The study was anchored on three objectives, namely, to ascertain the media practitioners' knowledge level of climate change and artificial intelligence

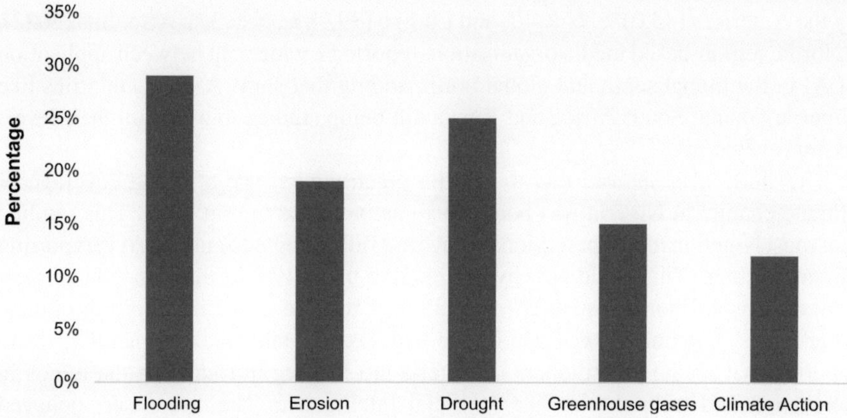

Fig. 20.5 Percentage distribution of climate change terminologies mostly reported by respondents using AI

Table 20.2 Knowledge level of AI and climate change among genders

		t	Df	p	Cohen's d
Knowledge level	Equal variances	−1.13	4	0.32	0.98
	Unequal variances	−1.43	3.78	0.23	1.24

in Nigeria, to assess the media use of artificial intelligence in coverage of climate change in Nigeria, and to find out the media practitioners' extent of use of artificial intelligence in disseminating climate change hazards in Nigeria. The data in Fig. 20.1 showed that the media practitioners' knowledge of climate change was fair (43%), but their knowledge level of AI was poor (57%). Table 20.2 (hypothesis result) further confirmed this as the knowledge level of the male and female media practitioners towards climate change and Artificial Intelligence was found not to be statistically significant ($t(4) = 1.13$, $p = 32$ at 95% confidence interval of −20.7, 8.7). This implied that the media practitioners should strive towards improving their knowledge level of both climate change and Artificial Intelligence with a view to contributing effectively to climate action and sustainability. Martinez (2023) also emphasised this while discussing various ways the media practitioners need the knowledge of AI tools in reporting climate change to combat fake news and promote facts in the world. Little wonder the UNESCO (2017) reiterated the needs for capacity building of media practitioners on climate change. In support of this, Onuegbu et al. (2024) urged the media to equally partner with the farmers for increased knowledge level and dissemination of climate mitigation and adaptation strategies. However, the findings indicated an improvement compared to Tagbo (2010) who reported poor knowledge of climate change among the media in Nigeria, and Bello et al. (2023) that reported high levels of awareness of AI among the journalists. No wonder de-lima-Samtos and Ceron (2022) found that AI was mostly applied in the newsrooms by the Americas (43.01%, N = 40) and Europe (39.78%, N = 37). Also, Ims (2022), a Nordic region-based media organisation, reported a wider gap between application of AI in the global south and global north, adding that some African countries like Nigeria, Ghana, South Africa and Kenya are being studied to ascertain their levels of improvements.

The study also showed that the media practitioners' use of AI in coverage of climate change in Nigeria was poor (14%) as indicated by Fig. 20.2. This implied that most Nigerian media practitioners have not fully adopted or utilise AI in reporting climate change. This could be why in Fig. 20.3 up to 49% of the respondents were neither satisfied nor dissatisfied (neutral) over the use of AI in reporting climate hang issues. The finding was also in line with Okocha and Ola-Akuma (2022) who reported that media practitioners in Nigeria are lagging in following the trends in innovation and technologies like Artificial Intelligence. The researchers believed that the knowledge of AI have revolutionise how things should be done in the world to encourage sustainability, reduce stress as well as create numerous opportunities for the media and entire world. Also, Lewis et al. (2024) posited that AI can assist the world in discerning, understanding and addressing climate change effectively. In other words, the media have to embrace or utilise the recommendations of the United Nations (2023) which at the end of World Economic Forum (WEF) highlighted various ways AI could be used to mitigate climate change. This could be in fashion, food and agriculture, carbon neutrality, pollution tracking, weather, among other disaster detention and management. Also, in Maher et al. (2022), 87% of the respondents agreed that AI tools were helpful in the fight against climate change in the world. Kaack et al. (2022) found the effects of AI in greenhouse gases, while Masterson (2024) uncovered how the UN are utilising AI to fight climate change in Sudan, Chad and Burundi, a project that could serve as media sources of data in reporting climate mitigation and adaptation.

It was also found that media practitioners' extent of use of artificial intelligence in disseminating climate change hazards in Nigeria was very little (46%) as contained in Fig. 20.4. In Fig. 20.5, flooding appeared to be the most climate change hazard the respondents have reported using AI (29%). This entailed that the Nigerian media are far behind their colleagues across the world. Some effects of this in Nigeria and other parts of the world are spreading of fake news, deep-fakes and misinformation which a journalist with adequate knowledge of could easily dictate and dismiss using AI tools like Climate AI, ChatGPT, among others (Dimock, 2022). Thus, Shafer (2024) and Simon (2024) reported that AI is already reverberating, transforming and revolutionising the newsroom especially in the Global North where it helps mostly in information generation, management, climate mitigation, processing and elimination of falsehood. Others benefits of AI to climate change reporting includes data generation, incorporation of solar and renewable energy, accelerate oil and gas exploration and extraction (Frueh, 2024). Therefore, it may be understatement to say that Nigeria can do more in mitigating the effects of climate change, including promoting climate adaptation strategies and other opportunities if the media practitioners learn and utilise knowledge of AI in their reportage. Also to be deduced from the finding is that the media practitioners in Nigeria are not completely living up to their expectations since the Sect. 22 of the constitution of the Federal Republic of Nigeria equally empowered them to gain access to many things and hold the public officers accountable. Besides, Borchardt (2023) added that reporting climate change should be at the heart of every journalist because climate journalism is not an option but an ethical responsibility every media practitioner should uphold, promote and dedicated to.

20.8 Conclusions, Recommendations and Limitations

The study focused on Artificial Intelligence and Media coverage of climate change in Nigeria. It was based on recommendations by the UN and other experts that climate change could easily be fought and conquered or mitigated using AI. Hence, the finding showed that the knowledge level of climate change was fair but that of Artificial Intelligence was poor. This was in line with some of the extant literatures cited in the work. The study concludes that Artificial Intelligence is not well utilised in coverage of climate change in Nigeria.

The researchers therefore advocated for training of media practitioners to be acquainted with the knowledge of use of Artificial Intelligence so they can improve in its usage coverage and reportage of climate change in Nigeria. They also suggested more studies to be carried out in the areas of AI and media coverage of climate change using content analysis, In Depth interview or focused group. Also, similar topics could be studied across countries or combinations of Africa or Global South and Global North.

The study, however, was limited by the methodology (descriptive survey), resources and data on media use of AI in coverage of climate change. It was also limited by the scope. Lack of funding, time and other resources also limited the researchers' ability to cover all the 36 States of Nigeria and Federal Capital Territory (FCT).

Acknowledgements Authors expressed their profound gratitude to the members of Nigeria Union of Journalists (NUJ) for their cooperation during the course of this study.

Funding This study attracted no funding.

Data Availability Statement The authors have confirmed that all the data generated or analysed during this study were duly acknowledged and referenced hereunder. The corresponding is also ready to provide any other relevant information if requested in due time.

Ethical Clearance Not applicable.

Conflict of Interest Authors hereby declared that there is no conflict of interest expressed in this study.

References

Abiola, R. (2023). *Once again, the Nigerian media is late to the artificial intelligence party, when is Our time?* Retrieved from https://technext24.com/2023/11/14/nigerian-media-artificial-intelligence/

Andrade, C., Menon, V., Ameen, S., & Praharaj, S. K. (2020). Designing and conducting knowledge attitude, and practice surveys in psychiatry: Practical guidance. *Indian Journal of Psychological Medicine, 42*(5), 478–481. https://doi.org/10.1177/0253717620946111

Atani, M. (2017). *UN environment supports capacity building of environmental journalists in Africa.* Retrieved February 3, 2017, from United environment programme. https://www.unep.org/news-and-stories/story/un-environment-supports-capacity-building-environmental-journalists-africa

Bello, A., Ishola, A. C., & Umeaku, P. C. (2023). Survey of awareness and adoption of Artificial Intelligence Journalism among Lagos and Kwara States Journalists in Nigeria. *Indonesian Journal of Communication Studies, 16*(2), 95–105. https://doi.org/10.31315/ijcs.v1i2

Bechett, C. (2023). *How AI is generating change in newsrooms worldwide.* Retrieved September 20, 2023, from https://blog.google/technology/ai/how-ai-is-generating-change-in-newsrooms-worldwide/

Becker, K. B., Simon, F. M., & Crum, C. (2023). *Policies in parallel? A comparative study of journalistic AI policies in 52 Global News Organisations.* Retrieved from https://doi.org/10.31235/osf.io/c4af9

Bimber, B. (1990). Karl Marx and the three faces of technological determinism. *Social Studies of Science, 20*(2), 333–351. https://doi.org/10.1177/030631290020002006

Borchardt, A. (2023). *Why climate change should be at the heart of modern journalism.* Retrieved October 9, 2024, from https://www.bbc.com/beyondfakenews/trusted-news-initiative/why-climate-change-at-heart-of-modern-journalism

British Geological Survey. (2021). What causes the Earth's climate to change? *British Geological Survey.* Retrieved June 7, 2021, from https://www.bgs.ac.uk/discovering-geology/climate-change/what-causes-the-earths-climate-to-change/#:~:text=These%20have%20been%20caused%20by,thousands%20or%20millions%20of%20years

Brown, T. (2024). *Global warming.* Retrieved February 21, 2024, from: https://education.nationalgeographic.org/resource/global-warming/

Campbell, E., Uppalapatim, S. S., Kotcher, J. & Maibach, E. (2023). Communication research to improve engagement with climate change and human health: a review. *Frontiers in public health,* 10(1), 1–8. https://doi.org/10.3389/fpubh.2022.1086858

Caswell, D. (2023). *AI and journalism: what is next?* Retrieved September 19, 2023, from https://reutersinstitute.politics.ox.ac.uk/news/ai-and-journalism-whats-next

Comery, A. L., & Lee, H. B. (1992). A first course in factor analysis cited in Wimmer, R.D., & Dominick, J.R (2003). *Mass media research: An introduction*, 7th ed. Thomson & Wadworth.

Copeland, B. J. (2024). *Artificial Intelligence.* Retrieved April 7, 2024, from https://www.britannica.com/technology/artificial-intelligence

Cools, H., & Diakopoulos, N. (2023). *Towards guidelines for guidelines on the use of Generative AI in newsrooms.* Retrieved June 10, 2023, from https://generative-ai-newsroom.com/towards-guidelines-for-guidelines-on-the-use-of-generative-ai-in-newsrooms-55b0c2c1d960

Dafoe, A. (2015). On Technological Determinism: A typology, scope conditions, and a mechanism. *Science, Technology, & Human Values, 40*(6), 1047–1076. https://doi.org/10.1177/0162243915579283

de-lima-Santos, M. & Weron, W. (2022). Artificial intelligence in news media: current perceptions and future outlook. *Journalism and media*, 3(1), 13–26. https://doi.org/10.3390/journalmedia3010002

Dimock, W. C. (2022). *What AI can do for climate change, and what climate change can do for AI.* Retrieved April 5, 2022, from https://environment.harvard.edu/news/what-ai-can-do-climate-change-and-what-climate-change-can-do-ai#main-content

Dunne, D. (2020). *The carbon brief profile: Nigeria.* Retrieved from https://www.carbonbrief.org/the-carbon-brief-profile-nigeria/

Frueh, S. (2024). *Exploring the Implications of AI for Climate.* Retrieved August 2, 2024, from https://www.nationalacademies.org/en/news/2024/08/exploring-the-implications-of-ai-for-climate

Guanah, S. J., Obi, I., & Ginikachukwu, A. C. (2020). Artificial Intelligence and its reportage in select Nigerian newspapers: A content Analysis. *International Journal of Language and Literary Studies,* 2(2), 45–61. https://doi.org/10.36892/ijlls.v2i2.298

Hanson, R. E. (2017). *Mass communication: Living in a Media World.* 6th ed., Sage publications Inc.

Haider, H. (2019). *Climate change in Nigeria: Impacts and responses. K4D Helpdesk Report 675.* Brighton, UK Institute of Development Studies. https://opendocs.ids.ac.uk/opendocs/handle/20.500.12413/14761

Haines, A. (2022). *Expert comment—over 40% of the world's population are highly vulnerable to climate.* Retrieved from https://www.lshtm.ac.uk/newsevents/news/2022/expert-comment-over-40-worlds-population-are-highly-vulnerable-climate#:~:text=In%20all%20regions%2C%20but%20particularly,circumstance%20in%20which%20they%20live.

Hase, V., Mahl, D., Schäfer, M. S., & Keller, T. R. (2021). Climate change in news media across the globe: An automated analysis of issue attention and themes in climate change coverage in 10 countries (2006–2018). *Global Environmental Change, 70,* 1–3. https://doi.org/10.1016/j.gloenvcha.2021.102353

Hulme, M. (2018). Gaps in climate change knowledge: Do they exist? Can they be filled? *Environ Humanit, 10*(1), 330–337.

Hurst, L. (2023). *Robot reporters? Here's how news organisations are using AI in journalism.* Retrieved August 24, 2023, from https://www.euronews.com/next/2023/08/24/robot-reporters-heres-how-news-organisations-are-using-ai-in-journalism

Ibm Data. & Ai, Team. (2023). *Understanding the different types of artificial intelligence.* Retrieved October 12, 2023, from https://www.ibm.com/blog/understanding-the-different-types-of-artificial-intelligence/

Ims. (2022). *How AI in newsrooms supports good journalism in the Global South.* Retrieved November 9, 2023, from https://www.mediasupport.org/news/how-ai-in-newsrooms-supports-good-journalism-in-the-global-south/

Islam, M. S., & Kieu, E. (2021). Sociological perspectives on climate change and society: A review. *Climate, 9*(7), 1–14. https://doi.org/10.3390/cli9010007

Jaiyeola, T. (2023). *Generative AI to help Nigeria's fight against climate change.* Retrieved November 28, 2023, from https://punchng.com/generative-ai-to-help-nigerias-fight-against-climate-change/#google_vignette

Joshi, N. (2019). *7 types of artificial intelligence.* Retrieved June 19, 2019, from https://www.forbes.com/sites/cognitiveworld/2019/06/19/7-types-of-artificial-intelligence/?sh=1c379a88233e

Kaack, L. H., Donti, L. P., Strubel, E., Kamiya, G., Greutzig, F., & Rolnick, D. (2022). Aligning artificial intelligence with climate change mitigation. *Nature Climate Change, 12*(1), 518–527. https://doi.org/10.1038/s41558-022-01377-7

Kaliyaperumal, K. (2004). Guideline for conducting a knowledge, attitude and practice study. *AECS Illumination, 4*(1), 7–9. Retrieved from https://v2020eresurce.org/content/files/guideline_kap_Jan_mar04.pdf

Kwak, C., Seo, Y. J., Park, K. H., & Han, W. (2022). Analysis of the knowledge, attitudes, and practice model of healthcare professionals on hearing loss at elderly dementia residences in Korea. *Heathcare (Basel), 10*(5), 792.

Lee, H., & Romero, J. (2022). *Summary for policymakers. in climate change 2023: Synthesis Report contribution of working groups I, II and III to the sixth assessment report of the Intergovernmental Panel on Climate Change* (pp. 1–34). Geneva, Switzerland. https://doi.org/10.59327/IPCC/AR6-9789291691647.001

Lewis, J. I., Toney, A., & Shi, X. (2024). Climate change and artificial intelligence: Assessing the global research landscape. *Discovery Artificial Intelligence, 4*, 64. https://doi.org/10.1007/s44163-024-00170-z

Liao, X., Nguyen, T. P. L., & Sasaki, N. (2022). Use of the knowledge, attitude, and practice (KAP) model to examine sustainable agriculture in Thailand. *Regional Sustainability, 3*(1), 41–52. https://doi.org/10.1016/j.regsus.2022.03.005

Maher, H., Meinecke, H. & Gromier, D. (2022). *AI is essential for solving the climate crisis.* Retrieved from https://www.bcg.com/publications/2022/how-ai-can-help-climate-change

Mann, M. E. (2024). *Global warming.* Retrieved March 19, 2024 from . https://www.britannica.com/science/global-warming

Martinez, M. (2023). *ChatGPT for climate reporting: What journalists should keep in mind.* Retrieved April 11, 2023, from https://ijnet.org/en/story/chatgpt-climate-reporting-what-journalists-should-keep-mind

Masterson, V. (2024). *9 ways Ai is helping tackle climate change.* Retrieved from. https://www.weforum.org/agenda/2024/02/ai-combat-climate-change/

Matias, Y., & Brandt, K. (2023). *Accelerating climate action with AI.* Retrieved from https://blog.google/outreach-initiatives/sustainability/report-ai-sustainability-google-cop28/

Mike, S., Willie, S., Cameron, S. & Teresa, C. (2013). *Climate change in Africa: a guide for journalists.* Retrieved from https://unesdoc.unesco.org/ark:/48223/pf0000225451

National Geographic Society. (2023). *The greenhouse effect and our planet.* Retrieved October 19, 2023, from https://education.nationalgeographic.org/resource/greenhouse-effect-our-planet/

Nasa. (2023). *What is climate change?* https://climate.nasa.gov/what-is-climate-change/?trk=publi_post_comment-text

Nwanyanwu, N. C., & Nwanyanwu, M. (2021). Utilization of artificial intelligence in journalism in Nigeria. *KIU Journal of Social Sciences, 7*(2), 205–212.

Ofori, B. Y., Ameade, E. P. K., Ohemeng, F., Musah, Y., Quartey, J. K., & Owusu, E. H. (2023). Climate change knowledge, attitude and perception of undergraduate students in Ghana. *Plos Clim, 2*(6). https://doi.org/10.1371/journal.pclm.0000215

Okere, N. (2021). *Reality of Climate Change.* Retrieved July 20, 2021, from https://thenationonlineng.net/reality-of-climatechange

Okiyi, G., & Nsude, I. (2019). Adopting Artificial Intelligence to journalistic practices in Nigeria: Challenges and way forward. *International Journal of Communication, an Interdisciplinary Journal of Communication, 1*, 141–162.

Okocha, D. O., & Ola-Akuma, R. O. (2022). Journalistic metamorphosis: robot journalism adopttion in Nigeria in a digital age. *Igwebuike: African Journal of Arts and Humanities, 8*(1).

Onuegbu, O. C., Wogu, J. O., Okaiyeto, S. A., &Wilson, A. O. (2024). Media practitioners' knowledge and coverage of climate change and rural farmers adoption in Nigeria. *International Journal of Agriculture and Biosciences, 13*, 390–396. https://doi.org/10.47278/journal.ijab/2024.135

Rozul, C. D. A., Gregorio, E. R., & Chiong, C. M. (2020). Baseline knowledge, attitudes, and practices of healthcare practitioners in Rizal province, Philippines toward implementing the universal newborn hearing screening program. *Acta medica Phillipina, 15*, 134–149. https://doi.org/10.47895/amp.v54i2.1504

Pona, H. T., Xiaoli, D., Ayantobo, O. O., & Tetteh, N. D. (2021). Environmental health situation in Nigeria: Current status and future needs. *Heliyon, 7*(1), 2405–8440.

Shafer, J. (2024, February 27). *How AI is already transforming the news business.* Politico. https://www.politico.com/news/magazine/2024/02/27/artificial-intelligence-media-00143508

Simon, F. (2024). *Artificial Intelligence in the news: How AI retools, rationalizes, and reshapes journalism and the public arena.* Retrieved February 6, 2024, from https://www.cjr.org/tow_center_reports/artificial-intelligence-in-the-news.php

Snow, J. (2019). *How artificial intelligence can tackle climate change.* Retrieved July 18, 2019, from https://www.nationalgeographic.com/environment/article/artificial-intelligence-climate-change

Tagbo, E. (2010). *Media coverage of climate change in Africa a case study of Nigeria and South Africa*. Retrieved from. https://reutersinstitute.politics.ox.ac.uk/sites/default/files/research/files/Media%2520Coverage%2520of%2520Climate%2520Change%2520in%2520Africa%2520A%2520Case%2520Study%2520of%2520Nigeria%2520and%2520South%2520Africa.pdf

Unesco. (2017). *Unesco strategy for action on climate change*. Retrieved from https://unesdoc.unesco.org/ark:/48223/pf0000259255

Unfccc. (1992). *United Nations Framework Convention on Climate Change*. https://unfccc.int/resource/ccsites/zimbab/conven/text/art01.htm#:~:text=2.,observed%20over%20comparable%20time%20periods.

United Nations. (2023). *What is climate change?* Retrieved from https://www.un.org/en/climatechange/what-is-climate-change

UN News (2023). Explainer: How AI helps combat climate change. Retrieved November 3, 2023, from https://news.un.org/en/story/2023/11/1143187

Wang, Y.-J., Chen, Y., Hewitt, C., Ding, W.-H., Song, L.-C., Ai, W.-X., Han, Z.-Y., Li, X.-C., & Huang, Z.-L. (2021). Climate services for addressing climate change: Indication of a climate livable city in China. *Advances in Climate Change Research, 12*, 744–751. https://doi.org/10.1016/j.accre.2021.07.006

Willing, N. (2024). *Top 10 most popular AI tools that you need to use in 2024*. Retrieved March 15, 2024, from https://www.techopedia.com/top-10-most-popular-ai-tools

Part III
Agriculture

Chapter 21
Gender and Media Coverage of Climate Change in Nigeria

Okechukwu Christopher Onuegbu, Joseph Oluchukwu Wogu, and Ogonna Wilson Anunike

Abstract Climate change is a global discourse due to its impact on earth. But there is paucity of scholarly findings on its impacts among genders. This study was aimed at bridging the knowledge gap. It specifically assessed the gender and media coverage of climate change in Nigeria. It adopted a purposive sampling procedure, while applying cognitive dissonance and agenda setting theories. Questionnaire prepared on Google form was distributed across 37 WhatsApp groups but only 390 journalists who filled it were chosen as the sample size as recommended by Glenn (1992). Its key objectives were to assess the media coverage of climate change in Nigeria; identify the place of gender in media coverage of climate change; and ascertain the influence of location and sources to media reporting of gender-based climate change stories in Nigeria. The findings showed that gender is not given adequate consideration in media reports of climate change in Nigeria.

Keywords Climate change · Gender · Journalists · Media coverage · Nigeria

21.1 Introduction

Climate change is gaining global attention due to its impact on humans, agriculture and food security (Naazie et al., 2024). In the African continent, and more specifically in Nigeria, evidence of climate change includes rising temperatures, changes in precipitation patterns, sea-level rise, flooding, erosion, extreme weather events, desertification, drought, land degradation, and biodiversity loss (Okere, 2021). Human activities significantly contribute to these climate issues through domestic and economic activities, leading to poverty, loss of lives, and essential resources, including homes, livestock, agricultural yields, freshwater fish, land, and

O. C. Onuegbu (✉) · J. O. Wogu
Department of Mass Communication, University of Nigeria, Nsukka, Nigeria
e-mail: oonuegbu3@gmail.com

O. W. Anunike
Department of Mass Communication, Federal Polytechnic, Oko, Nigeria

R. U. Onyeneke et al. (eds.), *Energy Transition, Climate Action and Sustainable Agriculture*, https://doi.org/10.1007/978-3-031-83165-2_21

economic trees (Besser et al., 2021). Natural factors, such as volcanic eruptions and solar fluctuations, also play a role (Newburger, 2021).

However, several individuals and institutions, including the United Nations (UN) women argue that the impact of climate change differs among genders (United Nation Women, 2022). They believed that climate change inflicts more pains and agony on women than men. This includes the agricultural sector where women contribute more than men, violence women face while seeking comfort, stillbirth and other excruciating pains, among others (Kareem, 2023). Religion, culture and tradition, socioeconomic factors and social conditioning also restrict the women's movements and access, as well as influence how they reason and relate with society (World Bank, 2021).

Therefore, researching gender and media coverage of climate change is very important, especially in developing countries like Nigeria where culture and tradition wield strong influence. This could equally contribute towards integrating gender related issues into climate change adaptation for easy interventions of policymakers (Ume et al., 2021). Similarly, the Constitution of the Federal Republic of Nigeria does not only guarantee equality to all genders but also empowers the media to hold the government accountable, inform, educate, and entertain the populace (Section 15 subsection 2, Section 42 subsection 1 and Section 22). Also, it is through the media that people often learn more about climate change and its inherent opportunities, and knowledge of climate action undertaken by the government, international organisations, individuals, non-governmental organisations and civil society groups (Mahl & Guenther, 2023). In other words, climate change should also be a point of focus for the mass media (Abdellatif, 2023) for the overall good of all. Unfortunately, there are little or no scholarly findings in that regard available on the Internet. Most studies available on the Internet focused on media framing of climate change, impact of climate, journalists' knowledge of climate change, amongst others. This study has filled the gap in knowledge. But it specifically assessed the gender and media coverage of climate change in Nigeria under the following study objectives:

1. Assess the media coverage of climate change in Nigeria.
2. Identify the place of gender in media coverage of climate change in Nigeria.
3. Ascertain the influence of location and sources to media reporting of gender-based climate change stories in Nigeria.

Research Questions

1. To what extent does the media cover climate change in Nigeria?
2. What is the place of gender in the media coverage of climate change in Nigeria?
3. Does location and sources influence media reporting of gender-based climate change stories in Nigeria?

21.2 Literature Review

Experts believe that there is a link between gender and climate change (Adpc, 2021). They argue that climate change affects both genders, but women felt the impact more because of sociocultural and socioeconomic backgrounds (United Nation Women, 2022). There are reports of how climate change is causing gender-based violence in developing countries, rural communities, among women and girls with disabilities, conflict prone areas, internally displaced persons camps, amongst others (Agrawal et al., 2023; Desai & Mandal, 2021; United Nations Human Right, 2022). For example, in regions or locations where there is intense heat, men could be seen sleeping outside or walking on streets without shirts but women who try it could be raped, stigmatised or punished. Also, some women have equally been restrained from access to social gatherings and religious homes due to their choice of dressings not minding the climatic condition of the location (Framework Convention on Climate Change, 2022; Jeffs, 2022; United Nation Women, 2022).

As a result, stakeholders, especially women are agitating for gender inclusion and equality in climate action, including in policy making, climate mitigation and adaptation among countries (United Nation Women, 2022). The United Nations Framework Convention on Climate Change (2022) also believed that women could play a pivotal role in mitigating the effects of climate change as planned when their suggestions, recommendations and presence are sought for and implemented accordingly. At the Conference of the Parties (COP25, COP27 and COP28), gender and women inclusion were top issues. This includes the needs for parties (institutions, organs and countries) to uphold, respect and promote the rights of everyone in climate change action. The media is among the institutions of the United Nations Framework Convention on Climate Change expected to be gender sensitive with regards to climate change coverage in the world. They are also expected to provide advocacy interventions for the fight to mitigate the impact of climate change in Nigeria and the rest of the world (Binuyo et al., 2022). This seems to be among the reasons many scholars should commit their time and resources towards that direction. The media are trusted stakeholders in climate change mitigation because they increase public knowledge and change people's behaviour towards the subject matter (Schäfer, 2015).

Unfortunately, several studies on media coverage of climate change and gender in the world have shown that gender related climate issues are not given favourable reportage. For instance, a comparison of the media coverage of climate change in 10 countries in the global North and the global South between 2006 to 2018 (Hase et al., 2021) showed there are dissimilarities in focus, content and attention due to scientific, political and economic reasons. The study was conducted in New Zealand, Australia, Germany, Canada, the United States of America (USA), the United Kingdom (UK), Namibia, India, Thailand and South Africa. The study analysed reports on climate change, global warming and greenhouse effects published by two leading national dailies in each of the countries noting that the qualities and quantities of the reports were not enough to address the climate change. This could be why a study conducted in Nigeria in 2021 (Ume et al., 2021) discovered that land tenure, lack of public awareness, poor access to credit and loan, among others constituted the gender issues affecting female household farmers in Nigeria. The study involving focused group discussion and in-depth interviews, also found that women are more vulnerable to climate change owing to cultural and unwarranted assumptions. A similar study earlier conducted in 2013 (Batta et al., 2013) showed that press coverage of climate change issues in Nigeria had implications for public participation opportunities. The study, which covered January 2007 to December 2009, focusing on four national dailies in Nigeria, discovered that out of 438 issues published only 134 items centred on climate change.

21.3 Theoretical Framework

This study adopted two theoretical frameworks namely, cognitive dissonance theory and agenda setting theory. The two were adopted given to the urgency of climate change and to prevent any possible lapses that could result from using one theory.

21.3.1 Cognitive Dissonance Theory (CDT)

The cognitive dissonance theory was propounded in 1957 by Leon Festingers, an American social psychologist. The theory came into existence to correct an impression that attitudes and beliefs were learned and could be relearned or unlearned through rewards and punishments (Brehm, 2007). Festinger argued that inconsistent cognitions such as knowledge, opinion, or belief about the environment, oneself, or one's behaviour could generate an uncomfortable motivating feeling (David & Bran, 2017). In other words, climate change knowledge could redefine how the media report or project its impact among genders. The proponent also countered the assumptions that an organisation or people could use messages about the impact of an action, behaviour or attitude to influence or achieve some positive behaviour in people's lives.

CDT proposed four assumptions namely, consequences of decisions, forced compliance, exposure to information, and the role of social support. It presupposed that influence to do something or change people's attitudes are rather driving by the individual's intrapersonal decisions or what occurs when incongruence between attitude and behaviour creates a tension that is resolved by altering either the belief or behaviour (Dainton, 2004). The theory gained prominence among the researchers even though it generated a lot of controversies. Some oppositions to the theory argued that if people change their attitudes in dissonance experiments it probably means that what they experienced had aversive effect (Zanna & Cooper, 1974). Others believed that motivations to change attitudes were aimed at rendering the consequences of a person's behaviour non-aversive (Cooper, 2019). This led to the emergence of more theories.

21.3.2 Agenda Setting Theory

Agenda setting theory was developed formally by Maxwell McCombs and Donald Lewis Shaw (1968) after attempts by Walter Lippmann in the 1920s and Bernard Cohen in 1960s. The theory argued that the mass media can make their audience start thinking towards the directions they (the media) want (Rogers & Dearing, 1988). The media do this through giving some news or reports prominence over others by appealing to the audience's minds, reporting them constantly, promoting them in front pages, as editorials, and so on. By this, the media can equally confer legitimacy, acceptance or quality to some products, discussions, objects or subjects (Mueller, 2017). In other words, it believes that the media has the power or ability to influence public opinion or always set an agenda for the masses.

Some scholars argued that the best or easiest way the media can set an agenda to the public is by frequently reporting critical issues like gender and climate change that appeal to their consciousness. This, according to them, would create pressure or influence the government and other stakeholders to take appropriate action. Scholarly findings also showed that agenda setting makes the audience or society form opinions and shaping of minds (Stacks et al., 2015). However, some controversies surrounding the applicability of this theory is that some theorists believed that the media cannot set an agenda for any course they do not understand or have positive attitudes towards. Also, some criticised the theory over its inability to prove cause and effect and if it truly sets an agenda for the public. This is because many studies have proven that policy makers often use lobbying and public relations to get the public's attention, set agenda or create certain policies (Kwansah-Aidoo, 2005).

21.4 Methodology

21.4.1 Research Design

This study was conducted between January 12 to March 12, 2024. It employed a survey as a research design using semi-structured questionnaires prepared on Google Form and distributed to the respondents (members of the Nigeria Union of Journalists, NUJ) across 37 Councils (chapters).

21.4.1.1 Study Population

The study population is 390 out of 15,000 media practitioners working in recognised print, electronic and new media platforms according to data obtained from the NUJ National Secretariat, Abuja. This was as recommended by Glenn (1992) sample size Table 21.2 which posited that the number was enough if they were only those who responded to online distributed questionnaires.

21.4.1.2 Sampling Procedure

A purposive sampling procedure was applied in the distribution of the questionnaires through 37 WhatsApp groups owned by the 37 NUJ councils. At the end of three months, only 390 responses were obtained and were chosen as the study's sample size. One hypothesis was also generated and tested using T-Test. Test–retest method was equally used to ascertain the reliability of the research instruments. The data were analysed using a simple table and percentages.

21.5 Results and Discussion

The Fig. 21.1 shows the respondents' gender distribution. According to the data, 51% of the respondents were males while 49% were females. This indicated that male respondents were more than females with 1%.

The data in Fig. 21.2 describes the respondents age brackets according to percentages. The data shows that those within the ages of 18 to 30 were 16%, while those between 31 to 42 years were 28%. Also, the respondents aged 43 to 54 years were 32% just as 24% others were between 55 years and above. This shows that most of the respondents were between 43 to 54 years old.

The data in Fig. 21.3 reveals the marital status of the respondents. The data shows that 20% of the respondents were not married (single). It further reveals that 74% of the respondents were married, while 5% others were already widows and widowers (widowed). Also, 1% of the respondents were divorces. This entails that the percentages of the respondents who were married were more in number.

The data in Fig. 21.4 shows the respondents academic or education qualifications. According to the data, the respondents whose highest academic qualification is National Certificate Education (NCE), and Ordinary National Diploma were just 2%. Also, 75% of the respondents have acquired Higher National Diploma (HND) or first degree such as Bachelor of Science (B.Sc.), Bachelor of Technology (B.Tech.)

Fig. 21.1 Percentage distribution of respondents according to gender

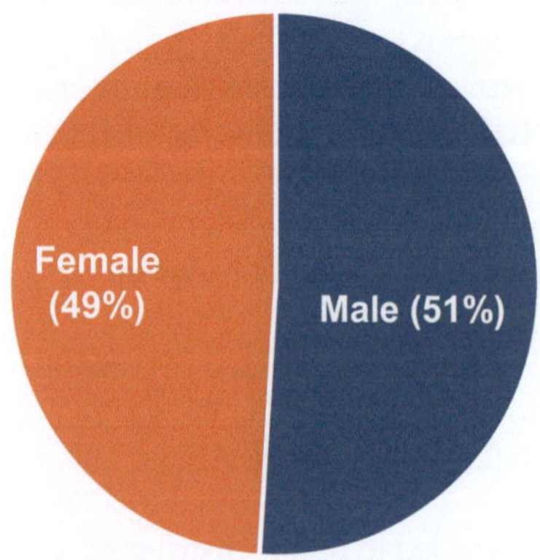

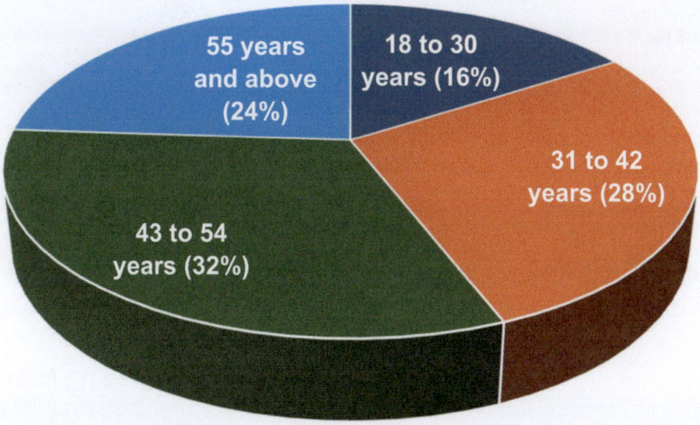

Fig. 21.2 Percentage distribution of respondents according to age

Fig. 21.3 Percentage distribution of respondents according to marital status

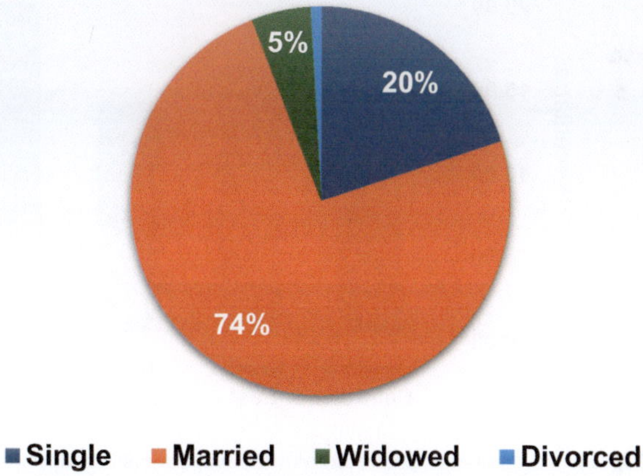

and Bachelor of Art (BA). The data also reveals that 14% of the respondents have acquired Master of Science (M.Sc) and 9% others have attained the level of Doctoral Degree (PhD.). This indicates that most of the respondents are first degree holders.

The data in Fig. 21.5 data presents the respondents geopolitical zones or regions in Nigeria. According to the data, the respondents from the North Central were 17% while those from Northeast were 15%. Also, 13% of the respondents hailed from the Northwest just as the Southeast respondents accounted for 19%. Similarly, 15% of the respondents came from the South South and 21% others were from the Southwest. Therefore, it could be deduced from the data that Southwest Nigeria had more respondents than others.

Figure 21.6 shows that the respondents were affiliated or members of various religious institutions in Nigeria. The data indicated that 39% of the respondents who attempted the questionnaire were Moslems or Islamic religious adherents. Similarly, there were 32% of the respondents who admitted being Christians according

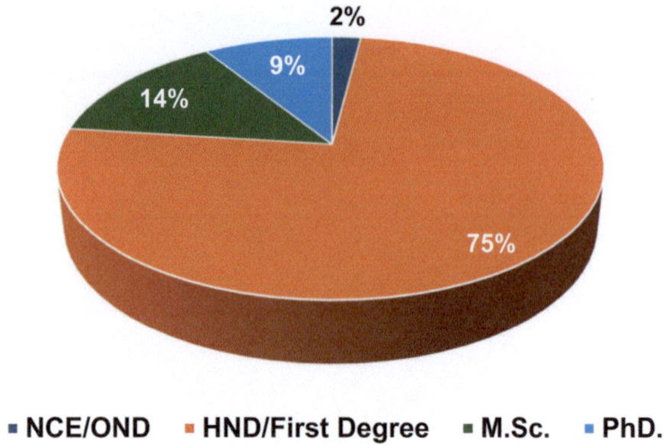

Fig. 21.4 Percentage distribution of the respondents' level of education

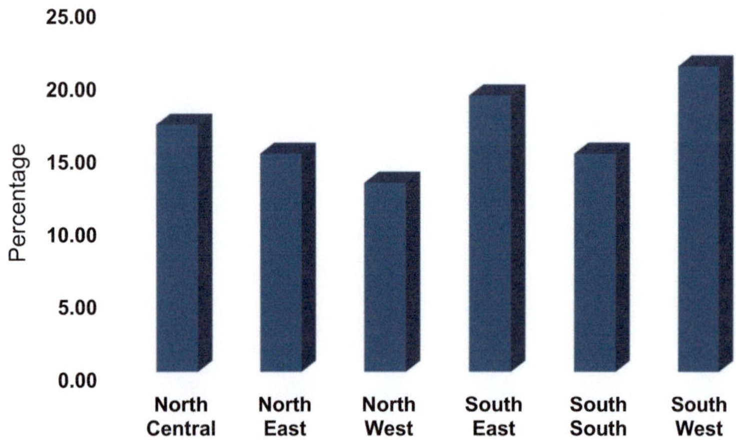

Fig. 21.5 Percentage distribution of the respondents' geopolitical zones

to the data. It further maintained that 11% of the respondents were members of African traditional religion, while 18% of others belonged to other religious institutions not mentioned in the questionnaire. This indicated that more Islamic religious worshippers attempted the question.

Media Coverage of Climate Change in Nigeria

Table 21.1 reveals that 99% of the respondents are aware of climate change, while a handful (1%) do not know about it. The respondents also disclose the following as causes of climate change; human activities-(59%), natural occurrence-(21%), and punishment from God (20%). Also, 48% of the respondents have covered climate change, while 52% have not done so. On why the respondents do not like covering or reporting climate change; 24% said they do not have the requisite training or skills to cover it, 21% stated that their medium will not broadcast/publish it, 17% lacked the financial capacities to execute it. Other reasons they gave were: it is against the

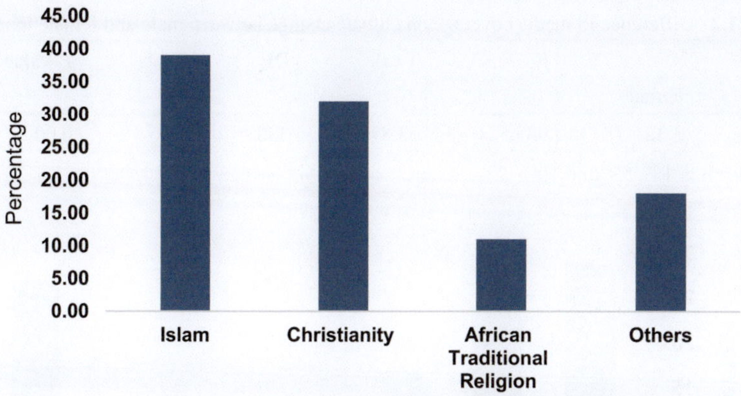

Fig. 21.6 Percentage distribution of the respondents according to religion

interest of our audience (14%), no experts or sources to validate or balance my reports (10%), it is against my religious belief (9%), and it is against the interest of the owner of my medium (7%).

The data in Fig. 21.7 identified the effects of climate change in Nigeria. This includes flooding (27%), erosion menace (22%) and water/food scarcity (15%).

Table 21.1 Media coverage of climate change in Nigeria

Variables and responses	Frequency	Percentage
Do you know about climate change?		
Yes	386	99
No	4	1
What do you think that causes climate change?		
It is a natural occurrence	81	21
It is a punishment from God	77	20
Human activities like bush burning, deforestation, etc	228	59
Have you covered climate change?		
Yes	187	48
No	199	52
If no, why don't you like covering climate change?		
My medium will not broadcast/publish it	41	21
I do not have the requisite training or skills to cover climate change it	47	24
It is against the interest of our audience	27	14
It is against the interest of the owner of my medium	13	7
There are no experts or sources to validate or balance my reports	20	9
I do not have the financial resources to execute it	33	17
It is against my religious belief	18	8

Table 21.2 Difference in media coverage on climate change between male and female journalist

Mean		F	T-cal.	DF	T tab	*P* Value
Male	Female					
2.75	3.32	985.52	33.87*	388	1.96	0.00

*Significant at 1%, 5% and 10%

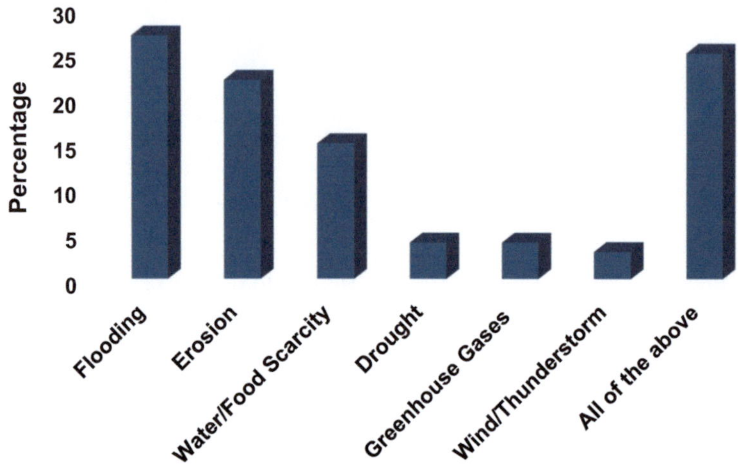

Fig. 21.7 Percentage distribution of effects of climate change on the respondents

Others were drought (4%), Greenhouse gases (4%) and Wind/Thunderstorm (3%). Also, 27% of the respondents identified all (all of the above) as effects of climate change in the country. This indicates that climate change has various effects on Nigeria.

The data in Fig. 21.8 shows that 33% of the respondents cover climate change related events or issues sometimes (not always). Similarly, 31% of the respondents rarely do that, while 29% of others give coverage to climate change issues always. Unfortunately, only 7% of the respondents give coverage to climate change related issues often, while 0% or none of the respondents chose never. This entails that all the respondents cover climate change related issues but there are disparities in how they do it.

The Fig. 21.9 above displays the areas media reports on climate change addressed most. According to the data, 39% of the respondents reports on climate change are mere breaking news and press statements about issues surrounding climate change. Also, 30% of the climate change reports centred on the disasters and risks involved in climate change. Similarly, 11% of the reports talked about causes of climate change incidents, while 13% and 7% addressed issues of climate change adaptation/ mitigation and climate change opportunities respectively. This entails that most of

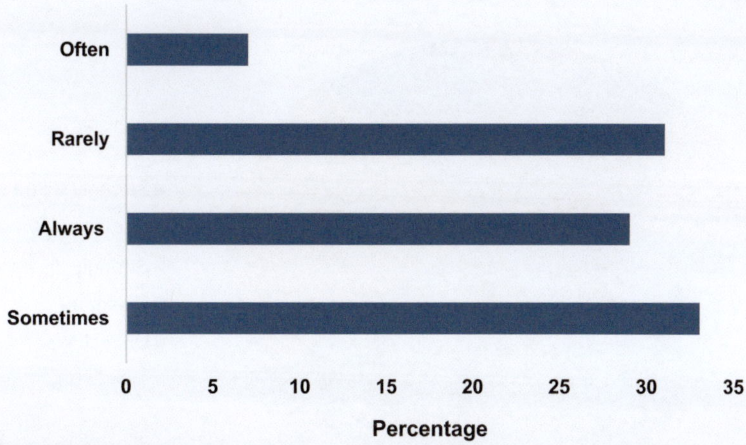

Fig. 21.8 Percentage distribution of respondents' frequency in coverage of climate change

the respondents write more breaking news and press statements centring on climate change than they do to other aspects of climate change.

The data in Fig. 21.10 shows the respondents level of agreements on impact of climate change between the males and females' genders. It revealed that 12% of the respondents agreed totally (strongly agree) that impact of climate change differs between the males and females' genders while 25% others did not totally agree (agree). There are also 19% others who strongly disagreed that climate change impact is more on any particular gender than another. Similarly, 17% of the respondents disagreed that the climate change impact differs between genders just as 27% others neither agreed nor disagreed (undecided).

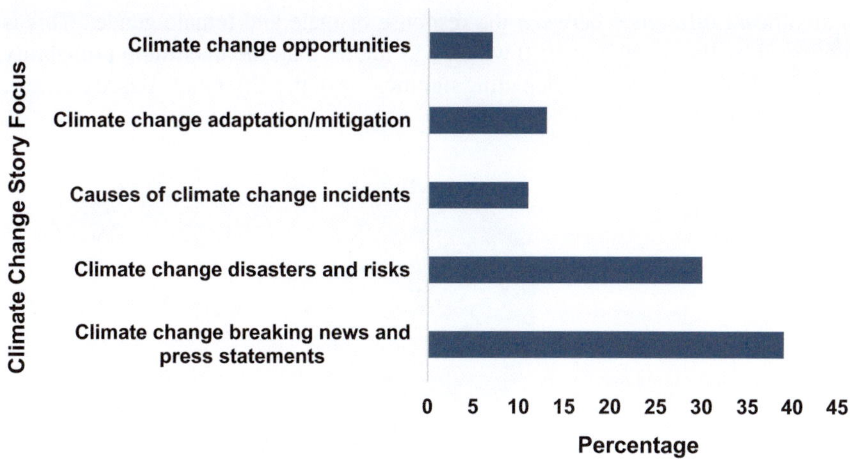

Fig. 21.9 Percentage distribution of climate change story focus

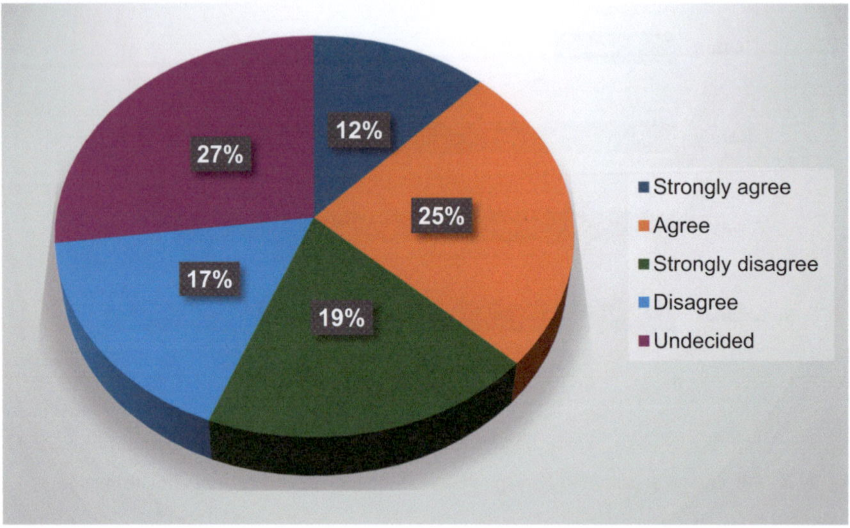

Fig. 21.10 Percentage distribution of the impact of climate change among genders

The data in Fig. 21.11 above shows the percentages of the respondents who are gender sensitive or have ever considered gender in their reportage of climate change issues or events. According to the data, most of the respondents (47%) are not sure if they have done it since they started writing on climate change. Also, a good number of the respondents (28%) admitted not to have considered gender in reporting climate change. The data further revealed that only 25% of the respondents are gender sensitive while reporting climate change.

The hypothesis result presented in Table 21.2 showed that the T calculated value of 33.87 was higher than the tabulated value of 1.96. This indicates the presence of significant difference between the response of male and female gender. This is confirmed by the p value of 0.00 which was higher than the maximum probability value for significance of 0.05, depicting significance of the difference. Based on these,

Fig. 21.11 Percentage distribution of gender consideration in climate change reporting

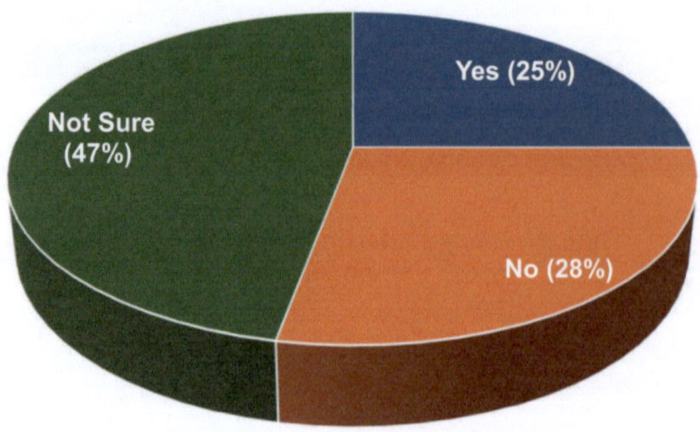

the study rejected the null hypothesis and concluded that there exists a significant difference in the media knowledge and coverage of climate change across gender.

The result in Table 21.3 showed that the correlation coefficient for whether there is relationship between gender and media coverage of climate change in Nigeria or not was 0.76 which means a very high and positive relationship. Also, the probability value of 0.00 depicts that the relationship was also significant. Therefore, the study accepted the null hypothesis.

Table 21.4 also provides the place of gender in the media coverage of climate change. On why the respondents do not consider gender importance in their report: 10 (19%) do not think it is necessary, 19 (37%) claimed that men are more disposed to address the media on climate change than women, 7 (13%) said their interest were usually on who and what is available while reporting climate change, 13 (25%) all of the above, and 3 (6%) none of the above. The respondents reflect climate change in their report in the following: consulting and citing more female sources/experts than males in their reports (1 or 2%), 6 or 13% give voices to both male and female experts or victims of climate change impact in equal proportion in their reports, 13 (28%) use gender sensitive language to avoid stereotyping or discrimination in their reports, 10 or 21% promote climate change opportunities and issues among genders in their reports, all of the above (15 or 32%), and none of the above – 2 or 4%.

The Fig. 21.12 above shows some of the gender-based climate change issues reported by the respondents since they started writing on climate change. According to the data, 28% of the respondents have reported how climate change is contributing to gender-based violence. Also, 13% of the respondents have reported how economic inclusion and exclusion are contributing to impact of climate change among genders. Similarly, 19% of the respondents have written on how political inclusion and exclusion are affecting how the various genders respond to climate change impact. Again, 6% of the respondents have used their media reports to address how equal and unequal access to education are contributing to impact of climate change among genders. There are also 26% of the respondents who have discussed in their media about how cultural and religious discriminations have impacted different genders with regards to climate change. Unfortunately, only 2% of the respondents have reported on climate change technology and businesses available to different genders just as 6% others have reported gender-based climate change mitigations and adaptation strategies.

The data in Fig. 21.13 shows the rate or frequency at which respondents report climate change gender-based issues. According to the data, 17% of the respondents report climate change gender-based issues always and very often respectively. Also, 36% of the respondents report gender-based climate change issues whenever they

Table 21.3 Spearman rank correlation result for relationship between gender and the media coverage of climate change in Nigeria

Pearson correlation coefficient	Probability value	N
0.76	0.00	390

*Significant at 1%, 5% and 10%

Table 21.4 Respondents gender-based climate change reports

Variables and responses	Frequency	Percentage
Why do you not consider gender in your climate change reportage?		
I do not think it is necessary to consider gender in my climate change reports	10	19
Men are more disposed to address the media on climate change than women	19	37
My interest is usually on who and what is available while reporting climate change, not minding the gender	7	13
All of the above	13	25
None of the above	3	6
Why do you reflect gender in your climate change reportage?		
I consult and cite more female sources/experts than males in my climate change reports	1	2
I give voices to both male and female experts or victims of climate change impact in equal proportion in my reports	6	13
I use gender sensitive language to avoid stereotyping or discrimination in my climate change reports	13	28
I promote climate change opportunities and issues among genders in my reports	10	21
All of the above	15	32
None of the above	2	4

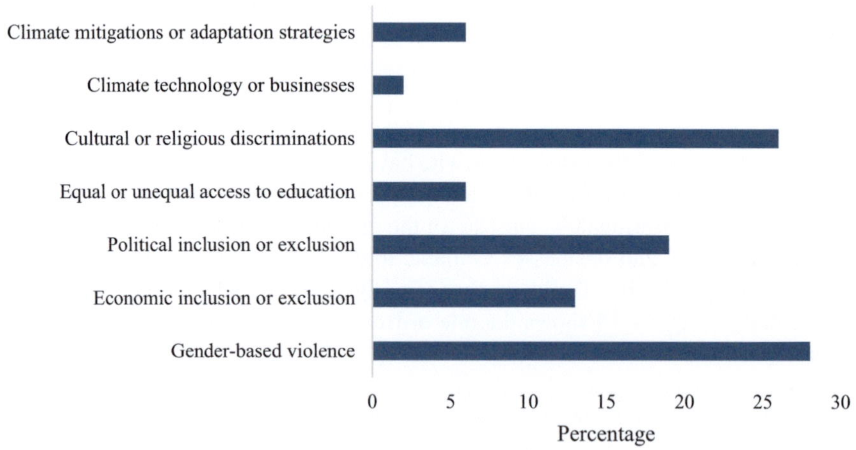

Fig. 21.12 Percentage distribution of gender-based climate change issues reported by the media

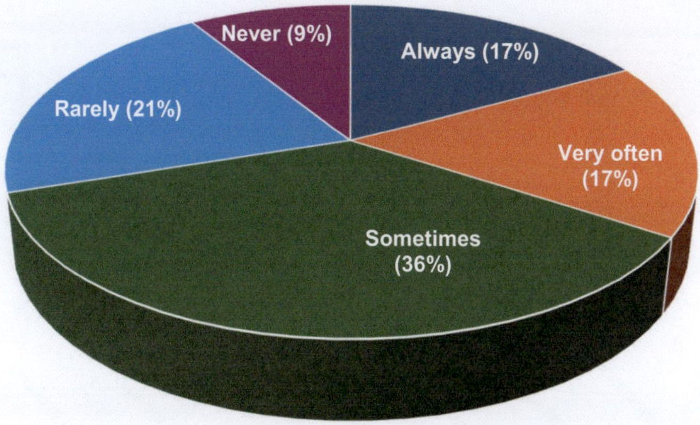

Fig. 21.13 Frequency in reportage of climate change gender-based issues

like (sometimes). There are also 21% of the respondents who rarely report them and 9% others who have never done that at all (never).

The data in Fig. 21.14 presents the location and sources of climate change gender-based stories to Nigerian media. It revealed that 24% of the respondents have sourced their climate change gender-based stories from rural communities just as 13% of the respondents have sourced theirs from urban communities. Similarly, there are 8% of the respondents who have sourced their climate change gender-based stories from semi-urban communities and 3% of the respondents who have sourced similar stories through religious institutions or gatherings. Also, 7% of the respondents have sourced their climate changed gender-based stories from the experts (Individuals, NGOs or corporate institutions), while 27% of the respondents have sourced theirs from government agencies (MDAs, NIMET, etc.). There are also 18% of the respondents who have accessed climate change gender-based stories through all of the above listed options.

The data in Fig. 21.15 indicate that location and sources can influence the extents media reports climate change gender-based stories in various ways. According to the data, 28% of the respondents said it can happen to a great extent (28%), while 25% of the respondents believed that it could happen sometimes or somewhat (25%). There are also 26% of the respondents who believed that it has very little influence and 21% others who insisted that they have little or no influence (Not at all).

The data in Fig. 21.16 revealed some reasons behind the respondents' choice of climate change gender-based stories with regards to locations and sources. It showed that 17% of the respondents' do that because of their media house-style. Again, 26% of the respondents maintained that they do that because of transportation challenges, while 11% of the respondents do that because of security challenges. There are also 17% of the respondents who do that because of their choice of audience and 8% of the respondents who do that because of religion. Again, 1% of the respondents do that because of their media outreach/coverage area and 16% of the respondents who that based on all of the reasons mentioned above (All of the above).

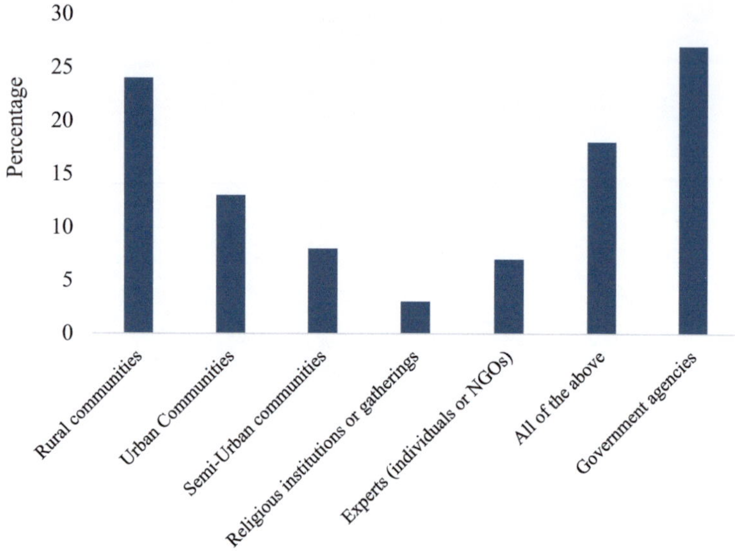

Fig. 21.14 Percentage distribution of respondents' gender-based climate change sources and locations

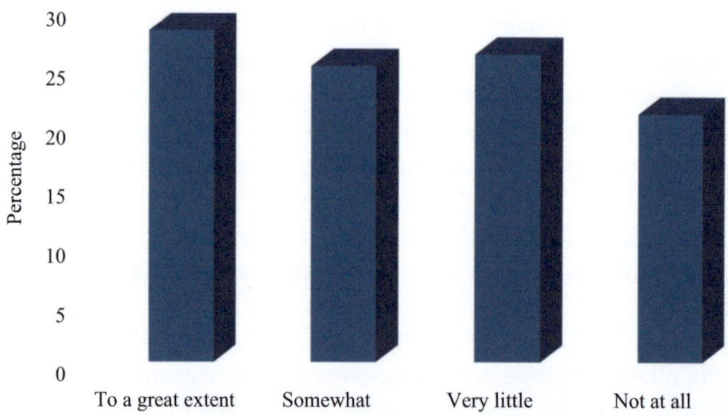

Fig. 21.15 Percentage distribution of extents location and sources influence climate change gender-based stories

The data in Fig. 21.17 showed that there are hope that media coverage of climate change can mitigate the impact among genders. According to the data, 41% of the respondents agreed that media coverage of climate change can mitigate the impact to a great extent. Also, 24% of the respondents said it can happen somewhat and 11% of the respondents said it is less likely to happen (Very little). Also, a very good number of the respondents (24) do not believe that it can happen (Not at all).

Figure 21.18 reveals how the media can improve in their climate change gender-based reportage. According to the data, 27% of the respondents believed that it

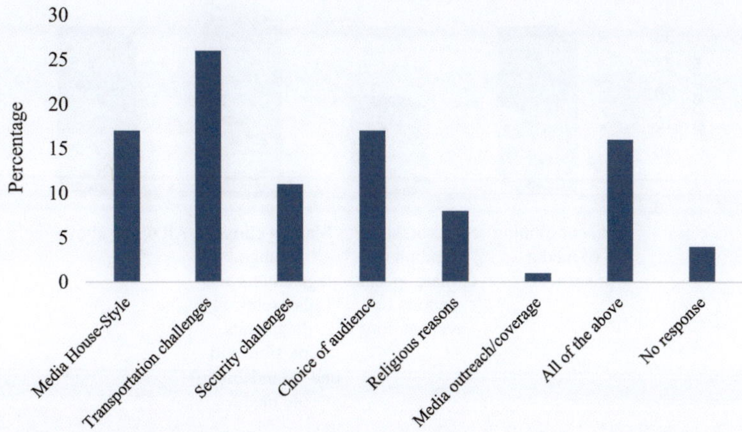

Fig. 21.16 Percentage distributions of some reasons locations and sources determine media choice of climate change gender-based stories

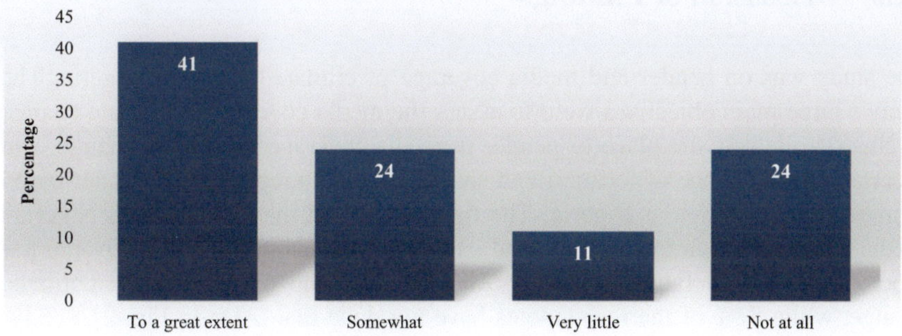

Fig. 21.17 Percentage distribution of extents media coverage of climate change can mitigate the impact among genders

could happen through frequent training of the media practitioners. Also, 15% of the respondents maintained that it can only happen through sponsorship or funding of climate special reports or investigations. Furthermore, most of the respondents (31%) emphasised that it can happen making climate change legislations and other relevant documents accessible and understandable to the media. Also, a good number of the respondents (27%) believed that all the options can contribute towards improving the media reportage of climate change gender-based stories.

Fig. 21.18 Percentage distribution of how media can improve gender-based climate change reportage

21.6 Discussion of Findings

The study was on gender and media coverage of climate change in Nigeria. The study's three main objectives were to assess the media coverage of climate change in Nigeria, identify the place of gender in media coverage of climate change, and ascertain the influence of location and sources to media reporting of gender-based climate change stories in Nigeria. The finding showed that even though Nigerian media are aware of climate change and its impact, the majority neither covered nor reported about it. In Table 21.1, only 48% of the respondents have covered climate change, while 52% of the respondents have not done that. Also in Fig. 21.8, 33% of the respondents cover climate change sometimes, 31% do that rarely, while 29% do that always. On why there is poor reportage or coverage of climate change in Nigeria, most of the respondents confirmed that they lack the requisite training or skills (24%), and their medium will not broadcast or publish the reports (21%). This finding was not entirely different from the previous studies in Nigeria, Ghana and other parts of the world (Danaa, 2018; Gadzekpo et al., 2018; Galle, 2019; Tagbo, 2010).

The study also found that gender is not given adequate consideration in media coverage of climate change in Nigeria. Figure 21.10 shows that only 37% of the respondents admitted that climate change impact differs among genders, 36% did not agree, while 27% were undecided. This is in line with the result of hypothesis in Table 21.3 which showed that the correlation coefficient for whether there is relationship between gender and media coverage of climate change in Nigeria or not was 0.76 depicting a very high and positive relationship. This was also in line with a previous study (Floro et al., 2019) where it was found that there is disparity in gender and perception of climate change awareness in Ethiopia. Also, most of the respondents do not consider gender in their climate change report as 28% admitted no as against 25% that responded yes and 47% who were not sure if they had done so (in Fig. 21.11). This could be why the hypothesis result presented in Table 21.2 shows that the T calculated value of 33.87 was higher than the tabulated value of 1.96. This indicates the presence of significant difference between the response of male and female gender. This is confirmed by the *p* value of 0.00 which was higher than the maximum probability value for significance of 0.05, depicting significance of the difference. Based on these, the study rejected the null hypothesis and concluded that there exists a significant difference in the media knowledge and coverage of climate change in Nigeria across gender. On why the respondents do not consider gender importance in their report, 37% of the respondents claim that men are more disposed to address the media on climate change than women, 19% said they do not think it is necessary, 13% said they were only interested on who and what is available while reporting climate change, while 25% of the respondents believed or selected all of the above (in Table 21.4). The finding was in line with some previous studies (Adpc, 2021; Herzhoff & Shanahan, 2010) where women are not adequately covered in climate change reports and the media were charged to consider all genders in their report with a view to improve people's knowledge and behaviour towards climate mitigation and adaptation. A study by Patronella (2021) also observed that journalistic norms and understanding of key issues influence public opinion and coverage of climate change.

The study also found that location and sources influence media reporting of gender-based climate change stories in Nigeria. In Fig. 21.14, the majority of the respondents (27%) identified government agencies (MDAs, NIMET, etc.), 24% valued rural communities more and 18% chose all of the above. Also, location or source can influence respondents' choice of gender-based climate change stories to a great extent (28%), Somewhat (25%), Very little (26%) and Not at all-39 (21%) as contained in Fig. 21.15. The reasons why location and sources could determine respondents' choice of gender-based climate change stories mostly are because of transportation or logistics challenges (26%), media house-style (17%) and choice of audience (17%) as recorded in Fig. 21.16. However, in Fig. 21.17, most of the respondents (41%) agreed that media coverage of climate change can mitigate the impact among genders to a great extent. On how the media can improve in their climate change gender-based reportage, 27% of the respondents in Fig. 21.18 advocated for frequent training of the media practitioners, 15% of the respondents' urged sponsorship or funding of climate special reports or investigations, 31% asked for making climate change legislations and other relevant documents accessible and understandable to the media and 27% said they stood for all (All of the above). Some scholars (Ugwu et al., 2021) also identified similar strategies to be adopted for effective implementation of climate change policies and legislations in Nigeria to include training, frequent radio and television broadcasts, feedback, norm appeal, panel discussion, cooperative learning and giving out free climate change literature and other incentives to the public. Hence, another scholar (Elia, 2021) also found that there is a direct relationship between professional journalism education and understanding of climate change, while recommending frequent training for journalists to make them report the issues surrounding it well.

21.7 Conclusion, Recommendations and Limitations

This study was focused on gender and media coverage of climate change in Nigeria. It had no intention of exploring all the areas of climate change or gender inclusions because of its restrictions on media coverage of the subject matter. The study was also limited by scope, methodology, financial resources and others. The cognitive dissonance and agenda setting theories were applied in the study. The findings showed among other things that media coverage of climate change with regards to gender was poor. This was in line to some of the extant literature cited in the work which implied that there is a need for gender to be given adequate consideration by the media while covering climate change in Nigeria and the rest of the world. This, according to our findings, would help to boost the fight to mitigate the impact of climate change in the world. Therefore, the researchers are recommending training and retraining of media practitioners with the knowledge of climate change and importance of gender inclusion in mitigating its impact. The researchers are also encouraging future scholars to explore this topic at regional or local levels, including applying other research methodologies like focus group discussions, in-depth interviews or cross sectional.

Acknowledgements Authors wish to express profound gratitude to the Head, Department of Mass Communication, University of Nigeria Nsukka, Professor Michael Ukonu, including Dr. Chidiebere Nwachukwu and Dr. Ijeoma Ajaero of the same Department for some useful advice they offered us at course of this study. Also, those whose works were cited in this study are highly appreciated.

Funding This article received no external funding.

Data Availability Statement Authors declare and affirm that the data supporting the findings of this study are available within the paper, its supplementary information file, and references. But individual respondents' data are available with the corresponding author who can only provide them on reasonable request.

Ethical Clearance

Conflict of Interest There is no conflict of interest expressed in this study by the authors.

References

Abdellatif, M. A. M. (2023). Flaming climate change on tv news websites: A study of Nile news, sky news Arabia, and CGTN Arabic. *Arab Media & Society, 34*, 5–10.

Adpc. (2021). *Applying a gender lens to climate actions: Why it matters, climate talks series care for South Asia.* Project Bangkok, Thailand

Agrawal, P., Post, L. N., Glover, J., Hearsay, D., Oberoi, P. & Biroscak, B. (2023). The interrelationship between food security, climate change and gender-based violence: A scoping review with system dynamics modelling. *PLOS Glob Public Health, 3*(2). https://doi.org/10.1371%2Fjournal.pgph.0000300

Batta, H. E., Ashong, A. C., & Bashir, A. S. (2013). Press coverage of climate change issues in Nigeria and implications for public participation opportunities. *Journal of Sustainable Development, 6*(2), 56–69. https://doi.org/10.5539/jsd.v6n2p56

Besser, H., Dhaouadi, L., Hadji, R., Hamed, Y. & Jemmali, H. (2021). Ecologic and economic perspectives for sustainable irrigated agriculture under arid climate conditions: an analysis based on environmental indicators for southern Tunisia. *Journal of African Earth Sciences, 177*(1), 1–15. https://doi.org/10.1016/j.jafrearsci.2021.104134

Binuyo, O. A., Bamgboye, O. E., & Adeola, G. A. (2022). Media advocacy in climate actionshowcasing best practices in West Africa. *Review Article Journal of Sustainability and Environmental Management, 1*(4), 419–424. https://doi.org/10.3126/josem.v1i4.50007

Brehm, J. W. (2007). A brief history of dissonance. *Social and Personality Psychology Compass, 1*(1), 381–391. https://doi.org/10.1111/j.1751-9004.2007.00035.x

Cooper, J. (2019). Cognitive dissonance: Where we've been and where we're going. *International Review of Social Psychology, 32*(1), 1–11. https://doi.org/10.5334/irsp.277

Dainton, M. (2004). *Introduction to communication theory*. Sage publication.

Danaa, Z. K. (2018). *Assessing the knowledge and perception of climate change among environmental journalists in Ghana*. (PhD Thesis). University of Ghana.

David, C. V., & Bran, A. (2017). *Cognitive dissonance theory*. Oxford bibliographies. https://doi.org/10.1093/OBO/9780199828340-0156

Desai, B. H., & Mandal, M. (2021). The role of climate change in exacerbating sexual and gender-based violence against women: A new challenge for international law. *Environmental Policy and Law, 51*(3), 137–157. https://doi.org/10.3233/EPL-210055

Elia, E. F. (2021). Journalists' awareness and understanding of climate change in Tanzania. *International Journal of Communication, 15*(19), 22–40. index.php/ijoc./article/view/10558

Floro, M., Yesuf, M. & Woldesenbet, T. (2019). Gender and perception of climate change in Ethiopia. *The International Journal of Climate Change: Impacts and Responses, 11*(2), 21–39. https://doi.org/10.18848/1835-7156/CGP/v11i02/21-39

Framework Convention on Climate Change. (2022). *Dimensions and examples of the gender-differentiated impacts of climate change, the role of women as agents of change and opportunities for women*. Retrieved from https://unfccc.int/sites/default/files/resource/sbi202207.pdf

Gadzekpo, A., Tietaah, G. K. M., & Segtub, M. (2018). Mediating the climate change message: Knowledge, attitudes and practices (KAP) of media practitioners in Ghana. *African Journalism Studies, 39*(3), 1–23. https://doi.org/10.1080/23743670.2018.1467838

Galle, U. A. (2019). News agenda setting and climate change in Nigeria: an appraisal. *International Journal of Innovative Research in Education, Technology & Social Strategies, 6*(1), 17–29. Retrieved from http://www.internationalpolicybrief.org/images/2019/APRIL/IJIRETSS/ARTICLE2.pdf

Glenn, I. D. (1992). Determining sample size. *Fact Sheet PEOD-6, Florida cooperative extension services, institute of food and agricultural sciences*. University of Florida.

Hase, V., Mahl, D., Schafer, M. S., & Keller, T. R. (2021). Climate change in news media across the globe: An automated analysis of issue attention and themes in climate change coverage in 10 countries (2006–2018). *Global Environmental Change, 70*, 1–12. https://doi.org/10.1016/j.gloenvcha.2021.102353

Herzhoff, N. & Shanahan, M. (2010). *Raising the profile of gender and generation: The role of the climate change media partnership, 1*, 1–12. Retrieved from https://www.iied.org/sites/default/files/pdfs/migrate/G03096.pdf

Jeffs, N. (2022). *Why women's leadership is key to climate action*. Retrieved February 1, 2022, from https://www.weforum.org/agenda/2022/02/women-gender-equality-climate-change-leadership/

Kareem, K. (2023). *Why Nigeria must prioritise climate change awareness for women*. Retrieved May 31, 2023, from https://www.dataphyte.com/latest-reports/climate/why-igeria-must-prioritise-climate-change-awareness-for-women/

Kwansah-Aidoo, K. (2005). Prospects for Agenda-Setting research in the 21st century in topical issues in communications and media research. In Aidoo, K.K (Ed.) (pp. 40–41). Nova Science Publishers.

Mahl, D., & Guenther, L. (2023). Content analysis in the research field of environmental & climate change coverage. *Standardised Content Analysis in Communication, 2*, 203–2012. https://doi.org/10.1007/978-3-658-36179-2_18

Mueller, T. S. (2017). Consumer perceptions of electric utilities: Insights from the center for analytics research & education project in the United States. *Energy Research & Social Science, 26*, 34–39.

Naazie, G. K., Agyemang, I., & Tampah-Naah, A. M. (2024). Exploring climate change adaptation strategies of crop and livestock farmers in urban Ghana. *Discover Environment, 2*, 53. https://doi.org/10.1007/s44274-024-00084-8

Newburger, E. (2021). *Fires in the arctic, record sea levels: NOAA report details effects of climate change in 2020*. Retrieved August 26, 2021, from https://www.cnbc.com/2021/08/26/noaa-2020-climate-change-report-record-sea-levels-greenhouse-gases.html

Okere, N. (2021, July 20). *Reality of climate change*. Retrieved July 10, 2021, from https://thenationonlineng.net/reality-of-climate-change/

Patronella, A. M. (2021). Covering climate change: A sentiment analysis of major newspaper articles from 2010–2020. *Inquiries Journal, 13*, 1–10.

Rogers, E., & Dearing, J. (1988). Agenda-setting research: Where has it been, where is it going? *Communication Yearbook, 11*, 555–594.

Scha¨fer, M. S. (2015). Climate change and the media. *International Encyclopaedia of the Social and Behavioural Sciences, 2*, 853–859. https://doi.org/10.1016/B978-0-08-097086-8.91079-1

Stacks, D. W., Cathy-Li, Z., & Spaulding, C. (2015). Media effects. *International Encyclopaedia of the Social & Behavioural Sciences, 2*, 29–34.

Tagbo, E. (2010). *Media coverage of climate change in Africa a case study of Nigeria and South Africa*. Retrieved from https://reutersinstitute.politics.ox.ac.uk/sites/default/files/research/files/Media%2520Coverage%2520of%2520Climate%2520Change%2520in%2520Africa%2520A%2520Case%2520Study%2520of%2520Nigeria%2520and%2520South%2520Africa.pdf

Ugwu, N. F., Onyekwere, O. K., Adekoya, A. F., Anibueze, A. U., & Ibeneme, C. B. (2021). Strategies for enhancing climate change knowledge, attitude and practices: Examining the views of experts. *Nigerian Journal of Health Promotion, 14*, 128–138. https://journals.aphriapub.com/index.php/NJHP/article/view/1666

Ume, C. O., Opata, P. I., Onyekuru, A. N. J. (2021). Gender and climate change adaptation among rural households in Nigeria. In N. Oguge, D. Ayal, L. Adeleke, I. da Silva (Eds.), *African handbook of climate change adaptation*. Springer. https://doi.org/10.1007/978-3-030-45106-6_182

United Nation Women. (2022). *Explainer: How gender inequality and climate change are interconnected*. Retrieved February 28, 2022, from https://www.unwomen.org/en/news-stories/explainer/2022/02/explainer-how-gender-inequality-and-climate-change-are-interconnected

United Nations Human Right. (2022). *Climate change exacerbates violence against women and girls*. Retrieved from https://www.ohchr.org/en/stories/2022/07/climate-change-exacerbates-violence-against-women-and-girls

World Bank. (2021). *Gender dimensions of disaster risk and resilience*. Report, World Bank.

Zanna, M. P., & Cooper, J. (1974). Dissonance and the pill: An attribution approach to studying the arousal properties of dissonance. *Journal of Personality and Social Psychology, 29*, 703–709. https://doi.org/10.1037/h0036651

Chapter 22
A Comparative Analysis of Cattle Rumen Content and Anaerobic Mono-and Co-digestion of Food Waste with Evidence from Microbial Communities, Biogas Yield, and Redox Potential

Afeez Oladeji Amoo, Ahmed Sabo, Adamu Haruna, Adeniyi Olarewaju Adeleye, Nureni Babatunde Amoo, Emmanuel Madu Ijanu, and Catherine Iyabo Asaju

Abstract Anaerobic digestion is a pivotal waste-to-energy technology that converts organic waste into biogas, offering notable environmental and economic benefits. This research investigates the comparative efficacy of mono- and co-digestion processes involving cattle rumen content (CRC) and food waste (FW), providing insights into their potential for enhancing biogas production and supporting sustainable waste management. The substrates physicochemical properties were characterized, microbial communities analysed, and biogas production, TOC, and redox potential monitored for 40 days. FW had higher moisture (71.2%) and volatile solid content (89.3%) than CRC (68.4%). Co-digestion increased microbial diversity and resulted in higher biogas yields (16,205.81 mL/kgVS) compared to mono-digestion (CRC: 13,425.33 mL/kgVS, FW: 11,077.04 mL/kgVS). TOC decreased significantly from 32,280.8 to 9912.5 mg/kg for FW, 26,920.4 to 8143.3 mg/kg for CRC, and 29,600.6 to 2649.9 mg/kg during digestion, indicating organic matter degradation. Redox potential decreased from − 53.2 to − 190.7 mV for FW, − 130.4 to − 276.2 mV for CRC, and − 110.6 to − 308.4 mV for co-substrates, reflecting anaerobic conditions favourable for biogas production. Correlation analysis showed significant relationships between biogas production, TOC removal, and redox potential, highlighting the importance of substrate composition and microbial activity. This study provides

A. O. Amoo (✉) · A. O. Adeleye · E. M. Ijanu · C. I. Asaju
Department of Environmental Sciences, Federal University Dutse, Jigawa State, Nigeria
e-mail: afeezoladeji@fud.edu.ng

A. O. Amoo · A. Sabo · A. Haruna
Department of Environmental Management Technology, Abubakar Tafawa Balewa University, Bauchi, Bauchi State, Nigeria

N. B. Amoo
Department of Surveying and Geo-Informatics, Federal Polytechnic Ede, Osun State, Nigeria

R. U. Onyeneke et al. (eds.), *Energy Transition, Climate Action and Sustainable Agriculture*, https://doi.org/10.1007/978-3-031-83165-2_22

insights for optimizing AD processes using CRC and FW, with implications for waste management, renewable energy generation, and greenhouse gas mitigation.

Keywords Biogas yield · Microbial communities · Waste-to-energy · Redox potential · Environmental benefits

22.1 Introduction

Anaerobic digestion (AD) has garnered significant attention as a sustainable method for converting organic waste materials into valuable biogas while mitigating environmental pollution (Rasapoor et al., 2020). The digestion of a single substrate, and co-digestion, which entails the simultaneous digestion of multiple substrates, offer distinct advantages and challenges (Karki et al., 2021). Mono-digestion simplifies operational control and substrate management but may lead to process inefficiencies due to substrate imbalance (Zhang et al., 2019). In contrast, co-digestion presents opportunities for synergistic interactions between substrates, potentially enhancing biogas yield and process stability through improved substrate balance and microbial diversity (Ferdes et al., 2023; Rabii et al., 2019). Biogas production has attracted substantial attention in recent years due to its potential to generate renewable energy while simultaneously minimizing greenhouse gas (GHG) emissions (Amoo et al., 2023a; Burg et al., 2018; Koirala et al., 2021; Kumar et al., 2022; Siddiki et al., 2021). According to Mason et al. (2022), the global potential for biogas production from organic wastes is 400 billion cubic meters per year, or 20% of global natural gas consumption. By lowering household energy and manure management costs while marketing more biogas and digestate, a nutrient-rich fertilizer, which improves soil health (Lee et al., 2021; Wang et al., 2021) and biogas production can as well boost the economy (Amoo et al., 2023b; Lu & Gao, 2021; Naveen et al., 2020).

Microbial communities are essential to the AD processes by catalysing the breakdown of organic matter into carbon dioxide and methane (Amoo et al., 2023c; Li et al., 2022; Mutungwazi et al., 2021). For instance, Lactobacillus sp. plays a vital role to both acetogenesis and acidogenesis during the anaerobic digestion of food waste, according to Li et al. (2021), while Zhang et al. (2023) emphasized the role of *Clostridium* sp. in cellulose and hemicellulose degradation during anaerobic digestion of cow rumen content. *Methanobrevibacter* sp. was identified by Ma et al. (2020) as a key methane producer from food waste, and *Acetobacterium* sp. was linked to acetate production by Oladejo et al. (2020). Chen et al. (2023), identified dominant species including *Lactobacillus* sp., *Clostridium* sp., *Methanobacterium* sp., *Methanoculleus* sp. and *Methanosarcina* sp. during single digestion and co-digestion of chicken manure, food waste, and corn straw. Metagenomic studies, such as Ju et al. (2021), revealed novel bacteria, like Candidatus Cloacamonas acidaminovorans, involved in protein degradation. Microbial additives, like *Bacillus* sp. (Ma et al., 2019), and microbial electrochemical systems (METs) (Zhu et al., 2021) have also shown promise in enhancing biogas production.

Despite the growing body of literature on anaerobic digestion, there exists a gap in understanding the comparative effects of co-digestion as well as mono-digestion strategies on microbial community dynamics, production of biogas, and redox potential, particularly concerning food waste and cattle rumen content substrates. This study aims to address this gap by analysing anaerobic mono-and co-digestion methods using cattle rumen content and food waste as the substrates. It investigates how these digestion strategies affect microbial community structure, yield of biogas, and redox potential in anaerobic digesters.

22.2 Methodology

22.2.1 Study Area

Dutse, the capital city of Jigawa State, Nigeria, serves as the administrative centre for the state 28 local government areas. It is situated at 11.76° North latitude and 9.34° East longitude, with an elevation of 460 m above sea level, this city has a population of more than 17,129 residents (Adeleye et al., 2019; Amoo et al., 2024) as presented in Fig. 22.1. Dutse ultra-modern (new) and old market (famously known as Yantifa) were used for this study. The Yantifa market which was a farmland covered with grasses was established after the birth of Jigawa state in 1991, and effort to upgrade and relocate the Yantifa market resulted to the establishment of the Dutse ultra-modern market in 2011. Arrays of vegetables, fruits, slaughtering of livestock (cattle, sheep, goat, chicken), and food products are sold in these markets.

22.2.2 Collection and Pretreatment of the Waste Substrates

Two main processes were carried out in the collection and pretreatment of the waste substrates: the development of rumen water and the treatment of waste substrates for anaerobic digestion. The waste substrates used for anaerobic digestion comprised food waste and rumen content. The cattle-derived rumen content (CRC) was sourced from the Dutse Central Abattoir in Dutse urban, as depicted in Fig. 22.1. After collection, the rumen content was immediately transported to the experimental site in an air-tight, non-transparent 15 L-capacity plastic container. Food wastes such as orange peels, watermelon peels, vegetables remnants, spoiled water lettuce, and pineapple peels were collected at their source of generation (motor parks and market) in Dutse. The collected food waste was immediately transported to the experimental site where it was blended and homogenized using an electric mixer, following the baseline properties outlined by Amoo et al. (2023d), to create the substrates used for the biogas production process.

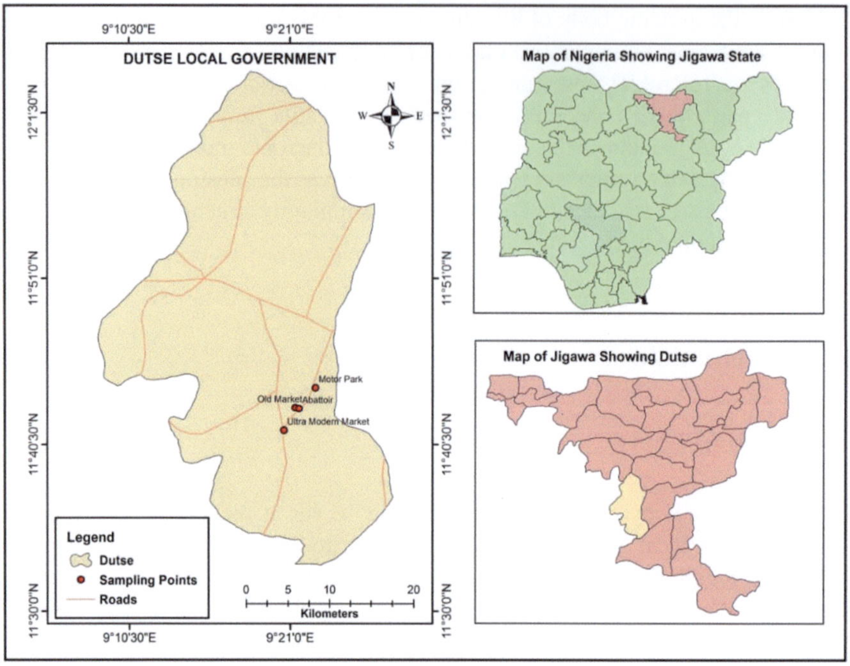

Fig. 22.1 Map of the study area

22.2.3 Experimental Design for Anaerobic Digestion of Selected Organic Substrates

Table 22.1 outlines the experimental setup for the AD of cattle rumen content and food waste. Three AD systems were considered: AD_{FW} for food waste (FW), AD_{CRC} for cattle rumen content (CRC), and $AD_{Co\text{-}S}$ for the combined substrates (Co-S), comprising both FW and CRC. It provides data on wet solid, water content, total solid, volatile solid, rumen fluid, and total digester content, offering insights into substrate composition and quantities for each AD system as described by Amoo et al. (2023c).

Development of rumen fluid (RF): To establish a methanogenic microbial community for use as a microbial inoculum source, rumen fluid was prepared via anaerobic digestion of rumen content (Meyer et al., 2022). Initially, 1 kg of rumen content was sourced from an abattoir, and this content was then mixed with 4 kg of tap water and supplemented with D-glucose to encourage microbial growth. The mixture was loaded into a 7-L capacity anaerobic digester and monitored for biogas production for a month. Following this incubation period, the resulting liquid digestate, termed rumen fluid, was obtained. This rumen water served as the microbial inoculum source for the subsequent anaerobic digestion of the waste substrates.

Table 22.1 Experimental design for anaerobic digestion of organic substrates

Anaerobic digestion (AD) system	Substrate	Wet solid (kg)	Water content (kg)	Total solid (kg)	Volatile solid (kg)	Rumen fluid (kg)	Total digester content (kg)
AD_{FW}	FW	1.0	0.230	0.760	0.963	1.0	2.0
AD_{CRC}	RC	1.0	0.217	0.780	0.726	1.0	2.0
$AD_{Co\text{-}S}$	Co-S	1.0	0.219	0.702	0.714	1.0	2.0

Food Waste = FW, Cattle Rumen Content = CRC, and Combined Substrates (Co-S) = FW + CRC

22.2.4 Physico-Chemical and Microbiological Characterization of the Substrates

The physical properties of the substrates, such as total solids, volatile solids, moisture content, and ash content, were measured using the Gravimetric Method as described by Mecha and Kiplagat (2023) and Jain et al. (2019). This method allowed for accurate measurement of the physical composition of the substrates. Chemical properties, including total organic carbon (TOC) content and total nitrogen content (TNC) of the substrates, were evaluated using the High-Temperature Combustion Method (Erdogan et al., 2015) and the Kjeldahl Method (Qiu et al., 2023), respectively. These methods provided insights into the substrates' organic carbon and nitrogen contents, which are essential for understanding their biodegradability and nutrient content as described by Singh et al. (2015).

22.2.5 Configuration and Operation of the Anaerobic Digester

A plastic digester with a cylindrical shape was utilized. The digester had a volume capacity of 2 L, out of which 1.5 L were used for loading the substrate, leaving a headspace volume of 0.5 L. The digester was equipped with an inlet for substrate addition, an outlet for digestate removal, and an outlet for biogas collection as detailed in Fig. 22.2.

22.2.6 Monitoring and Determination of Selected Process Parameters

The 300 mL gas syringe that was attached was used to continuously measure the output of biogas during the 40-day anaerobic digestion process. Every five days, at regular intervals, the volume of biogas produced was recorded. Following biogas

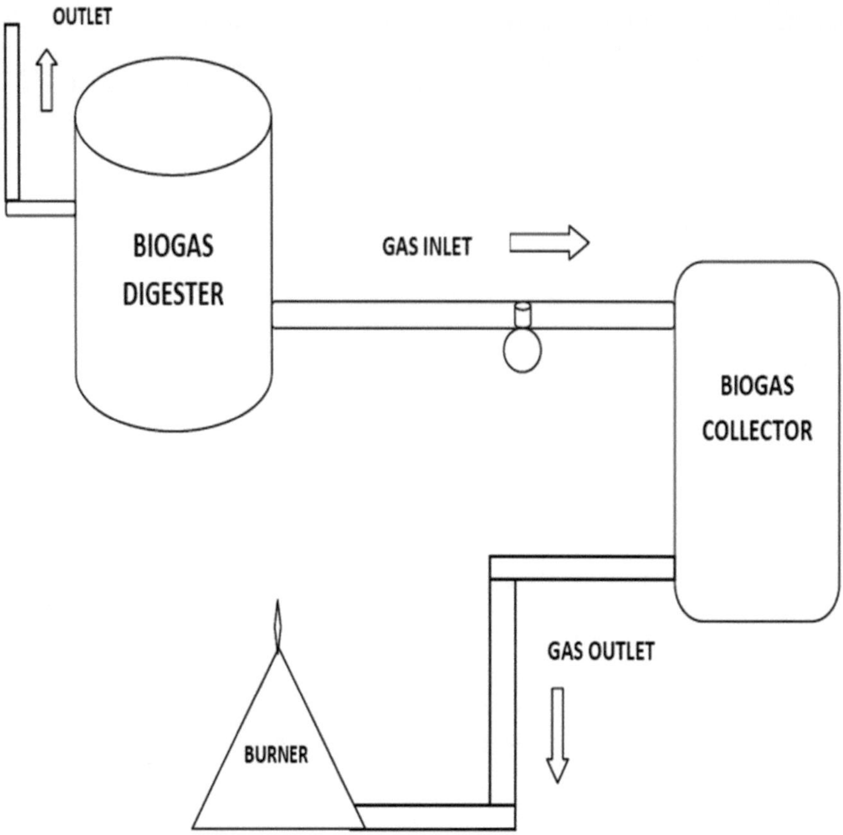

Fig. 22.2 Schematic experimental set-up of the anaerobic digester for biogas production

measurement, the generated biogas was evacuated from the digester and stored in a biogas bag. Total organic carbon (TOC) changes were monitored over time using standard techniques as described by Erdogan et al. (2015). This study provided insight into the development of anaerobic digestion and microbial activities within the system.

To ensure the anaerobicity of the digester contents, the redox potential was measured using a redox electrode and accompanying meter (inPro3250i/SG/120, Mettler-Toledo, Germany) according to the standard procedure outlined by Vongvichiankul et al. (2017). This analysis provided insights into the prevailing anaerobic conditions within the digester, ensuring optimal conditions for anaerobic digestion processes.

A digital pH meter with a long probe (inPro3250i/SG/120, Mettler-Toledo, Germany) was used to monitor the pH of the anaerobic digestion process. Hydrochloric acid or potassium hydroxide was used to adjust the pH and keep it stable at 7.2 throughout the experiments (Montalvo-Rodriguez et al., 2022). To foster the mixing of their contents, the bio-digesters were manually agitated three times a day (Zhang et al., 2021).

22.2.7 Bacteria Species Isolation

Three different test tubes were filled with 9 mL of sterile distilled water, and 1 g of each type of bacteria was suspended in the water, ensuring full mixing, in order to separate the facultative and strict anaerobic bacteria from food waste, rumen from cattle, and composite digestates. Following this, rumen material, food waste, and composite digestates were used to identify facultative and anaerobic bacterial species using the agar roll-tube spreading technique. These were then cultured at 30 °C for two to seven days (Singh et al., 2015). For short-term storage at 4 °C, the isolated colonies were re-cultured on pre-reduced agar slants and incubated for 2–7 days at 30 °C.

22.2.8 Culturing Media Preparation

The culture medium adopted from Ogbonna et al. (2018) was modified to isolate strict anaerobic and facultative anaerobic bacteria from the substrates and composite digestate samples. The following reagents were added to 1 L Erlenmeyer flasks with butyl rubber corks as the culture medium for stringent anaerobic bacteria: 1.0 g of NH_4Cl, 2.0 g of NaCl, 5.0 g of $NaHCO_3$, 0.3 g of KH_2PO_4, 0.3 g of K_2HPO_4, 0.16 g of $MgCl_2.6H_2O$, 0.01 g of $CaCl_2.2H_2O$, 12.5 mL of oligoelement solution, 2.0 g of yeast extract, 1.0 mL of resazurin solution (1% w/v), 5.5 g of D-glucose, 3.0 g of sodium acetate, 0.5 g of sodium thioglycolate, and 15 g of agar were all added. The final pH was 7.2. The facultative anaerobes were isolated using the same media, with the exception that 6 g of agar was utilized rather than the 15 g in the stringent anaerobic culture medium (Ogbonna et al., 2018).

22.2.9 Identifying and Characterizing Different Species of Bacteria

The bacterial isolates were identified using morphological and biochemical techniques, following the guidelines set out in Bergey's Manual of Determinative Bacteriology (Holt et al., 1994) and Bergey's Manual of Systematics of Archaea and Bacteria (Whitman et al., 2012). The identification of bacterial isolates involved a series of tests including Gram staining, acid-fast staining, motility assays, oxygen requirement assessments, and biochemical tests such as oxidase, catalase, coagulase, citrate, indole, urease, methyl red, and voges-proskauer tests. Additionally, fermentation tests with glucose, mannitol, sucrose, salicin, cellobiose, mannose, melezitose, raffinose, sorbitol, and cellulose hydrolysis, as well as gelatin and esculin hydrolysis were conducted, following the methods described by Barrow and Feltham (1993).

22.3 Results and Discussion

22.3.1 Physico-Chemical Properties of the Substrates

The substrates of (food waste and rumen content from cattle) physico-chemical characteristics are shown in Table 22.2. FW has a higher moisture content (71.2%) compared to CRC (68.4%). Rumen content has a higher total solid content (78.3%) than food waste (76.8%). FW exhibits a higher volatile solid content (96.3%) compared to rumen content (72.6%). Rumen content has a higher ash content (27.4%) compared to FW (3.7%). TOC content is higher in FW (32,280.8 mg/kg) compared to CRC (26,920.4 mg/kg). Total nitrogen (TN) content is slightly higher in CRC (1850.6 mg/kg) compared to FW (1926.5 mg/kg). The carbon to nitrogen (C/N) ratio is higher in FW (29.44) compared to CRC (22.48).

These results align with previous studies characterizing food waste as having high moisture content, promoting rapid decomposition due to its organic composition (Ho & Chu, 2019). Similarly, the elevated volatile solid content in food waste is consistent with its organic-rich nature reported in prior research (Palaniveloo et al., 2020). Moreover, the higher ash content in rumen content compared to food waste corresponds with studies on the mineral composition of animal-derived substrates, which typically contain more inorganic matter due to the animal's diet and digestive processes (Fonoll et al., 2021). Additionally, variations in the C/N ratio between food waste and rumen content are consistent with previous findings, indicating differences in carbon and nitrogen composition among waste substrates that influence their suitability for anaerobic digestion (Wang et al., 2023).

Table 22.2 Physico-chemical properties of the waste substrates

Physico-chemical properties	Food waste (FW)	Cattle rumen content (CRC)
Moisture content [%]	71.2	68.4
Total solid [%]	76.8	78.3
Volatile solid [%]	96.3	72.6
Ash content [%]	3.7	27.4
Total organic carbon (TOC) [mg/kg]	32,280.8	26,920.4
Total nitrogen content (TNC) [mg/kg]	1850.6	1926.5
C/N ratio	29.44	22.48

These findings provide valuable information about the composition of food waste and rumen content substrates, which is essential for evaluating their effectiveness in anaerobic digestion. The higher moisture content in food waste suggests increased microbial activity during digestion, enhancing organic matter decomposition and biogas production (Mu et al., 2020). Conversely, the elevated total solid and ash content in rumen content may impact process efficiency, necessitating adjustments in operating parameters to optimize biogas production (Sarker et al., 2019). The differences in organic carbon, nitrogen content, and the C/N ratio highlight the distinct biochemical compositions of the substrates, influencing microbial dynamics and biogas yield (Lee et al., 2020). Understanding these distinctions is pivotal for designing effective digestion strategies, maximizing biogas production, and ensuring process stability.

22.3.2 Biochemical Characteristics of Bacteria Isolated from the Substrates and Their Digestates

In total, sixteen distinct bacterial species across twelve genera were identified in the rumen product, composite digestates, and food waste. As detailed in Tables 22.3, 22.4, and 22.5, these species include *Acetobacterium* sp., *Bacteroides* sp., two types of *Clostridium* sp., *Enterobacter* sp., *Escherichia coli*, two types of *Lactobacillus* sp., two types of *Pseudomonas sp.*, *Ruminococcus* sp., *Staphylococcus* sp., *Streptococcus* sp., *Syntrophomonas* sp., and *Syntrophobacter* sp.

The diverse microbial communities found in rumen content, food waste, and their composites digestates provide crucial insights into the intricate microbial dynamics during anaerobic digestion. Genera such as *Clostridium, Lactobacillus, Pseudomonas, Bacteroides, Enterobacter, Escherichia coli, Streptococcus,* and *Ruminococcus* found in both substrates and digestates align with their established roles in organic matter degradation during anaerobic digestion (Kumar et al., 2021; Li et al., 2022). Among the microbes identified in both substrates and digestates, *Clostridium, Bacteroides, Ruminococcus, Pseudomonas,* and *Enterobacter* species play pivotal roles in hydrolysis and acidogenesis, crucial initial steps in breaking down complex organic compounds into soluble forms (Li et al., 2021; Zhang et al., 2021). These processes facilitate subsequent metabolic pathways, supporting efficient substrate degradation.

Furthermore, the identification of *Lactobacillus, Streptococcus,* and *Ruminococcus* species in the digestates indicates active acidogenesis, wherein soluble organic compounds are converted into volatile fatty acids (VFAs) and alcohols (Chen et al., 2021; Cui et al., 2020). These intermediates serve as precursors for acetogenesis and methanogenesis, driving biogas production (Detman et al., 2021). *Acetobacterium, Syntrophomonas,* and *Syntrophobacte*r species in the digestates suggest robust acetogenesis and methanogenesis activities (Amin et al., 2021; Zhang et al., 2023). *Acetobacterium* species facilitate acetate production from VFAs, while

Table 22.3 Bacteria isolated from the AD and their biochemical properties (IS 1–6)

Biochemical Tests	IS 1	IS 2	IS 3	IS 4	IS 5	IS 6
Gram stain	–	–	+	+	–	–
Shape	Rod	Rod	Rod	Rod	Rod	Rod
Arrangement	Single	Single	Single	Single	Single	Single
Spore	+	–	+	+	–	–
Acid fast	–	–	–	–	–	–
Motility	+	+	+	–	+	+
O_2 requirement	FA	OA	OA	OA	FA	FA
Oxidase	+	–	–	–	–	–
Coagulase	–	–	–	–	–	–
Citrate	+	–	–	–	–	+
Catalase	+	–	+	–	+	+
Indole	+	–	–	–	+	–
Urease	+	+	–	–	–	–
H_2S Production	+	+	–	–	–	–
Nitrate red	+	–	–	–	+	+
Methyl red	–	+	+	+	+	–
Voges Proskauer	–	–	–	–	–	+
Ornithinedecarboxylase	–	–	–	–	+	+
D-glucose	+/+	+/+	+/+	+/+	+/+	+/+
D-mannitol	+/+	+/+	–	+/+	+/+	+/+
D-sucrose	+/+	+/+	+/+	+/+	+/+	+/+
Lactose	–	+/+	+/+	+/+	+/+	+/+

(continued)

Table 22.3 (continued)

Biochemical Tests	IS 1	IS 2	IS 3	IS 4	IS 5	IS 6
D-maltose	+/+	+/+	+/+	+/+	+/+	+/+
D-xylose	–	+/+	+/+	+/+	+/+	–
L-arabinose	–	+/+	+/+	+/+	+/+	+/+
Salicin	–	+/+	+/+	+/+	+/+	+/+
Cellulose	–	–	+/+	+/+	–	–
Starch	+/+	+/+	±	±	–	–
Gelatin	+	–	–	±	–	–
Esculin	–	±	±	–	±	–
Glycerol		–	–	+/+	+/+	–
D-cellobiose	–	–	+/+	+/+	–	+/+
D-mannose	–	+/+	+/+	+/+	+/+	–
D-melezitose	–	–	–	–	–	–
D-raffinose	–	+/+	–	+/+	–	+/+
D-sorbitol	–	–	–	+/+	–	–
L-rhamnose	–	+/+	–	+/+	+/+	+/+
D-trehalose	–	–	–	+/+	–	–
Probably identify	Acetobacterium sp.	Bacteroides sp.	Clostridium sp.	Clostridium sp.	Escherichia coli	Enterobacter sp.

Key Obligate anaerobe (OA), Aerobe (A), Facultative anaerobe (FA), Acid and gas production (+/+), Acid production without gas production (±), No fermentation (–)

Table 22.4 Bacteria isolated from the AD and their biochemical properties (IS 7–12)

Biochemical Tests	IS 7	IS 8	IS 9	IS 10	IS 11	IS 12
Gram stain	+	+	+	+	+	+
Shape	Rod	Rod	Rod	Rod	Cocci	Cocci
Arrangement	Chain	Chain	Single	Single	Pair	Pair
Spore	–	–	–	–	–	–
Acid fast	–	–	–	–	–	–
Motility	+	+	+	+	–	–
O_2 requirement	OA	OA	FA	A	OA	OA
Oxidase	–	–	+	+	–	–
Coagulase	–	–	+	–	–	–
Citrate	+	–	+	+	–	–
Catalase	–	–	+	+	–	–
Indole	–	–	–	–	–	–
Urease	+	–	+	+	+	–
H_2S Production	–	+	–	–	+	–
Nitrate red	–	–	+	–	–	–
Methyl red	–	–	–	–	–	–
Voges Proskauer	–	–	–	–	–	–
Ornithine decarboxylase	–	–	–	–	–	–
D-glucose	+/+	+/+	–	±	+/+	+/+
D-mannitol	+/+	+/+	±	+/+	+/+	–
D-sucrose	+/+	+/+	–	–	+/+	+/+
Lactose	+/+	+/+	–	+/+	+/+	+/+

(continued)

Table 22.4 (continued)

Biochemical Tests	IS 7	IS 8	IS 9	IS 10	IS 11	IS 12
D-maltose	+/+	+/+	–	+/+	+/+	+/+
D-xylose	+/+	–	–	–	+/+	+/+
L-arabinose	+/+	+/+	–	–	+/+	+/+
Salicin	+/+	–	–	–	+/+	+/+
Cellulose	–	–	–	–	+/+	+/+
Starch	–	–	–	–	–	±
Gelatin	–	–	±	±	–	–
Esculin	–	–	±	–	±	±
Glycerol	–	–	+/+	–	–	–
D-cellobiose	+/+	+/+	–	–	+/+	+/+
D-mannose	–	+/+	–	–	+/+	+/+
D-melezitose	–	–	–	–	+/+	–
D-raffinose	–	–	–	–	+/+	+/+
D-sorbitol	+ / +	–	–	–	±	+/+
L-rhamnose	–	+/+	–	–	+/+	+/+
D-trehalose	–	–	–	–	+/+	+/+
Probably identify	Lactobacillus sp.	Lactobacillus sp.	Pseudomonas sp.	Pseudomonas sp.	Ruminococcus sp.	Ruminococcus sp.

Key Obligate anaerobe (OA), Aerobe (A), Facultative anaerobe (FA), Acid and gas production (+/+), Acid production without gas production (±), No fermentation (–)

Table 22.5 Bacteria isolated from the AD and their biochemical properties (IS 13–16)

Biochemical Tests	IS 13	IS 14	IS 15	IS 16
Gram stain	+	+	–	–
Shape	Cocci	Cocci	Rod	Rod
Arrangement	Cluster	Chain	Single	Single
Spore	–	–	–	–
Acid fast	–	–	–	–
Motility	+	–	+	–
O_2 requirement	FA	FA	OA	OA
Oxidase	+	+	–	–
Coagulase	+	+	–	–
Citrate	–	–	–	–
Catalase	+	–	–	–
Indole	–	–	–	–
Urease	+	+	–	–
H_2S Production	–	–	+	+
Nitrate red	–	–	+	–
Methyl red	–	+	–	–
Voges Proskauer	–	+	–	–
Ornithine decarboxylase	–	–	–	–
D-glucose	+/+	+/+	±	+/+
D-mannitol	±	±	±	–
D-sucrose	+/+	+/+	±	–
Lactose	+/+	+/+	±	–
D-maltose	+/+	+/+	±	–
D-xylose	–	–	±	–
L-arabinose	–	+/+	–	–
Salicin	–	–	–	–
Cellulose	–	–	–	–
Starch	–	±	±	–
Gelatin	±	–	–	–
Esculin	±	±	–	–
Glycerol	–	–	–	–
D-cellobiose	–	–	±	–
D-mannose	–	+/+	–	–
D-melezitose	–	–	–	–
D-raffinose	–	+/+	–	–

(continued)

Table 22.5 (continued)

Biochemical Tests	IS 13	IS 14	IS 15	IS 16
D-sorbitol	–	–	–	–
L-rhamnose	–	–	–	–
D-trehalose	–	+/+	±	–
Probably identify	Staphylococcus sp.	Streptococcus sp.	Syntrophomonas sp.	Syntrophobacter sp.

Key Obligate aerobes (OA), Obligate anaerobes (OAN), and facultative anaerobes (FA) – = No fermentation;+/+ = Acid and gas production; ± = Acid production but no gas production

Syntrophomonas, Syntrophobacter, and *Syntrophus* species contribute to syntrophic acetate oxidation and hydrogenotrophic methanogenesis, crucial processes for methane generation (Harirchi et al., 2022).

Moreover, the observed differences in microbial species composition between mono-digestion and co-digestion systems underscore the influence of substrate diversity on microbial community structure, potentially influencing process stability and biogas production efficiency (Li & Huang, 2024; Pan et al., 2022). Understanding these microbial dynamics is paramount for optimizing anaerobic digestion processes, maximizing biogas yield, and ensuring operational stability.

22.3.3 Dynamics of Biogas Produced by the Substrates

Figure 22.3 presents the cumulative biogas production (B_{pro}) over time from three different anaerobic digesters treating different substrates. In the anaerobic digester treating the food waste (AD_{FW}), rumen content (AD_{RC}), and the combined substrates ($AD_{Co\text{-}S}$), cumulative biogas increased steadily from 0.0 mL at day 1 to 2846.8 mL, 3020.7 mL, and 3905.6 mL to 40 days. Overall, the biogas yield in the anaerobic digester which treated the food waste (AD_{FW}), rumen content (AD_{RC}), and the combined substrates ($AD_{Co\text{-}S}$) reached 11,077.04 mL/kgVS, 13,425.33 mL/kgVS, and 16,205.81 mL/kgVS respectively after 40 days.

The cumulative biogas production and yields in this study are consistent with previous research on anaerobic digestion of food waste, rumen content, and their co-digestion (Ihoeghian et al., 2022; Sumardiono et al., 2021). These findings indicate the effectiveness of anaerobic digestion processes in biogas generation, with increasing biogas production observed over time. Additionally, our results support the trend of higher biogas yields in co-digestion systems compared to mono-digestion, as demonstrated in studies by Buragohain et al. (2021). This suggests that combining diverse substrates, like food waste and rumen content, enhances biogas production efficiency potentially due to improved substrate balance (resulting from complementary biochemical compositions) and improved microbial diversity (Chen et al.,

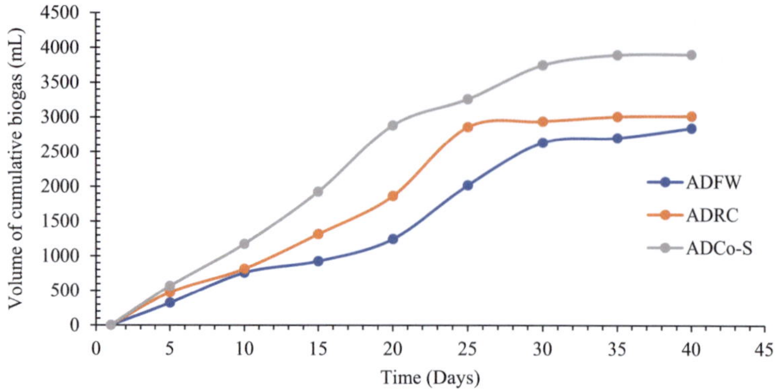

Fig. 22.3 Cumulative biogas produced by the substrates

2023; Zhao et al., 2022). In general, these findings emphasize the importance of substrate diversity in anaerobic digestion processes to maximize biogas production efficiency. They also have implications for the design and optimization of anaerobic digestion systems, highlighting the benefits of co-digestion strategies in enhancing biogas yields and overall process sustainability.

22.3.4 Dynamics of TOC Inside the Anaerobic Digesters

Figure 22.4 shows the dynamics of total organic carbon (TOC) content during anaerobic digestion of different substrates. In the anaerobic digester treating food waste (AD_{FW}), rumen content (AD_{RC}), and the combined substrates ($AD_{Co\text{-}S}$), the TOC content decreased from 32,280.8 mg/kg, 26,920.4 mg/kg, and 29,600.6 mg/kg at day 1 to 9912.5 mg/kg, 8143.3 mg/kg, and 2649.9 mg/kg at day 40 respectively. Overall, the results indicate a progressive reduction in TOC content during the AD process for all substrates, suggesting the decomposition of organic matter into biogas.

The progressive reduction in TOC content throughout anaerobic digestion indicates the effective biodegradation of organic matter into biogas, highlighting the potential of both mono-digestion and co-digestion processes for biogas production (Borth et al., 2022; Rocamora et al., 2020). The more substantial TOC reduction observed in the co-digestion system suggests that combining food waste and rumen content enhanced the efficiency of organic matter degradation. Previous studies have reported similar trends in TOC reduction during anaerobic digestion. Research by Ibro et al. (2022) revealed that digestion of two or more substrates often results in enhanced organic matter degradation compared to mono-digestion. This could be attributed to synergistic interactions between different substrates and the diverse microbial populations they support. The implications of these findings are significant for optimizing anaerobic digestion systems for biogas production. Co-digestion

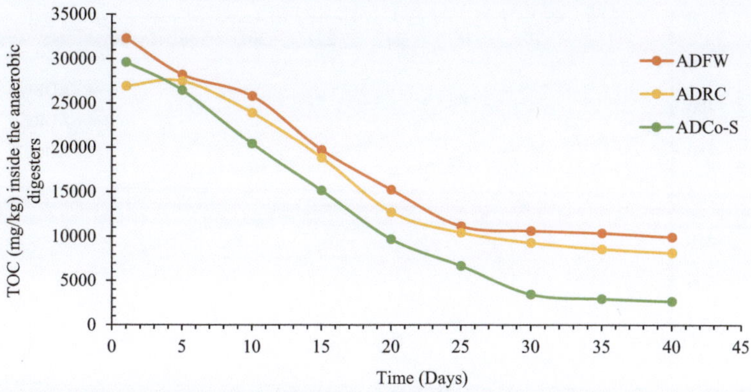

Fig. 22.4 Dynamics of TOC during anaerobic digestion of the substrates

strategies that incorporate diverse feedstocks, such as food waste and rumen content, could improve process efficiency and biogas yields by promoting more extensive organic matter degradation. Moreover, understanding the dynamics of TOC reduction provides valuable insights into the biodegradation kinetics during anaerobic digestion, aiding in the design and operation of efficient biogas production systems.

22.3.5 Dynamics of Redox Potential (RP) During Anaerobic Digestion of the Substrates

Figure 22.5 presents the dynamics of redox potential (RP) during the anaerobic digestion of different substrates in the anaerobic digester treating food waste (AD_{FW}), rumen content (AD_{RC}), and the combined substrates ($AD_{Co\text{-}S}$), the redox potential started at − 53.2 mV, − 130.4 mV, and − 110.6 mV on day 1 and gradually decreased to − 190.7 mV, − 276.2 mV, and − 308.4 mV respectively by day 40. Overall, the results indicate a consistent decline in redox potential throughout the anaerobic digestion process for all substrates, suggesting the establishment and maintenance of anaerobic conditions conducive to biogas production.

The progressive decrease in RP throughout AD reflects the establishment and maintenance of anaerobic conditions necessary for efficient biogas production (Shen et al., 2021). The more significant reduction in RP observed in the co-digestion system suggests that combining food waste and rumen content accelerated the transition to anaerobic conditions. Previous studies have reported similar patterns of RP decline during anaerobic digestion (Ibro et al., 2022; Shen et al., 2021; Wang et al., 2019). This could be attributed to synergistic interactions between the diverse substrates and microbial communities present in co-digestion systems, facilitating faster microbial adaptation and metabolic activity. Research by Ao et al. (2021) and Zhao et al. (2021) demonstrated that anaerobic conditions are essential for optimal biogas production,

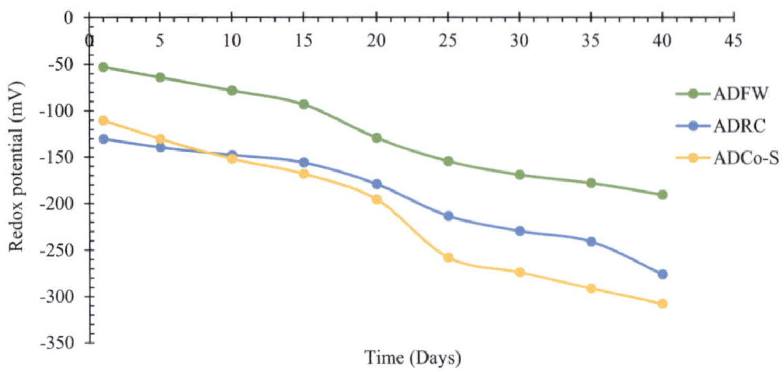

Fig. 22.5 Dynamics of redox potential (RP) during AD of the substrates

as they facilitate the activity of methanogenic microorganisms. The consistent decline in RP observed in both mono-digestion and co-digestion systems aligns with these findings, indicating the establishment of anaerobic environments conducive to biogas generation. The implications of these findings are significant for optimizing anaerobic digestion processes for biogas production. Co-digestion strategies that incorporate diverse feedstocks, such as food waste and rumen content, can enhance RP reduction and promote anaerobic microbial activity, ultimately improving biogas yields. Moreover, monitoring RP dynamics provides valuable insights into the establishment of anaerobic conditions during anaerobic digestion, aiding in process optimization and ensuring the efficient conversion of organic matter into biogas.

22.3.6 Temporal Relationship Between Biogas Production, Redox Potential and TOC

Figure 22.6 shows the correlations between biogas production (B_{Pro}), redox potential (R_{Pot}), and total organic carbon (TOC) during anaerobic digestion of the substrates. For the anaerobic digester treating food waste (AD_{FW}), there is a strong negative correlation between biogas production and TOC (– 0.9486), indicating that as TOC decreases, biogas production increases. There is also a strong negative correlation between biogas production and redox potential (– 0.9871), suggesting that as redox potential decreases (indicating more anaerobic conditions), biogas production increases. Additionally, there is a strong positive correlation between TOC and redox potential (0.9712), implying that as TOC decreases, redox potential also decreases.

Similarly, for the anaerobic digester treating rumen content (AD_{RC}), there are strong negative correlations between biogas production and both TOC (– 0.9801) and redox potential (– 0.9829). There is also a strong positive correlation between TOC and redox potential (0.9557). For the anaerobic digester treating the combined substrates ($AD_{Co\text{-}S}$), the correlations follow a similar pattern, with strong negative

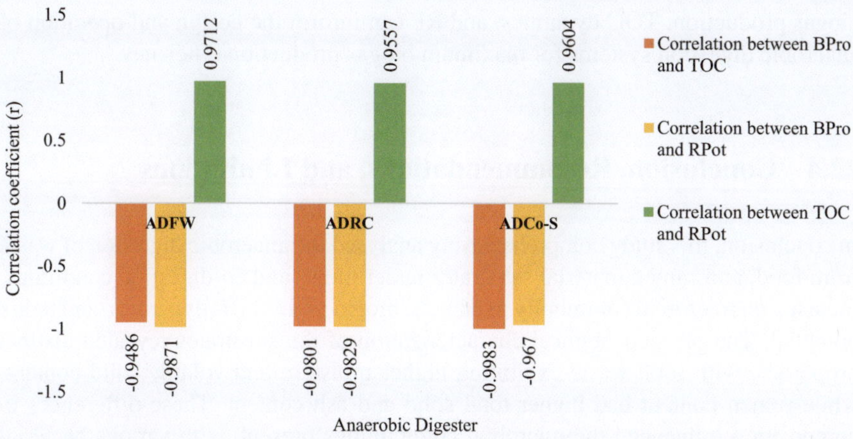

Fig. 22.6 Relationship between biogas production, redox potential and TOC during anaerobic digestion of the substrates

correlations between biogas production and both TOC (− 0.9983) and redox potential (− 0.9670), as well as a strong positive correlation between TOC and redox potential (0.9604).

Overall, the results suggest that there is a significant temporal relationship between biogas production, TOC, and redox potential during anaerobic digestion, with increased TOC removal and lower redox potential correlating with higher biogas production.

In the result above, the strong negative correlations between biogas production and both TOC and RP indicate that efficient biogas production is associated with the depletion of organic carbon and the establishment of anaerobic conditions. This underscores the importance of organic matter degradation and microbial activity in biogas generation. The positive correlation between TOC and RP further highlights the relationship between substrate degradation and the establishment of anaerobic environments conducive to biogas production. Previous studies by Mao et al. (2019) and Wu et al. (2019) have reported similar correlations between biogas production, TOC dynamics, and RP dynamics during anaerobic digestion. These studies emphasized the importance of TOC removal and the establishment of anaerobic conditions for efficient biogas production. The consistency of these correlations across mono-digestion and co-digestion systems reinforces the fundamental principles governing anaerobic digestion processes.

These results have significant implications for optimizing anaerobic digestion processes. Monitoring TOC dynamics and RP during anaerobic digestion can serve as indicators of process efficiency and biogas production potential (Ibro et al., 2022). Strategies aimed at enhancing TOC removal and maintaining low RP levels, such as optimizing feedstock composition and reactor conditions, can improve biogas yields and process stability. Additionally, understanding the temporal relationships between biogas production, TOC dynamics, and RP can inform the design and operation of anaerobic digestion systems for maximum biogas production efficiency.

22.4 Conclusion, Recommendations, and Limitations

In conclusion, this study comprehensively analysed the anaerobic digestion of waste from food, and rumen material substrates under mono-and co-digestion conditions, focusing on microbial community dynamics, biogas yield, TOC dynamics, and redox potential. The physicochemical characterization of the substrates revealed distinct properties, with food waste exhibiting higher moisture and volatile solid content, while rumen content had higher total solid and ash content. These differences in composition influenced the microbial communities present, with various bacterial species identified across both substrates and their digestates. The findings underscored the importance of microbial diversity and activity in anaerobic digestion processes. Regarding biogas production, co-digestion systems consistently outperformed mono-digestion, yielding higher cumulative biogas production and biogas yields. This trend aligned with previous studies, emphasizing the synergistic effects of combining diverse substrates on biogas production efficiency. The temporal relationships between biogas production, TOC dynamics, and redox potential further elucidated the factors influencing anaerobic digestion efficiency. Lower TOC content and redox potential correlated with higher biogas production, indicating the importance of organic matter degradation and the establishment of anaerobic conditions.

Based on the findings of this study, it is recommended to prioritize the combination of diverse substrates, as results in higher biogas yields due to synergistic effects on microbial community dynamics and substrate breakdown. Furthermore, monitoring key parameters such as total organic carbon dynamics and redox potential during digestion can help optimize the process for improved efficiency. Industrial-scale anaerobic digestion systems should incorporate substrate diversity to maximize energy recovery from organic waste. Understanding the complex interactions between microbial communities, substrate composition, and process parameters is crucial for enhancing biogas yield, process stability, and overall efficiency in waste treatment and renewable energy production.

Additionally, the study did not explore the long-term stability or scalability of the co-digestion systems for industrial applications. Likewise, there was no detailed investigation into the potential inhibitory effects of specific substrate components on biogas production.

Acknowledgements The authors would like to acknowledge the staff of Soil Science, and Microbiology Laboratory both of Federal University Dutse, Jigawa State, Nigeria for their role in physico-chemical and biochemical analysis of the study.

Funding The authors received no financial support for the research, authorship, and/or publication of this manuscript.

Data Availability Statement The data supporting the findings of this study are available from the corresponding author upon reasonable request. The physico-chemical properties of the waste substrates, biochemical characteristics of bacteria isolated from the substrates and their digestates, as well as the biogas production data, were generated during the current research. All relevant datasets used or analysed are included in the manuscript and its supplementary materials, where applicable. Additionally, any raw data or additional materials related to the microbial dynamics, TOC analysis, and cumulative biogas production can be provided to qualified researchers interested in further studies.

Ethical Clearance The study involves laboratory analysis of food waste collected from (Dutse-ultra modern market, old market, and motor parks), and cattle rumen content obtained from Dutse central slaughterhouse. No human or live animal subjects are involved in the research, and the samples are processed in a controlled laboratory environment. The handling and disposal of all waste samples are conducted in accordance with institutional biosafety guidelines. All laboratory procedures follow standard protocols to minimize environmental impact, and the lead researcher is trained in waste management and laboratory safety.

Competing Interests The authors declare that they have no known competing financial interests or personal relationships that could have appeared to influence the work reported in this paper.

Declarations This declaration ensures transparency about authorship, competent interest, funding, ethical clearance, and data availability statement, which are important aspects for publication in your reputable journal (Springer Nature).

References

Adeleye, A. O., Yalwaji, B., Shiaka, G. P., Amoo, A. O., & Udochukwu, C. C. (2019). Bacteriological quality of fresh-cut ready-to-eat vegetables sold in Dutse Market North-west, Nigeria. *FUW Trends in Science and Technology Journal (FTSTJ), 4*(1), 54–57.

Amin, F. R., Khalid, H., El-Mashad, H. M., Chen, C., Liu, G., & Zhang, R. (2021). Functions of bacteria and archaea participating in the bioconversion of organic waste for methane production. *Science of the Total Environment, 763*, 143007.

Amoo, A. O., Ijanu, E. M., Haruna, A., Adeleye, A. O., & Sabo, A. (2023b). Review on the pre-treatment advancements of biogas production barriers. *UMYU Journal of Microbiology Research,* 8(1), 6–30. https://doi.org/10.47430/ujmr.2381.002

Amoo, A. O., Ahmed, S., & Haruna, A. (2023d). The impact of process variables on the quantity and quality of biogas generated from anaerobic digestion of food waste and rumen contents. *Industrial and Domestic Waste Management,* 3(1), 27–37. https://doi.org/10.53623/idwm.v3i1.196

Amoo, A. O., Adeleye, A. O., Bate, G. B., Ijanu, M. E., Adamu, S. B., Asaju, C. I., & Olaitan, I. I. (2024). Microbiological analysis of hawked-cooked food: Evidence from ready-to-eat food vendors in Dutse Ultra-Modern Market, Northwestern, Nigeria. *Tropical Environment, Biology, and Technology, 2*(1), 23–33. https://doi.org/10.53623/tebt.v211.423

Amoo, A. O., Adeleye, A. O., Ijanu, E. M., Haruna, A., & Ahmed, S. (2023). Maximizing efficiency in biogas production: A comprehensive review of operational parameters. *Nigerian Research Journal of Engineering and Environmental Sciences, 8*(1), 48–63. https://doi.org/10.5281/zenodo.8093879

Amoo, A. O., Ahmed, S., & Haruna, A. (2023). Combinatorial effect of process parameters on the rate of biogas production and rate of substrate degradation following anaerobic digestion of food waste and rumen content. *Journal of Applied Science and Environmental Management, 27*(32), 445–455.

Ao, T., Chen, L., Zhou, P., Liu, X., & Li, D. (2021). The role of oxidation-reduction potential as an early warning indicator and a microbial instability mechanism in a pilot-scale anaerobic mesophilic digestion of chicken manure. *Renewable Energy, 179*, 223–232.

Barrow, G. I., & Feltham, R. K. A. (1993). *Cowan and Steel's Manual for the Identification of Medical Bacteria* (3rd ed.). Cambridge University Press.

Borth, P. L. B., Perin, J. K. H., Torrecilhas, A. R., Lopes, D. D., Santos, S. C., Kuroda, E. K., & Fernandes, F. (2022). Pilot-scale anaerobic co-digestion of food and garden waste: Methane potential, performance, and microbial analysis. *Biomass and Bioenergy, 157*, 106331.

Buragohain, S., Mahanta, P., & Mohanty, K. (2021). Biogas production from anaerobic mono- and co-digestion of lignocellulosic feedstock: Process optimization and its implementation at a community level. *Environmental Technology & Innovation, 24*, 101981.

Burg, V., Bowman, G., Haubensak, M., Baier, U., & Thees, O. (2018). Valorization of an untapped resource: Energy and greenhouse gas emissions benefits of converting manure to biogas through anaerobic digestion. *Resources, Conservation and Recycling,* 136, 53 – 62.https://doi.org/10.1016/j.resconrec.2018.04.004

Chen, R., Zhou, J., Zheng, X., Jiang, L., & Duan, N. (2023). Unveiling the synergy of Chlorella sp. and cattle manure co-digestion under high feeding load. *Energy,* 270, 126877.

Chen, Y., Yan, X., Wu, S., Zhang, L., Xie, S., Huang, H., & Zou, D. (2021). Effects of trace elements on anaerobic co-digestion of food waste and sewage sludge: Biogas production and microbial community. *Bioresource Technology, 331*, 125073.

Cui, L., Wang, X., He, L., Wang, J., & Zhang, Y. (2020). The effect of temperature on the abundance and diversity of *Clostridium* sp. and other acetogenicbacteria in biogas digesters treating dairy manure. *Applied Microbiology and Biotechnology*, 104(1), 405–414.

Detman, A., Bucha, M., Treu, L., Chojnacka, A., Pleśniak, Ł, Salamon, A., & Sikora, A. (2021). Evaluation of acidogenesis products' effect on biogas production performed with metagenomics and isotopic approaches. *Biotechnology for Biofuels, 14*(1), 125.

Erdogan, E., Atila, B., Mumme, J., Reza, M. T., Toptas, A., Elibol, M., & Yanik, J. (2015). Characterization of products from hydrothermal carbonization of orange pomace including anaerobic digestibility of process liquor. *Bioresource Technology, 196*, 35–42.

Ferdeș, M., Paraschiv, G., Ionescu, M., Dincă, M. N., Moiceanu, G., & Zăbavă, B. Ș. (2023). Anaerobic co-digestion: A way to potentiate the synergistic effect of multiple substrates and microbial diversity. *Energies, 16*(5), 2116.

Fonoll, X., Shrestha, S., Khanal, S. K., Dosta, J., Mata-Alvarez, J., & Raskin, L. (2021). Understanding the anaerobic digestibility of lignocellulosic substrates using rumen content as a cosubstrate and an inoculum. *ACS ES&T Engineering, 1*(3), 424–435.

Harirchi, S., Wainaina, S., Sar, T., Nojoumi, S. A., Parchami, M., & Taherzadeh, M. J. (2022). Microbiological insights into anaerobic digestion for biogas, hydrogen or volatile fatty acids (VFAs): A review. *Bioengineered, 13*(3), 6521–6557.

Ho, K. S., & Chu, L. M. (2019). Characterization of food waste from different sources in Hong Kong. *Journal of the Air & Waste Management Association, 69*(3), 277–288.

Holt, J. G., Krieg, N. R., Sneath, P. H. A., Staley, J. T., & Williams, S. T. (1994). *Bergey's manual of determinative bacteriology*. Williams & Wilkins Publisher UK.

Ibro, M. K., Ancha, V. R., & Lemma, D. B. (2022). Impacts of anaerobic co-digestion on different influencing parameters: A critical review. *Sustainability, 14*(15), 9387.

Ihoeghian, N. A., Amenaghawon, A. N., Ajieh, M. U., Oshoma, C. E., Ogofure, A., Erhunmwunse, N. O., & Martin, A. D. (2022). Anaerobic co-digestion of cattle rumen content and food waste for biogas production: Establishment of co-digestion ratios and kinetic studies. *Bioresource Technology Reports, 18*, 101033.

Jain, M. S., Paul, S., & Kalamdhad, A. S. (2019). The interplay of physical and chemical properties during in-vessel degradation of sewa upge sludge. *Waste Management, 98*, 58–68.

Ju, F., Beck, K., Yin, X., Munk, P., Zhang, X.-X., & Zhuang, G. (2021). Metagenomic analysis reveals novel Candidatus Cloacamonas acidaminovorans involved in biogas production from food waste. *Bioresource Technology, 319*, 124186.

Karki, R., Chuenchart, W., Surendra, K. C., Shrestha, S., Raskin, L., Sung, S., & Khanal, S. K. (2021). Anaerobic co-digestion: Current status and perspectives. *Bioresource Technology, 330*, 125001.

Koirala, S., Alam, S., Poudel, P., & Hoon Shin, S. (2021). A review on anaerobic digestion technology for biogas production. *Renewable and Sustainable Energy Reviews, 139*, 110673.

Kumar, S., Sivaprasad, S., & Das, D. (2022). Municipal solid waste management using biogas technology: A review. *Waste Management, 137*, 15–27.

Kumar, V., Nabaterega, R., Khoei, S., & Eskicioglu, C. (2021). Insight into interactions between syntrophic bacteria and archaea in anaerobic digestion amended with conductive materials. *Renewable and Sustainable Energy Reviews, 144*, 110965.

Lee, J., Hong, J., Jeong, S., Chandran, K., & Park, K. Y. (2020). Interactions between substrate characteristics and microbial communities on biogas production yield and rate. *Bioresource Technology, 303*, 122934.

Lee, M. E., Steiman, M. W., & Angelo, S. K. S. (2021). Biogas digestate as a renewable fertilizer: Effects of digestate application on crop growth and nutrient composition. *Renewable Agriculture and Food Systems, 36*(2), 173–181. https://doi.org/10.1017/s1742170520000186

Li, Y., Wang, J., Hu, J., & Zhou, Z. (2021). Isolation and identification of Lactobacillus sp. from an anaerobic digester treating food waste and its contribution to acidogenesis and acetogenesis during biogas production. *Journal of Cleaner Production*, 279, 123604.

Li, J., & Huang, C. (2024). Anaerobic co-digestion of corn straw, sewage sludge and fresh leachate: Focusing on synergistic/antagonistic effects and microbial mechanisms. *Bioresource Technology, 395*, 130414.

Li, Y., Chen, Z., Peng, Y., Huang, W., Liu, J., Mironov, V., & Zhang, S. (2022). Deeper insights into the effects of substrate-to-inoculum ratio selection on the relationship of kinetic parameters, microbial communities, and key metabolic pathways during the anaerobic digestion of food waste. *Water Research, 217*, 118440.

Lu, J., & Gao, X. (2021). Biogas: Potential, challenges, and perspectives in a changing China. *Biomass and Bioenergy, 150*, 106127. https://doi.org/10.1016/j.biombioe.2021.106127

Ma, G., Ndegwa, P., Harrison, J. H., & Chen, Y. (2020). Methane yields during anaerobic co-digestion of animal manure with other feedstocks: A metaanalysis. *Science of the Total Environment, 728*, 138224.

Ma, X., Yu, M., Yang, M., Gao, M., Wu, C., & Wang, Q. (2019). The synergistic effect from anaerobic co-digestion of food waste and *Sophora flavescens* residues at different co-substrate ratios. *Environmental Science and Pollution Research, 26*, 37114–37124.

Mao, C., Wang, Y., Wang, X., Ren, G., Yuan, L., & Feng, Y. (2019). Correlations between microbial community and C:N: P stoichiometry during the anaerobic digestion process. *Energy, 174*, 687–695.

Mason, P., Kougias, P., Schmidt, J. E., & Angelidaki, I. (2022). The potential of biogas from anaerobic digestion in the circular economy. *Sustainable Production and Consumption, 30*, 174–182.

Mecha, A. C., & Kiplagat, J. (2023). Characterization of kitchen and municipal organic waste for biogas production: Effect of parameters. *Heliyon, 9*(5), e16360.

Meyer, G., Okudoh, V., & Van Rensburg, E. (2022). A rumen-based anaerobic digestion approach for lignocellulosic biomass using barley straw as feedstock. *South African Journal of Chemical Engineering, 41*, 98–104.

Montalvo-Rodriguez, R., Garcia-Pena, E. I., Garcia-Gonzalez, M. C., Torres-Gonzalez, L., & Poggi-Varaldo, H. M. (2022). Development and optimization of a bioprocess for the degradation of pharmaceutical waste using anaerobic digestion. *Bioresource Technology, 342*, 126046.

Mu, L., Zhang, L., Zhu, K., Ma, J., Ifran, M., & Li, A. (2020). Anaerobic co-digestion of sewage sludge, food waste and yard waste: Synergistic enhancement on process stability and biogas production. *Science of the Total Environment, 704*, 135429.

Mutungwazi, A., Ijoma, G. N., & Matambo, T. S. (2021). The significance of microbial community functions and symbiosis in enhancing methane production during anaerobic digestion: A review. *Symbiosis, 83*(1), 1–24.

Naveen, K. S., Srinivasa, R., & Reddy, P. S. (2020). An overview of anaerobic digestion technology and its role in sustainable rural development. *Renewable and Sustainable Energy Reviews, 127*, 109848.

Ogbonna, C. B., Stanley, H. O., & Abu, G. O. (2018). Effect of seasonal variation on anaerobic treatment of organic municipal solid waste-II: Population dynamics of bacteria and archaea communities. *Applied Microbiology, 4*(3), 11–17.

Oladejo, O. S., Dahunsi, S. O., Adesulu-Dahunsi, A. T., Ojo, S. O., Lawal, A. I., Idowu, E. O., & Ajayi, O. E. (2020). Energy generation from anaerobic co-digestion of food waste, cow dung and piggery dung. *Bioresource Technology, 313*, 123694.

Palaniveloo, K., Amran, M. A., Norhashim, N. A., Mohamad-Fauzi, N., Peng-Hui, F., Hui-Wen, L., & Razak, S. A. (2020). Food waste composting and microbial community structure profiling. *Processes, 8*(6), 723.

Pan, S., Zabed, H. M., Li, Z., Qi, X., & Wei, Y. (2022). Enrichment and balancing of nutrients for improved methane production using three compositionally different agro-livestock wastes: Process performance and microbial community analysis. *Bioresource Technology, 357*, 127360.

Qiu, J., Guo, H., Xue, Y., Liu, Q., Xu, Z., & He, L. (2023). Rapid detection of chemical oxygen demand, pH value, total nitrogen, total phosphorus, and ammonia nitrogen in biogas slurry by near-infrared spectroscopy. *Analytical Methods, 15*(31), 3902–3914.

Rabii, A., Aldin, S., Dahman, Y., & Elbeshbishy, E. (2019). A review of anaerobic co-digestion with a focus on the microbial populations and the effect of multi-stage digester configuration. *Energies, 12*(6), 1106.

Rasapoor, M., Young, B., Brar, R., Sarmah, A., Zhuang, W. Q., & Baroutian, S. (2020). Recognizing the challenges of anaerobic digestion: Critical steps toward improving biogas generation. *Fuel, 261*, 116497.

Rocamora, I., Wagland, S. T., Villa, R., Simpson, E. W., Fernández, O., & Bajón-Fernández, Y. (2020). Dry anaerobic digestion of organic waste: A review of operational parameters and their impact on process performance. *Bioresource Technology, 299*, 122681.

Sarker, S., Lamb, J. J., Hjelme, D. R., & Lien, K. M. (2019). A review of the role of critical parameters in the design and operation of biogas production plants. *Applied Sciences, 9*(9), 1915.

Shen, Y., Yu, Y., Zhang, Y., Urgun-Demirtas, M., Yuan, H., Zhu, N., & Dai, X. (2021). Role of redox-active biochar with distinctive electrochemical properties to promote methane production in anaerobic digestion of waste-activated sludge. *Journal of Cleaner Production, 278*, 123212.

Siddiki, S. Y. A., Uddin, M. N., Mofijur, M., Fattah, I. M. R., Ong, H. C., Lam, S. S., & Ahmed, S. F. (2021). Theoretical calculation of biogas production and greenhouse gas emission reduction potential of livestock, poultry and slaughterhouse waste in Bangladesh. *Journal of Environmental Chemical Engineering, 9*(3), 105204. https://doi.org/10.1016/j.jece.2021.105204

Singh, R., Singh, B., & Marchaim, P. (2015). Anaerobic bacteria: Isolation techniques. In *Essentials of clinical infectious disease* (pp. 313–317). Springer Publisher Switzerland.

Sumardiono, S., Adisukmo, G., Hanif, M., Budiyono, B., & Cahyono, H. (2021). Effects of pretreatment and ratio of solid sago waste to rumen on biogas production through solid-state anaerobic digestion. *Sustainability, 13*(13), 7491.

Vongvichiankul, C., Deebao, J., & Khongnakorn, W. (2017). Relationship between pH, oxidation reduction potential (ORP) and biogas production in mesophilic screw anaerobic digester. *Energy Procedia, 138*, 877–8882.

Wang, R., Gu, J., Wang, Q., Jiang, S., Wu, Z., Wang, J., & Gong, X. (2023). Enhancing methane production in dry anaerobic digestion of ruminant manures through substrate ratio regulation for strengthened microbial interactions. *Environmental Technology & Innovation, 32*, 103389.

Wang, S., Ma, F., Ma, W., Wang, P., Zhao, G., & Lu, X. (2019). Influence of temperature on biogas production efficiency and microbial community in a two-phase anaerobic digestion system. *Water, 11*(1), 133.

Wang, X., Li, Y., Li, W., Li, X., & Zhang, Z. (2021). The effects of biogas fertilizer on crop growth and soil properties: A review. *Renewable and Sustainable Energy Reviews, 148*, 111474.

Whitman, W. B., Goodfellow, M., Kämpfer, P., Busse, H. J., Trujillo, M. E., Ludwig, W., & Suzuki, K. (2012). *Bergey's manual of systematic bacteriology* (Vol. 5). Springer Publisher Switzerland.

Wu, D., Li, L., Zhao, X., Peng, Y., Yang, P., & Peng, X. (2019). Anaerobic digestion: A review on process monitoring. *Renewable and Sustainable Energy Reviews, 103*, 1–12.

Zhang, J., Mao, L., Nithya, K., Loh, K. C., Dai, Y., He, Y., & Tong, Y. W. (2019). Optimizing mixing strategy to improve the performance of an anaerobic digestion waste-to-energy system for energy recovery from food waste. *Applied Energy, 249*, 28–36.

Zhang, Y., Li, C., Yuan, Z., Wang, R., Angelidaki, I., & Zhu, G. (2023). Syntrophy mechanism, microbial population, and process optimization for volatile fatty acids metabolism in anaerobic digestion. *Chemical Engineering Journal, 452*, 139137.

Zhang, Y., Li, Y., Chen, Z., Wu, Y., & Chen, M. (2021). Application of anaerobic digestion technology in the treatment of food waste: A review. *Waste Management, 123*, 3–13.

Zhao, S., Chen, W., Liu, M., Lv, H., Liu, Y., & Niu, Q. (2022). Biogas production, DOM performance and microbial community changes in anaerobic co-digestion of chicken manure with Enteromorpha and green waste. *Biomass and Bioenergy, 158*, 106359.

Zhao, S., Chen, W., Luo, W., Fang, H., Lv, H., Liu, R., & Niu, Q. (2021). Anaerobic co-digestion of chicken manure and cardboard waste: Focusing on methane production, microbial community analysis and energy evaluation. *Bioresource Technology, 321*, 124429.

Zhu, X., Wang, Y., Wang, X., Ma, S., & Liu, X. (2021). A microbial electro-chemical system (MES) for biogas production from food waste: Performance and microbial community analysis. *Bioresource Technology, 329*, 124903.

Chapter 23
Addressing Climate Change Vulnerability Issues Through Financial Risk Management in Cassava Farming: Evidence from the Niger Delta Region, Nigeria

Adeyinka Richard Aroyehun, Anthony Ojonimi Onoja, and Vivian Chinelo Ugwuja

Abstract Cassava farming in the Niger Delta region of Nigeria faces significant challenges due to climate change, including increased vulnerability to droughts, floods, and unpredictable weather patterns. These challenges threaten food security and the livelihoods of smallholder farmers who depend on cassava production. This study focuses on how the vulnerability of the farmers can be reduced through financial risk management strategies. It relied on primary data sourced from 260 cassava farmers after which a Multivariate Probit model was used to analyse the data after its validation. It was found that off-farm jobs (0.153, $p < 0.05$), credit availability (0.175, $p < 0.05$) and farmers' age (0.007, $p < 0.05$) significantly determined the decision to adopt income diversification strategies as a way of coping with climate risks; but the adoption of contract farming was significantly determined by the cassava farm size (0.067, $p < 0.05$), educational level (− 0.019, $p < 0.05$). The adoption of cautious saving was determined by the flood (− 0.120, $p < 0.01$) and extension contacts (− 0.242, $p < 0.05$). The adoption of credit facilities was influenced by high rates of pests and diseases (0.062) and credit availability (0.541), which were significant at the 0.05 and 0.01 levels, respectively. Moreover, government policies promoting climate-smart farming techniques and improving access to financial services are

A. R. Aroyehun (✉) · A. O. Onoja · V. C. Ugwuja
Agricultural Economics and Agribusiness Management Department, Faculty of Agriculture, University of Port Harcourt, Port Harcourt, Nigeria
e-mail: aroyehun_adeyinka@uniport.edu.ng

A. O. Onoja
e-mail: anthony.onoja@uniport.edu.ng

V. C. Ugwuja
e-mail: vivian.ugwuja@uniport.edu.ng

A. O. Onoja
Department of Agriculture and Animal Health, College of Agriculture and Environmental Sciences, University of South Africa (UNISA), Johannesburg, South Africa

R. U. Onyeneke et al. (eds.), *Energy Transition, Climate Action and Sustainable Agriculture*, https://doi.org/10.1007/978-3-031-83165-2_23

critical in strengthening farmers' adaptive capacity. The study concludes with policy recommendations that emphasize the need for multi-stakeholder collaboration in implementing sustainable financial risk management frameworks to support cassava farmers, financial literacy of farmers, improved access to affordable credit, enterprise diversification, reduced dependence on a single commodity, and financial resource accessibility to be resilient. These recommendations are vital for ensuring long-term agricultural resilience and food security in the face of increasing climate variability.

Keywords Cassava production · Climate change · Financial risk · MVP · Vulnerability

23.1 Introduction

The diverse economic and biophysical conditions in which farming is practiced expose agricultural operations to a broad spectrum of risks and uncertainties. According to Ullah et al. (2015), risk is typically connected with the uncertainty surrounding outcomes that entail some kind of misfortune or loss that has a detrimental impact on an individual's well-being. Farming enterprises are linked to several risk factors. Five primary categories of risk in agriculture were identified by USDA (Ullah et al., 2015): production risk, marketing risk, financial risk, legal and environmental risk, and risk related to human resources. The primary risk that farmers encounter is the production risk. Risks associated with agricultural production can harm output levels and result in significant losses. In agriculture, weather is widely acknowledged as a primary cause of output risk. Extreme natural disasters including storm surges, hailstorms, droughts, floods, and cyclones are usually unpredictable and leave farmers with little control over them. Farmers must be able to identify and control production risk since their primary source of income comes from production output (Soulibouth et al., 2021; Parshley, 2023; United Nations Climate Change, 2021). Therefore, controlling risks and uncertainties in agriculture, especially in the production of cassava, is essential as it influences other economic sectors and is often seen as a critical issue in the decisions made by farmers and the policies that influence these decisions. Cassava farmers have a variety of options to manage farm risks, and many of them utilize these risk management tools concurrently (Omodara et al., 2023).

Cassava farmers have also been working to improve their financial status to finance measures for adapting to climate change (Izuogu et al., 2023; Omodara et al., 2023). Some of these tactics include diversifying their sources of income outside of cassava farming. A few examples include working off the farm, managing transportation services, agro-processing, having a retail location, small- to large-scale marketing, adding value to agricultural products, and diversifying farm revenue. On-farm and off-farm diversification tactics are the two categories into which diversification may be divided. Redistributing resources for cassava cultivation to other uses, such as raising fish, poultry, or small ruminants like goats, sheep, or cattle, is known as "on-farm diversification" One strategy that cassava farmers may use to minimize income inequality and financial risk is through on-farm revenue diversification (Adnan et al., 2020). For example, cultivating in combination with other crops, such as cassava, or sharing acreage with another agricultural production firm (Izuogu et al., 2023). Furthermore, the practice of cassava farmers seeking to increase their revenue or manage additional financial risks associated with the crop might be referred to as "on-farm earnings diversification" Off-farm diversification, on the other hand, is the use of available agricultural labour in a variety of non-farm businesses to enhance farmers' livelihood and reduce the impact of financial risk.

As one of the main causes of risk associated with cassava farming, global climate change and its repercussions have drawn attention in recent decades (Anjali et al., 2019). Producers of cassava face risks and uncertainty regarding their output because of the destructive consequences of changing climatic conditions, including floods, droughts, high temperatures, low humidity, and heavy rains (National Geographic, 2022). Thus, the issue of global climate change is upsetting and depressing. The issues faced by cassava growers include deplorable crop failures throughout both the growing and harvesting stages, a decrease in the agricultural sector's GDP contribution, an increase in famine, malnutrition due to nutritional deficiencies, and infections (Otitoju, 2013). In light of the country's changing environment, it is concerning and concerning that Nigeria is producing less cassava to feed its growing population. Concerns regarding the effects of climate change on cassava production in Nigeria, particularly in the Niger Delta area, have been raised by recent studies (Anjali et al., 2019).

Virlics (2013); Ullah et al. (2015); Sarfraz et al. (2018); Adnan et al. (2020); Rahman et al. (2020), among many others, are among the previous studies on financial risk management confirmed the significance of financial risk management strategies concerning agricultural output. It was also observed that the use of any risk system device might lead to the use of other risk scheme devices at the same time. Information on the impact of financial risk management initiatives is scarce, particularly in the Nigerian Niger Delta. As a result, the present study looks at financial risk management techniques and cassava farmers' plans for the effects of climate change to bridge the previously noted gap. This should accelerate Nigeria's efforts to meet the Sustainable Development Goals (SDGs) of ending poverty, ending hunger, and addressing climate change. It should also offer evidence for policy on the effective use of financial resources to increase cassava farmers' productivity and resilience in a changing climate in Nigeria, particularly in the Niger Delta region.

This research aims to explore the climate change vulnerability issues through financial risk management in cassava farming in the Niger Delta region, Nigeria thus accelerating the pace of Nigerian's drive to achieve the Sustainable Development Goals; one SDG stated as no to prevalence to poverty; second SDG stated as zero to hunger; and third SDG stated as take climate action. Furthermore, cassava farmers used credit facilities, contract farming, and precautionary savings in addition to diversification as a self-insuring strategy for coping with the challenges posed by climate change (Adnan et al., 2020; Ullah et al., 2015). The conceptual framework

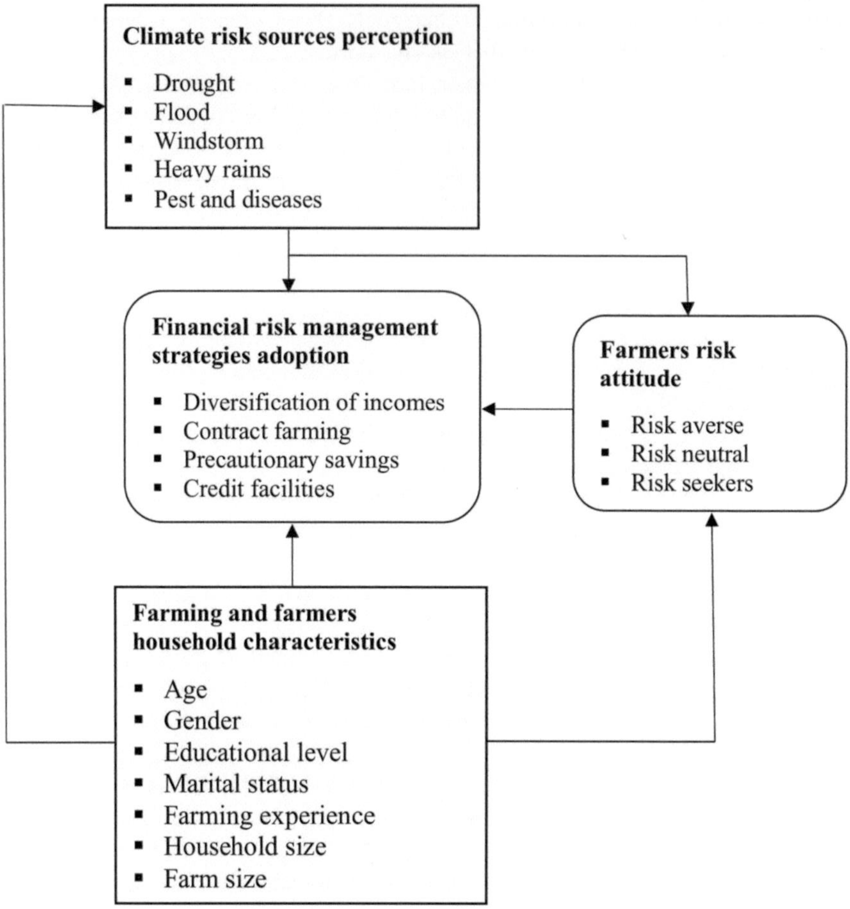

Fig. 23.1 Conceptual framework for financial risk management and climate change adaptation strategies

for the cassava farmers' financial risk management and climate change adaptation plans is depicted in Fig. 23.1.

23.2 Literature Review

Cassava is one of the root crops that supplies starchy food for above 500 million populaces in the tropical regions, and the sixth most vital crop globally. The importance of cassava is steadily rising in the pharmaceutical and brewery-like beer industries because of its starchy requirement (Omondi & Yermiyahu, 2021). Still, cassava global production stands at 262.6 million tonnes. Cassava (*Manihot esculenta*) was

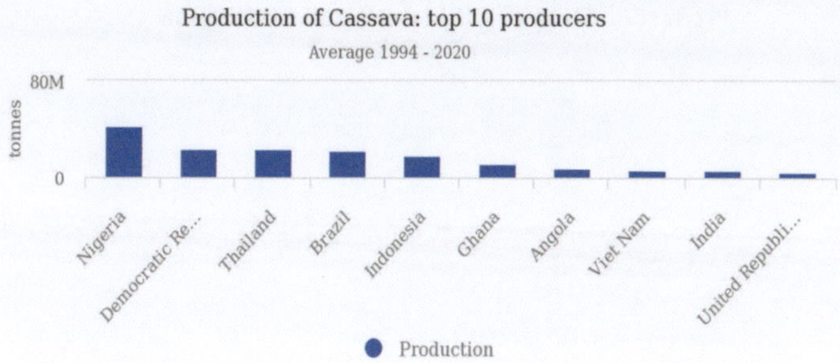

Fig. 23.2 Cassava global production trends. *Source* Authors computation (2023)

introduced to Tropical Africa, belonging to the *Euphorbiaceae* family. It is a staple food and source of food security root crop as well as a source of income for the majority of the African populace (Henri-Ukoha, 2020). Population increase is not measurable to production of food crops particularly cassava in sub-Saharan Africa resulting in starvation and malnutrition which can only be filled by importations. These deficiencies between the increase in population and cassava food yields are connected with the ineptitude and incapacitation of crops to attain their potential for food security (Omondi & Yermiyahu, 2021). They associated this with several factors such as erratic rainfall patterns, lack of finance for proper adaptation technologies, inadequate farm inputs, poor soil environment, and influence of climate change among others.

Cassava production proportion by region is as follows; Africa 56.5%, Asia 29.7%, Americas 13.7%, and Oceania 0.1%. This implies that Africa is the largest producer of cassava globally and Nigeria is the largest producer of cassava in Africa as shown in Fig. 23.2. The data used is from the FAOSTAT database.

Nigeria's cassava production trends are shown in Fig. 23.3. The aggregate yield of 59.5 million MT of cassava produced in Nigeria is based on the year 2018 estimations. Nigeria as a nation has the economic possibility of realizing income of up to US $427.3 and $2.98 billion for local value utilization and exports of cassava with its products respectively (Princewaterhouse Coopers Limited PwC, 2020). Cassava and its derivative products have huge and vast potential as outlined in Fig. 23.4. The data used is from the FAOSTAT database.

23.2.1 Theoretical Framework

Risk management theory provides a basis for identifying, assessing, analysing, and managing risks associated with a firm like cassava production. Utilizing identifying possible risks, cassava farmers can develop strategies to avoid or minimize the risks associated with finance and climate change. Cassava farmers hedge to mitigate the effect of financial risk on investment. Risk management aids in alleviating the impact

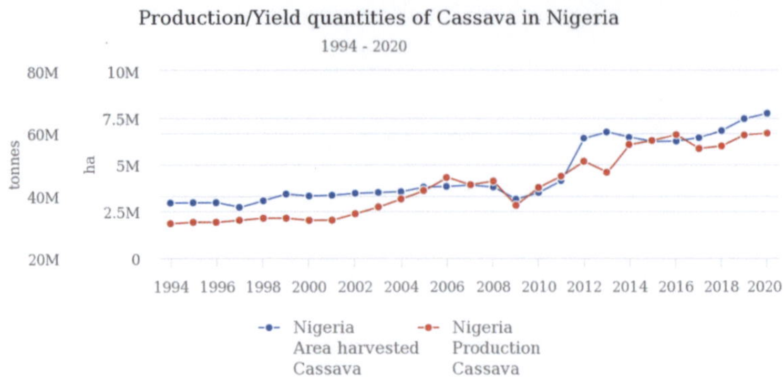

Fig. 23.3 Cassava production trends in Nigeria. *Source* Authors computation (2023)

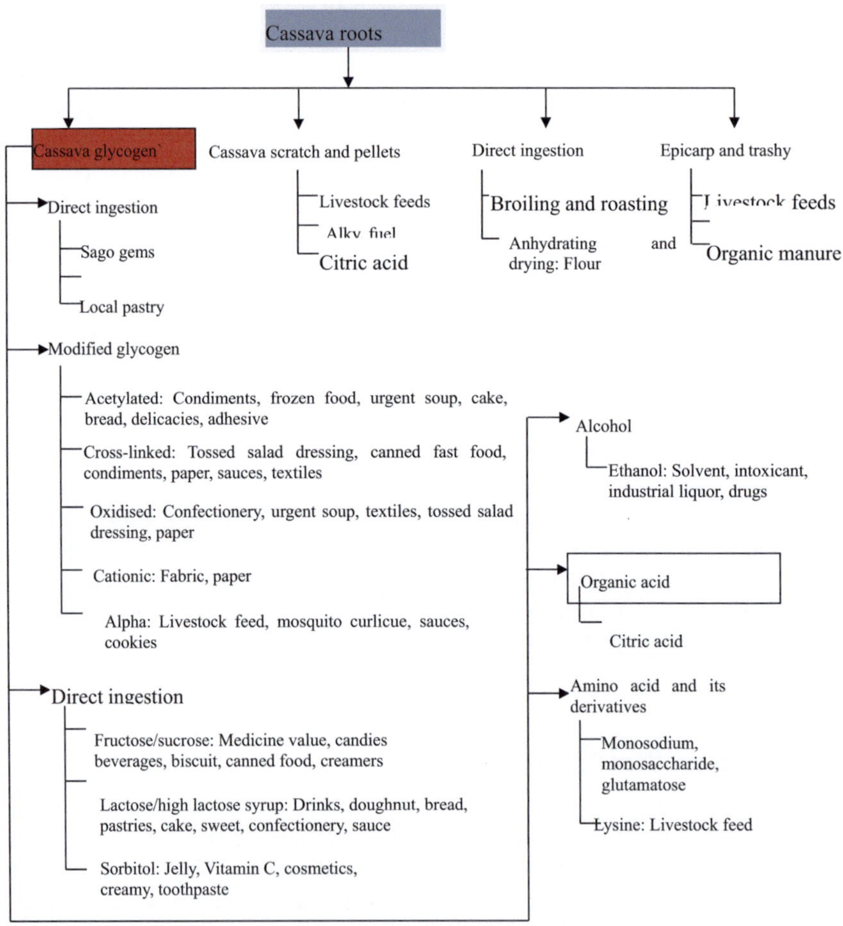

Fig. 23.4 Cassava and its derivatives. *Source* Authors computation (2023)

of financial risk rationing as it decreases the volatility of cash flows that can be used to finance adaptation strategies in cassava production where access to finance is limited or not available (Giambona et al., 2018). Hence, the risk management approach will contribute to increasing the possibility of productivity of cassava and minimizing both the likelihood of cassava failure and the uncertainty of achieving cassava farmers' overall objectives of profitability.

The risk management theory is built on the basic concepts of utility, regression, and diversification (Ajupov et al., 2019). Hence, risk management could be valuable for cassava farmers who desire to reduce the risk exposure of finance and climate change. Effective risk management can help to reduce the probability of finance and climate change risks occurrence; minimize the impact of hazards that do occur as a result of climate change and finance; improve cassava farmers' resilience to climate change; as well as increase the cassava farmers' productivity and profitability.

23.3 Methodology

The Nigerian Niger Delta area served as the study's site. The Niger River Delta, which is located in Nigeria next to the Gulf of Guinea on the Atlantic Ocean, is known as the Niger Delta area (Umoh et al., 2022). The region is made up of nine States and they are Ondo, Edo, Delta, Bayelsa, Imo, Abia, Rivers, Akwa-Ibom, and Cross River State (Udotong et al., 2017). The Niger Delta is a coastal area that is primarily below sea level, at around three meters. According to Umoh et al. (2022), the annual precipitation usually ranges from 2000 to 4000 mm or even more than 4500 mm, and the relative humidity is around 70% (Matemilola et al., 2019). There are two distinct seasons in the delta (wet and dry). Typically, the rainy season lasts eight months, from March to October, with ten to twenty-one days of dry weather occurring around August (also known as the August break). The August break's overall idea is not well defined. Nonetheless, the southwest monsoon winds are said to be the reason, since they transport moisture from the ocean into the hinterland (Matemilola et al., 2019). Climate change has been blamed for the absence of the August break in recent years. Beginning in November and lasting until February is the dry season. Between December and February, the area additionally experiences harmattan, which is defined by chilly, dry winds that are caused by dusty, tropical air masses coming from the north. This coastal region receives intense monsoon rains and is extremely sensitive to the effects of climate change, such as increases in sea levels (Matemilola et al., 2019). The estuary region experiences a range of air temperatures, with a mean of 25.4 °C. During the flood (wet) season, the temperature ranges from around 22–29 °C, and during the dry season, it varies from 30 to 37 °C, with an average temperature of 33 °C (Umoh et al., 2022).

Because the region is riverine, the fishing business is crucial to the residents' way of life. Sea snails, oysters, a variety of fish species, crabs, and periwinkles are just a few of the marine edibles that may be found nearby. Goats, cattle, and poultry of various sorts, including quail, turkeys, and broilers, are among the livestock that may be found in the Niger Delta region. The Niger Delta's environment does, however, also support the growth of cash crops, including rubber, coconut, pears, cocoa, cashew, oil palm, kolaunt, gum Arabic, sesame, and coconut. The region grows rice, cassava, maize, melon, yams, cocoyam, and sweet potatoes among other arable crops. However, the primary emphasis of this research was on cassava (*Manihot* spp.) producers.

The interview schedule and questionnaire were used as tools for gathering data. The questionnaire was intentionally and consciously created with logic and clarity in mind, to obtain responses from the cassava producers. As specified in the questionnaire, the interview schedule was employed to get precise information from the cassava producers.

The selection of respondents (cassava farmers) involved the use of multi-stage sampling procedures. First, out of the nine states in the Niger Delta, three states (Akwa-Ibom, Bayelsa, and Rivers State) were carefully chosen using a simple random technique. For administrative reasons, the Agricultural Development Programme (ADP) split each State into agricultural zones (Umoh et al., 2015). Second, every State's agricultural zones were chosen, making a total of twelve (12) agricultural zones. Thirdly, one (1) Local Government Areas (LGAs) were selected from each agricultural zone using simple random technique, making a total of twelve (12) Local Government Areas (LGAs) in all. Fourthly, three communities were selected from each LGA using a simple random technique, making a total of thirty-six (36) communities. Last, a list of cassava farmer households from each community was assembled with the assistance of local extension agents, and ten (10) and five (5) cassava farmers were chosen. The selection of 10 and 5 cassava farmers depended on the total number of agricultural zones in the State and the goal of having an equal number of 90 cassava producers from each State. There are six agricultural zones in Akwa-Ibom State, three in Rivers, and three in Bayelsa State. Thus, the two hundred and seventy (270) cassava growers that were chosen for the study constitute the overall sample size, as shown in Table 23.1. The sample size required for the study was estimated using a sample size estimator. The Kuder-Richardson, KR-20 test was utilized to gauge the questionnaire's reliability, and a multivariate probit (MVP) model was employed to evaluate the collected data.

Using a 90% (1.65) confidence level, 0.5 standard deviation, and 5% (0.05) margin error, the sample size proportion estimated by Andrew Fisher and utilized by Kibuacha (2021) was used. The estimation of sample size is stated as:

$$n = \frac{Zs^2 * P(1 - P)}{e^2} \tag{23.1}$$

where n is the sample size needed; Zs is the confidence level (z-score); P is the standard deviation; and e is the margin error.

$$n = \frac{(1.65)^2 * 0.5(1 - 0.5)}{(0.05)^2} = 272.25$$

Table 23.1 Sample frame and sampling technique of the study area

State	LGA	Population of the farmers	Total of samples	Total of samples retrieved
Rivers	Etche	569	30	29
	Abua/Odual	355	30	28
	Tai	418	30	30
Bayelsa	Nembe	303	30	29
	Southern Ijaw	354	30	30
	Ekeremor	336	30	29
Akwa-Ibom	Etim Ekpo	325	15	14
	Eastern Obolo	322	15	13
	Etinan	320	15	14
	Essien Udim	319	15	15
	Mbo	312	15	14
	Uruan	314	15	15
3	12	4247	270	260

Source Authors Survey, 2023

An estimated 272.25 sample size was obtained, making 6.41% of the total population of cassava farmers from the studied communities. Consequently, 270 sample sizes were employed and 260 samples making up 96.3% of the survey that was distributed were recovered for the analysis.

The primary data were collected based on a random process using an open-ended questionnaire; the confidentiality of the survey participants was strictly and unwaveringly respected. The information for the data gatherer was fully and accurately obtained by the questionnaire. The findings were applied to both intervention and, more broadly, to society at large. Face validity was therefore employed for the questionnaire instrument to be utilized to have high potentiality for the inquiry and the findings to also have high validity. Enumerators were trained for effective and efficient data collection. Data collection instruments were cautiously cantered on the financial risk management strategies adopted by farmers, socio-economic, fundamental farm-specific features, and other vital and relevant information that was needed for the study. The retrieved questionnaire and interview schedule were cross-checked for accuracy to avert any deviation and abnormality.

To determine the consistency between the measuring equipment, and the questionnaire parts, reliability measurement is essential. The reliability of the questionnaire using Kuder-Richardson, KR-20 as utilized by Chinda et al. (2023) and adopted, stated as:

$$\text{KR - 20} = \left(\frac{n}{n-1}\right)\left(1 - \frac{\sum_{i=1}^{n} p_i q_i}{\sigma^2}\right) \tag{23.2}$$

where: KR-20 is the reliability Kuder-Richardson; n is the number of questionnaires to be examined; Σ is the sum; pi is the fraction of the test answering 'yes' to an item (1); qi is the fraction of the test answering 'no' to an item (0); and σ^2 is the variance of the test-score.

Decision rule: The higher the value of KR-20 close to 1 (say between 0.7 and 0.85) the better the measurement instrument. Where n is 10, $\sum_{i=1}^{n} p_i q_i$ is 2.1097 and σ^2 is 7.5755.

$$\mathrm{KR}-20 = \left(\frac{10}{10-1}\right)\left(1-\frac{2.1097}{7.5755}\right) = 0.8017$$

The measuring instrument is good for evaluating the data, as indicated by the Kuder-Richardson (KR-20) which is equal to 0.80. Because of this, the researchers may rely on the study's findings for their future research because of its high validity and reliability.

23.3.1 Model Specification

The multivariate probit model (MVP) for financial risk management strategies as used by Adnan et al. (2020), Adhikari and Khanal (2021) expressed as:

$$Y_{ij}^{*} = X_{ij}{\prime}\beta_i + \varepsilon_{ij} \quad (23.3)$$

where, Y_{ij} ($j = 1, \ldots, m$) indicates the risk management strategies ($n = 4$ in the study) encountered by the ith cassava farmer (i = 1, …, m), $X_{ij}{\prime}$ is 1 × k vector of experiential variables that influence the risk management adaptation decision, β_j is k × 1 vector of unidentified parameters to be assessed, and e_{ij} is the unobserved error term. Expressed MVP equations as:

$$Y_{11}^{*} = \alpha_{11} + X\beta_{11} + \varepsilon_{11}(\text{Diversification of income}) \quad (23.4)$$

$$Y_{21}^{*} = \alpha_{21} + X\beta_{21} + \varepsilon_{21}(\text{Contract farming}) \quad (23.5)$$

$$Y_{31}^{*} = \alpha_{31} + X\beta_{31} + \varepsilon_{31}(\text{Precautionary savings}) \quad (23.6)$$

$$Y_{41}^{*} = \alpha_{41} + X\beta_{41} + \varepsilon_{41}(\text{Credit facilities}) \quad (23.7)$$

where, Y_{11}^{*}, Y_{21}^{*}, Y_{31}^{*}, and Y_{41}^{*} are the variables for each risk management strategy chosen by the cassava farmers; $Y_{ij} > 0$, $Y_{ij}^{*} = 1$ and 0, otherwise.

The parameter estimations as regards the multivariate probit (MVP) model provide direction of the effect of the explanatory variables on the dependent that's adopted optimal variable, and expressed as:

$$\begin{aligned} Y_{i=0,1,\ldots n} = {} & \beta_0 + \beta_1 X_1 + \beta_2 X_2 + \beta_3 X_3 + \beta_4 X_4 + \beta_5 X_5 \\ & + \beta_6 X_6 + \beta_7 X_7 + \beta_8 X_8 + \beta_9 X_9 + \beta_{10} X_{10} \\ & + \beta_{11} X_{11} + \beta_{12} X_{12} + \beta_{13} X_{13} + \beta_{14} X_{14} + \beta_{15} X_{15} \end{aligned} \tag{23.8}$$

Table 23.2 shows the dependent and independent variables as well as their descriptions and measurements.

Partial eta squared in the MVP model is expressed as:

$$\delta_p^2 = \frac{SS_{effect}}{SS_{effect} + SS_{error}} \tag{23.9}$$

where

δ_p^2 Partial eta squared
SS Sum of squares

Partial eta squared was employed to examine the effect of independent variable(s) on the dependent variable(s). Rule of thumb:

$\delta_p^2 = 0.01$; indicates a small effect
$\delta_p^2 = 0.06$; indicates a medium effect
$\delta_p^2 = 0.14$; indicates a large effect

23.3.2 Justification of the Choice of the Independent Variables

The choice of independent variables in this study captured a wide range of socioeconomic, demographic, environmental, and institutional factors as well as productivity, income, financial risk management, or financial decisions that could influence cassava farming outcomes. Each variable has theoretical and empirical support in agricultural economics and risk management literature, providing a strong basis for their inclusion.

Age is a crucial factor in agricultural decision-making, as it influences a farmer's experience, willingness to adopt innovations and risk tolerance. Older farmers tend to have more experience but may be less inclined to adopt new farming technologies due to risk aversion, while younger farmers may be more innovative but less experienced in farming practices (Brown et al., 2019; Rizzo et al., 2023).

Gender can influence access to land, credit, and agricultural inputs. Male farmers often have better access to resources and decision-making power compared to

Table 23.2 Description of variables

Key	Variables	Description and measurement
Dependent variables		
Y_1	Diversification of income	Dummy = 1 if the farmer uses diversification of income, 0 otherwise
Y_2	Contract farming	Dummy = 1 if the farmer uses contract farming, 0 otherwise
Y_3	Precautionary savings	Dummy = 1 if the farmer uses precautionary savings, 0 otherwise
Y_4	Credit facilities	Dummy = 1 if the farmer uses credit facilities, 0 otherwise
Independent variables		
X_1	Age	Cassava farmer's age (in years)
X_2	Gender	Dummy: if cassava farmer is a male = 1, 0 otherwise
X_3	Marital status	Level: single = 1; married = 2; widow/widower = 3; divorced = 4
X_4	Farming experience	Number of years spent in farming cassava
X_5	Educational level	Number of years spent in schooling to obtain the level of educational
X_6	Household size	Total number of household members
X_7	Off-farm job	Dummy = 1 if farmer engaged in off-farm job, 0 otherwise
X_8	Farm size	Size of cassava farm cultivated (in hectares)
X_9	Flood	Dummy = 1 if the farmer experiences a flood, 0 otherwise
X_{10}	Drought	Dummy: 1 if the farmer experiences drought, 0 otherwise
X_{11}	Heavy wind	Dummy: 1 if the cassava farmer experiences heavy wind, 0 otherwise
X_{12}	Pest and diseases	Level: highly perceived = 1; moderately perceived = 2; less perceived = 3; not sure = 4
X_{13}	Access to credit/financial facilities	Dummy: 1 if cassava farmer has access to credit/finance, 0 otherwise
X_{14}	Extension contact	Dummy: 1 if farmer has contact with extension agent within one year, 0 otherwise
X_{15}	Farm association membership	Dummy: 1 if the cassava farmer belongs to any farmer's association, 0 otherwise

female farmers, particularly in patriarchal societies. Gender differences in agricultural productivity are well-documented, with female farmers often facing constraints in accessing resources (Brown et al., 2019; Food and Agriculture Organization FAO, 2023; Rizzo et al., 2023).

Marital status can affect household labour availability and financial stability. Married farmers may have more family labour and shared financial resources, while widowed or divorced farmers may face greater challenges in managing farm activities. Marital status also plays a role in risk management, with different family structures influencing financial decisions (FAO, 2023; Rizzo et al., 2023).

Experience in farming is critical for developing expertise in crop management, risk mitigation, and financial decisions. More experienced farmers are likely to have better knowledge of farming practices, pest and disease control, and climate adaptation strategies, as well as make informed decisions about input use, crop diversification, and farm management practices. This can influence their productivity, risk management behaviour, and the decision to access credit or participate in farmer associations (Adnan et al., 2020; FAO, 2023).

Education improves a farmer's ability to understand and adopt new technologies, make informed decisions, and manage risks more effectively. More educated farmers are also better equipped to access financial services and navigate market dynamics, which can influence their farm productivity and resilience to shocks (Brown et al., 2019; Rizzo et al., 2023).

Household size affects labour availability for farming activities, especially in smallholder systems, consumption needs, and the financial burden on the farming household. Larger households may have more labour available for farming but also more mouths to feed, which could affect the farmer's risk management decisions and financial needs. Household size can also influence participation in off-farm employment and credit access (Adnan et al., 2020; Rizzo et al., 2023).

Engagement in off-farm employment is a risk diversification strategy that provides additional income. Farmers with off-farm jobs can supplement their income, making them less vulnerable to agricultural shocks such as cassava crop failure or market price fluctuations. This variable also captures the farmer's overall financial resilience and the ability to invest in farm inputs or access credit (FAO, 2023; Rizzo et al., 2023).

Farm size is a critical determinant of agricultural productivity and income. Larger farms may have higher production potential but also require more resources and capital to manage effectively. Farm size can also influence a farmer's ability to access credit, use extension services, and participate in risk management practices such as crop insurance (Adnan et al., 2020; Brown et al., 2019; Rizzo et al., 2023).

Floods can cause significant damage to crops, infrastructure, and livelihoods. Including this variable helps assess the impact of natural disasters on farmers' productivity, income, and risk management strategies. It also captures the need for financial facilities like credit and insurance to recover from such events (Adnan et al., 2020; Jabbar et al., 2020).

Like floods, drought is a major risk for agricultural production. Drought can lead to crop failures and reduced income, making it essential to consider its impact when analysing farmers' behaviour and financial needs. This variable captures the effect of climate variability on cassava production and the role of risk management strategies in coping with it (Adnan et al., 2020; Jabbar et al., 2020).

Heavy winds can damage crops, particularly in regions where cassava is vulnerable to such conditions, which can lead to significant financial losses. Including this variable helps assess the extent of exposure to extreme weather events and how this influences farmers' risk management behaviours and decisions to access financial facilities or engage in precautionary savings (Adnan et al., 2020).

Pest and disease outbreaks are significant risks in cassava farming. Farmers' perceptions of these risks can influence their willingness to invest in pest control measures, seek technical assistance, or adopt resistant varieties. This variable reflects the farmer's assessment of crop health risks and their management choices (Jabbar et al., 2020).

Access to credit and financial facilities is crucial for farm investment, purchasing inputs, adopting new technologies, and managing risks. Farmers with access to credit are more likely to adopt improved farming practices, invest in crop protection, and cope with shocks like floods or droughts, which could improve their productivity and resilience to risks such as climate shocks and market fluctuations. This variable directly captures the financial capacity of the farmer to manage risks (FAO, 2023; Rizzo et al., 2023).

Extension services provide farmers with technical advice, information on best practices, training, technologies, and access to resources that can improve productivity, resilience, and risk management. Farmers who have contact with extension agents are more likely to adopt improved farming practices, adopt appropriate risk management strategies, and access resources like credit and insurance. This variable captures the importance of information and advisory services in farming decisions (Brown et al., 2019; FAO, 2023).

Membership in a farmer association can provide farmers with access to collective resources, market information, credit, and input subsidies. It also enhances their bargaining power and ability to manage risks through cooperative support. This variable captures the social capital and network effects that can influence a farmer's financial and risk management decisions (Brown et al., 2019; FAO, 2023).

Therefore, these variables are justified based on their relevance to key factors influencing a cassava farmer's productivity, financial risk management strategies, and access to financial resources, as well as providing a comprehensive view of how these variables shape farmers' decisions and financial behaviour.

23.3.3 Model Validation

The goal of the model validation method is to ensure that the models utilized in the research are appropriate and do not go against the theoretical presumptions. According to Kerr & Goethel (2014), model validation determines if the model employed in research accurately represents the system's performance. The statement went on to say that model validity might be evaluated conceptually (by determining if the model's underlying theory and assumptions are upheld) as well as operationally (by determining whether the model's output agrees with the data under examination).

The models' multicollinearity was tested using the Variance Inflating Factor (VIF) (GeeksforGeeks, 2021). Regression analysis models are tested for multicollinearity using the Variance Inflation Factor (VIF). VIF is expressed in the regression model as:

$$Y = \beta_0 + \beta_1 X_1 + \beta_2 X_2 + \cdots + \beta_k X_k \quad (23.10)$$

$$VIF = \frac{1}{1 - R^2} \quad (23.11)$$

where

$$R^2 = \frac{\sum (y_{cal} - \overline{y})^2}{\sum (y_{given} - \overline{y})^2} \quad (23.12)$$

Decision rules:

A VIF value equal to one (1) implies no correlation; that is multicollinearity does not exist.

A VIF value in the range 1 and 5 specifies relative correlation; that is a low level of multicollinearity exists.

A VIF value greater than five (5) shows extreme correlation; that is a high level of multicollinearity exists.

The inverse of VIF is known as Tolerance and expressed as:

$$TOL = \frac{1}{VIF} = \left(1 - R^2\right) \quad (23.13)$$

Hence, when R^2 is equal to zero (R2 = 0), it indicates that no collinearity exists, then the Tolerance is high (that's equal to 1). The results of VIF and Tolerance are shown in Table 23.4.

23.4 Results and Discussion

23.4.1 Socioeconomic Characteristics of the Cassava Farmers

The descriptive statistics of the socioeconomic characteristics of the cassava farmers of this research are displayed in Table 23.3. It could be seen from the Table that the high rate (72%) of diversification of income suggests that most cassava farmers engage in multiple income-generating activities to manage risk. This is consistent with existing literature, which highlights that income diversification is a common strategy among smallholder farmers to cope with the uncertainty of agricultural income due to price volatility and environmental risks (Adnan et al., 2020). Only 20% of farmers participate in contract farming, indicating low adoption. The literature suggests that contract farming can provide farmers with stable markets and income. However, low participation may imply barriers such as the perceived risks associated with contracts, lack of access to buyers, or limited infrastructure. According to Ton et al., (2018), contract farming can enhance farmers' resilience by providing a guaranteed market and inputs, but success often depends on strong institutional support. The low adoption may suggest the need for better contract enforcement mechanisms or farmer education on the benefits of contracts. The high rate of precautionary (personal) savings (81%) shows that personal savings are a critical financial risk management strategy for farmers. Savings allow farmers to self-insure against future income shocks. However, reliance on personal savings could indicate limited access to formal insurance mechanisms or safety nets. The importance of precautionary savings is noted in the literature as a key risk management tool among farmers (Adnan et al., 2020). Furthermore, the low percentage of farmers using credit facilities (25%) points to a gap in access to formal financial services, despite its potential to improve productivity and risk resilience. This is a concern, as access to credit is often linked to higher farm investments and improved risk management. The finding is consistent with Rizzo et al. (2023) and FAO (2023) who reported that access to credit remains a critical bottleneck for smallholder farmers. The limited use of credit could be due to high interest rates, collateral requirements, or lack of financial literacy among farmers.

The average age of farmers is 49.44 years, indicating an aging farming population. This could have implications for technology adoption and productivity, as older farmers may be less likely to adopt innovative farming techniques or diversify their production (Brown et al., 2019; Rizzo et al., 2023). The gender distribution shows that 40% of cassava farmers are male, implying that the majority (60%) are female. Gender plays a significant role in access to resources, with literature highlighting that women often face more constraints in accessing credit, inputs, and markets (Rizzo et al., 2023; FAO, 2023). Gender disparities in agriculture are widely documented (FAO, 2023). Women, who make up a significant proportion of farmers, often lack access to resources that could improve their productivity. Most farmers are married (mean, 2.07), which may influence household decision-making, labour

Table 23.3 Descriptive statistics of the variables

Variables	Descriptive	Mean	S.D
Diversification of incomes	Dummy = 1 if the farmer uses diversification of income, 0 otherwise	0.72	0.45
Contract farming	Dummy = 1 if the farmer uses contract farming, 0 otherwise	0.20	0.40
Precautionary (personal) savings	Dummy = 1 if the farmer uses precautionary savings, 0 otherwise	0.81	0.40
Use of credit facilities	Dummy = 1 if the farmer uses credit facilities, 0 otherwise	0.25	0.44
Age (in years)	Cassava farmer's age (in years)	49.44	10.30
Gender	Dummy: if cassava farmer is a male = 1, 0 otherwise	0.40	0.49
Marital status	Level: single = 1; married = 2; widow/widower = 3; divorced = 4	2.07	0.52
Farmers experience	Number of years spent in farming cassava	23.01	13.55
Education level	Number of years spent in schooling to obtain the level of educational	7.83	5.23
Household (in number)	Total number of household members	6.00	1.88
Off-farm job opportunities	Dummy = 1 if farmer engaged in off-farm job, 0 otherwise	0.69	0.48
Cassava farm size	Size of cassava farm cultivated (in hectares)	1.2527	1.03
Flooding of farmlands	Dummy = 1 if the farmer experiences a flood, 0 otherwise	0.74	0.44
Drought	Dummy: 1 if the farmer experiences drought, 0 otherwise	0.66	0.48
Heavy winds	Dummy: 1 if the cassava farmer experiences heavy wind, 0 otherwise	0.60	0.49
High rate of pest and diseases incidence	Level: highly perceived = 1; moderately perceived = 2; less perceived = 3; not sure = 4	1.87	0.87
Access to credit/ finance facilities	Dummy: 1 if cassava farmer has access to credit/finance, 0 otherwise	0.45	0.50

(continued)

Table 23.3 (continued)

Variables	Descriptive	Mean	S.D
Extension contact	Dummy: 1 if farmer has contact with extension agent within one year, 0 otherwise	0.26	0.44
Farmer association	Dummy: 1 if the cassava farmer belongs to any farmer's association, 0 otherwise	0.24	0.43

Note S.D means Standard Deviation
Source Field Survey (2023)

Table 23.4 Collinearity diagnostics of multivariate probit (MVP) model used for financial risk management strategies analysis by cassava farmers

Variables	Collinearity statistics	
	Tolerance	VIF
Age (in years)	0.441	2.268
Gender	0.837	1.194
Marital status	0.578	1.731
Farmers' experience in crop farming (in years)	0.611	1.638
Education level	0.760	1.315
Household (in number)	0.850	1.176
Off-farm job opportunities	0.904	1.107
Farm size (in ha)	0.811	1.232
Flooding of farmlands	0.771	1.297
Drought	0.653	1.532
Heavy winds	0.737	1.357
High rate of pest and disease incidence	0.891	1.122
Access to credit/finance facilities	0.660	1.515
Extension contact in the last one (1) year	0.784	1.275
Farmer association	0.681	1.469

Note VIF means variance inflating factor
Source Field Survey (2023)

allocation, and financial risk management. Married individuals might have better access to family labour and more stable household structures. This finding agrees with Rizzo et al. (2023) who reported that marital status influences household labour availability and decision-making processes. Farmers in the study have on average, over 23 years of farming experience, which suggests they are likely well-versed in farming activities, however, the majority of the farmers were small-scale operators with an average (1.25 ha) size of farmed cassava farms. However, experienced farmers might be less inclined to adopt new financial risk management strategies like contract farming or credit use. As noted by Adnan et al. (2020); and FAO (2023), experienced farmers tend to rely on traditional methods of financial risk management and may resist change, even if new strategies could reduce risks more effectively. With only 45% of farmers having access to credit or financial services, there is a significant gap in formal financial inclusion. This is consistent with broader findings that access to rural finance is a challenge in developing countries (Birhanu et al., 2023). Lack of credit access hinders farmers' ability to invest in productivity that could enhance the management of financial risks effectively. Only 26% of farmers had contact with extension agents, indicating that extension services are not reaching the majority of farmers. This limits the dissemination of new technologies and financial risk management strategies. According to Brown et al. (2019) and FAO (2023), extension services are crucial in providing farmers with information on modern agricultural practices. Low extension contact reduces farmers' capacity to improve productivity and manage financial risks effectively. Furthermore, the low percentage (24%) of farmers belonging to associations suggests that collective action is underutilized. Farmer groups can facilitate access to credit, information, and appropriate financial risk management. FAO (2023) highlights that farmer associations can play a key role in enhancing farmers' access to inputs, markets, and collective bargaining power. Strengthening farmer groups could improve access to resources and mitigate financial risks more effectively (Rizzo et al., 2023).

23.4.2 *Multicollinearity Diagnostics Results for Validation of the Models*

It is possible to infer from Table 23.4 that there is no multicollinearity, indicating that there was no correlation between the variables. Given that the VIF values were greater than one (1) and within the acceptable range, and that the tolerance level (roughly equal to one, T = 1) was high as indicated by the tables, the models were, at most, accurate and suitable for measuring the data collected. One example of a variable with low multicollinearity is age (2.268).

23.4.3 *Correlation Coefficients of Different Financial Risk Management Strategy*

The pairwise correlation (Rho) coefficients of the four financial risk management strategies adoption choices of the cassava farmers were estimated, as shown in Table 23.5. The correlation coefficient relationships offer insight into how different risk management strategies can complement or substitute for one another, helping farmers balance financial risks through a combination of income sources, savings, and credit access. This finding is in contrast to Adnan et al. (2020) who reported

Table 23.5 Pairwise correlation (Rho) coefficient estimates in different pairs of financial risk management strategies

Financial risk management combination	Correlation (Rho) coefficients
Diversification of income and contract farming	0.143*
Diversification of income and precautionary savings	– 0.111*
Diversification of income and credit facilities	0.189**
Contract farming and precautionary savings	– 0.292**
Contract farming and credit facilities	0.260**
Precautionary savings and credit facilities	0.129*

Note **, * means significant at 1%, 5% respectively
Source Field Survey (2023)

a simultaneous relationship in the choices of risk management among farmers in Bangladesh.

23.4.4 Determinants of Financial Risk Management Strategies Adopted by Cassava Farmers

Table 23.6 shows the result of the MVP model of financial risk management strategies adopted by cassava farmers. Bartlett's test of sphericity Log likelihood chi-squared tests of 52.12 was obtained and significant at 1% level, null hypothesis that the residual covariance matrix is proportional to an identity matrix. The coefficient of multiple determination (R^2) shows that about 43% magnitude of financial risk management strategies adopted were explained by the independent variables included in the model, while Adjusted R^2 0.393 demonstrates the correct measurement of the model. Partial eta squared is used to measure the extent of the effect the independent variable(s) has on the dependent variable.

23.4.4.1 Diversification of Incomes

From Table 23.6, age with a coefficient of 0.007 (p-value 0.048), is significant at 5% and positively influences the probability of adopting diversification of incomes as financial risk management strategies by the cassava-based farmers. This implies that an increase in the age of the farmers could increase by using diversification incomes as a way of adapting to financial risk, with a partial eta squared of 0.016 this indicates a small effect on the use of diversification of incomes as financial risk management strategies adaptation. This finding is at variance with Ullah et al. (2015) and Adnan et al. (2020) who obtained a negatively correlated coefficient in their studies.

Table 23.6 Parameter estimates of the multivariate probit (MVP) model result showing cassava (Manihot Spp) based farmers financial risk management strategies adopted and their determinants in the study area

Variables	Diversification of incomes	Contract farming	Precautionary savings	Credit facilities
Intercept	0.221 (0.206) [0.005]	0.410** (0.179) [0.022]	1.111*** (0.187) [0.130]	– 0.247 (0.172)[0.009]
Age	0.007** (0.004) [0.016]	0.001 (0.003) [0.000]	– 0.002 (0.003) [0.002]	– 0.001 (0.003) [0.000]
Gender	– 0.041 (0.057) [0.002]	0.027 (0.050) [0.001]	– 0.091* (0.052) [0.013]	0.032 (0.048) [0.002]
Marital	0.081 (0.065) [0.007]	– 0.012 (0.057) [0.000]	– 0.087 (0.059) [0.009]	– 0.012 (0.054) [0.000]
FrmExp	– 0.002 (0.002) [0.004]	– 0.004* (0.002) [0.015]	0.005** (0.002) [0.019]	– 0.002 (0.002) [0.004]
EduLev	– 0.001 (0.006) [0.000]	– 0.019*** (0.005) [0.060]	0.005 (0.005) [0.003]	0.001 (0.005) [0.000]
HH	– 0.017 (0.015) [0.005]	– 0.015 (0.013) [0.006]	– 0.009 (0.013) [0.002]	0.013 (0.012) [0.005]
OffFrmJob	0.153*** (0.056) [0.030]	– 0.032 (0.507) [0.002]	0.035 (0.051) [0.002]	0.072 (0.047) [0.010]
CasFrmSiz	0.007 (0.027) [0.000]	0.067*** (0.023) [0.034]	0.030 (0.024) [0.006]	0.022 (0.022) [0.004]
Flood	0.019 (0.067) [0.000]	0.048 (0.059) [0.003]	– 0.120** (0.061) [0.016]	0.065 (0.056) [0.006]
Drought	– 0.006 (0.067) [0.000]	– 0.053 (0.058) [0.003]	– 0.005 (0.061) [0.000]	0.009 (0.056) [0.000]
HeavyWd	0.065 (0.061) [0.005]	– 0.056 (0.053) [0.005]	0.070 (0.055) [0.007]	0.071 (0.051) [0.008]
HighPestDis	– 0.041 (0.032) [0.007]	– 0.029 (0.027) [0.005]	0.001 (0.029) [0.000]	0.062** (0.026) [0.022]
AccesCred	0.175*** (0.062) [0.032]	0.332*** (0.054) [0.138]	– 0.049 (0.056) [0.003]	0.541*** (0.052) [0.314]
ExtCont	0.272*** (0.065) [0.069]	0.123** (0.056) [0.020]	– 0.242*** (0.059) [0.066]	– 0.077 (0.054) [0.008]
FarmAss	– 0.378*** (0.073) [0.101]	– 0.047 (0.063) [0.002]	0.073 (0.066) [0.005]	0.011 (0.061) [0.000]

Note ***, **, * means significant at 1%, 5%, and 10% respectively; the first figures are the betas, the bracket "()" is the standard errors, and the parenthesis "[ ]" are the partial eta squared; 0.01, 0.06 and 0.14 connotes small, medium and large effect of partial eta squared respectively; Bartlett's test of sphericity: Loglikehood Chi-square = 52.121 (0.000), $R^2 = 0.429$, Adjusted $R^2 = 0.393$; Sample size = 260

Source Field Survey (2023)

Off-farm job opportunities with a coefficient of 0.153 (*p*-value 0.007), are statistically significant at 1% and positively impact the probability of adopting diversification of incomes as financial risk management strategies. A unit increase in off-farm job opportunities would increase the chances of adopting diversification of incomes by cassava farmers by 15.3%, signifying the adaptation of off-farm job opportunities would improve the chances of cassava-based farmers adopting diversification of incomes as financial risk management strategies. Off-farm jobs avoid opportunities for the farmers to get extra income to combat the financial risks they might be facing. This finding agrees with the report of Adnan et al. (2020) who noted that farmers with a higher income have a greater possibility to diversify, which could invariably yield more benefits in diversification and obtain additional benefits. A partial eta squared of 0.030 was obtained, which indicates a small effect on the use of diversification of incomes as a financial risk management strategy adaptation.

Access to credit facilities with a coefficient of 0.175 (*p*-value 0.005), is statistically significant at 1% and positively influences the probability of adopting diversification of incomes as financial risk management strategies. A unit rise in access to credit could increase the chances of adopting diversification of incomes by cassava farmers by 17.5%, signifying the adaptation access to credit/financial would improve the chances of cassava-based farmers adopting diversification of incomes as financial risk management strategies. A partial eta squared of 0.29 was obtained for access to credit/ financial, which indicates a medium effect on the use of diversification of incomes as a financial risk management strategy adaptation. A partial eta squared of 0.032 was obtained, which indicates a small effect on the use of diversification of incomes as a financial risk management strategy adaptation.

Extension contact with a coefficient of 0.272 (*p*-value 0.000), is statistically significant at 1% and positively influences the probability of adopting diversification of incomes as financial risk management strategies. A unit increase in extension contact could increase the chances of adopting diversification of incomes by cassava farmers by 27.2%, signifying the adaptation of extension contact would improve the chances of cassava-based farmers adopting diversification of incomes as financial risk management strategies. A partial eta squared of 0.069 was obtained, which indicates a medium effect of diversification of incomes as financial risk management strategies adaptation. This implies that additional agricultural extension officer contacts could increase the prospect of better use of financial risk management strategies. This finding agrees with Adnan et al. (2020) who noted that extension services offer information and give accurate information about the sources of income, and risk sources and minimize losses.

Farming association with a coefficient of − 0.378 (*p*-value 0.000), is statistically significant at 1% and negatively influences the probability of adopting diversification of incomes as financial risk management strategies. A unit increase in farming associations could reduce the chances of adopting diversification of incomes by cassava farmers by 37.8%, signifying the adaptation of farming associations would reduce the chances of cassava-based farmers adopting diversification of incomes as financial risk management strategies. A partial eta squared of 0.101 was obtained, which indicates a large effect on the use of diversification of incomes as a financial risk management strategy adaptation. This implies that farmer association has a great influence on income diversification of the farmers. Farming association is expected to increase the rate of diversification of income of cassava farmers. Hence, the results of the multivariate probit (MVP) model provide sufficient indication for the complement or substitute adaptation choices.

23.4.4.2 Contract Farming

From Table 23.6, educational level with a coefficient of – 0.019 (*p*-value 0.000), is statistically significant at 1% and negatively impacts the probability of adopting contract farming as a financial risk management strategy. A unit increase in educational level could reduce the chances of adopting contract farming by cassava farmers by 1.9%, suggesting the adaptation educational level would improve the chances of cassava-based farmers adopting contract farming as a financial risk management strategy. A partial eta squared of 0.060 was obtained, which indicates a small effect on the use of contract farming as a financial risk management strategy adaptation. This implies that the years spent in acquiring the level of education will determine the level of exposure which invariably could probably influence the adaptation of contract farming as a means of financial risk management strategy. This finding agrees with Yirga et al. (2015) who reported that educational level was significant with arable crop farmers with formal education and especially, those who spent more than eight (8) years to attain the level of education among arable crop farmers in Ethiopian.

The size of the cassava farm with a coefficient of 0.067 (*p*-value 0.004), is statistically significant at 1% and positively impacts the probability of adopting contract farming as a financial risk management strategy. A unit increase in cassava farmland size could increase the chances of adopting contract farming by cassava-based farmers by 6.7%, suggesting the adaptation of cassava farm size would improve the chances of cassava-based farmers adopting contract farming as a financial risk management strategy. A partial eta squared of 0.034 was obtained, which indicates a small effect on the use of contract farming as a financial risk management strategy adaptation. This implies that farmers with large farm sizes could adapt to financial risk and minimize losses by contract farming. This result agrees with Adnan et al. (2020) who reported that farm size also influences contract farming as an adaptation to financial risk management.

Access to credit with a coefficient of 0.332 (*p*-value 0.000), is statistically significant at 1% and positively impacts the probability of adopting contract farming as a financial risk management strategy. A unit increase in access to credit could increase the chances of adopting contract farming by cassava-based farmers by 33.2%, suggesting that adaptation access to credit would improve the chances of cassava-based farmers adopting contract farming as a financial risk management strategy. A partial eta squared of 0.138 was obtained which indicates a large effect on contract farming as financial risk management strategies adaptation. This implies that access to credit facilities could aid or increase the rate of contract farming of cassava farmers.

Extension contacts with a coefficient of 0.123 (*p*-value 0.030), is statistically significant at 5% and positively impacts the probability of adopting contract farming as a financial risk management strategy. A partial eta squared of 0.020 was obtained which indicates a small effect on the use of contract farming as a financial risk management strategies adaptation. This shows the critical role agricultural extension officer contacts could play has regard to prospects on the use of financial risk management strategies to reduce losses. Hence, the results of the multivariate probit (MVP) model provide sufficient indication for the complement or substitute adaptation choices.

23.4.4.3 Precautionary (Personal) Savings

Table 23.6, shows that farming experience with a coefficient of 0.005 (*p*-value 0.034), is statistically significant at 5% and positively impacts the probability of adopting the use of precautionary (personal) savings as financial risk management strategies. A partial eta squared of 0.019 was obtained, which indicates a small effect on the use of precautionary (personal) savings as a financial risk management strategy adaptation. This implies that the more the years in farming or growing cassava the higher the rate of utilizing the use of precautionary (personal) savings as a means of financial risk management strategies adaptation.

Flood with a coefficient of − 0.120 (*p*-value 0.050), is statistically significant at 5% and negatively impacts the probability of adopting the use of precautionary (personal) savings as financial risk management strategies. A partial eta squared of 0.016 was obtained, which indicates a small effect on the use of precautionary (personal) savings as a financial risk management strategy adaptation.

Extension contacts with a coefficient of − 0.242 (*p*-value 0.000), is statistically significant at 1% and negatively impacts the probability of adopting precautionary (personal) savings as financial risk management strategies. A unit increase in contact with extension officers could reduce the chances of adopting precautionary (personal) savings by cassava farmers by 24.2%, suggesting the adaptation extension contact would improve the chances of cassava-based farmers adopting precautionary (personal) savings as financial risk management strategies. A partial eta squared of 0.066 was obtained which indicates a medium effect on the use of precautionary (personal) savings as a financial risk management strategy adaptation. This result agrees with Yirga et al. (2015) who reported that training arable crop farmers is crucial to sustainable potato production in Ethiopia.

23.4.4.4 Credit Facilities

From Table 23.6, the high rate of pest and disease incidence with a coefficient of 0.062 (*p*-value 0.020), is statistically significant at 5% and positively impacts the probability of adopting credit facilities as financial risk management strategies. This implies that cassava-based farmers could adopt credit facilities to prevent pest and disease infestation in cassava production. A partial eta squared of 0.022 was obtained, which indicates a small effect on the use of credit facilities as a financial risk management strategy adaptation.

Access to credit with a coefficient of 0.541 (*p*-value 0.000), is statistically significant at 1% and positively impacts the probability of adopting credit facilities as financial risk management strategies. A unit increase in contact with access to credit could increase the chances of adopting credit facilities by cassava farmers by 54.1%, suggesting the adaptation access to credit would improve the chances of cassava-based farmers adopting credit facilities as financial risk management strategies. A partial eta squared of 0.314 was obtained, which indicates a large effect on the use of credit facilities as a financial risk management strategy adaptation. This implies that credit is needed to aid the rate of financial risk management strategies adaptation using credit facilities.

23.5 Conclusions, Recommendations and Limitations

This research provides valuable insights into how financial risk management strategies are employed by cassava farmers in the Niger Delta Region, Nigeria, to address climate change vulnerability. The findings highlight the critical role of diversification of income, precautionary savings, and access to credit facilities in enhancing farmers' resilience to climate-related risks such as flooding, drought, heavy winds, and pest and disease outbreaks. However, the results also reveal several gaps in financial inclusion and institutional support, particularly in contract farming participation, access to credit, and extension services. Cassava farmers in the Niger Delta, who face significant climatic challenges, have developed adaptive strategies like diversification of income and precautionary savings.

In conclusion, addressing climate change vulnerability in cassava farming in the Niger Delta requires a multifaceted approach. This includes improving financial inclusion, promoting the use of contract farming, enhancing access to credit and extension services, and fostering the development of farmer associations. These interventions are vital to enabling cassava farmers to mitigate the financial risks posed by climate variability and secure their livelihoods in an increasingly uncertain climate.

The study primarily focuses on financial risk management strategies (diversification of income, contract farming, precautionary savings, and credit facilities), potentially overlooking other critical non-financial strategies that farmers use to adapt to climate change. Although the study provides a breakdown of gender, it does not delve deeply into how gender dynamics influence access to financial resources and climate adaptation strategies. While the study identifies gaps in access to credit and extension services, it does not deeply analyse the broader institutional or policy context that might explain why these gaps exist. Addressing these limitations could provide a more implication understanding of how cassava farmers in the Niger Delta and beyond manage financial risks in the face of climate change. The key policy areas the study recommends include:

i. Enhancing the financial literacy of farmers, particularly in the context of managing climate-related risks. Education and training programs should be developed to help farmers understand financial tools such as savings, credit, and insurance, and how to use them for financial risk management. Policies to integrate financial and climate risk education into existing agricultural extension services could improve overall resilience.
ii. Policies targeting improving access to affordable credit could potentially enhance cassava farmers' ability to mitigate risks. Credit availability is critical for resilience, and enhancing financial inclusion should be a policy priority.
iii. Extreme weather events including droughts, floods, and temperature swings are becoming more frequent in the area, and this has a big impact on cassava yield. Therefore, to help farmers effectively traverse the problems brought on by climate change, financial risk management measures including diversification, climate-smart practices, and access to financial resources must be coupled with capacity-building programmes. These would guarantee sustained cassava production and strengthen the adaptability of cassava producers.
iv. Establish robust monitoring and evaluation (M&E) systems to track the effectiveness of climate risk management policies and practices. Governments and NGOs should create M&E frameworks that assess how well financial risk management strategies reduce vulnerability in cassava farming. Furthermore, improving institutional frameworks to better coordinate efforts in addressing climate change risks in agriculture. Governments at all levels (local, regional, national) need to ensure that climate risk management in agriculture is integrated into national development plans.
v. Achievable plans and initiatives must be established to assist cassava growers in proactively adapting to the risks associated with climate change. Enhancing the accessibility and availability of low-interest loans for cassava farmers, for instance, might improve their adaption tactics. Other players, including NGOs and government organizations, might carry out this task.

Acknowledgements The authors are very grateful fund received from the Nigerian Conservation Foundation (NCF) under the Chief S.L. Edu PhD research grant for data collection. The authors also appreciate the respondents for their valuable time and information.

Funding Partial financial support was received from Nigerian Conservation Foundation (NCF) under the Chief S.L. Edu PhD research grant to support the field work in data collection.

Data Availability Statement The datasets used or analysed during the current study are available from the corresponding author on reasonable requests.

Conflict of Interest The authors declared that there is no conflict of interest.

Ethical Clearance The confidentiality of the survey participants was strictly and unwaveringly respected.

References

Adhikari, S., & Khanal, A. R. (2021). Economic sustainability and multiple risk management strategies: Examining interlinked decisions of small American farms. *Sustainability, 13*, 1741. https://doi.org/10.3390/su13041741

Adnan, K. M. M., Ying, L., Ayoub, Z., Sarker, S. A., Menhas, R., Chen, F., & Yu, M. M. (2020). Risk management strategies to cope catastrophic risks in agriculture: The case of contract farming, diversification and precautionary savings. *Agriculture, 10*(351), 1–16. https://doi.org/10.3390/agriculture10080351

Ajupov, A., Sherstobitova, A., Syrotiuk, S., & Karataev, A. (2019). The risk-management theory in modern economic conditions. In *E3S Web of Conferences* (Vol. 110, pp. 02040). https://doi.org/10.1051/e3sconf/201911002040

Anjali, C., Debashis, D., & Amar, D. (2019). Farm-level adaptation strategies to climate change in India: An overview. *Journal of Pharmacognosy and Phytochemistry, 8*(4), 1685–1690. Retrieved from: https://www.phytojournal.com/archives/2019.v8.i4.9189/farm-level-adaptation-strategies-to-climate-change-in-india-an-overview

Birhanu, M. Y., Kassie, G. T., & Dessie, T. (2023). Evolution and scope of rural and agricultural finance in developing countries: A review. *International Livestock Research Institute (ILRI) Research report 121*, Nairobi, Kenya. Retrieved from: https://www.researchgate.net/publication/377851498_Evolution_and_scope_of_rural_and_agricultural_finance_in_developing_countries_A_review

Brown, P., Daigneault, A., & Dawson, J. (2019). *Age, values, farming objectives, past management decisions, and future intentions in New Zealand agriculture*. Retrieved from: https://www.sciencedirect.com/science/article/am/pii/S0301479718311460

Chinda, C. I., Aroyehun, A. R., & Chinda, W. (2023). Instrument reliability determination. In N. Woun & P. Igwe (Ed.) *Research methods for university scholars* (pp. 189–202). Publisher.

Food and Agriculture Organization of the United Nations FAO. (2023). *The status of women in agrifood systems*. Rome. https://doi.org/10.4060/cc5343en

GeeksforGeeks. (2021). *Test of multicollinearity*. Retrieved from: https://www.geeksforgeeks.org/test-of-multicollinearity/

Giambona, E., Graham, J. R., Harvey, C. R., & Bodnar, G. M. (2018). The theory and practice of corporate risk management: Evidence from the field. *Financial Management, 47*(4), 783–832. https://doi.org/10.1111/fima.12232

Henri-Ukoha, A. (2020). Assessment of the cost-benefits of climate change adaptation strategies of cassava-based farmers in southern Nigeria. *International Journal of Environment and Climate Change, 10*(10), 99–110. https://doi.org/10.9734/IJECC/2020/v10i1030253

Izuogu, C. U., Njoku, L. C., Azuamairo, G. C., Atasie, C. M., & Ekumankama, O. O. (2023). Climate change adaptation among cassava farmers in Okigwe Agricultural zone of Imo State Nigeria. *International Journal of Agriculture System, 11*(1), 12–23. https://doi.org/10.20956/ijas.v11i1.2418

Jabbar, A., Wu, Q., Peng, J., Sher, A., Imran, A., & Wang, K. (2020). Mitigating catastrophic risks and food security threats: Effects of land ownership in Southern Punjab, Pakistan. *International Journal of Environmental Research and Public Health, 17*, 9258. https://doi.org/10.3390/ijerph17249258

Kerr, L. A., & Goethel, D. R. (2014). Simulation modeling as a tool for synthesis of stock identification information. In S. X. Cadrin, L. A. Kerr & S. Mariani (Eds.), *Stock identification methods* (2nd ed., pp. 501–533). https://doi.org/10.1016/C2011-0-07625-1

Kibuacha, F. (2021). *How to determine sample size for a research study*. Retrieved from: https://www.geopoll.com/blog/sample-size-research/

Matemilola, S., Adedeji, O. H., Elegbede, I., & Kies, F. (2019). Mainstreaming climate change into the EIA process in Nigeria: Perspectives from projects in the Niger Delta region. *Climate 7*(29). https://doi.org/10.3390/cli7020029.

National Geographic. (2022). *All about climate*. Retrieved from: https://education.nationalgeographic.org/resource/all-about-climate

Omodara, O. D., Ige, O. A., Oluwasola, O., Oyebanji, A. T., & Afape, O. O. (2023). Factors influencing cassava farmers' choice of climate change adaption practices and its effect on cassava productivity in Nigeria. *Heliyon, 9*(3), 1–13. https://doi.org/10.1016/j.heliyon.2023.e14563

Omondi, J. O., & Yermiyahu, U. (2021). *Improvement in cassava yield per area by fertilizer application* (pp. 1–10). IntechOpen. https://doi.org/10.5772/intechopen.97366

Otitoju, M. A. (2013). *The effects of climate change adaptation strategies on food crop production efficiency in Southwestern Nigeria*. Retrieved from: https://oer.unn.edu.ng/

Parshley, L. (2023). *Nothing's predictable: Extreme weather is ruining farmers' crops, and their finances*. Retrieved from: https://www.preventionweb.net/news/nothings-predictable-extreme-weather-ruining-farmers-crops-and-their-finances

Princewaterhouse Coopers Limited PwC. (2020). *Harnessing the economic potential of cassava in Nigeria*. Retrieved from: https://www.agriknowledge.org/file_downloads/xg94hp56z

Rahman, A., Jianchao, L., Adnan, K. M., Islam, M. D. I., Zhao, M., & Sarker, S. A. (2020). How indebted farmers perceive and address financial risk in environmentally degraded areas in Bangladesh. *Environmental Science and Pollution Research, 27,* 7439–7452. Retrieved from: https://doi.org/10.1007/s11356-019-07374-2

Rizzo, G., Migliore, G., Schifani, G., & Vecchio, R. (2023). Key factors influencing farmers' adoption of sustainable innovations: A systematic literature review and research agenda. *Organic Agriculture, 14*(1), 57–84. https://doi.org/10.1007/s13165-023-00440-7

Sarfraz, M., Qun, W., Hui, L., & Abdullah, M. I. (2018). Environmental risk management strategies and the moderating role of corporate social responsibility in project financing decisions. *Sustainability, 10*(2771), 1–17. https://doi.org/10.3390/su10082771

Soulibouth, L., Hwang, H. S., & Shin, D. H. (2021). The impact of flood damage on farmers, agricultural sector and food security in Laos: A regional case study of Champhone district, Savannaket province. *Journal of International Development Cooperation, 16*(2), 151–170. https://doi.org/10.34225/jidc.2021.16.2.151

Ton, G., Vellema, W., Disiere, S., Weituschat, S. C., & D'Haese, M. (2018). Contract farming for improving smallholder incomes: What can we learn from effectiveness studies? *World Development, 104*, 46–64. https://doi.org/10.1016/j.worlddev.2017.11.015

Udotong, J. I. R., Udoudo, U. P., & Udotong, I. R. (2017). Effects of oil and gas exploration and production activities on production and management of seafood in Akwa-Ibom state, Nigeria. *Journal of Environmental Chemistry and Ecotoxicology, 9*(3), 20–42. https://doi.org/10.5897/JECE2017.0405

Ullah, R. Shivakoti, G. P., Rehman, M., & Kamran, M. A. (2015). Catastrophic risks management at farm: The use of diversification, precautionary savings and agricultural credit. *Pakistan Journal of Agricultural Sciences, 52*(4), 1139–1147. Retrieved from: https://pakjas.com.pk/papers%5C2528.pdf

Umoh, N. E., Offiong, R. A., Ekpe, I. A., Yta, E. M., & Enya, E. A. (2015). Assessment of income poverty among farmers in Akwa Ibom State, Nigeria. *Journal of Emerging Trends in Economics and Management Sciences (JETEMS), 6*(8), 345–349. Retrieved from: https://www.researchgate.net/publication/335892605_Assessment_of_Income_Poverty_among_Farmers_in_Akwa_Ibom_State_Nigeria/citation/download

Umoh, U. U., Li, L., Wang, J., Kauluma, N., Asuquo, F. E., & Akpan, E. R. (2022). Glycerol dialkyl glycerol tetraether signatures in tropical mesotidal estuary sediments of Qua Iboe river, Gulf of Guinea. *Organic Geochemistry, 170*(104461), 1–17. https://doi.org/10.1016/j.orggeochem.2022.104461

United Nations Climate Change. (2021). *Climate change-related disasters a major threat to food security—FAO*. Retrieved from: https://unfccc.int/news/climate-change-related-disasters-a-major-threat-to-food-security-fao

Virlics, A. (2013). Investment decision-making and risk. In *International Economic Conference of Sibiu 2013 Port Crisis Economy: Challenges and Opportunities, IECS 2013. Procedia Economics and Finance* (Vol. 6, pp. 169–177). Retrieved from: https://www.sciencedirect.om

Yirga, C., Atnafe, Y., & AwHassan, A. (2015). Multivariate analysis of factors affecting adoption of improved varieties of multiple crops: A case study from Ethiopian highlands. *Ethiopian Journal of Agricultural Sciences, 25*(2), 29–45.

Chapter 24
Land Utilization and Food Security in Energy Transition: Role of Food Supply in Nigeria

Adeyinka Richard Aroyehun

Abstract The relationship between land utilization and food security in Nigeria is increasingly important as the country transitions to more sustainable energy sources. As the demand for land grows, driven by both agricultural and energy needs, the challenge of balancing food production with renewable energy infrastructure development becomes critical. This research explores the macroeconomic impact of the energy transition on Nigerian food security and supply from 2000 to 2021, using data from the World Development Indicators (WDI) and FAOSTAT. The dependent variables are land utilization (LdU), gross domestic product (GDP), energy transition (EgT), agricultural value added (AVA), and urbanization (Urb), while the independent variable is food supply (FSS). Descriptive statistics, unit root, correlation test, FMOLS, and Granger causality were employed to derive the research verdicts and evaluate the study's objectives. The unit root of stationarity utilizing both ADF and PP shows that all the variables were of the first difference order integration I(1). The FMOLS estimated results indicate that LdU (– 1439.20), GDP (– 33.11), AVA (6.87), and Urb (85.71) were statistically significant at a 1% level influencing food security and supply. On the other hand, food security and supply are only slightly impacted by the energy transition. The Granger causality test demonstrates that some of the variables are subject to short-run causality. Therefore, this research suggests policies to minimize land degradation and integrate food security, energy transition, and sustainable agriculture into national policies.

Keywords Agricultural products · Energy transition · Food security · GDP · Urbanization

---

A. R. Aroyehun (✉)
Department of Agricultural Economics and Agribusiness Management, Faculty of Agriculture, University of Port Harcourt, Port Harcourt, Nigeria
e-mail: aroyehun_adeyinka@uniport.edu.ng

R. U. Onyeneke et al. (eds.), *Energy Transition, Climate Action and Sustainable Agriculture*, https://doi.org/10.1007/978-3-031-83165-2_24

24.1 Introduction

There are various ways to define food security. At what time do all and sundry have physical, social, as well as financial access to ample nutritious foodstuff that meets their dietetic requirements and preferences for a lively, active, and healthful lifestyle, this is commonly referred to as food security (Food and Agriculture Organization FAO, 2002 as cited in Zhuang et al., 2022). Hence, food security's associated pillars comprise of availability, usability, accessibility, and stability of food. About local food production as well as the distribution, of food supply, food accessibility means how effortlessly people have access to dietary and nutritious sources of food. Hence, food accessibility is the capability of an individual or a household to obtain easily available food. According to Pinstrup-Andersen (2009), food choice is integral to food availability because the capacity to acquire food does not at all times translate into a definite acquisition of food. Usability of food comprises food preparation, processing, treatment, and cooking of food. Food stability, on the other hand, refers to both the potential for and capacity to recover from shocks. Thus, it's critical to bear in mind that, under this situation, an individual's food security is an essential nevertheless insufficient precondition for adequate sustenance. According to the UNICEF's guideline on nutritional deficiency (Black et al., 2008), nutrition status is determined by the link between food in addition to nutrient ingestion and fitness with vigour.

Food security programmes were given top priority by anthropogenic climate change authorities both spatially and temporally (Connolly-Boutin & Smit, 2015). Global food security, livelihood sustainability, and the energy transition are severely threatened by factors such as climate change, increasing rainfall irregularities, protracted droughts, and heavy flooding that occurs unexpectedly (Sa et al., 2017). To establish lasting solutions, these unavoidable variabilities demand a fundamental shift in human socioeconomic institutions. Nigeria must therefore create the mechanisms necessary to become a low-carbon economy, or decarbonised. Hassan et al. (2022) claim that doing so would assist save the environment by reducing the consequences of climate change. Achieving zero fossil fuel output and consumption is made possible by useful and effective ongoing approaches (Ajjur & Al-Ghamdi, 2022; Yoon et al., 2022).

Green energy is essential for food security, and the impact of the energy shift affects conventional energy utilization designs worldwide, which in turn sways and stimulates food systems. It has a significant impact on the production of food for everyone in a fossil fuel-dependent society (Strapasson et al., 2017). But there's a chance that this energy shift will have unintended societal repercussions. The societal outcomes incorporate more noteworthy consequences throughout the logical dissemination and allocation of unfavourable results on food security, humanoid livelihood, sustainable land use, and favourable prospects. By avoiding the resource-intensive paradigm, the logical energy transition may aid in the reduction of inequities in food security, which may support the current food security systems (Zhuang et al., 2022). A low carbon emissions economy is also ensured by "just energy transition," which also protects the livelihoods of workers, and the prospect of the societies involved in the energy transitions. Working-class people, unions, the government, and businesses' consultation with the community and civil society are all involved in this social discourse (Galgóczi, 2020). According to Heffron et al. (2021), the energy transition also highlights the decentralization of energy systems, their significance, and the demands of marginalized cultures and communities as top priorities. Furthermore, a fair energy transition aims to advance social resilience, wellness, economic sustainability, and environmental integrity, all of which are supported by stable and efficient democratic governance procedures. As a result, it expedites the mapping of the energy transition to ensure realistic results that are legitimately in line with the social and economic growth of affected neighbourhoods, towns, and regions. Similarly, having an effect atop the various mainstays and cores of food security, socioeconomic variables may also have an impact on how food security is measured. Cultural differences have a considerable impact on the food supply in addition to socioeconomic issues since they heavily influence the practices of intra-household food distribution.

The energy industry does a great job of increasing the food supply in response to growing worries about food security. Because so much traditional energy was used, environmental deterioration was the price paid for solving the problem of food insecurity. As a result, the energy transition has gained attention at summits discussing how energy will be used in human and commercial endeavours in the future. The demand for sustainable development is currently driving the energy shift, which is frequently tied to the adjustments that ought to be made to move towards renewable and environmentally friendly energy sources (Luciani, 2020). Numerous social, political, political, and economic concerns are discussed. The majority of previous bids to provide energy facilities to large populations have proved abortive, and the energy systems in place today are generally untenable and unsustainable. Therefore, for food to be sustainable, there must be an energy shift (Zhuang et al., 2022).

Developed countries have seen an energy transition as a result of rising incomes. The primary goal is to transition from conventional biomass to fossil fuels and then to natural gas as well as renewable energy sources for instance solar as well as wind power (Kim, 2019). The shift to renewable energy in countries that are still developing necessitates a substantial upsurge in the availability as well as the ability to afford energy services, which occasionally could lead to a decrease in the availability of food. These factors are caused by rising industrial production, urbanization, economic development, and the expansion of transportation networks. Several authors have given an opinion against centralized energy structures and are in favour of locally adaptive way out to socioeconomic challenges to achieve the present-day goals of the agenda for the Energy Transition (Goldthau & Hughes, 2020). Decentralized, local energy systems have emerged in this context as a novel strategy that enables cities and areas to become autonomous of energy (Adil & Ko, 2016). Domestic initiatives like capacity training and governance are essential for advancing the energy transition (Baker et al., 2014). Thus, the shift to a sustainable energy system depends critically on technology innovation as well as the market, pricing, economics, legislation, and policymakers. To steer the world energy system towards sustainable paths, policy action is necessary.

The main goal of this research is to identify the main obstacles to the food supply as well as investigate the tactics employed in Nigeria towards overcoming them. To close the knowledge gap, not a single study has attempted to look into how the Nigerian economy's energy shift is related to land use, food security, and supply. Furthermore, a plethora of case studies has been shown to expound upon the varied correlation between the availability of food and its associated factors. Additionally, the energy transition has not been regarded as a major factor influencing the availability of food worldwide, especially in Nigeria. For this reason, this study will add something new to the body of knowledge regarding the potential effects of the energy transition on food supply. This study considers the energy transition in addition to the other fundamental factors affecting the food supply, such as land use, gross domestic product, agriculture value-added, and urbanization. Socioeconomic variables are important and can have a big impact on the availability of food; contrariwise, the availability of food has a big impact on economies' ability to grow economically. Similarly, value added from agriculture also makes a weighty impact on the food supply. Examining the mild influence of the transition in energy on land usage, value of agricultural products added, as well as gross domestic product and their possible effects on the food supply system is another important addition to this research. It opens up new avenues for future research on how to examine the important role that energy plays in various sectors. We can determine whether the food supply is considerably wedged by and caused by the energy transition in land utilization, on the agriculture sector as well as economic growth by this research. It is recognized that the three food components listed above have a direct impact on the energy sector of the economy and this could lead to an upsurge in the availability of food. Unlike earlier case investigations, this current study attempts to explore the aforementioned goals using several innovative estimators. The most dependable unit roots—the Augmented Dickey-Fuller and Philips-Perron—as well as pairwise correlation with the fully modified least squares (FMOLS) model, were used in this study for this reason. Lastly, to look into the causative relationship between a few chosen variables, the current study used pairwise panel Granger causality tests.

24.2 Literature Review

According to the United Nations (2015), one of the seventeen Sustainable Development Goals (SDGs) is to end hunger. Despite the massive and exceptional study that has been conducted over the preceding years to form and assemble procedures as well as strategies to guarantee worldwide food security, one out of about ten individuals globally now experience austere food insecurity. Essential factors include population expansion, continuous urbanization, and unmaintainable utilization of non-renewable resources coupled with changes in climatic factors and modifications to food ingestion forms (for instance an upsurge in aggregate caloric intake or a shift in the parts of diets) (Abd-Elmabod et al., 2020). Extensive resource exploitation could probably lead to land deprivation and a reduction in soil efficiency, while malnutrition is expected to rise in conjunction with population growth (Hall et al., 2017). An increase in the intensity of pests and diseases caused by climate change and increasing in acute incidents for instance droughts and flooding may perhaps result in crop failure or total loss (Richardson et al., 2020). Furthermore, the demand for additional agricultural produce and varying consumption patterns are escalating the utilization of resources of land and water, diminishing these resources, and increasing problems of food security (Ruben et al., 2018). Hence, the worldwide food security strategy demands speedy and prompt global combat and complete international food insurance, particularly in Nigeria.

Food security is determined by the quantity of food that is readily accessible for ingestion by the populace in a certain place during a given period; this is accomplished through imports from areas where there is an excess of food or through home-grown production from agriculture (Kumar et al., 2017). Food security is also defined by Headey et al. (2022) as the state in which a nation has ample food to distribute to either meet local requirements vial home-growing output or imports. Nonetheless, a few contemporary instance investigations also attempted to illustrate the significance of the energy transition and discovered that it had a major impact on several industries. Shah et al. (2020) attempted to scrutinize the influence of energy transition on environmental value in the Asian continent. The long-term results demonstrate how the energy shift has a favourable impact on environmental deterioration. Naqvi et al. (2021) case study on income groups also looked into the significance of the energy shift to environmentally friendly durability. Utilizing the mean groups (MG), augmented mean groups (AMG) as well as common correlated effects mean groups (CCE-MG), they discovered a noteworthy impact on eco-friendly value in both high- and upper-middle-income nations. In their attempt to expound the energy transition in eco-friendly worth, Zhang et al. (2022) also used fully modified ordinary least squares (FMOLS) to demonstrate the important contributions of nuclear and wind energy. Zhuang et al. (2022) also examine land usage and food security in the context of energy transition. They use FMOLS with dynamic ordinary least squares (DOLS) to analyse the role of food supply. Their findings indicate a negative relationship between land usage and food security, energy shift, GDP, and value of agricultural products added play a role in the supply of food, urbanisation plays a negative role in the food supply, additionally, and the impact of the energy transition is found to be significant on food supply. Furthermore, previous research has not examined Nigeria, which is the focus of this study; this will add fresh knowledge to the body of literature already in existence. Lastly, not a single study attempts to predict or make recommendations on whether or not the future food supply will decrease. Therefore, by including such food supply forecasts in future publications, this study will also contribute. Therefore, local businesses' activities are necessary for a sustainable food supply.

24.2.1 Theoretical Background

Numerous ingenuities and plans were successfully implemented globally to tackle the increasing need for energy resources devoid of utilizing—running out of fossil fuels that produce diverse harmful sectors (Qadir et al., 2021). Zhuang et al. (2022) claim that Governments were required to subsidize green energy technology under the 1997 Kyoto Protocol. Following the 2015 Paris Agreement/COP21 (Conference of the Parties), the listed energy plans and policies were established: green energy technologies, like solar, wind as well as biomass must be developed further (Imteyaz & Tahir, 2019); sustainable renewable energy (RE) involves offering replacements that differentiate energy sources as well as promote energy security; moreover, renewable energy may perhaps facilitate the development of social and economic spheres, encourage energy accessibility, and, most importantly, lessen how climate change affects the environment and the health of individuals (Panwar et al., 2011). For this reason, switching to RE from fossil fuels is crucial to a greener future opportunity. The switch from energy sources that depend on fossil fuels to energy from renewable sources is referred to as the "renewable energy transition" (RET), according to Li et al. (2020). However, it may probably take an enormous volume of resources to execute an efficient RET. Despite new technology innovations greatly decreasing the total expenses related to renewable energy production, there has not been an equivalent increase in outlay (Irena & Desa, 2019). As a result, funding the RET may be one of Nigeria's and other nations' biggest challenges since financiers are less enthusiastic to take on financial risk owing to fluctuations in regulations as well as the required quantity of cash.

Environmental changes can occur at any point in the process of producing, storing, processing, distributing, selling, and consuming food. These changes can be small-scale variations around average conditions (like the climate, nutrient, and water cycle), large-scale anomalous disruptions, or slow changes in average conditions. With the increasing global interconnectedness of food supply chains, it is critical to understand the potential special risks associated with how environmental variability is dispersed or repressed within food systems. The interaction between producers and customers has become more complex and efficient due to the rise of multinational food companies and continuous globalization. There is still much to learn about how major events (such as economic downturns, restrictions, and epidemics) may intensify these outcomes, how variability in the environment spreads via local and international chains of food supply as a result of these dynamics, and what these factors subsequently imply for the accessibility and reliability of diets high in nutrients. Reducing the threat to the security of nutrition and food posed by recurrent food shortages and boosting resilience in the system need an understanding of the fundamental actors and processes in the supply chain of food as well as how they interact with and are vulnerable to economic and variability in the environment. Human activity has brought about major changes to the natural environment. Growing populations and rising food demands accelerate changes in land utilization and cover, requiring more farmland to produce the same amount of food.

24.3 Methodology

The panel data employed in this research encompasses the Nigerian economy from 2000 to 2021. Food supply, land utilization, energy transition, agricultural products value-added, GDP, as well as urbanization are all included in the panel data. On the other hand, the data dependability depends on the unit, definition, and source information. Thus, the food supply is determined by calculating Kcal/capita/day and is gathered from agricultural as well as food-related organizations. Similar to other core variables, land utilization is estimated by the aggregate of agricultural land in square kilometres, GDP per capita in US dollars, population density in urban areas, and agricultural value added in US dollars, as shown in Table 24.1. These data are gathered from the World Development Indicators (WDI) and FAOSTAT The primary focus of this study is Nigeria's energy transition, which is defined as the attempt of economies to replace their antiquated energy patterns with more contemporary and environmentally friendly ones. Thus, as a measure of the energy transition, the current study also included the percentage of RE use in overall energy utilization.

24.3.1 Model Construction and Specification

The prime target of this investigation is to examine the factors that influence the food supply and how they may impact the food supply system. The following models have been suggested for analysis in this research. However, the base model considers how land utilization, energy transition, the value of agricultural products added, urbanization, as well as economic expansion, affect the availability of food. For this reason, the base model utilized in Zhuang et al. (2022) research is as follows:

$$FSs_{i,t} = f\left(\beta_0, LdU_{i,t}^{\beta 1}, EgT_{i,t}^{\beta 2}, AVA_{i,t}^{\beta 3}, GDP_{i,t}^{\beta 4}, Urb_{i,t}^{\beta 5}\right) \tag{24.1}$$

where;

FSs Food supply
LdU Land utilization
EgT Energy transition
AVA Agricultural products value-added
GDP Gross domestic products
Urb Urbanization
i Figure of cross-sections
t Time in years

It is possible to convert Eq. (24.1) into the double-log model, which is expressed as follows:

Table 24.1 Description of the variables

Variables	Variable code	Indicators	Source of data	Definition	Link
Food supply	FSs	Per capita food supply variability (kcal/cap/day)	FAOSTAT	Food supply is the amount of food commodities entering the market, providing information on changes in food consumption over time and estimated nutrient content. This metric monitors the fluctuations in the average daily caloric supply accessible to individuals over a specific period, serving as an indicator of the stability and reliability of food availability in a region	https://www.fao.org/faostat/en/#data
Land utilization	LdU	Agricultural land (km^2)	WDI	Agricultural land is the share of land area that is arable, under permanent crops, and permanent pastures. Land abandoned as a result of shifting cultivation is excluded	https://databank.worldbank.org/source/world-development-indicators
Energy transition	EgT	Renewable energy consumption (% of total final energy consumption)	WDI	Renewable energy consumption is the share of renewable energy in total final energy consumption. Energy transition is noncarbohydrate energy that does not produce carbon dioxide when generated. It includes hydropower and nuclear, geothermal, and solar power, among others	https://databank.worldbank.org/source/world-development-indicators

(continued)

Table 24.1 (continued)

Variables	Variable code	Indicators	Source of data	Definition	Link
Agricultural products value-added	AVA	Agriculture, forestry, and fishing, value added (constant 2015 US$)	WDI	Agricultural products value-added include forestry, hunting, and fishing, cultivation of crops, and livestock production. Value added is the net output of a sector after adding up all outputs and subtracting intermediate inputs. Data are in constant 2015 prices, expressed in U.S. dollars	https://databank.worldbank.org/source/world-development-indicators
Gross domestic products	GDP	GDP per capita (current US$)	WDI	GDP per capita is gross domestic product divided by midyear population. GDP is the sum of gross value added by all resident producers in the economy plus any product taxes and minus any subsidies not included in the value of the products. Data are in current U.S. dollars	https://databank.worldbank.org/source/world-development-indicators
Urbanization	Urb	Urban population	WDI	Urban population refers to people living in urban areas. It is calculated using World Bank population estimates and urban ratios from the United Nations World Urbanization Prospects	https://databank.worldbank.org/source/world-development-indicators

$$LnFSs_{i,t} = \beta_0 + \beta_1 LnLdU_{i,t} + \beta_2 LnEgt_{i,t} + \beta_3 AVA_{i,t} + \beta_4 GDP_{i,t} + \beta_5 Urb_{i,t} + \varepsilon_{i,t} \quad (24.2)$$

where;

Ln Natural logarithms
Ɛ error term

The present study anticipates a negative relationship between land utilization and food supply. This is because it makes sense that using agricultural land for infrastructure development and building construction would reduce the availability of food. Thus, the coefficient of land utilisation is predicted to be less than zero (β <

0). Equally, the energy transformation has the potential to enhance the condition and quality of food supply. This is predicated on the idea that traditional energy sources reduce productivity since they are extremely expensive, whereas green energy sources may result in lower total production costs and increased food supply. This research believes the coefficient of energy transition to be positive (β > 0) because of the positive effect that energy transition plays in food production. A further significant factor influencing the amount of food available is agricultural value added, which is predicted to raise the amount of food available. According to this research, the coefficient value is expected to be positive (β > 0). Nonetheless, the GDP likewise considers the expansion of the food supply, and the coefficient value is attributed to be positive (β > 0). In conclusion, the food supply may be reduced as a result of urbanization, and the coefficient value is attributed to be lesser than zero (β < 0). Similar to how the major influence of the chosen variables might not respond effectively, this research also attempts to look into the energy transition's adequate function on the GDP, land utilization, and agricultural products value-added, as well as how these factors affect the panel's food supply.

As employed by Pedroni (2000), the Fully Modified Ordinary Least Squares (FMOLS) estimator can be expressed as follows:

$$\hat{\beta}FMOLS = \left(\sum_{t=1}^{T} X_t' P_t X_t\right)^{-1} \left(\sum_{t=1}^{T} X_{t=1}' P_t Y_t\right) \tag{24.3}$$

where;

$\hat{\beta}FMOLS$	Vector of estimated coefficients using the FMOLS estimator.
X_t	Matrix of independent variables for time period t.
P_t	Matrix of weighting coefficients for time period t to address serial correlation.
Y_t	Vector of dependent variable(s) for time period t.
T	Number of time periods in the panel data.

The Granger causality test as used by Wang et al. (2022) and adopted is estimated as:

$$X_t = \alpha_o + \sum_{j=1}^{k} \alpha 1sXt - s + \sum_{j=1}^{k} \alpha 2sXt - m + \varepsilon_{1t} \tag{24.4}$$

$$Y_t = \beta_o + \sum_{j=1}^{k} \beta 1sXt - j + \sum_{j=1}^{k} \beta 2sXt - h + \varepsilon_{1t} \tag{24.5}$$

It can be inferred from Eqs. (24.4) and (24.5) that there is no correlation between 1 and 2t. Should the calculated coefficient α2i be statistically significant, then Y Granger causes in X if α2i ≠ 0. Only when there is a relationship in the coefficient amidst the two variables that is greater than zero, or when c2h ≠ 0, is β2h statistically

Table 24.2 Summary of descriptive statistics

Variables	Mean	Median	Maximum	Minimum	Std. Dev.	Sum	Prob.
LFSS	3.347	3.480	4.277	1.792	0.712	73.635	0.357
LLDU	13.416	13.416	13.439	13.393	0.014	295.152	0.515
LEGT	6.607	6.608	6.666	6.552	0.030	99.110	0.949
LAVA	25.065	25.159	25.506	24.153	0.392	551.434	0.177
LGDP	7.694	7.783	7.893	7.288	0.184	169.276	0.205
LURB	18.098	18.112	18.577	17.572	0.313	416.243	0.481

significant. The importance of 2i and 2h indicates a relationship between two definite variables. In the event where both β2h and α2i are zero, Y and X variables would be autonomous terms.

24.4 Results and Discussion

24.4.1 Summary Statistics

The summary of descriptive statistics is shown in Table 24.2. For the chosen panel data, there is no discernible variance among the values of mean and median. Regarding the results, the provided data does not contain any outliers. By way of measuring the mean against the central tendency, using standard deviation, which indicates how wide the mean intercept is between the observations. The average divergence for every result from the mean is shown by the standard deviation. In a normal distribution, a small standard deviation indicates a distribution of observations close to the mean, while an excessive standard deviation indicates an overall departure of results from the mean. This means that the average observation may not be close to most of the observations if the standard deviation is large. Hence, all the variables clustered around the mean.

24.4.2 Stationarity Test

Before conducting any additional analysis, the order of integration of each variable was ascertained using unit root tests. Two separate unit root tests were conducted, as Table 24.3 illustrates the results for each variable for the unit root testing of the ADF and PP systems. Every test is conducted for both the starting point and the variable level. The null hypothesis, which states that variables assume both individual and common unit root processes, cannot be rejected in the case of variables. However,

Table 24.3 Stationarity unit root test

Variables	Unit root test for ADF		Unit root test for PP	
	Level	First difference	Level	First difference
LFSS	0.449 [0.508]	3.962*** [0.000]	0.356 [0.544]	4.630*** [0.000]
LLDU	11.328 [0.999]	11.868*** [0.000]	212.020 [0.999]	13.098*** [0.000]
LEGT	2.275 [0.192]	3.577*** [0.002]	0.827 [0.879]	3.585*** [0.002]
LAVA	6.019 [0.999]	2.982*** [0.005]	3.251 [0.999]	2.900*** [0.006]
LGDP	0.249 [0.748]	1.939** [0.052]	1.943 [0.984]	5.353*** [0.002]
LURB	0.804 [0.354]	1.927** [0.054]	26.078 [0.999]	3.389*** [0.002]

Note PP is for Phillips-Perron; *ADF* stands for Augmented Dickey-Fuller; significant levels are indicated by ***, **, and *, at the 1%, 5%, and 10% levels respectively

all the variables meet the conditions of the study when the first difference is applied. We can thus accept that they are stationary for the 99% confidence interval. Unit root testing is crucial, according to Roquez-Diaz and Escot (2018), since if variables are not integrated, regression may be erroneous and defective. The findings demonstrated that at the first difference, all pertinent sustainable growth metrics are stagnant. As a result, the integrated variables are all of the first difference order I(1).

24.4.3 Pairwise Correlation Between the Dependent and Independent Variables

The pairwise correlation values are displayed in Table 24.4. The findings demonstrate a strong positive relationship between food security and supply (LFSS) and land utilization (LLDU), energy transition (LEGT), agricultural product value added (LAVA), gross domestic product (LGDP), and urbanization (LURB). Similarly, energy transition (LEGT) has a very durable positive relationship with land utilization (LLDU), and a very durable positive relationship with gross domestic product (LGDP), urbanization (LURB), and agricultural value added (LAVA). It also has a positive correlation with the explained variable (LFSS). As anticipated, LFSS has a positive correlation with each of the sustainable growth metrics.

Table 24.4 Pairwise correlation test

Variables	LFSS	LLDU	LEGT	LAVA	LGDP	LURB
LFSS	1					
LLDU	0.725	1				
LEGT	0.648	0.735	1			
LAVA	0.727	0.947	0.748	1		
LGDP	0.710	0.988	0.743	0.982	1	
LURB	0.726	0.999	0.735	0.949	0.989	1

24.4.4 Determinants of Food Security

Table 24.5 displays the long-run co-integration vector using time-series FMOLS techniques amongst food supply and the aforementioned determining factor. A time-series data model estimate is used in the study to take deterministic inclinations and coefficients into consideration. To examine how land utilization (LLDU), energy transition (LEGT), agricultural products value added (LAVA), gross domestic product (LGDP), and urbanization (LURB) affect food security in Nigeria's time-series FMOLS estimator for all the economies that were chosen (FSS). The R-squared (R^2) of 0.816 is considered a good fit, indicating a large proportion of about 81.6% variance in the dependent variable is explained by the independent variables in the model.

Land utilization (LDU) is acknowledged as the first element of food security as well as reveals an inverse relationship with food supply (FSS). From the result of FMOLS in Table 24.5; a 1% increase in land utilization would result in a drop in the food supply by − 1439.20. Land utilization, conversely, is strongly knotted to agricultural and food systems and is seen as an essential component of the food supply (Stein, 2014; Zhuang et al., 2022). Hence, the land utilization analysis could result in a drop in output and its utilization in unfertile farmlands. Comparable, it could be a result of insecurity in some parts of Nigeria, though the land is productive. Land

Table 24.5 Fully modified least squares (FMOLS)

Variable	Coefficient	Standard Error	t-Statistic	Probability
Constant	17,833.22***	5059.374	3.525	0.008
LLDU	− 1439.200***	402.851	− 3.573	0.007
LEGT	2.087	1.675	1.246	0.248
LAVA	6.865***	0.883	7.776	0.000
LGDP	− 33.112***	3.679	− 9.001	0.000
LURB	85.705***	19.780	4.333	0.003
R-squared (R^2)	0.816	Adjusted R-squared (R^2)		0.701

Note Significant at 1%, 5%, and 10% are indicated by ***, **, and * respectively

utilization adjustments are significantly influenced by factors such as the growth in food demand and other political and economic factors. However, industrial and urban growth could be another factor accountable for a substantial loss in a high proportion of land utilization for agricultural purposes.

A one percent increase in agricultural value added (LAVA) would result in a 6.87% rise in food security since LAVA has a positive relationship with food supply. Refrigerators quickly became commonplace in restaurants and grocery stores because cold storage made it possible to preserve fresh meals while reducing food losses and waste safely. This upward trend may be the result of these two factors, the expansion of the agricultural sector and the adoption of new technology for home usage in food preservation. Moreover, according to Zhuang et al. (2022), it can be the consequence of the green revolution's adoption by several countries, including Nigeria, in the middle of the twentieth century to increase agricultural production. The phrase "green revolution" describes a broad campaign that was launched in the 1960s to encourage more ecologically friendly farming methods to boost the food supply in developing countries. Two essential strategies to accomplish this goal were to use higher-yielding plant varieties and to develop "friendlier" fertilizers. Industrial agriculture, which depends on scientific advancements for both expansion and survival, is one example of a science-based agricultural practice that may be responsible for this growing tendency, a nationwide transportation network, adequate machine technology, dependable and extensive irrigation systems, and the use of pesticides and synthetic fertilizers to improve crop productivity. Among other benefits, using this innovative agricultural method may result in increased crop yields, more productivity, less expensive food, profitability for large corporations, and increased export potential.

The findings show that the gross domestic product (LGDP) has a negative impact on food security in this model. This is particularly clear from Table 24.5, where a 1% increase in GDP would cause the food supply to decrease by $-$ 33.11 in the model, contrary to the previous expectation. The findings show that the gross domestic product (LGDP) has a negative impact on food security in this model. It can indicate that economic growth in the context of the model is not translating into improvements in access to food or equitable distribution of resources. This might happen due to factors like income inequality, poor agricultural investment, policy challenges, or market failures. This finding agrees with Timmer (2005) who reported that economic growth and food security, if concentrated in non-agricultural sectors, can negatively impact food security. He emphasizes that growth that excludes agricultural investment and rural development may increase food insecurity in poorer segments of the population. The finding also agrees with Cornia (2017) who shows that high levels of inequality in growing economies often lead to worsening food insecurity for poorer populations. As the wealthy benefit disproportionately from growth, the poor may face increasing food prices and reduced access to adequate nutrition. However, the finding is not consistent with Olorunmola et al. (2024) who reported that food security has a positive impact on economic growth and suggests that food availability, food accessibility, food utilization, and a stable inflation rate must combine to achieve economic growth. Therefore, it is essential to focus on inclusive growth, with an emphasis on agriculture, rural development, and strong social safety nets to ensure that economic growth benefits all segments of the population. This highlights the need for measures to increase food availability to achieve one of the 17 Sustainable Development Goals (SDGs) that the United Nations (UN) proposed in 2015 to end hunger in Nigeria. These measures could be taken by establishing an environment that encourages farmers to adopt and use new technologies that have lower costs but higher productivity. Therefore, higher income levels may increase food consumption, assisting Nigeria in achieving better economic growth and food security. The GDP may rise if consumers can buy enough food at home or in foreign markets. If a sizeable percentage of people in urban, semi-urban, and rural areas were involved in farming, the impact would be greater.

Urbanization (LURB) has a positive coefficient with food supply, an increase in LURB factor of 1% would result in an increase in food security by 85.71, which opposed the prior expectation. This implies that urban dwellers also engaged in product agricultural practices despite growing in urbanization. This finding agrees with Chengsheng et al. (2016) who reported that urbanization and food security have experienced steady development, yet there has not been an equivalent rise in their relationship. However, energy transition (LEGT) is not significant, but the coefficient (2.09) shows a positive relationship with food supply.

24.4.5 Granger Causality Test

There is no short-run causal relationship between the independent and dependent variables, according to the null hypothesis. While the alternative hypothesis stated that, there is a short-term causal relationship between the independent and dependent variables. The rule of thumb is to reject the null hypothesis and infer that there is causation if the probability is less than 0.05 ($p < 0.05$); if not, consider that there is no short-run causality (Maiti, 2020). This is employed to show the direction of causality. Granger causality thus examines whether a link is bidirectional, if there is no meaningful relationship at all, or whether one variable in a linear relation can be implicitly designated as a dependent variable and the other as an independent variable. Rejecting the null hypothesis shows that one variable, in reality, granger causes the other while accepting the null hypothesis concludes that there is no causality between variables at either 0.01, 0.05, or 0.10 significant level. FMOLS model produces a very notable results, on the other hand, the causal influence that each variable exerts on its own is not explicable by this model. Consequently, the granger causality test is used in this research to examine the causal relationship that exists between the variables. The results of pairwise granger causality tests at lag 2 is shown in Table 24.6. The remarks indicate the conclusion of the null hypothesis; accepted means there is causality going between the variables while rejected means there is no causality going from one variable to the other.

Table 24.6 Pairwise granger causality tests at Lag 2

S/N	Null Hypothesis (Ho)	F- Stat.	Prob.	Remarks
1	LLDU does not grant evidence LFSS	0.591	0.566	Accepted
	LFSS does not grant evidence LLDU	0.027	0.974	Accepted
2	LEGT does not grant evidence LFSS	3.750	0.071	Rejected
	LFSS does not grant evidence LEGT	0.163	0.852	Accepted
3	LAVA does not grant evidence LFSS	0.213	0.810	Accepted
	LFSS does not grant evidence LAVA	0.515	0.608	Accepted
4	LGDP does not grant evidence LFSS	0.339	0.718	Accepted
	LFSS does not grant evidence LGDP	2.821	0.091	Rejected
5	LURB does not grant evidence LFSS	4.025	0.040	Rejected
	LFSS does not grant evidence LURB	1.247	0.316	Accepted
6	LEGT does not grant evidence LLDU	0.166	0.850	Accepted
	LLDU does not grant evidence LEGT	2.837	0.117	Accepted
7	LAVA does not grant evidence LLDU	0.587	0.568	Accepted
	LLDU does not grant evidence LAVA	20.529	0.000	Rejected
8	LGDP does not grant evidence LLDU	0.033	0.968	Accepted
	LLDU does not grant evidence LGDP	3.601	0.053	Rejected
9	LURB does not grant evidence LLDU	0.138	0.873	Accepted
	LLDU does not grant evidence LURB	0.131	0.878	Accepted
10	LAVA does not grant evidence LEGT	1.316	0.321	Accepted
	LEGT does not grant evidence LAVA	0.005	0.995	Accepted
11	LGDP does not grant evidence LEGT	1.850	0.219	Accepted
	LEGT does not grant evidence LGDP	0.830	0.471	Accepted
12	LURB does not grant evidence LEGT	1.224	0.344	Accepted
	LEGT does not grant evidence LURB	8.760	0.010	Rejected
13	LGDP does not grant evidence LAVA	5.420	0.017	Rejected
	LAVA does not grant evidence LGDP	2.836	0.090	Accepted
14	LURB does not grant evidence LAVA	80.963	0.000	Rejected
	LAVA does not grant evidence LURB	0.256	0.778	Accepted
15	LURB does not grant evidence LGDP	9.566	0.002	Rejected
	LGDP does not grant evidence LURB	1.202	0.328	Accepted

24.5 Conclusions, Policy Recommendations and Limitations

To ensure that food security is not compromised during the energy transition, Nigeria must adopt policies that promote sustainable land use practices, protect smallholder farmers, and enhance food supply chains. Investment in agricultural technology, infrastructure, and social safety nets will also be essential to mitigate the risks posed by the land-energy competition. A balanced and holistic approach is vital to ensuring that energy development contributes to, rather than detracts from, the nation's food security objectives. Underinvestment in agriculture, further complicates the ability to balance energy development with food security needs. While transitioning to sustainable energy is critical for environmental and economic growth, this shift must be accompanied by integrated and well-coordinated policies that prioritize both energy development and agricultural productivity.

The lack of consideration data on inflation and types of renewable energy such as hydropower and nuclear, wind, ocean, bioenergy (biomass), geothermal, and solar power in this research serve as the limitation. The research therefore recommends the following policy:

i. Incorporate sustainable agriculture and food security into national policy and create targeted initiatives and plans for energy transition and economic growth for those who are most at risk from environmental shifts and food insecurity.
ii. Establish plans and methods for reducing land degradation and restoring deteriorated areas, with a focus on community-designed initiatives.
iii. Provide underprivileged farmers with the tools they need to boost the yield of a variety of suitable crops by defending their rights to land and water, expanding their access to markets, financing, and insurance, and building local capacity through farmer and community-based organizations.
iv. Create comprehensive laws and initiatives that minimize waste in food supply chains. Some examples include requiring distributors to separate and minimize food waste and using economic innovation to allow low-income producers to stockpile food during times of excess supply.

Acknowledgements The author is very grateful for the availability of data from WDI and FAOSTAT databanks.

Funding The author received no funds for this research.

Data Availability Statement The datasets used and/or analysed during the current study are available on WDI and FAOSTAT database sites.

Conflict of Interest The author declared that there is no conflict of interest.

References

Abd-Elmabod, S. K., Muñoz-Rojas, M., Jordán, A., Anaya-Romero, M., Phillips, J. D., & Jones, L. (2020). Climate change impacts on agricultural suitability and yield reduction in a Mediterranean region. *Geoderma, 374*, 114453. https://doi.org/10.1016/j.geoderma.2020.114453

Adil, A. M., & Ko, Y. (2016). Socio-technical evolution of decentralized energy systems: A critical review and implications for urban planning and policy. *Renewable and Sustainable Energy Reviews, 57*, 1025–1037. https://doi.org/10.1016/j.rser.2015.12.079

Ajjur, S. B., & Al-Ghamdi, S. G. (2022). Towards sustainable energy, water and food security in Qatar under climate change and anthropogenic stresses. *Energy Reports, 8*, 514–518. https://doi.org/10.1016/j.egyr.2022.02.099

Baker, L., Newell, P., & Phillips, J. (2014). The political economy of energy transitions: The case of South Africa. *New Political Economy, 19*, 791–818. https://doi.org/10.1080/13563467.2013.849674

Black, R. E., Allen, L. H., Bhutta, Z. A., Caulfield, L. E., De Onis, M., & Ezzati, M. (2008). Maternal and child undernutrition: Global and regional exposures and health consequences. *Lancet, 371*, 243–260. https://doi.org/10.1016/S0140-6736(07)61690-0

Chengsheng, Y., Yufei, Q., & Lin, H. (2016). Coupling relationship between urbanization and food security in China: An empirical study. *China Soft Science, 8*, 75–88. https://doi.org/10.1016/j.jclepro.2021.129957

Connolly-Boutin, L., & Smit, B. (2015). Climate change, food security, and livelihoods in sub-Saharan Africa. *Regional Environmental Change, 16*, 385–399. https://doi.org/10.1007/s10113-015-0761-x

Cornia, G. A. (2017). Income inequality trends in sub-Saharan Africa: Divergence, determinants, and consequences: Inequality levels, trends and determinants in sub-Saharan Africa: An overview of main changes since the early. UNDP Africa Reports 267642, United Nations Development Programme (UNDP). https://doi.org/10.22004/ag.econ.267642

Galgóczi, B. (2020). Just transition on the ground: Challenges and opportunities for social dialogue. *European Journal of Industrial Relations, 26*, 367–382. https://doi.org/10.1177/0959680120951704

Goldthau, A., & Hughes, L. (2020). *Protect global supply chains for low-carbon technologies.* Nature Publishing Group.

Hall, P.A., Lowe, C., & Vincent, C. (2017). Brain stimulation effects on food cravings and consumption: An update on Lowe et al. (2017) and a response to Generoso et al. (2017). *Psychosomatic Medicine, 79*, 839–842. https://doi.org/10.1097/PSY.0000000000000504

Hassan, W., Li, Y., Saba, T., Jabbi, F., Wang, B., & Cai, A. (2022). Improved and sustainable agroecosystem, food security and environmental resilience through zero tillage with emphasis on soils of temperate and subtropical climate regions: A review. *International Soil and Water Conservation Research, 10*, 530–545. https://doi.org/10.1016/j.iswcr.2022.01.005

Headey, D., Goudet, S., Lambrecht, I., Maffioli, E. M., Oo, T. Z., & Russell, T. (2022). Poverty and food insecurity during COVID-19: Phone-survey evidence from rural and urban Myanmar in 2020. *Global Food Security, 33*, 100626. https://doi.org/10.1016/j.gfs.2022.100626

Heffron, R., Halbrügge, S., Körner, M.-F., Obeng-Darko, N. A., Sumarno, T., & Wagner, J. (2021). Justice in solar energy development. *Solar Energy, 218*, 68–75. https://doi.org/10.1016/j.solener.2021.01.072

Imteyaz, B., & Tahir, F. (2019). Thermodynamic analysis of premixed and non-premixed oxy-methane combustion cycle with membrane-assisted oxygen separation. In *Paper Presented at the 8th Global Conference on Global Warming (GCGW)*, Doha, Qatar.

Irena, O., & Desa, U. (2019). *A new world: The geopolitics of the energy transformation.* Bonn.

Kim, J. E. (2019). Sustainable energy transition in developing countries: The role of energy aid donors. *Climate Policy, 19*, 1–16. https://doi.org/10.1080/14693062.2018.1444576

Kumar, A., Ahmad, M. M., & Sharma, P. (2017). Influence of climatic and non-climatic factors on sustainable food security in India: A statistical investigation. *International Journal of Sustainable Agricultural Management and Informatics, 3*, 1–30. https://doi.org/10.1504/IJSAMI.2017.082917

Li, H. X., Edwards, D. J., Hosseini, M. R., & Costin, G. P. (2020). A review on renewable energy transition in Australia: An updated depiction. *Journal of Cleaner Production, 242*, 118475. https://doi.org/10.1016/j.jclepro.2019.118475

Luciani, G. (2020). The impacts of the energy transition on growth and income distribution. In M. Hafner & S. Tagliapietra (Eds.), *The geopolitics of the global energy transition* (pp. 305–318). Cham: Springer.

Maiti, A. (2020). Fun with ARMA, VAR, and Granger Causality. Using Stata, R, and Python. Retrieved from https://towardsdatascience.com/fun-with-arma-var-and-granger-causality-6fdd29d8391c

Naqvi, S. A. A., Shah, S. A. R., Anwar, S., & Raza, H. (2021). Renewable energy, economic development, and ecological footprint nexus: Fresh evidence of renewable energy environment Kuznets curve (RKC) from income groups. *Environmental Science and Pollution Research, 28*, 2031–2051. https://doi.org/10.1007/s11356-020-10485-w

Olorunmola, D. R., Eggon, H. A., Ajidani, M. S., Ibbih, J. M., & Odonye, J.O. (2024). Impact of food security on economic growth in Nigeria. *Lafia Journal of Economics and Management Sciences, 9*(1), 137–152. Retrieved from https://lajems.com/index.php/lajems/article/download/291/222/

Panwar, N., Kaushik, S., & Kothari, S. (2011). Role of renewable energy sources in environmental protection: A review. *Renewable and Sustainable Energy Reviews, 15*, 1513–1524. https://doi.org/10.1016/j.rser.2010.11.037

Pedroni, P. (2000). Fully modified OLS for heterogeneous cointegrated panels. In L. Matyas & P. Sevestre (Eds.), *The econometrics of panel data: Fundamentals and recent developments in theory and practice* (pp. 93–130). https://doi.org/10.1007/978-94-015-9482-7_4

Pinstrup-Andersen, P. (2009). Food security: Definition and measurement. *Food Security, 1*, 5–7. https://doi.org/10.1007/s12571-008-0002-y

Qadir, S. A., Al-Motairi, H., Tahir, F., & Al-Fagih, L. (2021). Incentives and strategies for financing the renewable energy transition: A review. *Energy Reports, 7*, 3590–3606. https://doi.org/10.1016/j.egyr.2021.06.041

Richardson, R. A., Harper, S., Weichenthal, S., Nandi, A., Mishra, V., & Jha, P. (2020). Extremes in water availability and suicide: Evidence from a nationally representative sample of rural Indian adults. *Environmental Research, 190*, 109969. https://doi.org/10.1016/j.envres.2020.109969

Roquez-Diaz, A., & Escot, L. (2018). Relationship between trade openness and economic growth in Latin America: A causality analysis with heterogeneous panel data. *Review of Development Economics, 22*, 658–684. https://doi.org/10.1111/rode.12358

Ruben, R., Verhagen, J., & Plaisier, C. (2018). The challenge of food systems research: What difference does it make. *Towards Sustainable Global Food Systems, 11*, 171. https://doi.org/10.3390/su11010171

Sa, J. C., Lal, R., Cerri, C. C., Lorenz, K., Hungria, M., & de Faccio Carvalho, P. C. (2017). Low-carbon agriculture in South America to mitigate global climate change and advance food security. *Environment International, 98*, 102–112. https://doi.org/10.1016/j.envint.2016.10.020

Shah, S. A. R., Naqvi, S. A. A., Riaz, S., Anwar, S., & Abbas, N. (2020). Nexus of biomass energy, key determinants of economic development and environment: A fresh evidence from Asia. *Renewable and Sustainable Energy Reviews, 133*, 110244. https://doi.org/10.1016/j.rser.2020.110244

Stein, A. J. (2014). Rethinking the measurement of undernutrition in a broader health context: Should we look at possible causes or actual effects? *Global Food Security, 3*, 193–199. https://doi.org/10.1016/j.gfs.2014.09.003

Strapasson, A., Woods, J., Chum, H., Kalas, N., Shah, N., & Rosillo-Calle, F. (2017). On the global limits of bioenergy and land use for climate change mitigation. *GCB Bioenergy, 9*, 1721–1735. https://doi.org/10.1111/gcbb.12456

Timmer, C. P. (2005). Food security and economic growth: An Asian perspective. *Asian-Pacific Economic Literature, 18*(1), 1–17. https://doi.org/10.1111/j.1467-8411.2005.00155.x

United Nations. (2015). Transforming our world: The 2030 agenda for sustainable development A/RES/70/1. Retrieved from https://sustainabledevelopment.un.org/content/documents/21252030%20Agenda%20for%20Sustainable%20Development%20web.pdf

Wang, Y., Liu, M., Wang, S., Cui, X., Hao, L., & Gen, H. (2022). Assessing the impact of governance and health expenditures on carbon emissions in China: Role of environmental regulation. *Frontiers in Public Health, 10*, 949729. https://doi.org/10.3389/fpubh.2022.949729

Yoon, P. R., Lee, S.-H., Choi, J.-Y., Yoo, S.-H., & Hur, S.-O. (2022). Analysis of climate change impact on resource intensity and carbon emissions in protected farming systems using water-energy-food-carbon nexus. *Resources, Conservation and Recycling, 184*, 106394. https://doi.org/10.1016/j.resconrec.2022.106394

Zhang, Q., Shah, S. A. R., & Yang, L. (2022). An appreciated response of disaggregated energies consumption towards the sustainable growth: A debate on G-10 economies. *Energy, 254*, 124377. https://doi.org/10.1016/j.energy.2022.124377

Zhuang, D., Abbas, J., Al-Sulaiti, K., Fahlevi, M., Aljuaid, M., & Saniuk, S. (2022). Land-use and food security in energy transition: Role of food supply. *Frontier in Sustainable Food Systems, 6*, 1053031. https://doi.org/10.3389/fsufs.2022.1053031

Chapter 25
Willingness to Pay for Flood Insurance by Cassava-Based Farmers in Ikwerre Local Government Area Rivers State, Nigeria

Adanna Henri-Ukoha, Mercy Ebere Ndubueze-Ogaraku, Sochima Excel Obi, and Adeyinka Richard Aroyehun

Abstract This study investigates the willingness to pay (WTP) for flood insurance among cassava-based farmers in the Ikwerre Local Government Area (LGA) of Rivers State, Nigeria. The area, prone to seasonal flooding, presents significant risks to agricultural productivity, particularly for cassava farming, a primary livelihood source. Using descriptive and binary logit regression models with the aid of questionnaires and interview schedules, data were collected from 64 cassava farmers randomly selected across different communities within the LGA. The results show the majority (68.7%) of the cassava-based farmers were female, with a mean age of 41.7 years, an educational level mean of 9.8 years, a mean household size of 6 members, and 16.1 years as the mean of farming experience. The majority (71.9%) of the cassava-based farmers were willing to pay for flood insurance. The Logit regression model result shows that educational level and farming experience were statistically significant at 1% and 5% respectively determining the willingness of cassava-based farmers to pay for flood insurance. The major constraints are lack of awareness, affordability, limited access to credit, and institutional barriers. The study recommendation policy includes designing inclusive agricultural insurance products that improve farmers' resilience against climate-induced flooding, targeted subsidy programmes to make premiums more affordable, increased awareness campaigns on the benefits of flood insurance, and the introduction of flexible insurance schemes

A. Henri-Ukoha · M. E. Ndubueze-Ogaraku · S. E. Obi · A. R. Aroyehun (✉)
Agricultural Economics and Agribusiness Management Department, Faculty of Agriculture, University of Port Harcourt, Port Harcourt, Nigeria
e-mail: aroyehun_adeyinka@uniport.edu.ng

A. Henri-Ukoha
e-mail: adanna.henri-ukoha@uniport.edu.ng

M. E. Ndubueze-Ogaraku
e-mail: mercy.onu@uniport.edu.ng

S. E. Obi
e-mail: obisochima48@gmail.com

R. U. Onyeneke et al. (eds.), *Energy Transition, Climate Action and Sustainable Agriculture*, https://doi.org/10.1007/978-3-031-83165-2_25

tailored to the needs and capacities of small-scale farmers. These measures could enhance adoption rates and contribute to more sustainable cassava farming in the flood-prone area.

Keywords Adaptation · Cassava production · Flood · Insurance · Logit regression · Premium

25.1 Introduction

Rainfall, particularly in tropical areas, is an essential component of agricultural productivity. Research has indicated that crop output and agricultural productivity are impacted by variations in rainfall and shifts in climatic patterns (Ogunniyi et al., 2018). Nigeria has a large number of flood-prone areas, which can seriously harm infrastructure and agriculture (Mfon et al., 2022). Extreme weather events like floods and increasing rainfall variability have been linked to climate change. According to climate forecasts, Rivers State will have more intense and frequent rainfall, which would increase the frequency of flooding episodes (Egor et al., 2015). Hence, flooding can cause crop damage, economic loss, and farm and house displacement for cassava producers. The most frequent natural catastrophe in the world is flooding. Climate change has increased the frequency and severity of floods throughout time. In developing nations, agriculture and rural livelihoods are severely threatened by flooding. Extreme precipitation events can lead to floods, one of the most frequent and destructive natural catastrophes. Floods can inflict substantial harm to the environment, the economy, and society (Intergovernmental Panel on Climate Change, 2014). Floods have been linked to crop devastation, infrastructure loss, soil erosion, and human fatalities. Nigerian floods have harmed agricultural output.

Nigeria is an agriculturally based nation, and the country's economy is heavily impacted by climate change. The country has not been able to meet its food needs due to an increase in extreme weather events, such as floods. Floods have severely impacted Nigeria's production, causing crop damage, agricultural land losses, as well as loss of revenue for local farmers (Mfon et al., 2022). Adaptation strategies, like insurance against floods, could help mitigate the adverse impacts of climate variability on farming in the area (Aroyehun, 2023). Therefore, insurance is vital to transfer such losses to third parties.

An insurance policy is a contract in which two parties agree that, in exchange for the payment of a premium, one would provide coverage in the case of a defined event. Insurance offers a safety net to help people or businesses recover from unanticipated losses (Aroyehun, 2023). The ideas of risk management and financial protection are central to the insurance concept. A contract for insurance is a legal agreement between a person or organization (the policyholder) and an insurance provider (the insurer), under which the insurer agrees to pay monthly premiums in exchange for providing financial protection against particular risks (Maheshwari, 2023). The idea of insurance is to combine the resources of many policyholders. The insurer aggregates the premium payments from each policyholder to form an insurance pool, which is a fund. Subsequently, this fund is utilized to compensate policyholders for losses sustained during insured occurrences.

The insurance provider reimburses the policyholder for their monetary losses in the case of a covered loss. Usually, this compensation comes in the form of money returned, damaged property repaired or replaced, or the provision of services that are required and specified in the insurance policy (Life and General Insurance, 2023; Linnerooth-Bayer et al., 2019). Actuarial science is a tool used by insurance firms to estimate reserves, set rates, and evaluate risks. Actuaries assess the possibility and possible expense of various risks by examining statistical information, past loss trends, and other pertinent variables. This guarantees that insurance providers may continue to be financially stable while covering possible losses (Aroyehun, 2023). Providing people, corporations, and other entities with financial security and peace of mind is the main objective of insurance. It lessens the detrimental financial effects that unforeseen circumstances like accidents, sickness, property damage, or legal liabilities may have (Life and General Insurance, 2023). Insurance is therefore crucial to agriculture's efforts to reduce the risks brought on by unfavourable weather events.

Nigeria is not an exception to the major difficulties posed by climate change to nations worldwide. Nigeria is especially vulnerable to the effects of climate change, including a rise in the frequency and intensity of floods, due to its geographic position and poor ability to adapt (Egor et al., 2015; Mfon et al., 2022; Okon et al., 2021). Significant economic and social upheavals brought about by flooding have resulted in property and infrastructure devastation, fatalities, and relocation. In response to these difficulties, flood insurance has become a viable tactic for Nigerians to adapt to climate change and lessen the negative social and economic effects of flooding. This study examines flood insurance's usefulness and applicability in Nigeria as a tactic for coping with climate change. One possible tactic for farmers in flood-prone areas to adapt to climate change is flood insurance. The idea of flood insurance is to reimburse farmers for losses brought on by flooding. Among the earlier research on the effects of floods on agriculture are works by Ogunniyi et al. (2018). Previous research on insurance and farmers' willingness to pay for it has been done including Ayinde et al. (2016), and Adeoye et al. (2020) among others. Nevertheless, none of this research looked at the cassava growers' willingness to pay for flood insurance in the Ikwerre Local Government Area. A deficiency of understanding regarding farm insurance may provide obstacles to the creation and execution of successful flood insurance plans and initiatives. Therefore, it is very essential to examine empirically the willingness of cassava-based farmers in Ikwerre Local Government Area Rivers State, Nigeria to pay for flood insurance as an adaptation strategy to climate change. The specific objectives are to describe the socioeconomic characteristics of the cassava-based farmers in the study area; aascertain the willingness of cassava-based farmers to pay for flood insurance; determine the factors that influence cassava-based farmers' willingness to pay for flood insurance; and identify the challenges that cassava-based farmers face in accessing flood insurance.

25.2 Literature

Tropical rainfall is crucial because it has the potential to either save or end human life, depending on whether there is too much rain and floods or not enough rain and drought. Tropical rainfall is significant for the world's climate and weather in addition to its importance to human existence. Tropical regions get more than two thirds of the world's precipitation. This leads to a significant release of energy in the latent heat form in the low latitudes. In addition to counteracting radiation heat losses, the energy powers the atmosphere's worldwide circulation. Thus, a grasp of the regional distribution of rainfall is a prerequisite to understanding the worldwide distribution of heat sources that drive the globe's atmospheric heat regime. The most erratic factor in tropical climes is most likely the amount of rainfall. The annual total varies from location to location and from year to year. The distinctive seasonality and cyclical distribution, extent, length, and frequency of rainy days also serve as indicators of its temporal and geographical variability (Egor et al., 2015). Appropriately constructed insurance mechanisms can serve as a risk management inducement as the premium ought to decrease if the policyholder takes steps to reduce possible effects. This means that rates for flood insurance must be risk-based, meaning that they must fluctuate in proportion to the risk involved. However, flood insurance has to be purchased, and insurers need to be willing to supply the policy for flood insurance to offer this method of managing flood risk. For insurance plans, it is crucial to comprehend the structure of the willingness to pay (WTP) distribution. This is due to the requirement that rates charged and consumer willingness to pay overlap sufficiently for the development of a flood insurance market (Roder et al., 2019).

In several of the federation's states, including Rivers, severe flooding is wreaking havoc on agriculture and other economic sectors in Nigeria (Coker et al., 2014). This was brought on by both the Republic of Cameroun's discharge of water from Lagdo Reservoir and Nigeria's severe rains. Farmlands, villages, and other vital infrastructure like houses, marketplaces, bridges, highways, and power and communication systems were washed away. Floods may only last a short period, but if they are not adequately controlled, their economic repercussions can be disastrous. They are also unexpected and unpredictable. Agriculture's growth has been hampered by floods, particularly in regions that experience regular flooding. It has also been a significant barrier to the population of Africa's ability to escape poverty (Coker et al., 2014). There are several causes of floods, but the most frequent one is intense rain that raises the water table and river levels abruptly. Flooding may also happen when a stream overflows its banks and engulfs the surrounding land. However, a flood's potential to wreak damage varies depending on its intensity, duration, and—above all—the availability and dependability of the defenses the locals or government put in place to deal with the flood. According to Nwaobiala and Nwosu (2014), agriculture is one of the human endeavours that is most dependent on the weather, hence it suffers from climate change.

Many theoretical models influence cassava-based farmers' willingness to pay for flood insurance as a means of adapting to climate change. According to the theory of planned behaviour (TPB), a person's intention to carry out a behaviour is determined by their attitude toward the behaviour, subjective norms, and perceived behavioural control (Ryan & Worthington, 2021). According to the TPB, farmers' views regarding flood insurance, societal standards, and their perception of their control over their financial situation will all have an impact on how willing they are to pay for flood insurance. The Protection Motivation Theory (PMT), which contends that a person's perception of danger and coping mechanism affect their behaviour, is another theory that forms the basis of this research (Baghiani-Moghadam et al., 2015). The PMT indicates that farmers' perceptions of the threat that floods pose to their livelihoods and how successful they believe flood insurance is as a coping mechanism will affect their willingness to pay for flood insurance.

25.3 Methodology

The study was conducted in Nigeria's Rivers State, specifically in the Ikwerre Local Government Area (LGA). Its approximate geographic coordinates are 6° 14′ 30″ E and 7° 0′ 0″ E longitude and 4° 58′ 10″ N and 5° 14′ 30″ N latitude and covers an area of about 655 km^2 (National Population Commission [NPC], 2020). To the east, it is bounded by the LGAs of Etche, Obio-Akpor, and Port Harcourt City, to the south, by the LGAs of Emohua and Ogbakiri, and to the north, by the LGAs of Ahoada East and Ahoada West. The region is well-known for its rich soil and agricultural pursuits; one of the main crops grown there is cassava (Odum, 2016). The rainy season, which spans from April to October, and the dry season, which spans from November to March, are the two different seasons in the LGA's tropical climate.

A multistage sampling procedure was used to select cassava-based farmers. The first stage involved the random selection of four (4) districts in Ikwerre LGA. The second stage involved random selection of four (4) communities from each clan giving a total of 16 communities. The third stage involved the random selection of four (4) cassava-based farmers from each community giving a total of 80 cassava-based farmers. However, sixty-four (64) were retrieved for analysis. The study relied majorly on primary data collected through the administration of questionnaire to the cassava-based farmers. This research used a descriptive survey technique to obtain and collect relevant data on the relationships and importance of willingness to pay for flood insurance among cassava-based farmers, the independent variables consist of socioeconomic characteristics of the farmers, and factors influencing cassava-based farmers' willingness to pay for flood insurance as well as the constraints in accessing flood insurance by cassava-based farmers. The data was collected in 2023, and the authors administered the questionnaire/interviews. Objectives (i), (ii), and (iv) were achieved using descriptive statistics while the binary logit regression model was used to achieve objective (iii).

Model specification:

Logit regression model as used by Elum et al. (2018) and adopted is expressed as:

$$Ln(y_1/1 - y_1) = \beta_0 + \beta_1 X_1 + \beta_2 X_2 + \beta_3 X_3 + \beta_4 X_4 + \beta_5 X_5 + \beta_6 X_6 + u \quad (25.1)$$

where;

y_1	The probability that cassava-based farmer will be willing to pay for flood insurance.
$1 - y_1$	The probability that the cassava-based farmer will not be willing to pay for flood insurance.
β_0	Intercept.
$\beta_1 - \beta_6$	Estimated coefficients.
X_1	Gender (dummy; 1 = male, 0 = female)
X_2	Age (years)
X_3	Marital status (dummy; 1 = married, 0 = single)
X_4	Educational level (years)
X_5	Household size (numbers)
X_6	Farming experience (years)

Justification of the Choice of the Independent Variables

The justification for choosing the independent variables of gender, age, marital status, educational level, household size, and farming experience for examining cassava-based farmers' Willingness to Pay (WTP) for flood insurance can be explained by considering their socio-economic and farming-related characteristics that directly influence risk perception and financial decisions. Socioeconomic and Institutional Factors Influencing Willingness to Pay for Agricultural Insurance. These factors are crucial in understanding the adoption of flood insurance in farming contexts.

Gender is a key determinant in decision-making related to risk management, including insurance. Men and women have different levels of risk aversion, access to resources, and financial autonomy. Male farmers may have greater control over farm-related decisions and financial resources, which could influence their willingness to invest in insurance products. Studies in agriculture have shown that women, particularly in rural settings, tend to have less access to information and financial resources, which could lower their WTP for flood insurance (Elum et al., 2018).

Age affects a farmer's risk tolerance and capacity to invest in preventive measures like insurance. Younger farmers are often more willing to adopt innovative solutions, including insurance, as they are more future-oriented. Older farmers may rely on traditional coping mechanisms or be more cautious about new financial products like insurance (Ellis, 2017). Furthermore, older farmers may perceive themselves as having less time to benefit from long-term investments like insurance, reducing their WTP.

Marital status reflects household stability and access to additional resources. Married individuals often have greater financial security and support, which can increase their WTP for insurance, as they may need to protect their family's livelihood. Being married can also influence decision-making through shared responsibility, especially in financial matters like insurance (Elum et al., 2018).

Education improves awareness of financial products and their benefits, enhancing farmers' understanding of insurance and its role in mitigating flood risks. Farmers with higher education are more likely to recognize the long-term advantages of paying for insurance and are thus more likely to be willing to pay. Previous research shows a positive correlation between education and the adoption of risk-mitigating strategies like insurance (Ellis, 2017; Elum et al., 2018; Jensen et al., 2017).

Larger households may have higher financial obligations, which can either increase or decrease their WTP for insurance. On one hand, larger households have more to protect, making insurance more appealing. On the other hand, the financial burden of a larger household may limit disposable income for additional expenses like flood insurance. The impact of household size on WTP varies depending on the balance between labour availability and financial constraints (Ellis, 2017; Kebede et al., 2016).

Experienced farmers are more likely to have faced floods and other natural disasters, which can affect their risk perception. Farmers with more experience may have a deeper understanding of the potential financial losses caused by floods, making them more willing to pay for insurance. On the contrary, some experienced farmers might have developed traditional coping mechanisms over time and may rely less on formal insurance (Elum et al., 2018; Zhang et al., 2017).

25.4 Results and Discussion

25.4.1 Socioeconomic Characteristics of the Farmers

Table 25.1 shows that about 54.7% of the cassava-based farmers age range between 20 and 39 years and 39.9% ranged between 40 and 59 years, with a mean age of 41.7 years. This implies that the cassava-based farmers in the study area are majorly young adults or youths. Younger farmers may have a higher willingness to adopt new technologies and risk management strategies, including insurance, compared to older farmers who may be more conservative in their decision-making (Mendola, 2007). Majority (68.7%) of the cassava-based farmers were female. With majority (76.6%) of the cassava-based farmers married. This implies that married people have more responsibilities, and these led them to dominate in the field of cassava-based farming. According to Kebede et al. (2016), married farmers may have a higher capacity to afford insurance premiums and thus may be more willing to pay.

About 46.1% of the cassava-based farmers have a secondary education, and 20.6% have primary education, with a mean of 9.8 years spent to attain the level of education. This implies that the cassava-based farmers have enough level of formal education to make them understand the terms and conditions of insurance. According to Jensen

Table 25.1 Socioeconomic characteristics of the cassava-based farmers in the study area

Variables	Frequency	Percentage	Mean
Gender			
Male	20	31.3	
Female	44	68.7	
Age (years)			41.7
20–39	35	54.7	
40–59	26	39.9	
60–79	3	4.5	
Marital status			
Single	15	23.4	
Married	49	76.6	
Educational level (years)			9.8
No formal	7	10.9	
Primary	17	26.6	
Secondary	30	46.9	
Tertiary	10	15.6	
Household size (number)			6
1–5	37	57.8	
6–10	26	40.9	
11–15	1	1.6	
Farming experience (years)			16.1
1–10	31	48.6	
11–20	14	23	
21–30	12	18.9	
31–40	3	4.7	
41–50	2	3.1	
51–60	1	1.6	
Total	64	100	

Source Field survey, 2023

et al. (2017), higher levels of education among cassava-based farmers can positively influence their willingness to pay for flood insurance by enhancing their awareness, risk perception, decision-making abilities, access to information, and capacity for effective risk management. About 57.8% of the cassava-based farmers' household sizes fall within the range of 1–5 members, 40.9% within the range of 6–10 members, with a mean of 6 members. This shows that there is enough labour on the farm to enable them to pay for flood insurance. According to Kebede et al. (2016) the size of the household may influence farmers' willingness to pay for flood insurance. Larger households may have higher financial responsibilities, making the decision to purchase insurance more challenging due to limited financial resources. Conversely,

Table 25.2 Willingness to pay for flood insurance of the cassava-based farmers in the study area

Willingness to pay	Frequency	Percentage
No	18	28.1
Yes	46	71.9
Total	64	100

Source Field survey, 2023

larger households may also have a higher income-generating potential, increasing their ability to pay for insurance.

About 48.6% of the cassava-based farmers has farming experience between 1 and 10 years, 23% between 11 and 20 years with a mean of 16.1 years. This suggests that the cassava-based farmers in the study area have enough farming experience to enable them to understand the impact of the flood on the farmland and are probably willing to pay for flood insurance. This is following the study of Zhang et al. (2017) which stated that farmers with more farming experience may have a better understanding of the risks associated with their agricultural activities. Consequently, they may be more willing to invest in risk management strategies such as flood insurance.

25.4.2 Farmers' Willingness to Pay for Flood Insurance

Table 25.2 shows majority (71.9%) of the cassava-based farmers were willing to pay for flood insurance. This could likely be attributed to past experiences of flood on their farmland or risk perceptions. Farmers who are more aware of the risks associated with flooding may be more inclined to purchase insurance. According to Funkhouser (2019) this could be due to personal experiences with flooding, knowledge of the agricultural vulnerability to floods, or awareness of climate change impacts.

25.4.3 Factors Influencing Cassava-Based Farmers' Willingness to Pay for Flood Insurance

From Table 25.3, the Log-likelihood chi-square test statistics of 65.838 and the Hosmer and Lemeshow chi-square of 4.682 are significant showing that the model gives a good fit to the data. Educational level is statistically significant at 1%, with a positive coefficient of 0.105 is an important determinant of cassava-based farmers' willingness to pay for flood insurance in the study area. This implies that a 1% increase in educational level could increase the rate of cassava-based farmers to be willing to pay for flood insurance by 10.5%. Higher levels of education among cassava-based farmers can positively influence the probability of their willingness

Table 25.3 Factors influencing cassava-based farmers' willingness to pay for flood insurance in the study area

Variable	Coefficient	Standard error	P-value
Gender	0.262	0.686	0.703
Age	0.14	0.031	0.651
Marital status	0.363	0.774	0.639
Educational level	0.105	0.084	0.011***
Household size	0.079	0.142	0.075*
Farming experience	0.057	0.040	0.053**
Constant	-2.330	1.979	0.239

Note ***, **, * means significant at 1%, 5% and 10% respectively; Log-likelihood chi-square = 65.838; Hosmer and Lemeshow chi-square = 4.682
Source Field survey, 2023

to pay for flood insurance by enhancing their awareness, risk perception, decision-making abilities, access to information, and capacity for effective risk management (Jensen et al., 2017).

Farming experience is statistically significant at a 5% level of probability, with a positive coefficient of 0.057. This implies that the more years of farming experience, the greater the likelihood that the cassava-based farmers in the study area would be willing to pay for flood insurance. This finding agrees with Zhang et al. (2017) who stated that farmers with more farming experience may have a better understanding of the risks associated with their agricultural activities. Consequently, they may be more willing to invest in risk management strategies such as flood insurance. The household size is significant at 10%. This revealed that the larger the household size the more the probability that a farmer will be willing to pay for flood insurance. This is because larger household has enough labour on the farm and conversely may also have a higher income-generating potential, increasing their ability to pay for flood insurance (Kebede et al., 2016).

25.4.4 Constraints in Accessing Flood Insurance by Cassava-Based Farmers

Table 25.4 provides a breakdown of reasons that might hinder cassava-based farmers from engaging with insurance products. Lack of awareness (45.3%) is the most common barrier is a lack of knowledge or understanding about insurance products and their benefits. This indicates that many cassava-based farmers might not be fully informed about how insurance works or how it can benefit them financially. This finding agrees with Ngango et al. (2022) who reported low awareness insurance among crop farmers in Rwanda, and stated further that awareness of the crop insurance scheme does not automatically translate into acceptance by farmers. Awareness

Table 25.4 Constraints in accessing flood insurance by cassava-based farmers in the study area

Variables	Frequency	Percentage
Lack of awareness	29	45.3
Affordability	11	17.2
Lack of trust	4	6.3
Limited access to credit	10	15.6
Institutional barriers	6	9.4
Lack of data and risk assessment	4	6.3
Total	64	100

Source Field survey, 2023

campaigns and educational efforts are needed to address this gap. About 17.2% of cassava-based farmers cite affordability cost as a barrier. Insurance premiums can be a financial burden for those with lower incomes or limited disposable income. This shows that affordability is a critical factor, as individuals are less likely to purchase insurance if they perceive it as too expensive. This may be because the cost of flood insurance premiums may be excessively high for small-scale cassava-based farmers, making it financially challenging for them to purchase coverage. This finding agrees with Elum et al. (2018) who reported that the cost of insurance products is not affordable for poor farmers to cover their losses.

Lack of trust accounted for 6.3%, this shows that trust issues with insurance providers or the industry can prevent people from purchasing insurance. This could stem from concerns about policy transparency, claims processing, or past negative experiences. Farmers may be sceptical about the effectiveness and transparency of the claims process. They may fear that in the event of a flood, they will face difficulties in receiving timely and fair compensation. While about 15.6% have limited access to credit, farmers facing difficulties in accessing credit are less likely to purchase insurance. Credit access can be crucial for making premium payments, especially when households face liquidity constraints. Insurance often requires an upfront premium payment, which can be a significant financial burden for cassava-based farmers who are already facing challenges in accessing credit or loans for their farming activities. This finding agrees with Ellis (2017) who reported that access to credit is very vital in the willingness to purchase insurance among crop farmers in Ghana.

About 9.4% were faced with institutional barriers such as regulatory and institutional frameworks which may not be conducive to the development and dissemination of crop insurance products for cassava farmers. This includes unclear policies, regulatory challenges, bureaucratic hurdles, or inefficiencies in the insurance sector, and a lack of government support for agricultural insurance making it difficult for farmers to access insurance products. About 6.3% were faced with the constraints of lack of data and risk assessment which is a huge challenge because insurers often rely on historical data and risk assessment models to determine premiums. This finding agrees with Elum et al. (2018) who reported poor assessment of insurance. Insufficient data or inadequate risk assessments by insurers can hinder their ability to offer suitable products, especially for high-risk cassava-based farmers. This can lead to under-insurance in certain markets.

25.5 Conclusions, Recommendations and Limitations

In conclusion, the willingness to pay for flood insurance among cassava-based farmers in Ikwerre Local Government Area, Rivers State, Nigeria, reflects a significant concern about the risks posed by frequent flooding. The study reveals that while many farmers recognize the potential benefits of flood insurance in mitigating their losses, various socio-economic factors, including income levels, education, farm size, and previous flood experience, play crucial roles in influencing their willingness to participate. To enhance participation in flood insurance schemes, it is essential to create awareness and provide educational programs that clearly explain the benefits and operation of such insurance. Policymakers and insurance providers should also consider offering flexible payment plans or subsidies to make insurance more affordable for small-scale farmers. Overall, flood insurance can be an effective tool for enhancing the resilience of cassava farmers in Ikwerre against the growing threat of climate-induced flooding, but addressing the barriers to participation is crucial for widespread adoption. In conclusion, the findings provide insights for policymakers and insurance providers aiming to design inclusive agricultural insurance products that improve farmers' resilience against climate-induced flooding. The availability and accessibility of insurance services in rural areas like Ikwerre may be limited, which could affect the practicality of farmers purchasing insurance even if they express willingness. The study may not have fully explored these logistical challenges. The study did not consider the effect of flood on gender. These limitations suggest that further research should be done to develop a comprehensive understanding of farmers' willingness to pay for flood insurance. The study therefore recommends that:

i. Awareness campaigns and educational efforts should be intensified to reach farmers in rural areas to enlighten them on the importance of flood insurance and the benefits attached to it.
ii. Credit facilities should be made available as a support system to assist the cassava-based purchase insurance for their cassava farms.
iii. Policymakers and insurance providers should tailor strategies to address the specific barriers faced by cassava-based farmers ultimately improving insurance coverage, as well as making the insurance premium affordable for the farmers.

Acknowledgements The authors are very grateful to the respondents for their valuable time and information.

Funding The author received no funds for this research.

Data Availability Statement The datasets used and/or analyzed during the current study are available from the corresponding author upon reasonable request.

Ethical Clearance The confidentiality of the survey participants was strictly and unwaveringly respected.

Conflict of Interest The authors declared that there is no conflict of interest.

References

Adeoye, I. B., Oyinbo, O., Ojehomon, V., Adelowo, O., & Adeoye, F. B. (2020). Farmers' willingness to pay for crop insurance in Nigeria. Munich Personal RePEc Archive, 88701.

Aroyehun, A. R. (2023). Determinants of poultry egg farmers' participation in livestock insurance in Rivers State, Nigeria. *Black Sea Journal of Agriculture, 6*(6), 643–648. https://doi.org/10.47115/bsagriculture.1323005

Ayinde, O. E., Adejobi, A. O., & Sanda, M. A. (2016). Perception, determinants and farmers' willingness to adopt crop insurance policy in rural Nigeria. *Agro-Science, 15*(3), 41–48.

Baghiani-Moghadam, M., Seyedi-Andi, S., Shokri-Shirvani, J., Khafri, S., Ghadimi, R., & Parsian, H. (2015). Efficiency of two constructs called "fear of disease" and "perceived severity of disease" on the prevention of gastric cancer: Application of protection motivation theory. *Caspian Journal of Internal Medicine, 6*(4), 201–208. Retrieved from https://www.researchgate.net/publication/286372981_Efficiency_of_two_constructs_called_fear_of_disease_and_perceived_severity_of_disease_on_the_prevention_of_gastric_cancer_Application_of_protection_motivation_theory#pf2

Coker, A. A., Adebayo, C. O., Odoemena, B. C., Akogun, E. O., & Ezinne, C. G. (2014). Flood and cassava productivity in Kogi State, Nigeria: A quantitative analysis using cross-sectional data. *Ethiopian Journal of Environmental Studies & Management, 7*(6), 599–608. https://doi.org/10.4314/ejesm.v7i6.2

Egor, A. O., Osang, J. E., Uquetan, U. I., Emeruwa, C., & Agbor, M. E. (2015). Inter-annual variability of rainfall in some States of Southern Nigeria. *International Journal of Scientific & Technology Research, 4*(10), 134–154. Retrieved from https://www.researchgate.net/publication/326681877_Inter-Annual_Variability_of_Rainfall_in_Some_States_of_Southern_Nigeria/citations

Ellis, E. (2017). Farmers' willingness to pay for crop insurance: Evidence from Eastern Ghana. *International Journal of Agricultural Management and Development, 7*(4), 447–463. Retrieved from www.ijamad.iaurasht.ac.ir

Elum, Z. A., Nhamo, G., & Antwi, M. A. (2018). Effects of climate variability and insurance adoption on crop production in select provinces of South Africa. *Journal of Water and Climate Change, 9*(3), 500–511. https://doi.org/10.2166/wcc.2018.020

Funkhouser, D. (2019). Assessing agricultural insurance demand and willingness to pay among smallholder farmers in Mozambique. *The Journal of Development Studies, 55*(12), 2553–2569.

Intergovernmental Panel on Climate Change IPCC. (2014). Climate change 2014: Synthesis report. In R. K. Pachauri & L. A. Meyer (Eds.), *Contribution of Working Groups I, II and III to the Fifth Assessment Report of the Intergovernmental Panel on Climate Change* (pp. 1–151). IPCC, Geneva, Switzerland. Retrieved from https://www.ipcc.ch/site/assets/uploads/2018/05/SYR_AR5_FINAL_full_wcover.pdf

Jensen, N. D., Barrett, C. B., & Mude, A. G. (2017). Cash transfers and index insurance: A comparative impact analysis from Northern Kenya. *Journal of Development Economics, 129*(C), 14–28. https://doi.org/10.1016/j.jdeveco.2017.08.002

Kebede, E., Kagochi, J., & Joshi, P. K. (2016). Households' willingness to pay for weather index-based crop insurance in Ethiopia. *Climate Risk Management, 14*, 1–10.

Life and General Insurance. (2023). *Understanding the key objective of reinsurance in risk management*. Retrieved from https://lnginsurance.com/blog/objective-of-reinsurance-in-risk-management/

Linnerooth-Bayer, J., Surminski, S., Bouwer, L. M., Noy, I., & Mechler, R. (2019). Insurance as a response to loss and damage. In R. Mechler, L. Bouwer, T. Schinko, S. Surminski, J. Linnerooth-Bayer (Eds.), *Loss and damage from climate change. Climate risk management, policy and governance*. Cham: Springer. https://doi.org/10.1007/978-3-319-72026-5_21

Maheshwari, R. (2023). *Insurance: Definition, how it works and main types of policies*. Retrieved from https://www.forbes.com/advisor/in/insurance/what-is-insurance/

Mendola, M. (2007). Agricultural technology adoption and poverty reduction: A propensity-score matching analysis for rural Bangladesh. *Food Policy, 32*(3), 372–393.

Mfon, I. E., Oguike, M. C., Eteng, S. U., & Etim, N. M. (2022). Causes and effects of flooding in Nigeria: A review. *East Asian Journal of Multidisciplinary Research (EAJMR), 1*(9), 1777–1792. https://doi.org/10.55927/eajmr.v1i9.1261

National Population Commission NPC. (2020). Nigeria population projections and demographic indicators, national and states. Retrieved from https://factcheckhub.com/wp-content/uploads/2022/07/national-population-commission-Projection_2022.pdf

Ngango, J., Nkurunziza, F., & Ndagijimana, J. (2022). Assessing rural farmers' willingness to pay for crop insurance scheme: Evidence from Rwanda. *Cogent Economics & Finance, 10*, 2104780. https://doi.org/10.1080/23322039.2022.2104780

Nwaobiala, C. U., & Nwosu, I. E. (2014). Effect of climate change on cassava farmers' output in Cross River State, Nigeria. *International Journal of Agriculture and Rural Development, 17*(1), 1628–1634.

Odum, C. (2016). Cassava production in Nigeria: A brief overview. *Journal of Agriculture and Sustainability, 9*(1), 1–8.

Ogunniyi, L. T., Adesope, O. M., & Ogunniyi, A. O. (2018). Assessment of flood impact on cassava production in Nigeria. *Environmental Development, 25*, 82–89.

Okon, E. M., Falana, B. M., Solaja, S. O., Yakubu, S. O. Alabi, O. O., Okikiola, B. T., Awe, T. E., Adesina, B. T., Tokula, B. E., Kipchumba, A. K., & Edeme, A. B. (2021). Systematic review of climate change impact research in Nigeria: Implication for sustainable development. *Heliyon, 7*(9). https://doi.org/10.1016/j.heliyon.2021.e07941

Roder, G., Hudson, P., & Tarolli, P. (2019). Flood risk perceptions and the willingness to pay for flood insurance in the Veneto region of Italy. *International Journal of Disaster Risk Reduction, 37*, 101172. https://doi.org/10.1016/j.ijdrr.2019.101172

Ryan, M. J., & Worthington, A. K. (2021). Theory of planned behaviour. In A. K. Worthington (Ed.), *Persuasion theory in action: An open educational resource*. Retrieved from https://ua.pressbooks.pub/persuasiontheoryinaction/chapter/theory-of-planned-behavior/

Zhang, F., Chen, X., Fan, K., & You, L. (2017). Farmers' willingness to pay for agricultural insurance: Evidence from China. *Journal of Integrative Agriculture, 16*(4), 927–937.

Chapter 26
Rural-Farmers' Response to Climate Change in Nigeria—A Review

Chibuzo Uzoma Izuogu, Joy Obiageli Oparaojiaku, Loveday Chukwudi Njoku, Michael Olatunji Olaolu, Ndidiamaka Martha Ekweanya, Okorie David Amah, and Ojomugbokenyode Ebenehi

Abstract Climate change impact is extremely felt across continents with influence on agricultural production which is essential for rural livelihoods. The sub-Saharan Africa, identified as one of the most vulnerable continents to climate risks, faces severe food production challenges, especially among smallholder farmers in the rural areas. This study investigates the response of rural farmers in Nigeria to climate change, emphasizing their adaptation strategies in the face of increasing climate change related challenges. With the use of the PRISMA method for systematic review of empirical studies from various databases (2014–2024), the study analyses

C. U. Izuogu (✉) · L. C. Njoku · M. O. Olaolu
Department of Agricultural Extension, Alex Ekwueme Federal University Ndufu Alike, Ikwo, Ebonyi State, Nigeria
e-mail: chibuzo.izuogu@funai.edu.ng

L. C. Njoku
e-mail: chukzy4now1980@gmail.com

M. O. Olaolu
e-mail: michealolaolu@yahoo.com

J. O. Oparaojiaku
University of Agriculture and Environmental Sciences, Umuagwo, Imo State, Nigeria
e-mail: Joy.oparaojiaku@gmail.com

N. M. Ekweanya · O. D. Amah
Department of Agricultural Extension and Rural Development, Michael Okpara University of Agriculture, Umudike, Abia State, Nigeria
e-mail: ekweanyandidiamaka@mouau.edu.ng

O. D. Amah
e-mail: amahokorie@gmail.com

O. Ebenehi
Department of Agricultural Extension, Federal University Dutsin-Ma, Dutsin-Ma, Kastina State, Nigeria
e-mail: ebenehi.o@gmail.com

R. U. Onyeneke et al. (eds.), *Energy Transition, Climate Action and Sustainable Agriculture*, https://doi.org/10.1007/978-3-031-83165-2_26

69 articles focusing on climate change awareness, adaptation strategies, and socio-economic determinants. Findings reveal that farmers predominantly rely on indigenous production practices alongside contemporary innovations, with income, education, and extension services identified as critical determinants for successful adaptation. The review highlights the need for policies that incorporate local knowledge, enhance farmers' resilience, and promote inclusive engagement in climate adaptation strategies, thereby supporting rural households in mitigating climate challenges.

Keywords Climate change responses · Climate change awareness · Perception of climate change · Rural adaptation strategies

26.1 Introduction

Climate change impact has been felt across continents by both the natural and human structures, even though the outcome is more evident in the natural structure. These alterations in weather conditions have had detrimental consequences on human actions, especially agriculture which is the major occupation and anchor of rural livelihood (Soubry et al., 2020). According to the IPCC, Africa is among the most vulnerable continents to risks emanating from climate change given its socioeconomic and demographic characteristics. This is because of the low adaptive capacity of the continent amidst numerous stressors. IPCC further observed that most of the countries in Africa will suffer a severe reduction in food production by 50% by 2025 with smallholder farmers being the worst hit (IPCC, 2021). This threatens food security in a continent that is battling food insufficiency. FAO also lamented that climate change has retarded the successes recorded in the economic development of Africa given the huge amount that has been invested in the continent. There is every likelihood that the smallholder farmers, most of whom live in rural communities will be more affected thereby worsening their standard of living (Ifeanyi-Obi et al., 2017). Aside from its exacerbation of rural households' indebtedness, climate stress has made farmers commit suicide in India (Dattaa et al., 2022). Despite these challenges, rural farmers in Nigeria have continued to be the primary producers of food (Izuogu et al., 2024).

Agricultural production activities account for rural households' food, employment generation, income, and contributes about 24.17% of the national Gross Domestic Product (Butu et al., 2022; Izuogu et al., 2023). The impact of alterations in climatic indices on the nation is becoming more noticeable as there are more reports of floods, drought, extreme temperatures, and changes in rainfall regimes (Ifeanyi-Obi et al., 2017). The Nigerian Meteorological Agency (NIMET) reported that if these alterations continue unabated, the nation may experience more weather-related tragedies in the coming years (Ado et al., 2017).

The impact of climate change in Nigeria comes with more burden on rural farmers. Akyala et al. (2023) estimated that about 31 million hectares of small-scale cultivable land in rural areas are adversely affected. Reports from several research findings highlighted increasing temperature with a reduction of the number of months of rainfall in many rural communities (Ado et al., 2017). As a result of the problems of climate change in food production and other economic sectors, Nigeria has initiated many programs and models to support adaptation responses, especially in rural communities (Onyeneke et al., 2023). These adaptation strategies may include Indigenous (Agboola et al., 2024), non-indigenous (Jellason et al., 2022), and an integration of both. Although Jellason et al. (2022) reported the lack of consensus on what could be termed indigenous practices in climate change adaptation; indigenous adaptation strategies in this study include traditional values and belief patterns of the rural farmers that have been transmitted and transformed over the years. These strategies are products of the Indigenous farmers' reliance on their local surroundings, their consistent reflection on nature, and their analysis of the changes to take sustainable livelihood actions. In advocating for climate justice, these indigenous strategies cannot be ignored.

There are several studies on how climate change has influenced rural livelihood in Nigeria. The availability of much literature provides mixed outcomes on the variables that relate to rural farmers' reaction to climate change in Nigeria. This report seeks to galvanize the various results of the empirical research for a more focused policy-making and implementation. These studies identified several factors that determine rural farmers' response to climate change in Nigeria to include socio-economic characteristics (Onyeneke et al., 2023), levels of digitalization and digitization (Izuogu et al., 2023), previous experience (Magesa et al., 2023), etc. It is necessary to highlight the most important determinants to advance an ideal support and approach to improve rural farmers' resilience strategies as it regards their buffer capacity, self-organization, and adaptive capacity.

The existing studies also assessed the response of rural households to climate change by using various climate impact assessment tools, conceptual frameworks, and variables that influence the responses of rural farmers to climate change. However, research which affords an all-inclusive interpretation by initiating a systematic review of this aspect is not sufficient. Consequently, to bridge the knowledge gap, this review seeks to contribute to the current collection of ideas on the response of rural farmers to climate change through a current systematic review. Specifically, the study profiled the distribution of the studies by geographical zones; identified the spread of the studies across years; ascertained rural farmers' awareness and sources of climate change information; reviewed rural farmers' climate change adaptation strategies; described their perception of climate change; and highlighted the determinants of climate change adaptation across the studies.

In making these contributions to the existing body of knowledge, this review assists policymakers and other active participants in climate change-related discussions to pinpoint the most pressing issues affecting rural farmers' response to climate change and facilitate strategies that meet the felt needs of rural households. The review will also spotlight the prospective course of research for researchers and academics yet to be addressed.

26.2 Methodology

The studies in the review are from articles published in different areas of specialization that share a relationship with agriculture, rural development, environment, and climate change.

In conducting a comprehensive assessment of rural farmers' response to climate change, the first stage of a broad literature probe was done on Scopus, Web of Science, Directory of Open Access Journal (DOAJ), and ResearchGate from 2014 to 2024. Some of the advantages of these four databases include advanced search functions, wider spread (over 10,000 publishers), quality assurance, and diversification of disciplines.

The predefined keywords for the search conducted from 14th February to 30th July 2024 included 'rural farmers in Nigeria' and 'response to climate change'. To guarantee that studies are not obsolete and considering that climate change research needs about a minimum of ten years, a timeline range of 2013–2024 was set for articles to be included in the study. This search produced 168 articles. Criteria were set to screen the articles, only empirical, peer-reviewed articles were included in the study for quality assurance. Emphasis was placed on empirical studies given that they reveal the situation of alterations in Nigerian agriculture. Studies that were conducted in Nigeria were considered as well as other research publications within sub-Saharan Africa that presented results from Nigeria. To reduce the margin of error of misunderstanding, the study comprised only articles published in the English language. At this stage, seventeen articles were removed. Of the 151 articles that were left, four duplicated articles were deleted.

At the eligibility criteria, articles that fit properly into the research topic were selected. These articles were selected by reading their titles, abstracts, keywords, results, and discussion. At the end of this stage, thirty-four articles were eliminated from the 147 articles for the 113 articles that remained for quality evaluation. These 113 articles were vetted by five climate change research experts and 56 articles were classified as high, 13 as moderate, and 43 as low. For an article to be included in the data extraction, all the evaluators must agree that the quality of the paper is moderate. Only 69 articles came through the in-depth screening (Fig. 26.1).

26.3 Results and Discussion

26.3.1 Placement of Publications by Geographical Zones

In terms of the geographical spread of the articles, Fig. 26.2 shows that the number of publications grew from the north towards the southern region. The South-West zone was the maximum study geographical zone on the topic with 37.68% of the articles. The South-Eastern and North Central geographical zones had 26.08% and 14.49% of the articles respectively. The lowest percentage was in the North-Eastern

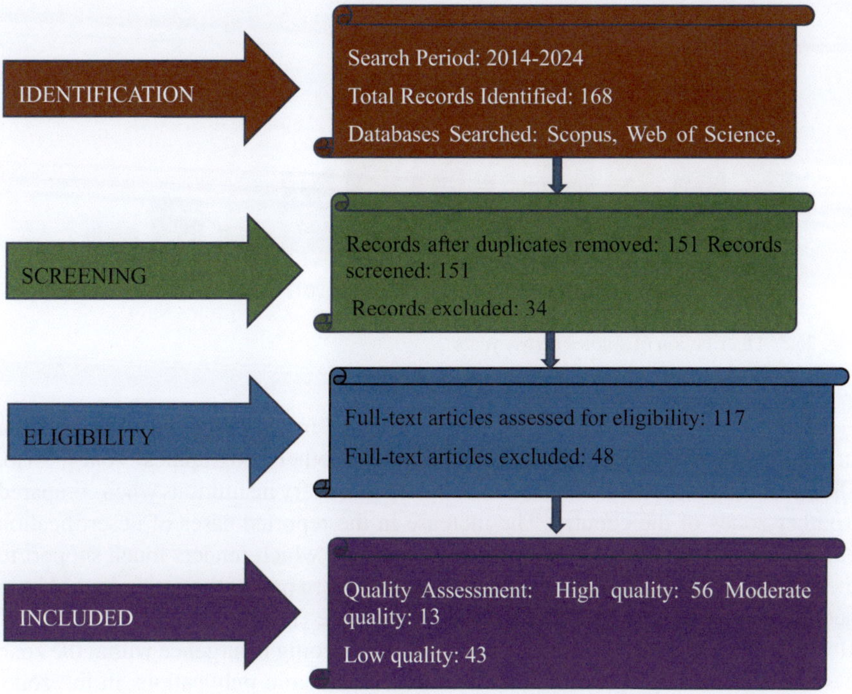

Fig. 26.1 PRISMA diagram of literature search on rural farmers' response to climate change in Nigeria

zone of the country with 4.34%. Given that the study did not set out to have an even distribution of articles as part of its inclusion criteria, the variation in the number of publications across the zone isn't a major issue of concern.

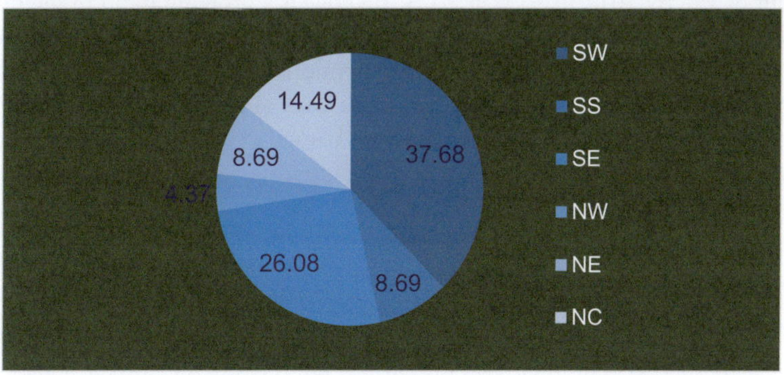

Fig. 26.2 Distribution of articles across geographical zone

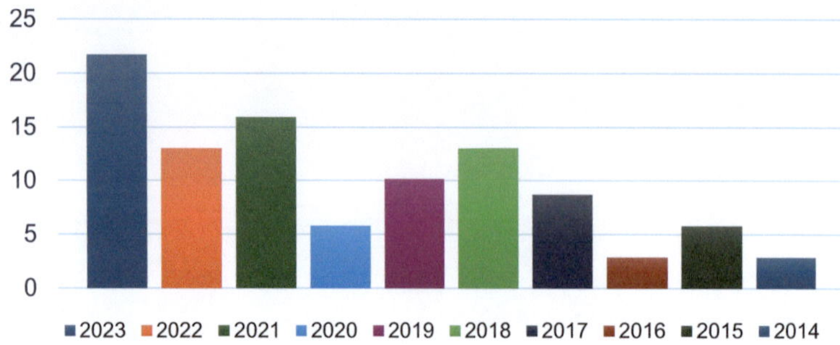

Fig. 26.3 Distribution of studies across years

That the majority of the studies were conducted in the southwest zone does not imply that climate change studies are ignored in other geographical zones. Also, Okon et al. (2021) reported that the zone has more tertiary institutions when compared to other zones of the country. The increase in the reported cases of desertification caused by drought and the shrinking of Lake Chad which renders much support to the livelihood of rural households in the North-Eastern part of the nation should have necessitated more research on the topic in the zone (Danso-Abbeam et al., 2021). However, Izuogu et al. (2021) disclosed that the ongoing insurgence within the zone has contributed largely to the low turn-out of academic publications in the zone. Also, the World Bank (2021) reports that the literacy rate in Nigeria is lower in the Northern region.

26.3.2 Distribution of Articles by Year of Publication

Figure 26.3 shows that among the reviewed articles, the number of publications grew out of 2% by 2014 to 15% in 2023. It showed a growing trend in the number of publications from 2016 which declined between 2019 and 2020. The drop could relate to the outbreak of the Covid-19 pandemic that gave rise to restrictions on movement before the total lockdown. Since the level of utilization of digital tools by researchers in developing nations is very low (Izuogu et al., 2023), such restrictions are expected to negatively influence research outputs.

26.3.3 Distribution According to Journals

Magesa et al. (2023) mentioned that the earliest publication on climate change in the African region was in 2001. Journals that had the major number of publications in the review were Heliyon (11.6%), Journal of Agricultural Extension (10.1%), and

Climate Services (8.7%). Others include Climate Change (7.2%), Journal of Environment and Earth Science (5.8%), Environmental Challenges (5.8%), and Bulletin of National Research Centre (4.3%) the rest of the relevant journals had less than less than three publications as shown on Table 26.1

26.3.4 Awareness and Source of Information

The majority (91.31%) of the articles used for the survey reported that farmers are enlightened of climate change as shown in Table 26.2, while 8.69% reported that farmers are not aware. Farmers are said to be conscious of climate change when they can identify its causes and effects. Ado et al. (2017), Gbolahan et al. (2020), Mustafa et al. (2023), Nwobodo et al. (2022), Onyeneke et al. (2023) reported that most rural farmers are aware of climate change but it's difficult to estimate the extent of their awareness. Rural farmers that are knowledgeable about of climate change and its adverse outcomes show more likelihood of adopting mitigation and adaptation strategies to counter its outcomes (Abdullahi, 2020; Tanimonure & Naziri, 2021; Onyekuru & Marchant, 2017; Oluwatayo and Ojo, 2016; Eta et al., 2023; Oyero et al., 2018).

With the increasing threat posed to agricultural production by climate change and increased research and publications on climate change impact, farming communities have become more informed of climate change and will recognize the alteration in climatic indicators such as rainfall, temperature, and relative humidity (Balogun & Onokerhoraye, 2022).

From Table 26.2, most of the articles reported that radio (79.76%), television (9.36%), and extension contact (4.70%) were the primary providers of climate advice across the nation. The capacity of phones to communicate climate change information among rural farmers has not been adequately utilized as farmers in rural areas mainly prefer using mobile phones to disseminate personal information (Adeloye et al., 2024; Gbolahan et al., 2020; Izuogu et al., 2023; Okpokiri et al., 2017).

Mustafa et al. (2023) agreed that climate change knowledge is an essential factor for adjustment among rural farmers. Climate change communication offers diverse options as it targets a reduction in environmental risks. For, as knowledge of climate change increases, agriculturist will presumably embrace environmentally friendly approaches to minimize the adverse effects of climatic elements. This ensures a sustainable utilization of resources given that rural farmers will become active participants in the agro-centred environmental programs. By this, a sense of possession is established among the rural farmers.

Table 26.1 Distribution according to journals

Journal	Frequency	Percentage
Heliyon	8	11.5942
Journal of Agricultural Extension	7	10.14493
Climate Services	6	8.695652
Climate Change	5	7.246377
Environmental Challenges	4	5.797101
Journal of Environment and Earth Science	4	5.797101
Bulletin of the National Research Centre	3	4.347826
Atmosphere	2	2.898551
Sarhad Journal of Agriculture	2	2.898551
African Population Studies	1	1.449275
Agricultura Tropica Et Subtropica,	1	1.449275
Agronomy	1	1.449275
Climate	1	1.449275
Climate Risk Management	1	1.449275
Current Research in Environmental Sustainability	1	1.449275
Ecological Economics	1	1.449275
Environ Dev Sustain	1	1.449275
Environment and Sustainability	1	1.449275
Environmental Development	1	1.449275
Financial Innovation	1	1.449275
Front. Vet. Sci	1	1.449275
Inter. Journ. of Advanced Academic Research Sciences, Technology & Eng	1	1.449275
Inter. Journ. of Env. Agric. and Biotech	1	1.449275
International Journal of Sciences: Basic & Applied Research	1	1.449275
Journal of Bangladesh Agricultural University	1	1.449275
Journ. of Environ. Sci. Toxicology and Food Tech	1	1.449275
Journal of Geography and Regional Planning	1	1.449275
Journal of Research, Policy & Practice of Teachers & Teacher Education	1	1.449275
Journal of Rural Studies	1	1.449275
Journal of Tropical Agriculture, Food, Environment and Extension	1	1.449275
Renew. Agric. Food Syst	1	1.449275
Science of the Total Environment	1	1.449275
The Journal of Developing Areas	1	1.449275
The Nigerian agricultural journal	1	1.449275
Weather and Climate Extremes	1	1.449275

(continued)

Table 26.1 (continued)

Journal	Frequency	Percentage
World Development	1	1.449275
World Development Sustainability	1	1.449275

Table 26.2 Awareness and sources of information

Variable	Percentage
Awareness	
Yes	91.31
No	8.69
Sources of information	
Radio	79.76
Television	9.36
Extension contact	4.70
Personal experience	2.18
Newspaper	2.02
Fellow farmers	1.08
Phone	0.80
Internet	0.10

26.3.5 Perception of Climate Change

Figure 26.4 shows the perception of climate change by rural farmers in Nigeria. The majority of the studies reported that farmers perceived climate change as an increase in rainfall (38.0%). Other indications of climate change as recognised by the farmers included a decrease in rainfall (25%), an increase in temperature (16%) early start of rain (9%), and desert encroachment (2.5%).

Of the 38.0 percent that reported an increase in rainfall as an indication of climate change, 70.5% were studies conducted in the Southern zone as shown in Fig. 26.5 (Ayanlade et al., 2017; Bakare et al., 2023; Onyeneke et al., 2021; Schroth et al., 2016). This points to the fact that the rainfall regime in Nigeria, especially the Southern zone has surged over the period. The difficulty in predicting the rainfall pattern was reported to have changed the indigenous crop cultivation calendar of the farmers (Agboola et al., 2024; Ekemini et al., 2019; Emediegwu et al., 2022; Oyebisi et al., 2023; Udeh & Ikpe, 2021).

Most of the studies (93.4%) that identified desert encroachment as an indication of climate change were conducted in the Northern zone as shown in Fig. 26.6. Izuogu et al. (2021) noted that farmers' understanding of climate change in Nigeria is multifaceted.

Farmers' perception across these studies gives empirical evidence that temperature has increased considerably across Nigeria (Onyeneke et al., 2023). Rural farmers may

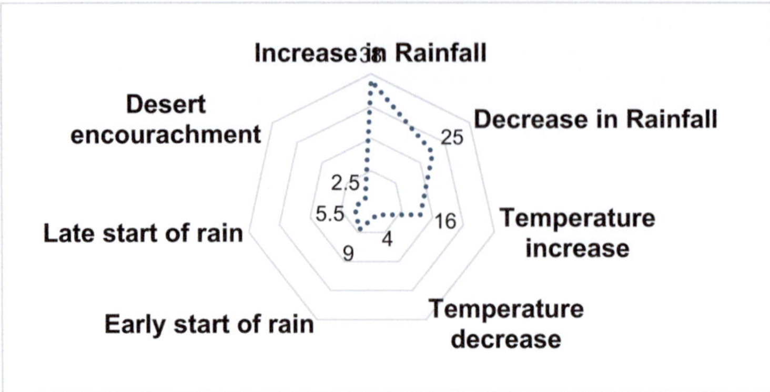

Fig. 26.4 Perception of climate change

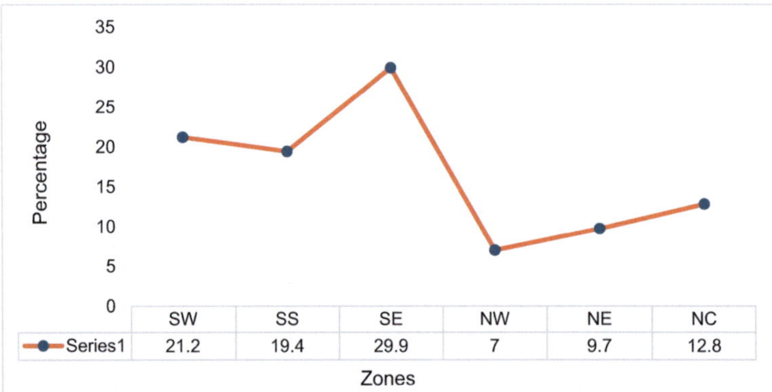

Fig. 26.5 Perception of climate change as temperature rise

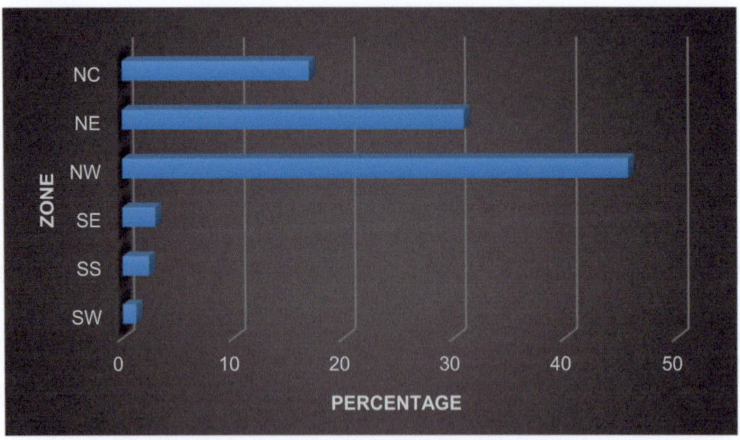

Fig. 26.6 Climate change perceived as desert encroachment

not perceive changes in meteorological variables immediately as this may require time before it manifests.

Knowledge of changes in these meteorological variables may assist rural farmers in inventing better responses to climate change. Nonetheless, the greater number of the studies were restricted to only alterations in major meteorological variables with very few of them reporting the details of how rural farmers perceive changes in other climatic indices like windstorm patterns, air density, solar radiation, etc.

Also, it's difficult to rule out the possibility of rural farmers' perceptions being influenced by the period when these studies were carried out. Jellason et al. (2022) reported that it was more probable that farmers will observe temperature increase during the summertime given that the atmospheric temperature will be high. Similarly, these farmers may more likely indicate that rainfall patterns are irregular during the dry season if their rainfall needs are not met, and their crops are experiencing water stress.

26.3.6 Climate Change Adaptation Strategies

The reviewed articles presented about 42 adaptation strategies by rural farmers in Nigeria. In line with Jellason et al. (2022), these were categorized into three major classes of adaptation strategies Indigenous (23.1%), non-indigenous (7.2%), and integrated (69.5%) as seen in Fig. 26.6.

Indigenous climate change adaptation strategies that were described in the publications include mulching (Adeagbo et al., 2021; Gbolahan et al., 2020; Mesike et al., 2015; Ogunleye et al., 2021; Yakubu et al., 2020), mixed cropping (Adeagbo et al., 2021), dry season cultivation, minimum tillage (Adeagbo et al., 2021), cover cropping (Mesike et al., 2015), using less water consuming crops (Abubakar et al., 2018), water harvesting (Abubakar et al., 2018), and prayers (Gbolahan et al., 2020) while the non-indigenous strategies included use of weather reports from NiMET (Ifeanyi-Obi et al., 2017), planting of early maturing varieties (Orifah et al., 2021; Izuogu et al., 2023), use of flood/drought resistant crops, changing of fertilizer (Abubakar et al., 2018; Adeagbo et al., 2021), improving irrigation (Abaje et al., 2015; Bakare et al., 2023; Ogunleye et al., 2021), and altering planting dates in keeping with weather reports (Bakare et al., 2023; Ogunleye et al., 2021). Improved certified crop varieties are innovative seeds that seek to increase the quality of agronomic advantages for diverse admirable traits in the seed sector (Danso-Abbeam et al., 2021; Izuogu et al., 2024; Onyeneke et al., 2023; Udeh & Ikpe, 2021). Magesa et al. (2023), (Duru et al., 2022), and Ifeanyi-Obi et al. (2017) reported that the utilization of improved seed varieties in the African region has been negatively influenced by their high prices which is not within the reach of farmers who lack access to credit facilities.

Leal et al. (2017) verified that poor farmers in rural areas are more exposed to risks relating to climate change with their utilization of indigenous adaptation strategies solely. Izuogu et al. (2021) however disagreed with the utilization of only non-indigenous strategies given its top-down approach and the assumption

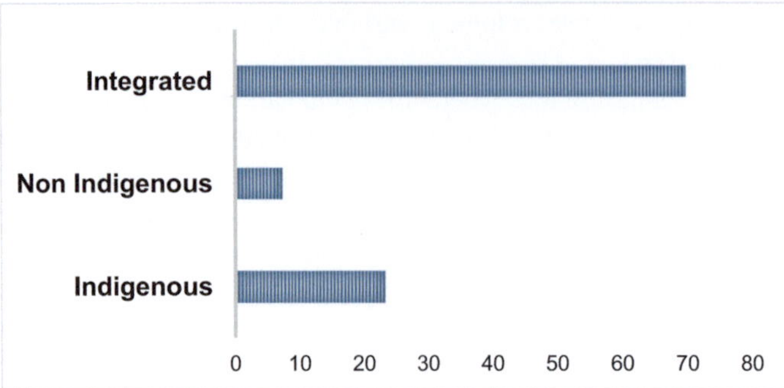

Fig. 26.7 Climate change adaptation strategies

that one strategy will suit all conditions. Indigenous adaptation strategies are critical for the sustainable livelihood of most rural farmers given that they offer them the capacity to understudy their locality, identify internal alterations, and advance various approaches to handle the changes thereby securing their livelihoods and survival. Also, the low level of involvement of rural farmers in the development of the non-indigenous strategy could be responsible for its reduced adoption. The use of indigenous adaptation strategy supports rural farmers' resilience since they will be able to make plans and adjust in line with their previous experiences of bad climatic conditions.

Figure 26.7 shows that 69.53% of the respondents integrated both indigenous and non-indigenous climate change adaptation strategies. An integrated approach to climate change adaptation among rural farmers' is an emerging strategy which seeks to complement non-indigenous strategies. This mix will facilitate and enhance the social education process by diversifying and enriching the picture of climate change matters (Agboola et al., 2024; Ekemini et al., 2019). This will make way for the effective transmission of innovative advantages to rural farm households with a bottom-up structured adaptation policy implementation. When this is realized, the achievements of the adaptation strategy are strengthened, and rural farmers' adaptive capacity will be boosted. Given that rural farmers depend on their environment for sustainable livelihood; climate change adaptation is essential to rural farmers to support their resilience.

26.3.7 Determinants of Choice of Adaptation Strategy

Figure 26.8 shows that most of the articles identified income (35.3%), extension contact (28.2%), education (21.4%), climate change awareness (8.0%), credit access (7.1%) as the major positive determinants of climate change adaptation among rural farmers in Nigeria. These factors may however not present the ideal construct

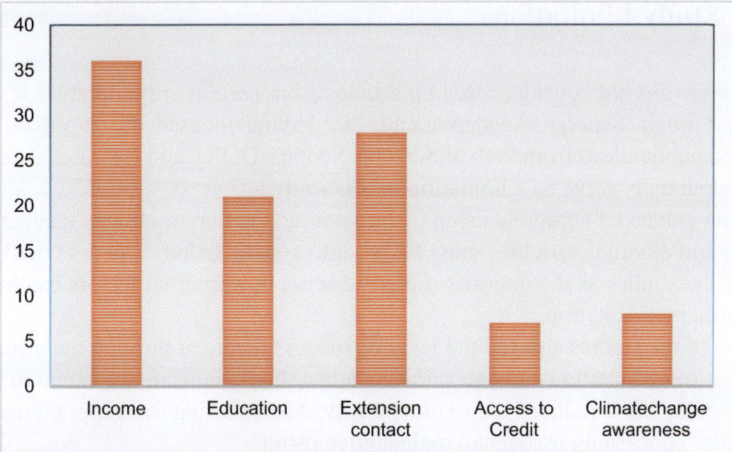

Fig. 26.8 Determinants of choice of adaptation

of the determinants of adaptation choice as a result of the non-utilization of in-depth research methods for more detailed investigation into rural farmers' choice of adaptation strategy.

Rural farmers' level of income proved to be a critical factor in determining their climate change adaptation strategy (Abraham & Fonta, 2018, Mulubrhan and Bedru, 2023). Low/non-availability of income implies that rural farmers will lack the financial support to secure climate change-related information (Ayanlade et al., 2017; Bakare et al., 2023; Nwobodo et al., 2022; Ogunleye et al., 2021; Orifah et al., 2021; Osuji et al., 2023; Owombo et al., 2014; Towolawi et al., 2023). It will also deter farmers from purchasing improved climate change resilient crop varieties (Adeagbo et al., 2021; Ikyoosu et al., 2015) engaging in soil conservation, water conservation, and discourage diversification of farm production (Iyilade et al., 2020; Yakubu et al., 2020).

Extension services occupy a central position in assisting rural farmers in making adaptation choices (Adeagbo et al., 2021; Ojo et al., 2021; Onyeneke, 2021; Onyeneke et al., 2023; Orifah et al., 2021; Osuji et al., 2023; Umunakwe et al., 2022). Beneficiaries of agricultural extension services can support climate change adaptation choices through social networking. However, Nwobodo et al. (2022) reported a dwindling role of agricultural extension services in Nigeria. This has resulted in low availability, access, and usage of extension services by farmers as majority of the farmers rely on other farmers for agricultural information.

Adequate knowledge of the variables that support rural farmers' adaptation to climate change is essential for the successful application of coping and adaptation strategies (Ayanlade et al., 2017).

26.4 Study Limitations

This review did not consider research articles from certain grey literature searches such as Google Scholar, Academia.edu, etc. having focused on empirical, peer-reviewed publications from Web of Science, Scopus, DOAJ, and ResearchGate. This may undoubtedly serve as a limitation of this study.

For an enhanced comprehension of farmers' adaptation to climate change, both internal and external variables must be put into consideration. This was missing in most of the studies as the majority did not discuss how internal factors can hamper rural farmers' adaptation.

None of the studies showed the level of success recorded through the utilization of the various adaptation strategies. Ideal studies should indicate the extent to which farmers; adaptation reduced their vulnerability. Also, nothing was reported in any of the studies concerning associated maladaptive results.

The study advocates for further studies that target rural farmers' willingness and challenges in adopting modern climate change adaptation schemes and the circumstances that influence them. Worthy of investigation also is the extent to which these adaptation models facilitate the attainment of the Sustainable Development Goals (SDGs).

26.5 Conclusion and Recommendations

The majority of rural farmers are aware of climate change and radio is the major source of information. Farmers' perception of climate change is multifaceted as a result of how these alterations in weather conditions affect them. A combination of indigenous and non-indigenous strategies has been adopted by farmers to mitigate the challenges of climate change. Income, education, and extension contact are the major determinants of the utilization of climate change adaptation strategies.

Policy making and implementation geared toward the mitigation of climate change adaptation for rural farming households should factor in the felt indicators of climate change among the beneficiaries. Chain actors in climate change adaptation should support the integration of indigenous and non-indigenous knowledge in policy development. Efforts should be made to increase farmers' income, access to education, and extension contact to boost their adaptation to climate change. Also, it is necessary to engage rural farmers in the designing of policies that relate to building their adaptive capacity since they are more aware and have encountered the alterations in their environment.

Acknowledgements The authors acknowledge the untiring efforts of Prof Chinedum Uzoma Nwajiuba aimed at encouraging young academics in their careers. The assistance of Prof Dennis Ikpe and the Hostclass team while preparing this work is also acknowledged.

Funding No external funds were received at any stage of the research including the decision to submit the article for publication.

Data Availability Statement The set of data analysed during the study are available through the corresponding author if the request is reasonable enough.

Conflict of Interest All the authors declare the absence of conflict of interest as it concerns the publication of this article.

Ethical Clearance The study does not require ethical clearance as the systematic reviews did not involve human subjects, animals, or any sensitive data.

References

Abaje, L. B., Sawa, B. A., Iguisi, E. O., & Ibrahim, A. A. (2015). Assessment of rural communities' adaptive capacity to climate change in Kaduna State, Nigeria. *Journal of Environment and Earth Science*, *5*(20), 14–23. Retrieved from http://www.iiste.org

Abdullahi, N. J. K. (2020). Raising awareness of climate change issues in Nigeria through education for policy and practice. *Journal of Research, Policy & Practice of Teachers & Teacher Education, 10*(1), 14–26.

Abraham, T. W., & Fonta, W. M. (2018). Climate change and financing adaptation by farmers in Northern Nigeria. *Financial Innovation, 4*(11), 2–17. https://doi.org/10.1186/s40854-018-0094-0

Abubakar, A., Maigari, A. I., Danhassan, S. S., Umar, S. I., Usman, M. K., & Gambo, J. (2018). Smallholder farmers' adaptation strategies to drought in the Sahelian zone of Jigawa State, Nigeria. *International Journal of Advanced Academic Research Sciences, Technology & Engineering*, *4*(10).

Adeagbo, O. A., Ojo, T. O., & Adetoro, A. A. (2021). Understanding the determinants of climate change adaptation strategies among smallholder maize farmers in South-west, Nigeria. *Heliyon*, 1–10. https://doi.org/10.1016/j.heliyon.2021.e06231

Adeloye, K. A., Okonkwo, V. C., & Afolayan, A. F. (2024). Rural cassava farmers' agro-climatic information needs in Osun State Nigeria. *Journal of Agricultural Extension, 28*(3), 60–69. https://doi.org/10.4314/jae.v28i3.7

Ado, A. M., Leshan, J., Savadogo, P., Bo, L., & Shah, A. A. (2017). Farmers' awareness and perception of climate change impacts: A case study of Aguie district in Niger. *Environment, Development and Sustainability*. https://doi.org/10.1007/s10668-018-0173-4

Agboola, A. F., Famakinwa, M., Williams, S. O., & Akerele, O. A. (2024). Indigenous climate change adaptation strategies among crop farmers in Osun State Nigeria. *Journal of Agricultural Extension, 28*(3), 134–143. https://doi.org/10.4314/jae.v28i3.14

Akyala, A., Ayanlade, A., Adeyeye, O., & Oluwatimilehin, I. A. (2023). Gender-responsive approaches to rapid climate warming among smallholder farmers. *Current Research in Environmental Sustainability, 5*, 100219. https://doi.org/10.1016/j.crsust.2023.100219

Ayanlade, A., Radeny, M., & Morton, J. F. (2017). Comparing smallholder farmers' perception of climate change with meteorological data: A case study from southwestern Nigeria. *Weather and Climate Extremes, 15*, 24–33. https://doi.org/10.1016/j.wace.2016.12.001

Bakare, A. Y., Ayodeji, S. O., & Ayodeji, D. K. (2023). Impacts of microcredit access on climate change adaptation strategies adoption and rice yield in Kwara State Nigeria. *World Development Sustainability, 2*, 100047. https://doi.org/10.1016/j.wds.2023.100047

Balogun, V. S., & Onokerhoraye, A. G. (2022). Climate change vulnerability mapping across ecological zones in Delta State, Niger Delta Region of Nigeria. *Climate Services, 27*, 100304.

Butu, H. M., Okeke, C. U., & Okereke, C. (2022). *Climate change adaptation in Nigeria: Strategies, initiatives, and practices*. Working Paper No. 3. Berlin, Germany: APRI

Danso-Abbeam, G., Ojo, T. O., Baiyegunhi, L. J. S., & Ogundeji, A. A. (2021). Climate change adaptation strategies by smallholder farmers in Nigeria: Does non-farm employment play any role? *Heliyon, 7*, e07162. https://doi.org/10.1016/j.heliyon.2021.e07162

Datta, P., Behera, B., & Rahut, D. B. (2022). Climate change and Indian agriculture: A systematic review of farmers' perception, adaptation, and transformation. *Environmental Challenges, 8*, 100543. https://doi.org/10.1016/j.envc.2022.100543

Duru, J., Aro, J., & Oladipo, R. E. (2022). The effects of climate change on the livelihood of rural women: A case study of Ilorin South, Nigeria. *Bulletin of the National Research Centre, 46*, 165. https://doi.org/10.1186/s42269-022-00834-9

Ekemini, Richard-Mbossoh, A., Adeolu, B., & Adelegan, J. O. (2019). *Factors influencing choice of climate change adaptation methods among underutilized indigenous vegetable farmers in Southwest Nigeria*. Invited paper presented at the 6th African Conference of Agricultural Economists, September 23–26, Abuja, Nigeria.

Emediegwu, L. E., Wossink, A., & Hall, A. (2022). The impacts of climate change on agriculture in sub-Saharan Africa: A spatial panel data approach. *World Development, 158*, 105967. https://doi.org/10.1016/j.worlddev.2022.105967

Eta, H. C., Idiku, F. O., Elemi, G. F., & Eremi, E. O. (2023). Crop farmers' access to e-information for climate-smart agriculture production in Cross River State Nigeria. *Journal of Agricultural Extension, 27*(3), 26–34. https://doi.org/10.4314/jae.v27i3.3

Gbolahan, O. O., Musiala, M. S., Olusola, O. O., Adisad, J. O., & Abasse, A. O. (2020). Adaptation practices among arable crop farmers against perceived effects of climate change in rural southwestern Nigeria. *International Journal of Sciences: Basic and Applied Research (IJSBAR), 52*(2), 191–202.

Ifeanyi-Obi, C. C., Togun, A. O., Lamboll, R., Adesope, O. M., & Arokoyu, S. B. (2017). Challenges faced by cocoyam farmers in adapting to climate change in Southeast Nigeria. *Climate Risk Management, 17*, 155–164. https://doi.org/10.1016/j.crm.2017.04.002

Ikyoosu, B. M., Ezihe, J. A. C., & Odoemenem, I. U. (2015). Climate change farm-level adaptation measures among soybean farmers in Benue State, Nigeria. *International Journal of Environment, Agriculture and Biotechnology (IJEAB), 2*(5). https://doi.org/10.22161/ijeab/2.5.12

IPCC. (2021). Climate change: The physical science basis. In *Working Group I contribution to the Sixth Assessment Report of the Intergovernmental Panel on Climate Change* (6th ed.). Cambridge, UK; New York, NY, USA: Cambridge University Press.

Iyilade, A. O., Alalade, O. A., Longe, M. P., Alokan, A. O., & Akinola-Soji, B. (2020). Factors influencing adoption of sustainable soil and water conservation practices among smallholder farmers in Kwara State Nigeria. *Journal of Agricultural Extension, 24*(4), 113–121. https://doi.org/10.4314/jae.v24i4.12

Izuogu, C. U., Olaolu, M. O., Azuamairo, G. C., Njoku, L. C., & Kadurumba, P. C. (2023). A review of the digitalization of agriculture in Nigeria. *Journal of Agricultural Extension, 27*(2), 47–64. https://doi.org/10.4314/jae.v27i2.5

Izuogu, C. U., Njoku, L. C., Azuamairo, G. C., Oparaojiaku, J. O., Orji, J. E., Njoku, J. I. K., Chinaka, I. C., & Ankrumah, E. (2024). Farmers' attitudes toward certified rice seeds in Ebonyi State. *Journal of Agricultural Extension, 28*(3), 124–133. https://doi.org/10.4314/jae.v28i3.13

Izuogu, C. U., Onyeneke, R. U., Njoku, L. C., Azuamairo, G. C., & Atasie, M. C. (2021). Repositioning Nigeria's agricultural extension system towards building climate change resilience. *Sarhad Journal of Agriculture, 37*(1), 180–189. https://doi.org/10.17582/journal.sja/2021/37.1.180.189

Jellason, N. P., Salite, D., Conway, J. S., & Ogbaga, C. C. (2022). A systematic review of smallholder farmers' climate change adaptation and enabling conditions for knowledge integration in Sub-Saharan African (SSA) drylands. *Environmental Development, 43*(100733), 1–11. https://doi.org/10.1016/j.envdev.2022.100733

Leal, W., Nzengya, D., Muasya, G., Chemuliti, J., & Kalungu, J. W. (2017). Climate change responses among the Maasai community in Kenya. *Climatic Change, 145*, 71–83.

Magesa, B. A., Mohan, G., Matsuda, H., Melts, I., Kefi, M., & Fukushi, K. (2023). Understanding the farmers' choices and adoption of adaptation strategies, and plans to climate change impact in Africa: A systematic review. *Climate Services, 30*, 100362.

Mesike, C. S., Ugwa, I. K., & Esekhade, T. U. (2015). Adaptation to climate change among rubber farmers in Delta State, Nigeria. *Climate Change, 1*(2), 98–104. www.discovery.org.in/climate_change.htm

Mulubrhan, A., & Bedru, B. (2023). Climate change, income sources, crop mix, and input use decisions: Evidence from Nigeria. *Ecological Economics, 211*(107892), 1–14. https://doi.org/10.1016/j.ecolecon.2023.107892

Mustafa, G., Alotaibi, B. A., & Nayak, R. K. (2023). Linking climate change awareness, climate change perceptions and subsequent adaptation options among farmers. *Agronomy, 13*, 758. https://doi.org/10.3390/agronomy13030758

Nwobodo, C. E., Nwokolo, B., Iwuchukwu, J. C., Ohagwu, V. A., & Ozioko, R. I. (2022). Determinants of ruminant farmers' use of sustainable production practices for climate change adaptation and mitigation in Enugu State, Nigeria. *Frontiers in Veterinary Science, 9*, 735139. https://doi.org/10.3389/fvets.2022.735139

Ogunleye, A., Kehinde, A., Mishra, A., & Ogundeji, A. (2021). Impacts of farmers' participation in social capital networks on climate change adaptation strategies adoption in Nigeria. *Heliyon, 7*, e08624. https://doi.org/10.1016/j.heliyon.2021.e08624

Ojo, A. O., & Oluwatayo, I. B. (2016). Awareness and adaptation to climate change among yam-based farmers in rural Oyo state Nigeria. *The Journal of Developing Areas, 50*(2), 97–108.

Ojo, T. O., Baiyegunhi, L. J. S., Adetoro, A. A., & Ogundeji, A. A. (2021). Adoption of soil and water conservation technology and its effect on the productivity of smallholder rice farmers in Southwest Nigeria. *Heliyon, 7*, e06433. https://doi.org/10.1016/j.heliyon.2021.e06433

Okon, E. M., Falana, B. M., Solaja, S. O., Yakubu, S. O., Alabi, O. O., Okikiola, B. T., Awe, T. E., Adesina, B. T., Tokula, B. E., Kipchumba, A. K., & Edeme, A. B. (2021). Systematic review of climate change impact research in Nigeria: Implication for sustainable development. *Heliyon*. https://doi.org/10.1016/j.heliyon.2021.e07941

Okpokiri, C. I., Agwu, N. M., & Anoliefo, D. C. (2017). Climate change awareness and its effects on the performance of agribusiness households in Abia State Nigeria. *The Nigerian Agricultural Journal, 48*(2), 204–210.

Onyekuru, N. A., & Marchant, R. (2017). Climate change perception, awareness and adaptation decision among forest communities in Nigeria. *Journal of Tropical Agriculture, Food, Environment and Extension, 16*(3), 51–62.

Onyeneke, C. J., Umeh, G. N., & Onyeneke, R. U. (2023). Impact of climate information services on crop yield in Ebonyi State, Nigeria. *Climate, 11*(7). https://doi.org/10.3390/cli11010007

Onyeneke, R. U. (2021). Does climate change adaptation lead to increased productivity of rice production? Lessons from Ebonyi State, Nigeria. *Renewable Agriculture and Food Systems, 36*, 54–68.

Onyeneke, R. U., Amadi, M. U., Njoku, C. L., & Osuji, E. E. (2021). Climate change perception and uptake of climate-smart agriculture in rice production in Ebonyi State Nigeria. *Atmosphere, 12*, 1503. https://doi.org/10.3390/atmos12111503

Orifah, M. O., Sani, M. H., Nasiru, M., & Ibrahim, A. A. (2021). Perceived effectiveness of adaptation strategies to climate change among rice farmers in Jigawa State, Nigeria: Implication for rice production. *Agricultura Tropica et Subtropica, 54*(OV), 122–135. https://doi.org/10.2478/ats-2021-0013

Osuji, E. E., Igberi, C. O., & Ehirim, N. C. (2023). Climate change impacts and adaptation strategies of cassava farmers in Ebonyi State Nigeria. *Journal of Agricultural Extension, 27*(1), 35–48. https://doi.org/10.4314/jae.v27i1.4

Owombo, P. T., Koledoye, G. F., Ogunjimi, S. I., Akinola, A. A., Deji, O. F., & Bolarinwa, O. (2014). Farmer's adaptation to climate change in Ondo State, Nigeria: A gender analysis. *Journal of Geography and Regional Planning, 7*(2), 30–35. https://doi.org/10.5897/JGRP12.071

Oyebisi, O. O., Adeola, O., & Bolarin, O. (2023). Resilience to climate shocks among rural households in Nigeria. *Tropentag.*

Oyero, O., Oyesomi, K., Abioye, T., Ajiboye, E., & Kayode-Adedeji, T. (2018). Strategic communication for climate change awareness and behavioural change in Ado-Odo/Ota Local Government of Ogun State. *African Population Studies, 32*(1).

Schroth, G., Läderach, P., Martinez-Valle, A. I., Bunn, C., & Jassogne, L. (2016). Vulnerability to climate change of cocoa in West Africa: Patterns, opportunities, and limits to adaptation. *Science of the Total Environment, 556*, 231–241. https://doi.org/10.1016/j.scitotenv.2016.03.024

Soubry, B., Sherren, K., & Thornton, T. (2020). Are we taking farmers seriously? A review of the literature on farmer perceptions and climate change, 2007–2018. *Journal of Rural Studies.* https://doi.org/10.1016/j.jrurstud.2019.09.005

Tanimonure, V. A., & Naziri, D. (2021). Impact of climate adaptation strategies on the net farm revenue of underutilized indigenous vegetables' (UIVs) production in Southwest Nigeria. *Resources, Environment and Sustainability, 5*(100029), 1–15. https://doi.org/10.1016/j.resenv.2021.100029

Towolawi, A. T., Oguntoke, O., Bada, B. S., Adejuwon, J. O., & Ibrahim, S. B. (2023). Adaptation strategies of rain-fed farmers to climate change in Nigeria. *Journal of Bangladesh Agricultural University, 21*(1), 109–123. https://doi.org/10.5455/JBAU.120899

Udeh, E. L., & Ikpe, E. (2021). Appraisal of farmers' understanding, perception and knowledge of climate change and adaptation strategies in Jema'a Local Government Area, Kaduna State, Nigeria. *Journal of Environmental Science, Toxicology and Food Technology, 15*(11), 45–54. www.iosrjournals.org

Umunakwe, P. C., Madukwe, M. C., & Aja, O. O. (2022). Capabilities of agricultural extension personnel to transfer climate change adaptation technologies to farmers in Imo State Nigeria. *Ife Journal of Agriculture, 34*(1), 43–56.

World Bank Group. (2021). *Population total for Nigeria.* www.data.worldbank.org. Retrieved July 10, 2024.

Yakubu, D. H., Akpoko, J. G., Akinola, M. O., & Abdulsalam, Z. I. (2020). Climate change adaptation practices and rice farmers' level of living in North-West Nigeria. *Journal of Agricultural Extension, 24*(3), 49–60. https://doi.org/10.4314/jae.v24i3.5

Chapter 27
Enhancing Adoption of Agroecological Practices Among Smallholder Farmers in Kenya

Brian Muthuri Kithinji, Bolly Kevin Echessa, Loraine Chivingo Kabaka, and Elijah Kakunza Bakari

Abstract Smallholder farmers play a pivotal role in sustaining Kenya's agrifood systems through producing food crops. Despite the agricultural sector's economic significance, over 95% of farming is rain-fed and susceptible to seasonal rainfall patterns and climate change. Agroecology as an approach to farming is gradually being embraced in Kenya as a form of sustainable agriculture that can withstand the intensifying impacts of climate change especially for the vulnerable smallholder farmers. However, adoption has been slow as the country lacks a clear policy framework that could support its rollout. This paper seeks to assess the current levels of adoption of agroecological practices among smallholder farmers in Kenya by examining the barriers, knowledge, and awareness gaps, and socio-economic impacts. The study sampled 200 smallholder farmers from Bomet, Bungoma, Isiolo, Kiambu and Kilifi Counties using qualitative methods involving surveys and interviews. The findings indicate that 56% of study participants do not hold positive perceptions on agroecology, while only 50% have experienced positive impacts on their personal income and soil health from practising agroecological principles such as crop diversification, organic farming, and agroforestry. Some of the barriers hindering adoption include resistance to change, access to land and capital, and climate vulnerability. The findings illustrate the importance for stakeholders to enhance access to training and hands-on capacity building programs for smallholder farmers, accelerate land reforms to enhance access and ownership, and improve access to markets for organic foods.

Keywords Smallholder farming · Agroecology · Crop diversification · Organic farming · Food security · Sustainable agriculture

B. M. Kithinji (✉) · L. C. Kabaka · E. K. Bakari
Policy Action Initiative, Nairobi, Kenya
e-mail: brian@policyactioninitiative.org

B. K. Echessa
Department of Spatial and Environmental Planning, Kenyatta University, Nairobi, Kenya

R. U. Onyeneke et al. (eds.), *Energy Transition, Climate Action and Sustainable Agriculture*, https://doi.org/10.1007/978-3-031-83165-2_27

27.1 Introduction

27.1.1 Background and Problem Statement

Agriculture is the backbone of Kenya's economy, contributing nearly 30% of the total gross domestic product and employing 45% of the national labour force, and 80% in the rural areas (Ministry of Agriculture & Livestock Development, 2023). Despite Kenya being the largest economy in Eastern Africa, the country relies on food imports from its neighbours Ethiopia, Uganda, and Tanzania to meet local consumption demands, in the process exacerbating food prices and inflation. Recent assessments by the Integrated Food Security Phase Coalition indicate that nearly 1.9 million Kenyans are facing acute food insecurity, while over 5.5 million (more than 10% of the population) are at serious risk of hunger (IPC, 2024). Furthermore, about 1 million children in the country are projected to be suffering the effects of malnutrition including stunted growths, protein deficiency, and weakened immune systems (UNICEF, 2024). With less than six years remaining before the elapse of the Sustainable Development Goals, Kenya is undeniably far away from achieving Goal 2 on Zero Hunger and its Constitutional provision of Article 43(1c) that guarantees the right of every Kenyan "*to be free from hunger and to have adequate food of acceptable quality.*"

Farming in Kenya is overwhelmingly rain-fed, and this has periodically affected food production as over 80% of the country's land mass is arid and semi-arid, and highly susceptible to droughts and seasonal rainfall patterns (Kithinji et al., 2023). Youths, who make up 29% of the labour force in the agricultural sector, lack meaningful opportunities to upscale their initiatives and the capacity to handle the escalating effects of climate change (Ministry of Agriculture, Livestock & Fisheries, 2017). Recent government-led efforts to upscale agricultural production in Kenya have largely centred on creating irrigational schemes such as Galana Kulalu in Tana River County but this strategy has proven to be unreliable and unsuccessful. Moreover, with agriculture contributing more than 75% of freshwater withdrawals in Africa, according to the United Nations Food and Agriculture Organization (FAO), there is a serious risk of water shortages for domestic household needs in the future (Van Huis, 2015). According to Macharia et al. (2020), the agricultural sector is responsible for 60% of Kenya's greenhouse gas (GHG) emissions and over 80% of the solid waste generated annually. If left unchecked, this may hinder the country achieving its target of reducing emissions by 32% by 2030, as outlined in the Nationally Determined Contributions which form part of Kenya's obligations as a signatory of the Paris Agreement (Ozor et al., 2020). Meanwhile, more than 2.6 million tons of pesticides with a market value of more than $25 billion U.S. Dollars are used annually across the world (Sharma et al., 2019). This excessive use of pesticides has affected the environmental biodiversity and soil health in Africa and runs contrary to the African Union's Climate Change and Resilient Development Strategy. While some mitigating solutions such as using high-yield crop varieties and synthetic fertilizers have been introduced in recent years, a new paradigm that is locally led and incorporates indigenous knowledge systems is required.

Agroecology is defined as the application of the principles of ecology to the management of agriculture and food production with the purpose of maintaining sustainability of the production systems. Contrary to intensive agriculture, which depends on factors such as synthetic inputs and single crops, agroecology enhances diversification, resource productivity and sustainability. Gliessman (2014) devoted a chapter to defining what agroecology is not—simply a package of practices, but a science, a movement, and a practice that focuses on the relationship between agriculture and its environment. With over 2 billion smallholder farmers globally operating more than 500 million small farms, these new systems could help elevate production, reduce emissions from the sector, and mitigate concerns with land tenure issues (Anderson et al., 2021).

Agroecology as an approach to farming is gradually being embraced in Kenya as a form of sustainable agriculture that can withstand the current natural challenges such as climate change and thus is a vital form of farming for the most vulnerable smallholder farmers. In the Kenyan context, the implementation of agroecological practices for food and income security is espoused based on the principles that Africans should sustainably use their assets, resources, knowledge, and ecosystems within their local environment. They also include social justice equity, where the practice of agroecology empowers farmers and communities. This approach supports the smallholder farmers' rights to use the agricultural inputs and management practices of their choice and oversee their resources as they participate in the global food economy at a time when they are being systematically neglected (Altieri, 2018). Agroecology facilitates decision-making at the grassroots level and scarce resources are used in ways that respect the farmer's knowledge and needs. In Kenya, where most of the farmers are smallholder farmers, agroecology is a decent way for them to develop sustainable agriculture that is less costly and environmentally friendly (Anderson et al., 2021). This paper seeks to assess the current levels of adoption of agroecological practices among smallholder farmers in Kenya by examining the barriers, knowledge and awareness gaps, and socio-economic and environmental impacts.

27.1.2 Objectives

The overall objective of this study is to assess the current levels of adoption of agroecological practices among smallholder farmers in Kenya. Specifically, the study aims to:

- Explore the socio-economic, cultural, and technical barriers hindering the adoption of agroecological practices by smallholder farmers.
- Evaluate the potential socio-economic and environmental benefits of agroecological practices to smallholder farmers.

- Assess the perceptions on agroecology and the level of knowledge and awareness among smallholder farmers about the benefits of agroecological practices.

27.2 Overview of Smallholder Farming and Agroecology in Kenya

27.2.1 Principles of Agroecology

Agroecological practices are organized in such a way that the farming systems can be productive, sustainable, and socially just. Some of these principles include diversification, interoperability, reliability, reusability, recycling, and finally, farmers' and community centricity. The policy of diversity is at the centre of agroecology owing to the complexity and multiplicity of farming practices and systems that bring into agri-food systems a diversity of crops, livestock, and other creatures (Altieri & Nicholls, 2017). Smallholder farmers in Kenya can benefit from crop diversification since crop failure is likely to be frequent due to the increasing prevalence of pests and diseases, or because of the changing seasonal rainfall patterns. Therefore, crop diversification can help in producing adequate food that can satisfy nationwide demand.

Closely related to the components is the interaction, or interdependence, of elements within the farming system, for example, crops, livestock, and natural resources in agroecology. These synergies improve the efficiency and effectiveness of the system as all the components of the system are in harmony with each other (Gliessman, 2014). For instance, the use of mixed farming, where cattle are included together with crops as in the case of rural Kenya, can help in replenishing the soil nutrients using manure and therefore raise household incomes. Additionally, it could reduce reliance on artificial fertilizers that have caused widespread soil degradation. Another principle is resource conservation in that agroecological practices require the prudent use of available resources, including water and soils, and therefore reduces the negative impacts of farming on the environment (Wezel et al., 2020).

Cooperation is an inherent feature of agroecological systems because maintaining farming systems and raising the resistance of the systems to negative external shocks is one of the main goals of the technology. This is especially so in Kenya where smallholder farmers are vulnerable to factors like climate risk, market prices, and availability of inputs. Recycling is another principle that deals with the use of nutrients, organic matter, and water, all in the farming system, to little or no extent relying on outside inputs (Gliessman, 2014). Last of all, agroecology has a rather significant focus on farmers and other members of communities who engage in farming, thus promoting decent working conditions, adequate recognition, and fair distribution of local resources. These scenarios are very evident in the Mt. Kenya and Rift Valley regions where social welfare is far institutionalized.

Smallholder farming is the most widespread agricultural system in Kenya. Of the total persons working in the agricultural sector, 75.2% are engaged in smallholder farming (IFAD, 2019). The majority of the smallholder farms in Kenya are below 2 hectares in size and the farmers are involved in low commercial and intensive farming of crops such as maize, beans, vegetables, and rearing livestock. Most smallholder farmers are involved in the production of the country's favourite staples, hence have a great contribution towards food security. However, success in agriculture and rural development depends more on the smallholder farmers' labour and these farmers encounter many problems that limit their productivity.

27.2.2 Smallholder Farming in Kenya

Smallholder farmers play a pivotal role in the global agrifood systems. There are more than 500 million smallholder farms that produce more than a third of the world's food, and about 80% of the food consumed in Asia and Africa (IFAD, 2023). In Kenya, most smallholder farmers live in rural areas and battle various constraints. Such farmers normally work within areas with low soil nutrient reserves and often with poor accessibility to water for irrigation, making them very sensitive to climate change and other manmade changes in the environment (FAO, 2017). Also, smallholders are credit-constrained and lack inputs and modern technology that could enable them to increase yield and adapt to the prevailing conditions.

Despite these hurdles, smallholder farming is still the dominant trend in Kenyan agriculture and there exists broad prospects for supporting and popularizing such sustainable approaches to agriculture through agroecological principles. In August 2024, the Ban Ki-moon Centre issued a Youth AgriChampions demand paper on locally led adaptation that called on world leaders to streamline and simplify procedures for accessing international climate finance, localize data collection and information dissemination, and ease access to land for young farmers (BKMC, 2024). The intensification of such calls for support to smallholder farmers could push governments worldwide to set strategies that can accelerate adoption of agroecological practices.

27.2.3 The Policy Framework on Agroecology in Kenya

In 2003, Kenya joined other African states in endorsing the Maputo Declaration on Agriculture and Food Security that mandates African governments to allocate at least 10 % of their national budgets to the agricultural sector. However, the country is yet to attain that goal, with the sector receiving just 3% in the 2024/2025 budget (The National Treasury & Economic Planning, 2024). In recent years there have been concerted efforts to strategize Kenya's agrifood systems as prices of key cash crops like coffee and tea fluctuated on the world market and food poverty increased.

In 2004, President Mwai Kibaki's administration launched the Strategy for Revitalizing Agriculture (SRA), whose vision was to transform Kenya's agricultural sector into a commercially oriented, profitable, regional and international competitive economic activity that provides high-quality, gainful employment to Kenyans. Six years later, the government followed up with the Agriculture Sector Development Strategy (ASDS), whose goal was to create a prosperous food-secure country. Despite these efforts, production of key staples such as maize, potatoes, and beans dropped as the strategy did not address the underlying issues plaguing smallholder farmers.

This situation forced a rethink and in 2019, the government launched the Agricultural Sector Transformation and Growth Strategy, with the goal of increasing small-scale incomes by 35%, agricultural output and value addition by 6%, and building household resilience (FAO, 2019). To sidestep the pitfalls of the past policy frameworks, this strategy seeks to strengthen community resilience through crop diversification, cultivating nutritious indigenous crops such as cassava, sorghum and millet, and introducing climate-smart natural resource management practices including soil health and water catchment. The animal health principle within the framework seeks to increase the provision of feeds, fodder, and roll out a livestock insurance program in the Northern region, where 60% of livestock in Kenya are found and over 90% of the population raise animals for beef and milk consumption (Ministry of Agriculture, Livestock, Fisheries and Cooperatives, 2020).

However, while this policy framework promotes some agroecological principles, Kenya lacks a specific strategy on agroecology which can anchor nationwide efforts for its adoption and scaling across the country. To mitigate this concern, earlier this year, the Ministry of Agriculture and Livestock Development introduced the National Agroecology for Food System Transformation Strategy (NAFSTS) 2024–2033. This strategy provides a comprehensive approach to include all actors in the agroecology space in transforming food and agricultural systems and addressing biodiversity loss, while providing benefits of climate adaptation, food and nutrition security, water conservation, sustainable livelihoods, ecosystem resilience and human rights. Additionally, unlike previous policies, it includes an explicit focus on boosting agribusiness, the socio-economic dimensions of food systems and the rights of youth, women, and indigenous communities, filling a vacuum that was left by the elapse of the Kenya Youth Agribusiness Strategy in 2022. Once the policy is formally adopted in the coming months, Kenya will join other African countries like Senegal and Uganda in having national policy frameworks on agroecology (Bottazzi & Boillat, 2021).

27.3 Methodology

This study employs a qualitative approach incorporating the use of surveys and interviews with small-holder farmers in Kenya. Participants for this study were selected through purposive sampling to ensure diversity in gender, age, and environment. The researchers identified 200 participants from across five counties in Kenya who had previously benefitted from agricultural support projects (Fig. 27.1).

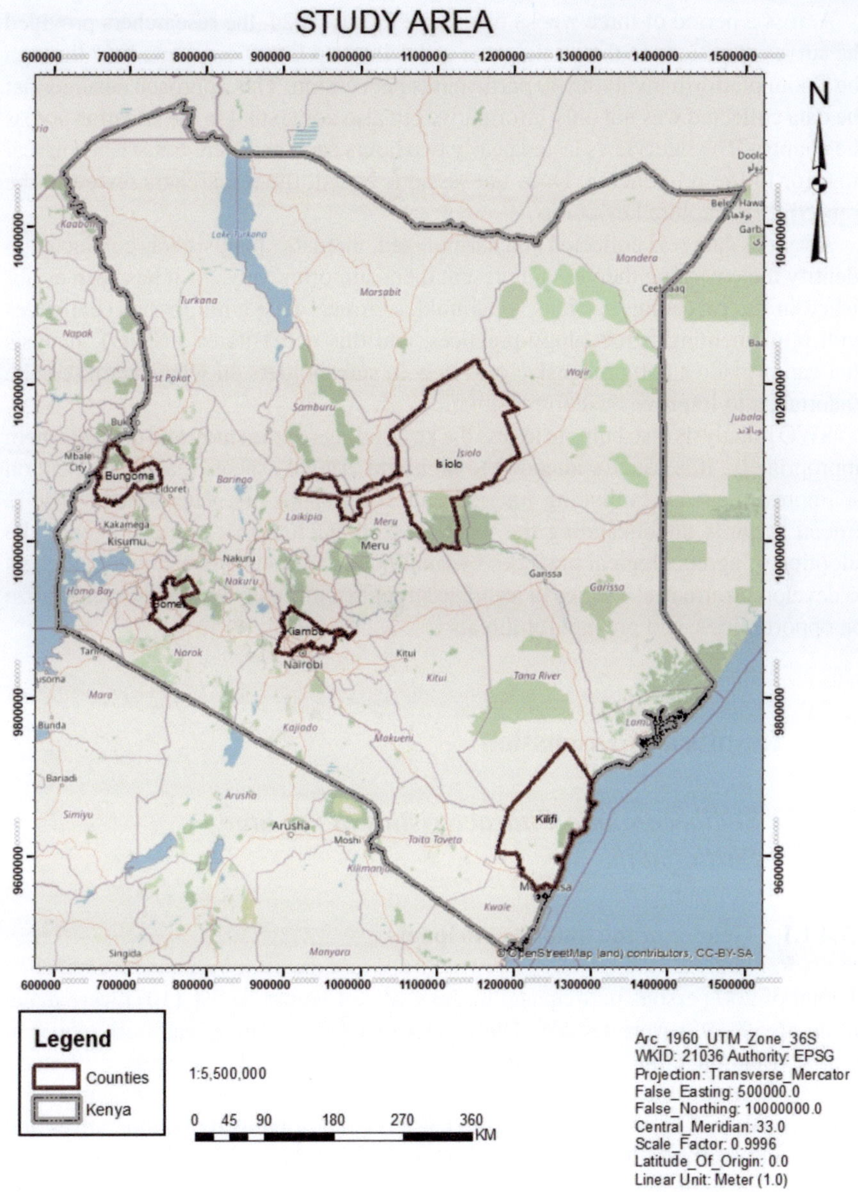

Fig. 27.1 The map shows the counties selected for this study. They include: Bungoma, Bomet, Isiolo, Kiambu, and Kilifi

Across a period of three weeks from June to July 2024, the researchers provided the survey questions to the participants and follow up interviews were held through the Zoom platform involving 40 participants per session. This approach ensured that the data collected was not only informative but also contextual to the situation across the country. The interviews lasted nearly two hours each and were transcribed in real time for future referencing. Once the sessions ended, the researchers reviewed the transcripts to capture key points,

After the data was collected and harmonized, thematic analysis was conducted to identify the emerging themes and present them authoritatively. As it has been established in the preceding sections, smallholder farmers face a number of challenges with implementing agroecology practices, and this necessitates a SWOT analysis that can provide a comprehensive overview to stakeholders on what actions can be undertaken to improve the current situation.

SWOT analysis also helps to assess the key findings of the study and presents them appropriately. It is usually intended to highlight critical aspects that need attention or improvement by evaluating internal and external factors. The holistic view is crucial towards understanding the wide range of factors that could influence the adoption of agroecological practices by smallholder farmers and enable researchers to develop informed strategies to leverage strengths, mitigate weaknesses, capitalize on opportunities, and prepare for threats.

27.4 Results and Discussion

27.4.1 Socioeconomic Characteristics of the Study Participants

27.4.1.1 Gender of the Study Participants

A total of 200 persons participated in the study, of whom 53.5% (107) were male, while 46.5% (93) were female. The gender distribution curve was fairly equal in Bomet, Bungoma, and Kilifi Counties. In Isiolo County, the percentage of study participants that were male shot up to 72.5% (29), while females were a mere 27.5% (11). Abdullahi (2016) investigated the participation of women in cooperatives and community welfare groups in Isiolo County, and found out that often, women are excluded due to customary beliefs that prioritize patriarchy across all levels of society. Isiolo County is also located in the Kenyan ASALs (Arid and Semi-Arid Lands), and it is dominated by pastoral production systems. In the 2019 Kenya Population and Housing Census, the county had a total population of 268,002 people, of whom 52.1% were males and 47.9% were females. The gender distribution in Isiolo County differs from the national average in Kenya, where women make up 50.5% of the country's population (KNBS, 2019). Recently, owing to pressure campaigns from the private sector and advocacy groups, the Isiolo County Government has introduced

Table 27.1 Gender representation of study participants

County	Males	Females	Percentage of males	Percentage of females
Bomet	18	22	45	55
Bungoma	21	19	52.5	47.5
Isiolo	29	11	72.5	27.5
Kiambu	15	25	37.5	62.5
Kilifi	21	19	53.5	47.5

Note The table shows the distribution of the 200 study participants across each county and their genders

a Gender Policy 2021–2015 that seeks to ensure the attainment of gender equality in the development process by redressing existing gender imbalances (KIPPRA, 2021).

However, in Kiambu County, it was the reverse. The study participants who identified as females were 62.5%, almost doubling their male counterparts. Cisco and Olungah (2016) noted that women take part actively in farming activities in Kiambu County, as men prefer white collar jobs and retail trade. Kiambu County is the second most populous county in Kenya after the capital Nairobi. In 2019, the county was recorded to have a population of 2,417,735, of whom 49.1% were males and 50.9% were females. It can therefore be construed from these results that gender imbalances exist in Kenya's agricultural sector, and this may have a correlating effect on the adoption of agroecological practices (Table 27.1).

27.4.1.2 Age Distribution of Study Participants

Figure 27.2 shows that 58.5% of the study participants were between the age bracket of 25–30 years, with 23% and 18.5% falling in the brackets of 30–40 years and 18–25 years respectively. This age distribution curve is an indication that most people begin to transition to work and other income generating activities at the age of 25 years—which is also the time when they complete studies. Additionally, this age group also makes up the largest percentage of Kenya's labour force (PwC, 2022), of whom an overwhelming majority work in the agricultural sector.

27.4.2 Thematic Analysis

27.4.2.1 Perceptions of Agroecology Among Smallholder Farmers

Figure 27.3 shows that while agroecology principles such as crop diversification, organic farming, vermicomposting and agroforestry have been practiced for decades in Kenya, 56% of smallholder farmers do not hold positive perceptions. This may

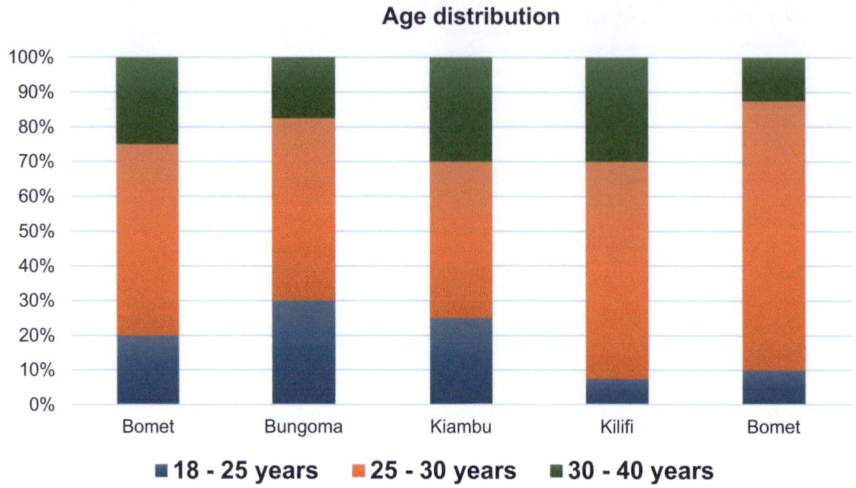

Fig. 27.2 Age distribution of study participants

result from a lack of understanding or awareness gaps on the benefits of agroecology principles such as crop diversification, organic farming, vermicompositing. There was also a common misconception that agroecology only refers to organic farming, because smallholder farmers naturally would prefer synthetic fertilizers to speedily increase soil nutrients ahead of the farming season. Kirui et al (2023) found out that 70% of farmers who had awareness on agroecological principles had a positive perception about their effectiveness. The proposed National Agroecology for Food System Transformation Strategy (NAFSTS) includes awareness and capacity-building as one of the priority areas to boost adoption of agroecology in Kenya.

On the other hand, only 44% of smallholder farmers surveyed expressed a positive view of agroecology, basing their positivity on personal experiences of its benefits in increasing yields and incomes. Those who expressed positive views on agroecology shared how participating in hands-on training projects enabled them to learn best practices, and for some, it was the financial support they received to launch their pilot projects. This illustrates the importance of increasing access to financial resources for locally led adaptation initiatives, particularly in the rural areas.

Overall, these findings are a wakeup call to stakeholders in the agricultural sector to take proactive measures to increase awareness on agroecology and its benefits. Community media stations can play a key role in challenging misinformation by speaking directly to smallholder farmers in a voice and language they can easily trust (Gatzweiler & Von Braun, 2016). The government can liaise with international development partners in providing one-off grants or zero-interest loans, alongside hands-on training from expert facilitators to smallholder farmers to implement agroecological practices.

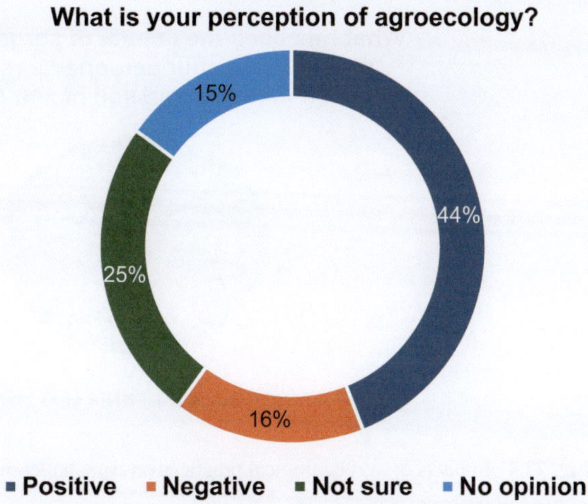

Fig. 27.3 Perception of agroecology among smallholder farmers in Kenya

27.4.2.2 Socio-economic and Environmental Impacts of Agroecology

Similarly to the findings on the perceptions of agroecology in Fig. 27.4, only half of the smallholder farmers who participated in the study reported having experienced positive financial or environmental benefits since they began implementing agroecological practices. The data in Fig. 27.4 illustrates the challenges that smallholder farmers face in trying to integrate this form of sustainable agriculture into their overstretched farming systems. According to the Kenya Demographic and Health Survey (2022), only 33% of women and 45% of men own land. Formal land ownership and documentation remains a significant challenge bearing from the legacies of colonial rule and the dispossession of natives from the highlands by white settlers. Successive governments have also failed to address the land question and institute sweeping reforms to the existing land tenure system, which has only served to entrench inequalities and accelerate the repurposing of agricultural lands for commercial real estates.

It is uneconomical to practice agroecology techniques on small tracts of land in which farmers need to maximize output across seasons. Agroecology also requires a significant initial investment that is an insurmountable barrier for smallholder farmers who lack access to credit finance and capital. Overall, these findings indicate that upscaling adoption of agroecology in Kenya requires addressing the underlying challenges facing smallholder farmers such as access to land, capital, extension services, and information.

27.4.2.3 Forms of Agroecology

From the responses gathered in the surveys, the researchers identified four main forms of agroecology that smallholder farmers are practising in their communities.

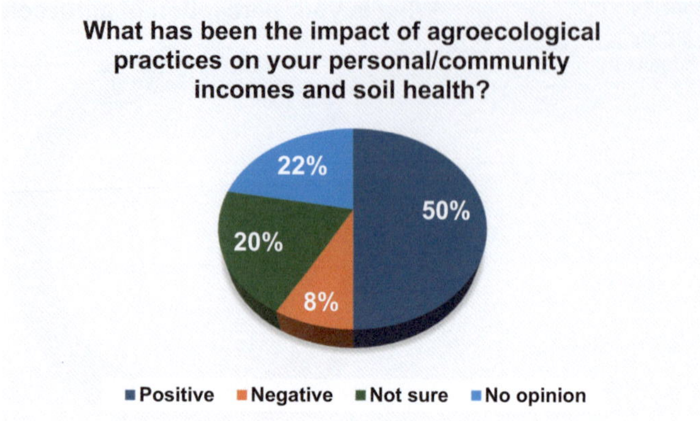

Fig. 27.4 Impacts of agroecological practices on personal/community incomes and soil health

Agroforestry

Smallholder farmers in Kenya have increasingly adopted this measure which involves integrating trees with crops and livestock and practicing alley cropping. Adoption of agroforestry has largely been fuelled by the launch of the nationwide tree planting program by the Ministry of Environment and Climate Change in November 2023. Through this program, the national government has set a goal of planting 15 billion trees in the next 10 years and increasing the country's forest cover to about 20%. In Kilifi County, agroforestry is viewed as strategy to tap into the carbon trading system following the roll out of the Carbon Markets Regulations (Muthuri et al., 2023). Due to its favourable location along the coastline, Kilifi has become a beehive for carbon trading projects led by multinationals that seek to leverage on its burgeoning mangrove forests to offset their carbon emissions.

Organic farming

The harmful effects of pesticides and chemical fertilizers on soil health has edged smallholder farmers to increasingly adopt vermicomposting and crop rotation. The former is particularly becoming popular in the grassroots owing to its simplicity and an avenue to recycle household waste. While organic waste accounts for about 80% of solid waste collected in Eastern Africa, just four per cent is recycled with the rest left to degrade in the open, pausing grave concerns to human and environmental health (Okot-Okumu, 2012). As the population in the region expands by nearly 10% year-on-year, annual bio-waste generation is projected to triple to 516 million tonnes by 2050. Additionally, the recent move by the government to scrap subsidies on synthetic fertilizers has increased prices of one 50 kg bag of nitrate fertilizers from Kshs. 2500 ($20 U.S. Dollars) to Kshs. 7500 (60 U.S. Dollars) and subsequently forced smallholder farmers to consider alternative cost-effective solutions to improve yields.

Indigenous farming

In Isiolo County, farmers work on communally owned lands and agriculture is a communal activity rather than a commercial venture like in Bomet, Bungoma, Kiambu and Kilifi Counties. Additionally, the methods applied are largely traditional, including seed saving and digging holes in the dry season to conserve ground water from the extreme heat. Researchers noted that these methods have been inherited from one generation to another and are also part of the community's strategy to withstand patterns of droughts and famine.

Crop diversification

This form of agroecology has become very popular among smallholder farmers in Kenya as it can help mitigate existing challenges with regards to access to land resources and quality seeds. Young farmers in Bomet, Bungoma, and Kiambu counties are now growing multiple crops such as beans, maize, and green grams on one-acre plots, unlike in the past where one single crop was grown on specific sections. The results of this method have helped increase yields as green grams and beans enrich the soil with nitrogen, which in turn eliminates the necessity to utilize nitrate fertilizers. Additionally, crop diversification has proven to be an effective method of pest control for cereal crops.

27.4.2.4 Barriers to Adoption of Agroecological Practices

Figure 27.5 indicates that 60% of the study participants in Isiolo County cited resistance to change as the main barrier preventing them from adopting agroecological practices. These findings can be interpreted by considering the cultural contexts within the county. Unlike most areas in Kenya, communities in Isiolo County primarily live traditional conservative lives where the man is the head of the household and community elders hold huge sway over daily life. In Bomet, Bungoma, Kiambu and Kilifi counties, resistance to change was the least cited barrier, indicating that awareness about agroecology is high in these areas and that smallholder farmers are very receptive to new information.

Access and ownership of land remains a major challenge that has largely been left unaddressed by successive governments since the end of colonial rule. Vast swathes of the rich fertile highlands in the Rift Valley and Mount Kenya regions that were grabbed by white settlers are yet to be equally redistributed among local communities. Figure 27.5 affirms this issue remains a pressing concern as 58% of smallholder farmers in Kiambu County (which is in the Mount Kenya region) felt that land tenure was their biggest barrier towards adopting agroecology. On the other hand, land barely registered as an issue in Isiolo County where most of the land is classified as community land. When farmers own lands they are able to embrace sustainable farming practices like agroforestry and soil conservation, which need secure tenure to invest in (Kinyangi et al., 2015).

Access to finance was mentioned as leading barrier by 46% of smallholder farmers in Kilifi County, though the results from the other four counties indicate that it may not hold considerable weight like land and resistance to change. In further

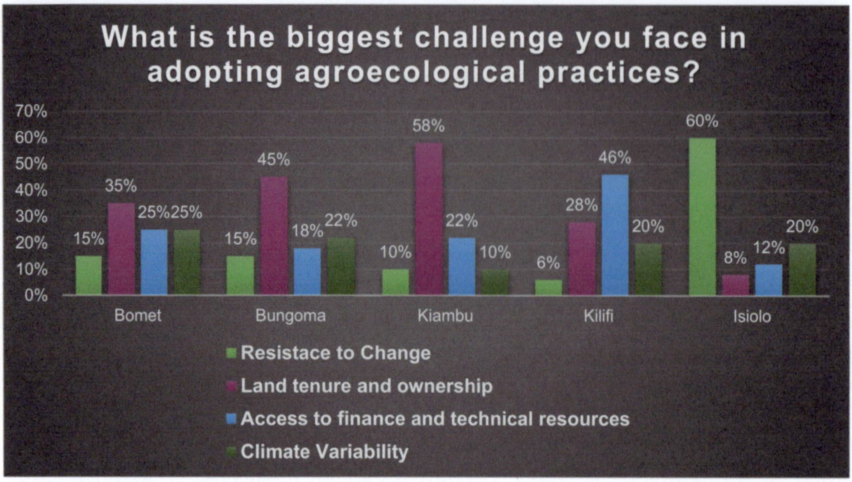

Fig. 27.5 Barriers to adoption of agroecological practices

interviews, the researchers established that the smallholder farmers believed that financial resources exist, but the processes of applying for these funds are often lengthy, require extensive paperwork and documentation that is not readily available to local communities, and knowledge of the English language, which is not commonly spoken in their environments. They called for the simplification and streamlining of processes to apply for private sector credit and other grant fundings, to enable them adopt agroecology.

Lastly, only 20% of the smallholder farmers surveyed in this study reported that climate variability poses the biggest barrier towards adoption of agroecological practices, showing concerns about climate change and the ability to adapt to its impacts is not seen as a pressing concern by smallholder farmers in Kenya. However, their vulnerability makes them more likely to suffer losses and damages from climate change.

27.4.3 SWOT Analysis of the Adoption of Agroecological Practices Among Smallholder Farmers in Kenya

Strengths

By promoting practices that enhance soil health and biodiversity, agroecology is beneficial to small holder farmers. The principles of agroecology can be tailored to the local environments and help the farmers in adapting to the intensifying impacts of climate change on seasonal rainfall patterns. For example, through vermicomposting, farmers in Kilifi County have been able to increase their soil nutrients and increase yields of green grams (mung beans) from two bags (roughly 90 kg) to four bags per acre (roughly 360 kg). Agroecological practices can increase income sources through waste recycling, sales of organic fertilizers to other farmers, while reducing input costs for seedlings.

Moreover, agroecology helps smallholder farmers reduce waste and curb land degradation. The increasing use of pesticides and synthetic fertilizers has infiltrated the water systems with dangerous chemicals and exhausted soil nutrients (Mahmood et al., 2016). Reversing these trends could have overall positive effects in human and animal health (Table 27.2).

Weaknesses

From the study findings in Fig. 27.3, it was noted that only 44% of smallholder farmers have positive perceptions about agroecology, illustrating the lack of understanding, awareness and extension services in grassroots communities. Since all the counties

Table 27.2 SWOT analysis of the adoption of agroecological practices among smallholder farmers in Kenya

Factor	Strengths	Weaknesses	Opportunities	Threats
Knowledge and awareness	Tailored practices Positive impacts (e.g., increased yields, reduced waste)	Lack of understanding, awareness, and extension services	National agroecology for food systems transformation strategy Increased research and technology	Low adoption rates due to limited knowledge and skills
Economic viability	Increased income sources Reduced input costs	Lack of markets for organic products Limited economic benefits for some farmers	Growing demand for organic food Potential for export markets	Rising inflation Fluctuating market prices
Policy environment	Potential for positive impact through National Agroecology for food systems trasformation strategy	Lack of a clear policy framework	Policy implementation Cross-sector collaboration	Slow implementation of land reform policies Limited access to capital
Land tenure and access	Potential for community-based approaches	Land ownership challenges Gender disparities	Land reform initiatives Promotion of community-based land management	Scarcity of arable land Land grabbing
Climate change and weather patterns	Resilience to climate change	Unpredictable weather events Extreme weather conditions	Climate-smart agroecological practices	Increased risks to crop yields and livelihoods

selected for the study are outside of the traditional food basket regions of Kenya, it is difficult to state whether this notion is widespread. However, it raises serious concerns whether smallholder farmers have access to the relevant technical expertise, training and capital to transition to agroecology.

Additionally, since Kenya lacks a policy framework on agroecology, there are no markets for organic or sustainably produced agricultural products that can incentivize its production among smallholder farmers. In the study, only 50% of smallholder farmers surveyed indicated having experienced positive economic or environmental impacts from practicing agroecology.

Opportunities

To accelerate the adoption of agroecological practices among smallholder farmers in Kenya, it is vital that the National Agroecology for Food System Transformation Strategy is quickly adopted and implemented. This policy framework provides clear pathways for integrating agroecology into the national agricultural strategies and a human-rights based approach that prioritizes the concerns of youth, women, and indigenous communities.

Advancements in research and technology could help refine agroecological practices and precision farming techniques. Moreover, it could attract more youths to consider careers in the agricultural sector, helping reduce the burden of high unemployment. Cross-sector collaboration between government agencies, non-governmental organizations, research institutions, private sector and international organizations will be key in enhancing knowledge sharing and resource mobilization to support agroecological pilot projects. Kenya's food insecurity burden also presents clear opportunities for smallholder farmers to meet increasing demand for favorite staples such as maize, beans, and potatoes, and in the process, increase incomes.

Threats

With Africa's population growing by 2% annually, there will be significant strains for national governments to feed the masses. Agriculture is therefore poised to remain the backbone for most economies, employing a majority of the 11 million youths entering the labor force annually (IFAD, 2019). This will strain the demand for arable lands, which are already scarce in Kenya where 80% of the landmass lies in the arid and semi-arid ecosystems (Kithinji et al., 2023).

While more than 65% of land in Kenya is classified as community lands, these are mostly undocumented, and majority of the residents lack formal documentation to prove ownership (Rights & Resources Initiative, 2020). Additionally, women only received 3% of land title deeds issued out in the last decade, exacerbating gender disparities and inequalities. As illustrated in the thematic analysis, land ownership is a big challenge in counties like Kiambu, which are located in the former "White Highlands" that were populated by colonial settlers' pre-independence. Meanwhile, only 8% of smallholder farmers that were surveyed in Isiolo County, where over 80% of the land is classified as community lands, listed access to land as their biggest barrier towards adopting agroecology.

There are some strategies that have the potential of resolving this challenge. Kenya enacted the Community Lands Act in 2016 which built upon the National Land Policy of 2009 (Alden, 2018). The Act promises land security for more than 10 million Kenyans living in rural communities. However, its implementation by the National Land Commission has been slow owing to the emotive nature of land as a sensitive political, cultural, and economic issue in Kenya. In 2010, the United Nations Economic Commission for Africa (UNECA) in collaboration with the African Union and the African Development Bank (AfDB) established a land policy framework to support African countries with land reforms, aligning itself with the World Bank's goal of expanding land ownership and rights by 2030 (UC-ECA-AfDB Consortium, 2010). More than a decade later, there has been limited progress across the continent in accelerating land surveying, titling, issuance of title deeds, and distribution; while conflicts over land have increased as more communities are forced to move due to climate change impacts.

In our study, it was noted that 25% of the smallholder farmers surveyed reported having challenges with accessing capital and credit facilities to implement agroecological principles. Youths looking for alternative earnings cannot fund agroecological projects such as agroforestry or set up structures for vermicomposting (Nasambu & Sugden, 2017).

Unpredictable weather patterns and extreme weather events will also pose threats towards the adoption of agroecological practices and can hinder long-term investment. Rising inflation and fluctuation of market prices may also scare away smallholder farmers from adopting agroecological practices that appear to be risky bets.

27.5 Conclusions, Policy Recommendations and Limitations

This study illustrates the perceptions, benefits and barriers of adopting agroecology among smallholder farmers in Kenya. Through a thematic and SWOT analysis, it was identified that agroecology holds promise for Kenya to achieve food security for her 54 million citizens, however, necessary policy frameworks are required that can support smallholder farmers to transition to these forms of farming.

The main limitation of the study was the parameters, and we recommend that in future, researchers seek to investigate each agroecological practice individually, and focus on youths and women, as these groups are not well represented in the agricultural sector. Also, further research is required to investigate the study's findings in other regions with different climates and cultural contexts.

Based on the observations from our study, the authors make the following recommendations:

Strengthen institutional support for agroecology

The National Agroecology for Food System Transformation Strategy should be quickly adopted and implemented as it provides a pathway for mainstreaming agroecology at the national and county level. The National Treasury should increase its budgetary allocations to the agricultural sector (in line with Kenya's commitments under the Maputo Declaration of 2003) to support funding of research and development that targets localizing agroecological practices by smallholder farmers.

Enhance access to training and hands-on capacity building

To scale up interest and adoption of agroecological practices, there is need for targeted capacity building programs for smallholder farmers. This can be implemented through development of farmer field schools, demonstration farms, and extension work on agroecological practices. The trainings should be continuous and targeted at increasing farmers' knowledge and skills on organic farming, crop diversification, soil fertility management and water conservation among others.

In the same vein, the government should engage local farmer cooperatives and community-based organizations to facilitate these training programs as they are able to speak to the smallholder farmers in a language they can easily understand. Moreover, the government should engage local non-governmental organizations and community-based organizations to facilitate these training programs and continuously monitor and evaluate their effectiveness.

Accelerate land reforms to enhance access and ownership by smallholder farmers

Land is a major factor of production, yet it is largely inaccessible to most smallholder farmers. This paper calls for expanding land access and ownership by acknowledging and redressing of past land injustices through comprehensive land reforms such as issuing title deeds, documentation, and surveying. Moreover, the National Land Commission should begin a national review of the National Land Policy of 2009 and the Community Lands Act, 2016, to create a unified land tenure system.

Improve access to markets

Smallholder farmers are frequently unable to find markets to sell their agroecological products at competitive prices. The Competition Authority of Kenya, in cooperation with county governments, should create local and regional markets for food that is organically grown, especially as an alternative to the imported genetically modified foods (GMOs). Additionally, banks should be supported to avail credit, financial services and insurance products tailored for the local needs of smallholder farmers.

Foster multi-stakeholder collaboration

Lastly, multi-stakeholder partnership is crucial for the adoption of agroecology. There is need for more collaboration between smallholder farmers, research and learning institutions, non-governmental organizations, the private sector, and international organizations to enable the sharing of expertise, inputs, and technologies, as well as to upscale the effectiveness of existing agroecological programs. Furthermore, these actors should be included in multi-stakeholder platforms aimed at policy and program coherence so they can harmonize their strengths to meet the needs of smallholder farmers.

Acknowledgements The authors would like to acknowledge the Kenya National Farmers' Federation for their technical support towards accomplishing this research project. We would also like to recognize Sylvia Wachira (Pan African Climate Justice Alliance) for her directional advice and strategic inputs.

Funding The authors of this study did not receive support from any organization for the submitted work.

Data Availability Statement The datasets used during this study are available from the corresponding author on reasonable requests.

Conflict of Interest The authors declare that they have no known competing financial interests or personal relationships that could have appeared to influence the work reported in this paper.

Ethical Clearance No ethical concerns were identified during the study.

References

Abdullahi, N. (2016). *Factors influencing performance of women owned development projects in Kenya: A case of Isiolo County* (Doctoral dissertation, University of Nairobi).

Alden Wily, L. (2018). The community land act in Kenya; opportunities and challenges for communities. *Land, 7*(1), 12.

Altieri, M. A. (2018). *Agroecology: The science of sustainable agriculture.* CRC Press.

Altieri, M. A., & Nicholls, C. I. (2017). Agroecology: A brief account of its origins and currents of thought in Latin America. *Agroecology and Sustainable Food Systems, 41*(3), 211–222.

Anderson, C. R., Bruil, J., Chappell, M. J., Kiss, C., Pimbert, M. P., Anderson, C. R., & Pimbert, M. P. (2021). Origins, benefits and the political basis of agroecology. In *Agroecology Now! Transformations towards more just and sustainable food systems*, 11–28.

BKMC. (2024). *Youth AgriChampions Demand Paper 2024.* Ban Ki-moon Centre for Global Citizens.

Bottazzi, P., & Boillat, S. (2021). Political agroecology in Senegal: Historicity and repertoires of collective actions of an emerging social movement. *Sustainability, 13*(11), 6352.

Cisco, O. B., & Olungah, C. O. (2016). Factors affecting rural women's participation in agriculture for development in Gatundu south sub-county. Kiambu county, Kenya. *International Review of Social Sciences and Humanities, 11*(2), 97–107.

FAO. (2017). *The state of food and agriculture: Climate change, agriculture, and food security.* FAO.

FAO. (2019). *Agricultural sector transformation and growth strategy.* Retrieved from https://faolex.fao.org/docs/pdf/ken189053.pdf

Gatzweiler, F. W., & von Braun, J. (2016). Innovation for marginalized smallholder farmers and development: An overview and implications for policy and research. In *Technological and institutional innovations for marginalized smallholders in agricultural development* (pp. 1–22).

Gliessman, S. R. (2014). *Agroecology: The ecology of sustainable food systems.* CRC Press.

IFAD. (2019). *Smallholder farming in Kenya: Challenges and opportunities.* IFAD.

IFAD. (2023). *Smallholders can feed the world.* IFAD.

Integrated Food Security Phase Coalition—IPC (2024). *IPC acute food insecurity and malnutrition analysis February–June 2024.* Retrieved from https://reliefweb.int/report/kenya/kenya-ipc-acute-food-insecurity-and-acute-malnutrition-analysis-february-june-2024-published-26-march-2024

Kinyangi, J., et al. (2015). *Land degradation, soil fertility decline, and climate change in Sub-Saharan Africa.*

KIPPRA. (2021). *Isiolo County Gender Policy 2021–2025.* Kenya Institute for Public Policy Research and Analysis. Retrieved from https://repository.kippra.or.ke/handle/123456789/4157

Kirui, E. C., Kidoido, M. M., Mutyambai, D. M., Okello, D. O., & Akutse, K. S. (2023). Farmers' knowledge, attitude, and practices regarding the use of agroecological-based pest management practices in crucifers and traditional african vegetable (TAV) production in Kenya and Tanzania. *Sustainability, 15*(23), 16491.

Kithinji, B., Mainye, N., Thini, P., & David, P. (2023). Building resilience for youth climate mobility in Africa. Kenya institute for public policy research and analysis (KIPPRA). Policy Brief No. 25 of 2023–2024.

KNBS. (2019). *2019 Kenya population and housing census results.* Kenya National Bureau of Statistics.

Macharia, J. M., Pelster, D. E., Ngetich, F. K., Shisanya, C. A., Mucheru-Muna, M., & Mugendi, D. N. (2020). Soil greenhouse gas fluxes from maize production under different soil fertility management practices in East Africa. *Journal of Geophysical Research: Biogeosciences, 125*(7), e2019JG005427.

Mahmood, I., Imadi, S. R., Shazadi, K., Gul, A., & Hakeem, K. R. (2016). *Effects of pesticides on environment. Plant, soil and microbes: Volume 1: Implications in crop science*, 253–269.

Ministry of Agriculture, Livestock and Fisheries, and Cooperatives. (2017). Kenya Youth Agribusiness Strategy 2017/2021.

Ministry of Agriculture, Livestock, Fisheries and Cooperatives. (2020). Sessional Paper No. 3 of 2020 on The Livestock Policy.

Ministry of Agriculture and Livestock Development. (2023). *National agroecology for food system transformation strategy, 2024–2033.*

Muthuri, C. W., Kuyah, S., Njenga, M., Kuria, A., Öborn, I., & van Noordwijk, M. (2023). Agroforestry's contribution to livelihoods and carbon sequestration in East Africa: A systematic review. *Trees, Forests and People*, 100432.

Nansambu, D., & Sugden, F. (2017). A generation on the move: Voices of youths in the context of climate change, migration, and livelihood transition. In *Session Report from the 11th International Community Based Adaptation Youth Conference*, Kampala, Uganda, 26–29 June 2017 (No. 615-2017-5856).

Okot-Okumu, J. (2012). Solid waste management in African cities—East Africa. In *Waste management—An integrated vision* (pp. 1–20).

Ozor, N., Nyambane, A., Onuoha, C. M., Makokha, M. O., & M'mboyi, F. (2020). Nationally determined contributions (NDCs) implementation index, monitoring and tracking tools for selected countries in Africa. Availabe at: https://atpsnet.org/wpcontent/uploads/2020/07/NDC-Implementation-Index-Report.pdf. Accessed on 7 November 2023.

PwC. (2022). *Agriculture in Kenya.* Price Waterhouse Coopers.

Rights and Resources Initiative. (2020). *Community land rights in Kenya: A summary of 2020 analyses.* RRI

Sharma, A., Kumar, V., Shahzad, B., Tanveer, M., Sidhu, G. P. S., Handa, N., Kohli, S. K., Yadav, P., Bali, A. S., Parihar, R. D., Dar, O.I., Singh, K., Jasrotia, S., Bakshi, P., Ramakrishnan, M., Kumar, S., Bhardwaj, R. & Thukral, A. K. (2019). Worldwide pesticide usage and its impacts on ecosystem. *SN Applied Sciences, 1*, 1–16.

The National Treasury and Economic Planning. (2024). Budget statement FY 2024/2025. Retrieved from https://www.treasury.go.ke/wp-content/uploads/2024/06/2024-Budget-Statement-for-the-FY-2024-25F.pdf

UC-ECA-AfDB Consortium. (2010). Land policy in Africa: A framework to strengthen land rights, enhance productivity and secure livelihoods. Retrieved from https://www.un.org/en/land-natural-resources-conflict/pdfs/35-EN-%20Land%20Policy%20Report_ENG%20181010pdf.pdf

United Nations Children's Fund—UNICEF. (2024). Humanitarian situation report No. 8. Available: www.unicef.org/appeals/kenya/situation-reports

Van Huis, A. (2015). Edible insects contributing to food security? *Agriculture & Food Security, 4*, 1–9.

Wezel, A., Herren, B. G., Kerr, R. B., Barrios, E., Gonçalves, A. L. R., & Sinclair, F., (2020). Agroecological principles and elements and their implications for transitioning to sustainable food systems.

MIX
Papier aus verantwortungsvollen Quellen
Paper from responsible sources
FSC® C105338

If you have any concerns about our products,
you can contact us on
ProductSafety@springernature.com

In case Publisher is established outside the EU,
the EU authorized representative is:
Springer Nature Customer Service Center GmbH
Europaplatz 3, 69115 Heidelberg, Germany

Printed by Libri Plureos GmbH
in Hamburg, Germany